Signals, Systems, and Controls

B. P. Lathi
Universidade Estadual de Campinas—São Paulo

THOMAS Y. CROWELL
HARPER & ROW, PUBLISHERS
New York Hagerstown Philadelphia San Francisco London

SIGNALS, SYSTEMS, AND CONTROLS

Library of Congress Cataloging in Publication Data

Lathi, Bhagwandas Pannalal.
  Signal, systems, and controls.
  (Intext series in circuits, systems, communications, and computers)
  Includes bibliographical references.
  1. System analysis.   2. Control theory.
I. Title.
QA402.L36   003   73-464
ISBN 0-7002-2431-9

Contents

Chapter 4: Frequency-Domain Analysis Using Generalized Exponentials

Chapter 5: Feedback and Control

Chapter 6: State-Space Analysis

Chapter 7: Discrete-Time Systems

Appendix A: Some Properties of Differential Operators

Appendix B: Partial-Fraction Expansion

Appendix C: Bode Plots

Appendix D: Vectors and Matrices

Appendix E: Second-Order System with a Zero

Appendix F: Nyquist Criterion for Stability

Bibliography

Index

Preface

In this book I have attempted to integrate some of the basic ideas of network and system theory, signal analysis and processing, and control systems and simulation in a fashion that will be accessible to undergraduate students taking a second course in networks or systems. The approach to time-domain and frequency-domain analysis is unified.

The two methods are presented not as different approaches, but as essentially the same approach employing different bases for input signal representation. Such a viewpoint is very effective in practical terms and has philosophical appeal. Thus, the frequency transforms (Fourier and Laplace) are introduced not as mechanical operators which aid in solving integrodifferential equations, but as tools for representing a signal as the sum of exponential signals with complex frequencies. The response of a linear system to any input signal is seen as the sum of the responses of the system to various exponential components of the input. This approach not only gives a deeper appreciation of interaction of signals with systems, but also allows one to integrate smoothly the basic concepts of signal analysis and processing with those of system analysis. It also unmasks frequency-domain analysis to reveal that it is in fact a time-domain analysis in disguise.

In the development of discrete-time systems, discrete-time signals are introduced first. The analysis of discrete-time systems then unfolds along lines similar to those in continuous-time systems, taking advantage of the parallel that exists between the two types of systems. Hybrid or sampled-data systems are then treated as special cases in which the techniques of discrete-time analysis can be applied conveniently. I believe that this approach greatly facilitates the learning of discrete-time systems as well as sampled-data systems.

The concept of the state of a system is introduced in the first chapter. Identification of the initial state with the initial conditions immediately dispels the veil of mystery surrounding the concept of the state and helps to clarify the meaning of the state to the student. The system response is then discussed in terms of zero-input and zero-state components.

As in all my previous books I have used mathematics not so much to prove abstract axiomatic theory, as to enhance physical understanding. Logical motivation is provided for introducing new concepts. Whenever possible, theoretical results are interpreted heuristically, supported by carefully chosen examples and illustrations. Written primarily for juniors and seniors, the material included is, I believe, the absolute minimum that a prospective graduate of electrical engineering or systems engineering must acquire. The book is self-contained, requiring only a modest background in calculus and in the elements of network theory or dynamic systems. It can therefore be used effectively for self-study by practicing engineers.

Thanks are due to Professors J. B. Cruz, W. D. Thayer, M. E. Van Valkenburg, and Mr. J. M. Elfelt for several suggestions. Discussions with Professor W. H. Huggins were especially helpful. A. Alonso and D. Sousa assisted in proofreading. Thanks are also due to John Wiley & Sons for allowing me to reproduce certain material from my earlier works.

Notes to the Instructor

The entire contents of this book can be covered in about 90 classroom hours. By judicious choice of topics, it can be used for a course lasting anywhere from 30 to 90 hours. It is therefore suitable as a text for a course lasting for one to three quarters or one to two semesters.

Linear Systems: For a course on signals and systems (or systems theory), any one of the following combinations of chapters should be appropriate.

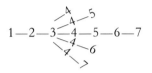

Chapter 4 gives two interpretations of the frequency domain: (1) the time-domain interpretation and (2) the conventional or transform interpretation. The instructor may omit the former interpretation (Secs. 4.1–4.8) without experiencing much discontinuity in the flow of the discussion. This, however, is not recommended unless there is no way of covering the desired material in a given time period.

Control Systems: The book can be used effectively as a text for a 30- to 45-hour course on control systems. For this purpose the following material is suggested. Chapters 1 and 2, part of Chapter 4 (Secs. 9–13, and 15–17 only), Chapter 5, and Appendixes A–F.

1

Introduction to Systems

1.1 INTRODUCTION

The dictionary gives several possible meanings of the word *system*. One of them is "A set of arrangement of things so related or connected as to form a unity or organic whole, as a solar system, irrigation system, or supply system." This rather broad definition includes all physical as well as nonphysical systems. Electrical, mechanical, electromechanical, hydraulic, acoustic, and thermal systems are examples of physical systems. Economic systems, political systems, traffic control systems, and industrial planning systems are examples of socioeconomic systems. The socioeconomic systems are made up of various components and their interactions. These are no more or less of systems than physical systems. As a result, the system theory applicable to physical and a class of nonphysical systems is emerging. Attempts are being made to apply system theory to such socioeconomic problems as education, transportation, public and private administration, and economic development. In this book we shall be concerned with physical systems in general, and with electrical, mechanical, and electromechanical systems in particular.

Each system performs some desired function (response or outputs) for a given set of driving functions (inputs). In electrical systems the driving functions are generally in the form of voltage and current sources, and the response will be voltages or currents at certain locations. For mechanical systems the inputs may be forces (or displacements), and the response may be displacement or velocity at some point. For a given system there may be several inputs (driving functions) acting simultaneously, and there may be several outputs (responses) of interest. Before defining any relationship between inputs (driving functions) and outputs (responses), let us consider a simple mechanical system consisting of a mass M which is acted upon by a force f as shown in Fig. 1.1.

In this case the driving function (input) is $f(t)$ and the output (response) is the velocity v. According to Newton's law, for a constant mass M the force f and the velocity v of the mass are related by

$$f = M \frac{dv}{dt}$$

or

$$v = \frac{1}{M} \int f(\tau) \, d\tau \tag{1.1a}$$

1

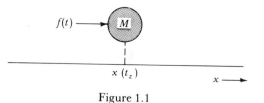

Figure 1.1

The velocity v is the response to the input force f. The velocity at any instant t is therefore the result of force f acting on M in the entire past. Hence the limits of integration in Eq. 1.1a are from $-\infty$ to t. Thus

$$v(t) = \frac{1}{M} \int_{-\infty}^{t} f(\tau)\, d\tau \tag{1.1b}$$

$$= \frac{1}{M} \int_{-\infty}^{0} f(\tau)\, d\tau + \frac{1}{M} \int_{0}^{t} f(\tau)\, d\tau \tag{1.1c}$$

From Eq. 1.1b (by letting $t = 0$) the first term on the left-hand side of Eq. 1.1c is seen to be $v(0)$. Hence

$$v(t) = v(0) + \frac{1}{M} \int_{0}^{t} f(\tau)\, d\tau \tag{1.1d}$$

Let us discuss the implications of Eq. 1.1d. From Eq. 1.1b, it is obvious that the velocity $v(t)$ of the mass at any instant t can be computed if we know the force that acted upon the mass in the entire past $(-\infty, t)$. In practice, however, it is impossible to keep record of force acting on a mass over the entire history of its existence. In such case the use of Eq. 1.1d proves very attractive. Suppose we know the force from some moment $t = 0$ onward we can still calculate $v(t)$ for $t \geq 0$ provided $v(0)$, the initial velocity (velocity at $t = 0$) was known. Thus $v(0)$ has all the relevant information about the entire past of the forces acting on M, that we need to calculate $v(t)$ for $t \geq 0$. The velocity $v(0)$ represents the value of the velocity at the initial moment $t = 0$ and is generally referred to as the *initial condition*. In the present case we arbitrarily chose the initial moment to be $t = 0$. We can, however, use $t = t_0$ as the initial instant.

Eq. 1.1d can be easily generalized as

$$v(t) = v(t_0) + \frac{1}{M} \int_{t_0}^{t} f(\tau)\, d\tau \tag{1.1e}$$

In the present situation we conclude that the response (velocity) for $t \geq t_0$ is a function of the initial condition $v(t_0)$ and the input $f(t)$ for $t \geq t_0$. This fact may be expressed as

$$v(t) = \phi[v(t_0), f(t)], \qquad t \geq t_0 \tag{1.2}$$

This result is true in general. A response of a system for $t \geq t_0$ is a function of the initial conditions at $t = t_0$ and the input(s) $f(t)$ for $t \geq t_0$.

In the present problem we needed only one initial condition. However, in general several initial conditions may be necessary. Consider again our problem of mass M acted upon by a force f. Let us determine the position x of the mass at some time t. We have

$$\frac{dx}{dt} = v$$

Hence

$$x(t) = \int_{-\infty}^{t} v(\tau)\, d\tau \qquad\qquad (1.3a)$$

$$= \int_{-\infty}^{0} v(\tau)\, d\tau + \int_{0}^{t} v(\tau)\, d\tau \qquad\qquad (1.3b)$$

From Eq. 1.3a (by letting $t = 0$), the first term on the right-hand side of Eq. 1.3b is seen to be $x(0)$. Hence

$$x(t) = x(0) + \int_{0}^{t} v(\zeta)\, d\zeta \qquad\qquad (1.3c)$$

where ζ is the dummy variable of integration. Substituting Eq. 1.1d in Eq. 1.3c we obtain

$$x(t) = x(0) + \int_{0}^{t} \left[v(0) + \frac{1}{M} \int_{0}^{\zeta} f(\tau)\, d\tau \right] d\zeta$$

$$= x(0) + v(0)t + \frac{1}{M} \int_{0}^{t} \int_{0}^{\zeta} f(\tau)\, d\tau\, d\zeta \qquad\qquad (1.3d)$$

It is obvious from Eq. 1.3d, that if the input force $f(t)$ is known from $t = 0$ onward, then to find the position $x(t)$ for $t \geq 0$, we need two initial conditions $x(0)$ and $v(0)$. Thus

$$x(t) = \phi[x(0), v(0), f(t)], \qquad t \geq 0$$

Note that the initial instant is arbitrarily chosen at $t = 0$. The results can be generalized for any value of the initial instant.

1.2   STATE OF A SYSTEM: THE VITAL KEY

The initial conditions at some $t = t_0$ collectively are called as the **state** of the system at $t = t_0$. Thus if a system has n initial conditions $x_1(t_0)$, $x_2(t_0)$, $\ldots$, $x_n(t_0)$, the state of the system at $t = t_0$ **(initial state)** is given by $x_1(t_0)$, $x_2(t_0)$, $\ldots$, $x_n(t_0)$. We may say that the state at some instant t_0 contains all the relevant information of the past history of the system that is needed to obtain the response for $t \geq t_0$ when the input is given for $t \geq t_0$. For the mass-force system in Fig. 1.1, the state of the system at $t = t_0$ is given by $x(t_0)$ and $v(t_0)$.

The state of a system at any time t_0 is the smallest set of numbers $x_1(t_0)$, $x_2(t_0)$, $\ldots$, $x_n(t_0)$ which is sufficient to determine the behavior of the system for all time $t \geq t_0$ when the input to the system is known for $t \geq t_0$.

In general there may be several inputs applied simultaneously at various points in a system and there may be several variables of interest which will be considered as response (output). For simplicity, we shall first consider the case of single-input, single-output system and then later extend the discussion to a general case of multiple-input, multiple-output system. A response $y(t)$ for $t \geq t_0$ of a system is a function of the state of the system at some initial instant $t = t_0$, and the input $f(t)$ for $t \geq t_0$. This can be expressed as

$$y(t) = \phi[x_1(t_0), x_2(t_0), \ldots, x_n(t_0), f(t)] \qquad t \geq t_0 \qquad (1.4a)$$

For the sake of convenience the initial state at $t = t_0$ represented by numbers $x_1(t_0), x_2(t_0), \ldots, x_n(t_0)$, will be denoted by $\{x(t_0)\}$. Using this notation, Eq. 1.4a can be expressed as

$$y(t) = \phi[\{x(t_0)\}, f(t)], \qquad t \geq t_0 \qquad (1.4b)$$

Figure 1.2 shows the block diagram representation of a system. A system is

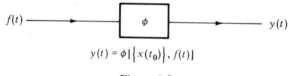

$f(t)$ $\qquad$ ϕ $\qquad$ $y(t)$

$$y(t) = \phi[\{x(t_0)\}, f(t)]$$

Figure 1.2

characterized by input(s), output(s) and the functional block diagram. The functional block diagram should be labeled by the input-output relationship (such as in Eq. 1.4b) for complete characterization of the system.

At this point we make an important observation. It was seen that the response $y(t)$ at any instant $t \geq t_0$ can be determined from the knowledge of initial state $\{x(t_0)\}$ and the input $f(t)$ over the interval (t_0, t). Let us consider the output y at $t = t_0$. From the above discussion it is evident that $y(t_0)$ can be determined from the knowledge of the initial state $\{x(t_0)\}$ and the input $f(t)$ over the interval (t_0, t_0). The latter is $f(t_0)$. Hence the response at any instant is determined completely from the knowledge of the state of the system at that instant (and the input at that instant). This result is also true for multiple-input, multiple-output systems. Every output (response) at any given instant t is completely determined by the state of the system (and the inputs) at that instant. Therefore the state of a system at some instant tells us everything about the system at that instant. It is evident that the state of a system is the single most important attribute of a system. It is the vital key to the system.

As an example, consider the electrical network shown in Fig. 1.3. We can easily show that if the capacitor voltage x_1 and the inductor current x_2 are known (along with the input) at any instant, then all voltages and currents in this network at that instant are known. From Fig. 1.3 it can be seen that

$$v_{R_1} = f(t) - x_1$$

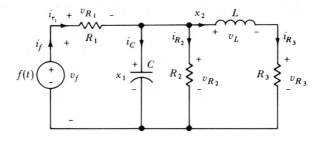

Figure 1.3

$$i_{R_1} = \frac{1}{R_1} v_{R_1} = \frac{1}{R_1} [f(t) - x_1]$$

$$v_{R_2} = x_1$$

$$i_{R_2} = \frac{1}{R_2} x_1$$

$$i_C = i_{R_1} - i_{R_2} - x_2 = \frac{1}{R_1} [f(t) - x_1] - \frac{1}{R_2} x_1 - x_2$$

$$v_C = x_1$$

$$i_{R_3} = x_2$$

$$v_{R_3} = R_3 x_2$$

$$v_L = v_{R_2} - v_{R_3} = x_1 - R_3 x_2$$

$$i_L = x_2$$

$$i_f = i_{R_1} = \frac{1}{R_1} [f(t) - x_1]$$

$$v_f = f(t)$$

$$(1.5)$$

It can be seen that if we know the capacitor voltage x_1, the inductor current x_2, and the input $f(t)$ at any instant, all the voltages and the currents in the network at that instant are determined. Consequently the state of this network at any instant t_0 is given by $x_1(t_0)$, $x_2(t_0)$. For electrical networks, in general, it can be shown that all voltages, and the currents at any instant are determined by the values of all inductor currents and all capacitor voltages (and the inputs) at that instant. Hence in electrical network the state of a system at any instant is given by all inductor currents and all capacitor voltages. It will also be seen that in mechanical systems all forces, displacements, and velocities at any instant are determined from the knowledge of positions, and velocities of all junctions† (along with inputs) at that instant. Hence the positions and velocities of all junctions represent a state of a mechanical system.

The concept of a state is very important. As the term implies, state of a system represents its position or status.

---

†A junction is the point where two or more elements are connected.

STATE OF A SYSTEM IS NOT UNIQUELY SPECIFIED

At this point we shall note that the state of a system can be specified in several ways. In the force-mass system (Fig. 1.1), for example, we may specify the state of a system by $x(t_0)$, $v(t_0)$, or we may define new variables w_1 and w_2 as

$$w_1 = a_{11}x + a_{12}v$$
$$w_2 = a_{21}x + a_{22}v$$

(1.6)

Solving these equations simultaneously, we can express x and v in terms of† w_1 and w_2. Therefore if w_1, and w_2 are known x and v are also known. Hence the response of the system can also be obtained from the knowledge of the input $f(t)$ and the initial conditions $w_1(t_0)$ and $w_2(t_0)$. Therefore, by definition $w_1(t_0)$, $w_2(t_0)$ also specify the state of the system.

In the electrical network shown in Fig. 1.3, one can easily show that the voltages and the currents in all the branches at any instant are determined from the knowledge of v_{R_1} and v_{R_3} (and the input) at that instant. Hence v_{R_1} and v_{R_3} also specify the state of the system. Other possible sets of variables which can specify the state of this network are (i_L, v_L), (i_C, v_C), (v_{R_1}, v_L), (i_C, v_L), and (i_C, v_{R_3}). It is left as an exercise for the reader to show that these sets can specify the state of the system in Fig. 1.3.

1.3 CLASSIFICATION OF SYSTEMS

The systems can be broadly classified into following categories: ‡

1. Linear and nonlinear systems.
2. Constant-parameter and time-varying-parameter systems.
3. Instantaneous and dynamic systems.
4. Lumped-parameter and distributed-parameter systems.
5. Continuous-time and discrete-time systems.

We shall now discuss the nature of these classifications. To begin with, we shall consider the single-input, single-output system and then extend the results to a general case.

1. LINEAR AND NONLINEAR SYSTEMS

Before defining a linear system it is necessary to understand the important concept of linearity.

Linearity Concept

The reader is no doubt familiar with the rudimentary notions of linearity. Broadly speaking, the linearity property implies two important concepts (i) *homo-*

---

† This is true provided $a_{11}a_{22} - a_{12}a_{21} \neq 0$. This condition is implicit here.

‡ There are few more classifications such as (1) discrete-state and continuous-state systems, and (2) deterministic and probabilistic systems. These classes, however, are beyond the scope of this book and will not be considered.

geneity and (ii) *superposition.* Homogeneity property implies that a k-fold increase in the input causes a k-fold increase in the output for any value of k (Fig. 1.4a). If $f(t)$ is the input and $y(t)$ is the corresponding response, then $ky(t)$ is the response when the input is $kf(t)$. This fact may be represented as

   If

$$f(t) \rightarrow y(t)$$

then

$$kf(t) \rightarrow ky(t) \tag{1.7}$$

The superposition property states that if there are several inputs (or causes) acting on a system, then the total response of the system due to all the inputs (causes) can be determined on installment basis by considering only one input (cause) at a time and by assuming the remaining inputs (causes) to be zero. The total output is the sum of all such output components computed by considering one input (cause) at a time. This property may be expressed as follows. If $y_1(t)$ is the response of a system to the input (cause) $f_1(t)$ and $y_2(t)$ is the response of the same system to the input (cause) $f_2(t)$, then if both inputs are acting simultaneously—that is, when the input is $f_1(t) + f_2(t)$—the response is $y_1(t) + y_2(t)$. This property may be expressed as

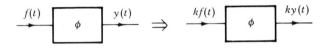

(a) Homogeneity property

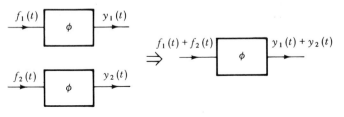

(b) Superposition property

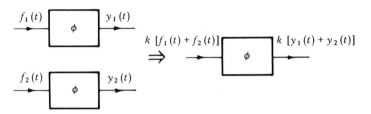

(c) Linearity property (homogeneity + superposition)

Figure 1.4

follows (see Fig. 1.4b):

If

$$f_1(t) \rightarrow y_1(t)$$
$$f_2(t) \rightarrow y_2(t)$$

then

$$f_1(t) + f_2(t) \rightarrow y_1(t) + y_2(t) \tag{1.8}$$

We can combine both the properties (Eqs. 1.7 and 1.8) into one equation as follows (see Fig. 1.4c):

If

$$f_1(t) \rightarrow y_1(t)$$
$$f_2(t) \rightarrow y_2(t)$$

then

$$k[f_1(t) + f_2(t)] \rightarrow k[y_1(t) + y_2(t)] \tag{1.9}$$

Note that Eq. 1.9 embodies the essence of both equations 1.7 and 1.8. Hence Eq. 1.9 represents the *linearity property* (homogeneity + superposition).

Definition of Linear and Nonlinear Systems

We shall now use the linearity concepts to define a linear system. As observed earlier, the output of a system depends upon not only the input $f(t)$ but also the initial state $\{x(t_0)\}$. We may view this as if the output (response) depends upon two different inputs or causes; $\{x(t_0)\}$ and $f(t)$. Consequently for a linear system we must demand that the output (response) should be given by a sum of two components, arising because of the two different causes. The component due to each cause is computed by assuming that only that cause is present and the other cause is zero. To be specific, the output of a linear system should be given by a sum of two components (i) the output of the system with the given initial state $\{x(t_0)\}$ and with zero input—that is, $f(t) = 0$ (*zero-input component*); and (ii) the output of the system with the given input $f(t)$ but with zero initial state—that is, $\{x(t_0)\} = 0$ (*zero-state component*). Thus the response $y(t)$ can be expressed as

$$\underbrace{y(t)}_{\substack{\text{total} \\ \text{response}}} = \underbrace{y_x(t)}_{\substack{\text{zero-input} \\ \text{response}}} + \underbrace{y_f(t)}_{\substack{\text{zero-state} \\ \text{response}}} \tag{1.10}$$

where $y_x(t)$ (the zero-input response) is a function of the initial state only and $y_f(t)$ is a function of the input $f(t)$ only. This property of a system which allows us to separate the components due to the initial state and the input is called the *decomposition property*. Thus the output of a linear system can be separated into two components. The first component (zero-input component) is obtained by letting the input be zero. This component of response is caused entirely by the initial conditions or the initial state. The second component (zero-state component) is obtained by letting the initial state be zero. This component of the response results due to the input alone. The decomposition property allows us to evaluate the two response components arising because of two different causes in a simpler way. Component due to each cause is computed as if only that cause is present and the other cause is zero. The total response is the sum of the responses due to each cause.

In the force-mass system in Fig. 1.1, the response $x(t)$ is given by

$$x(t) = \underbrace{x(0) + v(0)t}_{\substack{\text{zero-input response} \\ y_x(t)}} + \underbrace{\frac{1}{M} \int_0^t \int_0^\zeta f(\tau)\, d\tau\, d\zeta}_{\substack{\text{zero-state response} \\ y_f(t)}} \qquad (1.11)$$

It is obvious that the response can be separated into two components; the first component exists entirely due to initial conditions. If the initial state is zero—that is if $x(0) = 0$, $v(0) = 0$—this component is zero. Hence the first component is the zero-input response $y_x(t)$. The second component exists entirely because of the input $f(t)$ and is zero if $f(t) = 0$. Hence this is zero-state response $y_f(t)$. Obviously this system satisfies the decomposition property.

The decomposition property alone, however, is not enough for a system to qualify as a linear system. A linear system must exhibit linearity under all possible input conditions. Thus when the initial state is zero, the zero-state response $y_f(t)$ must exhibit linearity (superposition and homogeneity) with respect to various inputs (*zero-state linearity*). Similarly when the input $f(t) = 0$, the zero-input response y_x must exhibit linearity with respect to various initial states (*zero-input linearity*). Let us explain what we mean.

i. Zero-State Linearity.   When the initial state of the system is given to be zero, that is, $\{x(0)\} = 0$, the response is given by $y_f(t)$. For a linear system, we must demand that $y_f(t)$ exhibit linearity (homogeneity + superposition in Eq. 1.9) with respect to $f(t)$. Thus if $y_{f_1}(t)$ is the zero-state response due to input $f_1(t)$ and $y_{f_2}(t)$ is the zero-state response due to input $f_2(t)$, then the zero-state response due to input $k[f_1(t) + f_2(t)]$ must be $k[y_{f_1}(t) + y_{f_2}(t)]$. For the force-mass system this is certainly true because

$$y_f(t) = \frac{1}{M} \int_0^t \int_0^\zeta f(\tau)\, d\tau\, d\zeta$$

It is also true that

$$\frac{1}{M} \int_0^t \int_0^\zeta k[f_1(\tau) + f_2(\tau)]\, d\tau\, d\zeta$$

$$= \frac{k}{M} \left[\int_0^t \int_0^\zeta f_1(\tau)\, d\tau\, d\zeta + \int_0^t \int_0^\zeta f_2(\tau)\, d\tau\, d\zeta \right]$$

The left-hand side of this equation represents the zero-state response due to input $k[f_1(t) + f_2(t)]$. The right-hand side represents k times the sum of zero-state responses due to inputs $f_1(t)$ and $f_2(t)$, respectively. This is precisely the linearily property. Hence the system is zero-state linear.

ii. Zero-Input Linearity.   When the input $f(t) = 0$, the response of the system is given by zero-input response $y_x(t)$. For a linear system we must demand that $y_x(t)$ exhibit linearity (Eq. 1.9) with respect to initial state $\{x(0)\}$. Thus if $y_{x_a}(t)$ is the zero-input response of the system to initial state $\{x_a(0)\}$ and $y_{x_b}(t)$ is the zero-input response of the system when the initial state $\{x_b(0)\}$, then the zero-input response must be $k[y_{x_a}(t) + y_{x_b}(t)]$ when the initial state is

$$k[\{x_a(0)\} + \{x_b(0)\}].$$

For the force-mass system the initial state is $\{x(0), v(0)\}$ and $y_x(t)$ is given by (see Eq. 1.11)

$$y_x(t) = x(0) + v(0)t \tag{1.12}$$

Consider the initial state $\{x_a(0)\}$ given by

$$\{x_a(0)\} = \{a_1, a_2\}$$

that is,

$$x(0) = a \text{ and } v(0) = a_2$$

The zero-input response $y_{x_a}(t)$ is given by (see Eq. 1.12)

$$y_{x_a}(t) = a_1 + a_2 t$$

For initial state $\{x_b(0)\} = \{b_1, b_2\}$, the zero-input response $y_{x_b}(t)$ is given by (see Eq. 1.12)

$$y_{x_b}(t) = b_1 + b_2 t$$

If the initial state is $k[\{x_a(0)\} + \{x_b(0)\}] = \{k(a_1 + b_1), k(a_2 + b_2)\}$ the zero-input response is given by (see Eq. 1.12)

$$k(a_1 + b_1) + k(a_2 + b_2)t = k[(a_1 + a_2 t) + (b_1 + b_2 t)]$$
$$= k[y_{x_a}(t) + y_{x_b}(t)]$$

Hence this system is zero-input linear.

For a linear system we shall demand that it has a decomposition property, that it is zero-input linear and zero-state linear. All these three properties are the logical consequences of the linearity principle which allows us to evaluate the effect (response) due to several causes (inputs) by considering only one cause at a time. We shall now define a linear system as follows: *A system is linear if and only if it has a decomposition property and it is zero-input as well as zero-state linear.*

A system is nonlinear if it is not linear. The force-mass system in Fig. 1.1 satisfies all the three linearity conditions. Hence it is a linear system. As an example consider a linear system with two state variables x_1 and x_2.

(a) Let the zero-input response of this system be $2 + 3e^{-2t}$ when the initial state is given by $x_1(0) = 2$ and $x_2(0) = 1$—that is when the initial state is given by $\{2, 1\}$. If we increase the initial state five-fold—that is, if the initial state is $\{10, 5\}$, then the zero-input response will also increase five-fold, and will be given by $5(2 + 3e^{-2t})$.

(b) Let the zero-input response be $2 - 3e^{-2t}$ when the initial state is $\{1, 2\}$ and be $5 + 2e^{-2t}$ when the initial state is $\{4, 1\}$. If we form a new state by adding these two states—that is, if the new state is $\{5, 3\}$, then the zero-input response corresponding to this new state is the sum of the zero-input responses due to two component states, and is given by $(2 - 3e^{-2t}) + (5 + 2e^{-2t}) = 7 - e^{-2t}$.

(c) If the zero-state response for input e^{-t} is given by $2 - e^{-t} + e^{-2t}$, then the zero-state response for the input ke^{-t} is $k(2 - e^{-t} + e^{-2t})$.

(d) If the zero-state response for the input e^{-t} is $2 - e^{-t} + e^{-2t}$, and is $3 + 2e^{-2t} + e^{-5t}$ for the input e^{-5t}, then the zero-state response for the input $e^{-t} + e^{-5t}$ is $(2 - e^{-t} + e^{-2t}) + (3 + 2e^{-2t} + e^{-5t}) = 5 - e^{-t} + 3e^{-2t} + e^{-5t}$.

We shall give some examples of systems which do not meet all three linearity conditions. Consider the systems whose response $y(t)$ is given by

(a) $y(t) = \log x(0) + f^2(t)$

(b) $y(t) = x^2(0) \log f(t)$

(c) $y(t) = 3x(0) + f^2(t)$

(d) $y(t) = x^2(0) + \displaystyle\int_0^t f(\tau)\,d\tau$

System (a) has decomposition property but is neither zero-input linear nor zero-state linear. System (b) satisfies none of the three linearity conditions. System (c) has decomposition property and is also zero-input linear but it is not zero-state linear. System (d) has decomposition property, is zero-state linear, but it is not zero-input linear.

It will be shown later that if the input $f(t)$ and the output $y(t)$ are related by equation of the form

$$\frac{d^n y}{dt^n} + a_{n-1}\frac{d^{n-1}y}{dt^{n-1}} + \cdots + a_1\frac{dy}{dt} + a_0 y = b_m\frac{d^m f}{dt^m} + \cdots + b_1\frac{df}{dt} + b_0 f(t) \quad (1.13)$$

The system is a linear system. The coefficients a_i's and b_i's in this equation may be constants or functions of time.

Comment on Linear Systems

Analysis of linear systems is greatly simplified because of its properties mentioned earlier. First because of the decomposition property (Eq. 1.10) we may evaluate the two components of the output with ease. The zero-input component can be obtained by assuming zero input, and the zero-state component can be obtained by assuming the system in zero state. Furthermore, if we express $f(t)$ as a sum of simpler functions

$$f(t) = a_1\psi_1(t) + a_2\psi_2(t) + \cdots$$

then by virtue of zero-state linearity, the zero-state response is given by

$$y_f(t) = a_1 y_1(t) + a_2 y_2(t) + \cdots \quad (1.14)$$

where $y_k(t)$ is the zero-state response of the system to the input $\psi_k(t)$. This property appears only mildy valuable. In reality, however, it is extremely useful and opens many new avenues for analyzing linear systems. The methods such as convolution and frequency-domain analysis in linear systems (Chapters 3 and 4) are based on this property. In convolution analysis, the input $f(t)$ is expressed as a sum of impulse functions, whereas in frequency-domain analysis (Fourier and Laplace transforms) $f(t)$ is expressed as a sum of exponential functions of the form e^{st}, where $s = \sigma + j\omega$. These techniques will be discussed later in more details.

Nonlinear systems do not have the three above-mentioned properties of linear systems. This makes it difficult to analyze nonlinear systems. There are no straightforward methods of analysis and no general solutions available. Fortunately many nonlinear systems can be approximated by linear systems over a limited range of operation. For these reasons study of linear systems is important. Most of this book is devoted to linear systems.

2. CONSTANT-PARAMETER (TIME-INVARIANT) AND TIME-VARYING SYSTEMS

Systems whose parameters do not change with time are called *constant-parameter systems* or *time-invariant systems*. Most of the systems observed in practice are of this nature or can reasonably be approximated to this type.

If the system itself does not change with time, then it is apparent that for a given initial state the shape of the output will depend only on the shape of the input and not on the instant of application of the input. It is of course assumed that the initial state in both cases is identical. This property is expressed graphically in Fig. 1.5.

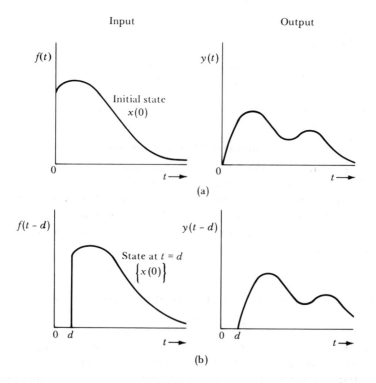

Figure 1.5

One can easily verify that the systems in Fig. 1.1 and 1.2 are cases of time-invariant systems. It will be seen later that all systems whose input-output relations are described by linear differential equations of the form in Eq. 1.13 where all the coefficients a_i's and b_i's are constants, are *linear time-invariant* systems. If the coefficients a_i's and b_i's are functions of time, the system is a linear time-varying system. All the networks composed of RLC networks and other commonly used active elements such as vacuum tubes and transistor (also gyrators) represent time-invariant systems. A familiar example of time-varying system is furnished by a carbon microphone where the resistance R is a function of the mechanical pressure (generated by sound waves) on carbon granules of the microphone. The equivalent circuit of the microphone is given in Fig. 1.6. The response is the

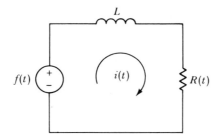

Figure 1.6

current $i(t)$. The equation describing this circuit is given by

$$L \frac{di}{dt} + R(t)i(t) = f(t)$$

One of the coefficients in this equation is $R(t)$, which is time-varying. Hence this circuit is an example of (linear) time-varying system.

3. INSTANTANEOUS AND DYNAMIC SYSTEMS

As observed earlier, the output of a system at any instant t depends upon the entire past of the input. In a special class of systems, however, the output at any instant t depends only on the input at that instant t. In resistive networks for example, any output in the network at some instant t depends only on the input at instant t. In these systems the past history is not relevant in determining the response at any particular instant. The response y can be expressed as

$$y = \phi(f)$$

where y and f are the values of the output and the input at some instant t. Such systems are said to be *instantaneous* or *memoryless* systems. More precisely, a system is said to be instantaneous (or memoryless) if its output any instant t depends at most on the strength of the sources (input signals) at the same instant but not on the past or future values of the sources. Otherwise the system is said to be *dynamic* (or system with memory). A system whose response at any instant t is completely determined by the input signals over the past T seconds [interval $(t - T)$ to t] is a finite-memory system and is said to have a memory of T seconds. The networks containing inductive and capacitive elements have, in general, infinite memory because, the response of such networks at any instant t is determined by the input signals over the entire past $(-\infty, t)$. The same is true of the force-mass system in Fig. 1.1.

In instantaneous or memoryless systems, the response (output) at any instant t depends only on the driving function at instant t. Initial conditions (or past history) do not play any role in determining the response. As a result there are no states variables for such systems. The input and the output are in general algebraically related.

4. LUMPED- AND DISTRIBUTED-PARAMETER SYSTEMS

A system is a collection of individual elements interconnected in a particular way. A lumped system consists of lumped elements. In a lumped model the energy in the system is considered to be stored or dissipated in distinct isolated elements (resistors, capacitors, inductors, masses, springs, dashpots, and the like). Also it is assumed that the disturbance initiated at any point is propagated instantaneously at every point in the system. This implies that the dimensions of the elements are very small compared to the wavelength of the signals to be transmitted. In a lumped-parameter electrical element the voltage across terminals and the current through it, are related through a lumped parameter. In contrast to lumped systems we have distributed systems such as transmission lines, waveguides, antennas, and mechanical shafts, where it is not possible to describe a system by lumped parameters. How may we describe a shaft which transmits the torque? The shaft has a mass (moment of inertia). It also acts as a (torsional) spring. But the effect of mass and spring is intermingled and are distributed uniformly throughout the length of the shaft. In an electrical transmission line, the resistance, inductance, and capacitance of the line are also continuously distributed along the line and cannot be isolated meaningfully by lumped elements. All three effects are present at each point along the line and are intermingled.

In addition, in distributed-parameter systems it takes a finite amount of time for a disturbance at one point to be propagated to the other point. We thus have to deal not only with the independent variable t (time), but also the space variable z. The equations describing the distributed-parameter systems are therefore partial differential equations.

In fact, all systems are distributed-parameter systems. However, we approximate these systems by lumped systems. Such a simplification is often justified if the dimensions of the system elements are small compared to the wavelengths of the input signals.

5. CONTINUOUS-TIME AND DISCRETE-TIME SYSTEMS

For any system the input and output signals are defined for all values of time. There are, however, cases where one is interested only in knowing what happens at some discrete instants of time. A familiar example is the digital computer, whose operation concerns us only in terms of outcomes at discrete instants. The response actually does exist between these instants but it is of no significance so far as the system model is concerned. In such cases the system analysis is considerably simplified by considering the model with input and output signals at discrete instants of time and finding the appropriate relationships. The inputs and outputs are represented as $f(t_k)$, $y(t_k)$ in contrast to $f(t)$, $y(t)$ for continuous systems. The relationship between discrete variables is given by difference equations. These systems are treated in Chapter 7.

1.4 MULTIPLE-INPUT, MULTIPLE-OUTPUT SYSTEMS

So far we have considered the systems with only one input and one output (response). In practice, there may be several inputs (applied at different points in the system) and there may be several outputs of interest. Such a system may be represented as shown in Fig. 1.7. The response $y_k(t)$ in general will

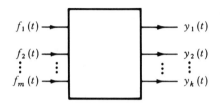

Figure 1.7

be a function of the initial state $\{x(t_0)\}$ of the system and all the inputs $f_1(t)$, $f_2(t)$, $\ldots, f_m(t)$. Thus we have

$$
\begin{aligned}
y_1(t) &= \phi_1[\{x(t_0)\}, f_1(t), f_2(t), \ldots, f_m(t)] \\
y_2(t) &= \phi_2[\{x(t_0)\}, f_1(t), f_2(t), \ldots, f_m(t)] \\
y_n(t) &= \phi_n[\{x(t_0)\}, f_1(t), f_2(t), \ldots, f_m(t)]
\end{aligned}
\tag{1.15}
$$

For multiple-input, multiple-output system, definition of linearity is a simple extension of the definition used earlier for the single-input, single-output case. We define a system to be linear if and only if each of the equations in Eq. 1.15 has a decomposition property

$$
y_k(t) = y_{x_k}(t) + y_{f_1}(t) + y_{f_2}(t) + \cdots + y_{f_m}(t)
\tag{1.16}
$$

where $y_{x_k}(t)$ is the zero-input component of the response $y_k(t)$ and $y_{f_j}(t)$ is the zero-state component of the response $y_k(t)$ caused solely by the input $f_j(t)$. This is the response of the system when the initial state and all the inputs but $f_j(t)$ are zero. In addition the function on $y_{x_k}(t)$ is linear with respect to initial state (zero-input linearity), and $y_{f_j}(t)$ is linear with respect to $f_j(t)$ for all $j = 1, 2, \ldots, m$ (zero-state linearity).

1.5   SYSTEM MODELING

A system is a collection of elements interconnected in a certain manner. Complete analysis of any system implies accurate prediction of the behavior of the system. For the purpose of analysis, the system must be described mathematically. This step, known as *modeling,* is crucial in the analysis problem.

When we study any system we first construct a mathematical model which approximates the system. The complexity of the model will depend upon the accuracy desired. As an example, consider a mass M acted upon by a force f, as shown in Fig. 1.1. A fairly accurate mathematical model of this system is provided by Newton's law:

$$f = M \frac{d^2x}{dt^2}$$

This model depicts the reality with fair accuracy when the mass velocity is small enough so that the air friction has a negligible effect. If the force f is of such a magnitude as to cause large velocity of the mass, we must account for the frictional force. Experimentally, it is found that the force due to air friction is proportional to the square of the velocity. Thus a better model in this case will be

$$f = M \frac{dv}{dt} + Kv^2$$

where v is the velocity of the mass. If the mass velocity approaches the velocity of light (3×10^8 m/sec), the theory of relativity predicts that the mass M cannot be assumed constant. In such case a more accurate model is given by

$$f = \frac{d}{dt}(Mv) + Kv^2$$

It is evident that, for the same system, there are different models depending on the circumstances and the degree of accuracy desired.

We shall give a few examples of modeling electrical, mechanical, and electromechanical systems.

ELECTRICAL NETWORKS

The basic elements used in modeling electrical systems are resistors, capacitors, inductors, voltage sources and current sources.† It is assumed that the reader is familiar with terminal properties of these elements. The network equations can be written by either using mesh or node equations. The network equations can be written in a more general framework of loop and cut-set equations. Yet another form (state equations) will be discussed in Chapter 6.

The network in Fig. 1.8 has two meshes. We define two-mesh currents i_1 and i_2 circulating along each of the meshes. Application of Kirchhoff's voltage law (KVL) along these meshes yields two-mesh equations as

---

†One may also add ideal transformers and gyrators.

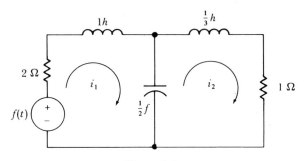

Figure 1.8

$$2i_1 + \frac{di_1}{dt} + 2 \int_{-\infty}^{t} (i_1 - i_2) \, dt = f(t)$$

$$2 \int_{-\infty}^{t} (i_2 - i_1) \, dt + \frac{1}{3} \frac{di_2}{dt} + i_2 = 0$$

(1.17)

These are two simultaneous integrodifferential equations in two variables i_1 and i_2, and can be solved by techniques discussed in Chapters 2, 3, and 4. The current and voltage associated with any element in the network can be expressed in terms of i_1 and i_2. So once i_1 and i_2 are determined, all the currents and voltages in the network are determined. The node method utilizes node voltages as unknowns.

MECHANICAL SYSTEMS

A planar motion can be resolved into translational (rectilinear) motion and rotational (torsional) motion. Translational motion will be considered first. We shall restrict ourselves to motions in only one dimension.

Translational Systems

The basic elements used in modeling translational systems are springs, masses and dashpots. A linear spring stretches by an amount proportional to the force applied (Hooke's law). The force required to stretch the spring by an amount x is given by

$$\text{Force} = Kx \qquad (1.18)$$

where K is a constant called the *stiffness of the spring*.

Similarly, for a linear dashpot, which operates by virtue of viscous friction, the force is proportional to the relative velocity $\dot{x}$ of one surface with respect to the other.

$$\text{Force} = D\dot{x} \qquad (1.19)$$

where D is a constant called the *damping coefficient*.

The mechanical system in Fig. 1.9 consists of a mass M acted upon by external force $f(t)$, two springs K_1 and K_2 and a dashpot D_2 (which operates on viscous friction).   There is also a viscous friction between the mass M and the ground. We shall now give a systematic procedure for writing system equations.

1. Determine all the junctions† and label the motion of each junction. In Fig. 1-9a there are 2 junctions whose motions are labeled x_1 and x_2 with reference directions.
2. Draw a free-body diagram at each junction.

The free-body diagram at junction 1 is shown in Fig. 1.9b. The mass M is

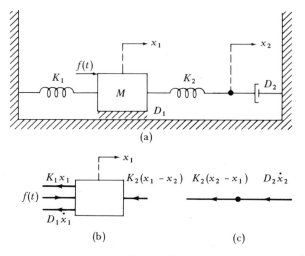

Figure 1.9

shown with all the forces acting on it.   These forces are (i) external force, $f(t)$, (ii) force, $K_1 x_1$, due to spring K_1, (iii) force, $K_2(x_1 - x_2)$, due to spring K_2, and (iv) force, $D_1 \dot{x}_1$, due to viscous friction.   The appropriate reference directions of these forces are shown in the free-body diagram. According to Newton's law, the net force in the positive reference direction x_1 must be equal to $M\ddot{x}_1$. Hence

$$M\ddot{x}_1 = f(t) - K_1 x_1 - K_2(x_1 - x_2) - D_1 \dot{x}_1 \qquad (1.20)$$

The free-body diagram at the second junction is shown in Fig. 1.9c. The net force at this junction is (mass) times $\ddot{x}_2$. Since the mass at the junction is zero, the net force must be zero. Hence

$$K_2(x_2 - x_1) + D_2 \dot{x}_2 = 0 \qquad (1.21)$$

Equations 1.20 and 1.21 are the two simultaneous differential equations in two variables, x_1 and x_2. These equations can be solved by techniques discussed in Chapters 2, 3, and 4.

_____

†A *junction*, as previously noted, is defined as a point where two or more elements are connected. The inertial frame of reference shown by hatched boundary is also a junction. All the motions are measured with respect to this junction.

Rotational Systems

Treatment of rotational systems is analogous to that of translational systems. For rotational systems we use torque instead of force, angular displacement instead of linear displacement, moment of inertia instead of mass, and rotational springs and dashpots instead of translational springs and dashpots. The terminal equations for various elements are analogous to those of translational elements.

ELECTROMECHANICAL SYSTEMS

There are a wide variety of electromechanical systems. Here we shall consider an example of conversion of electrical energy into mechanical energy using armature controlled shunt d-c motor. A field coil carrying current i_f generates a magnetic field in which the armature coil is free to rotate. The system is shown in Fig. 1.10a with a mechanical load. An external voltage $f(t)$ is applied across the armature coil with resistance R_a and inductance L_a (Fig. 1.10b). According to

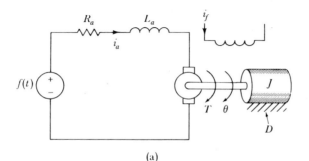

(a)

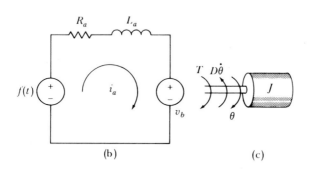

(b)   (c)

Figure 1.10

Faraday's law, when a coil rotates in a magnetic field, a voltage proportional to the angular velocity, $\dot{\theta}$, of the coil is generated in the coil. This voltage, v_b, called the *back emf*, is generated in the armature coil, and has such a polarity as to oppose its cause $f(t)$, as shown in Fig. 1.10b:

$$v_b = K_b \dot{\theta} \tag{1.22}$$

The mesh equation for the armature circuit in Fig. 1.10b is given by

$$R_a i_a + L_a \frac{di_a}{dt} + v_b = f(t)$$

Use of Eq. 1.22 yields

$$R_a i_a + L_a \frac{di_a}{dt} + K_b \frac{d\theta}{dt} = f(t) \tag{1.23}$$

When the armature current i_a passes through the armature coil placed in a magnetic field, a force is induced which is proportional to the product of the magnetic field and the armature current i_a. Since the magnetic field is proportional to the field current i_f, the torque generated is proportional to the product $i_f i_a$. Hence

$$\text{Torque} = K i_f i_a \tag{1.24}$$

For armature-controlled operation, the field current i_f is maintained constant. Hence the torque is proportional to the armature current:

$$T = K_t i_a \tag{1.25}$$

This torque is applied to the mechanical load. The free-body diagram for the mechanical system is shown in Fig. 1.10c. The rotating load with moment of inertia J experiences two torques (i) generated torque $K_t i_a$, (ii) viscous friction torque $-D\dot{\theta}$. Hence the net torque is $K_t i_a - D\dot{\theta}$. This according to Newton's law is $J\ddot{\theta}$. Hence

$$J \frac{d^2\theta}{dt^2} = K_t i_a - D \frac{d\theta}{dt} \tag{1.26}$$

Equations 1.23 and 1.26 are two simultaneous differential equations in two variables, i_a and θ. These equations can be solved by techniques discussed in Chapters 2, 3, and 4.

PROBLEMS

1.1.   One terminal of a spring (Fig. P-1.1) is fixed and the other terminal is moved with a velocity $v(t)$. Find the force $\phi(t)$ induced in the spring if the spring obeys Hooke's law (the induced force is proportional to the displacement, $\phi = Kx$). If the velocity $v(t)$ is known for $t \geq 0$, what initial condition(s) need be specified in order to find the force $\phi(t)$ for $t \geq 0$? Can the initial condition (initial state) be specified in more than one way?

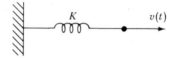

Figure P-1.1

1.2. Determine all the voltages and currents in the network in Fig. 1.3 at $t = 1$, if it is given that $x_1(1) = 5$, $x_2(1) = 2$, and the input $f(1) = 7$. The network parameters are

$$R_1 = 1, \qquad R_2 = 5, \qquad R_3 = 3$$

1.3. In an electrical network, inductor currents and capacitor voltage represent the state of the network. If we know the capacitor voltages and inductor currents (and the input) at some instant, then the voltages and currents in the entire network at that instant are determined. In the network shown in Fig. P-1.3, given the state at $t = t_0$ as

$$x_1(t_0) = 2, \qquad x_2(t_0) = 5$$

and input $f(t_0) = 10$. Determine voltages and currents associated with every branch of this network at $t = t_0$.

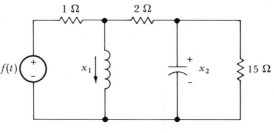

Figure P-1.3

1.4. For the network in Fig. 1.3, show that any of the following pairs of variables can specify the state of the network.

(a) i_L, v_L (b) i_C, v_C (c) v_{R_1}, v_L (d) i_C, v_L (e) i_C, v_{R_3}

1.5. State of a certain linear system is specified by two variables x_1, x_2. The input is $f(t)$, the output $y(t)$. It was observed that:

(i) When $f(t) = 0$, $\quad x_1(0) = 1$, $\quad x_2(0) = 0$, $\quad y(t) = 2e^{-t} + 3e^{-3t}$

$$\text{for} \quad t \geq 0$$

(ii) When $f(t) = 0$, $\quad x_1(0) = 0$, $\quad x_2(0) = 1$, $\quad y(t) = 4e^{-t} - 2e^{-3t}$

$$\text{for} \quad t \geq 0$$

Find $y(t)$ for $t \geq 0$, if $f(t) = 0$, $x_1(0) = 5$, and $x_2(0) = 3$.

1.6. An additional observation on the system in Prob. 1.5 showed that:

When $f(t) = u(t)$, $\quad x_1(0) = x_2(0) = 0$, $\quad y(t) = 2 + e^{-t} + 2e^{-3t}$ $\quad$ for $\quad t \geq 0$

Find $y(t)$ if $f(t) = 3u(t)$ and $x_1(0) = 2$, $x_2(0) = 5$.

1.7. For a certain system it was found that the response is $(6e^{-2t} - 5e^{-3t})u(t)$ when the input is $u(t)$ and the initial conditions are $x_1(0) = 1$ and $x_2(0) = 2$. The response is $(8e^{-2t} - 7e^{-3t})u(t)$ when the input is $3u(t)$ and the initial conditions are the same as before.

(a)   Find the response when the input is zero and the initial conditions are $x_1(0) = 1$, $x_2(0) = 2$.

(b)   Find the response when the input is $2u(t)$ and the initial conditions are zero.

1.8.   For a certain linear system with two initial condition variables x_1 and x_2, with input $f(t)$ and output $y(t)$, it was observed that:

(i) $y(t) = e^{-t}(7t + 5)$ when $x_1(0) = 5$, $x_2(0) = 2$,   and $f(t) = 0$
(ii) $y(t) = e^{-t}(5t + 1)$ when $x_1(0) = 1$, $x_2(0) = 4$,   and $f(t) = 0$
(iii) $y(t) = e^{-t}(t + 1)$ when $x_1(0) = 1$, $x_2(0) = 1$,   and $f(t) = u(t)$

Find

(a) $y(t)$ when $x_1(0) = 1$, $x_2(0) = 0$,   and $f(t) = 0$
(b) $y(t)$ when $x_1(0) = 0$, $x_2(0) = 1$,   and $f(t) = 0$
(c) $y(t)$ when $x_1(0) = 0$, $x_2(0) = 0$,   and $f(t) = u(t)$
(d) $y(t)$ when $x_1(0) = 2$, $x_2(0) = 1$,   and $f(t) = 3u(t)$

1.9.   Find $y(t)$ for the system in Prob. 1.7 if $x_1(0) = 1$, $x_2(0) = 2$, and $f(t)$ is as shown in Fig. P-1.9. Assume the system to be time-invariant.

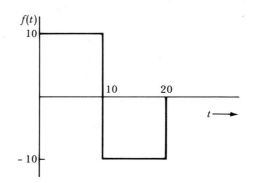

Figure P-1.9

1.10.   The response $y(t)$ and the input $f(t)$ for certain systems are related as follows [with $x(t_0)$ as initial condition]. State with reasons if these are linear or nonlinear systems.

(a) $y(t) = ax(t_0) + bf(t)$
(b) $y(t) = x^2(t_0) + 3t^3f(t)$
(c) $y(t) = x(t_0) \sin 5t + tf(t)$
(d) $y(t) = x(t_0) + 3t^3f(t)$
(e) $y(t) = x(t_0) + f(t)\,\dfrac{df}{dt}$
(f) $y(t) = 3x(t_0) + f^2(t)$
(g) $y(t) = x(t_0) + 2\displaystyle\int_0^t f(\tau)\,d\tau$

1.11.   Write system equations (mesh as well as node equations) for the network shown in Fig. P-1.11.

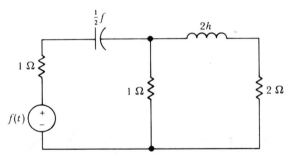

Figure P-1.11

1.12. Write system equations for the translational mechanical system shown in Fig. P-1.12.

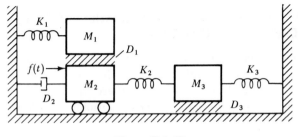

Figure P-1.12

1.13. Write system equations for field controlled dc motor shown in Fig. P-1.13.

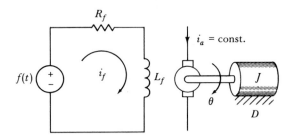

Figure P-1.13

The voltage $f(t)$ applied across the field coil which has a resistance of R_a ohms and inductance of L_f henries. The armature current i_a is maintained constant. The generated torque is therefore proportional to the field current i_f:

$$T = K_f i_f$$

2

Time-Domain Analysis

In this chapter we shall study one of the techniques of system analysis. This method deals directly with the time variable t. For this reason it is known as the *time-domain method*.

2.1 ELIMINATION OF VARIABLES IN SIMULTANEOUS INTEGRODIFFERENTIAL EQUATIONS

In the preceding chapter it was seen that equations describing a continuous-time system were a set of simultaneous differential equations. Here, we shall develop one of the techniques of solving these equations.

For convenience, we shall use the operational notation

$$p = \frac{d}{dt} \tag{2.1a}$$

and

$$\frac{1}{p} = \int_{-\infty}^{t} (\quad) \, dt \tag{2.1b}$$

Thus

$$px = \frac{dx}{dt}, \qquad p^n x = \frac{d^n x}{dt^n}$$

and

$$\frac{1}{p} x = \int_{-\infty}^{t} x \, dt$$

Note that

$$p \frac{1}{p} x = \frac{d}{dt} \int_{-\infty}^{t} x \, dt = x$$

or

$$p \frac{1}{p} x = x \qquad (2.2)\dagger$$

Using operational notation, the loop equations for the network in Fig. 2.1 can

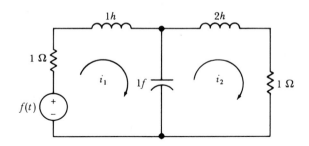

Figure 2.1

be written as

$$\left(p + 1 + \frac{1}{p}\right) i_1 - \frac{1}{p} i_2 = f(t) \qquad (2.3a)$$

$$-\frac{1}{p} i_1 + \left(2p + 1 + \frac{1}{p}\right) i_2 = 0 \qquad (2.3b)$$

The operational notation proves valuable in solving differential (or integrodifferential) equations.

It must never for a moment be forgotten that Eqs. 2.3 represent not algebraic equations but integrodifferential equations. For example, $(p^2 + ap + b)x$ represents

$$(p^2 + ap + b)x = \frac{d^2 x}{dt^2} + a \frac{dx}{dt} + bx$$

The term $(p^2 + ap + b)$ here is not an algebraic quantity multiplied by x but is an operator which transforms x into a new quantity shown on the right-hand side of the equation.

We shall now define an inverse operator as follows. If $D(p)$ is some polynomial in p, then by definition the equation

---

†However,

$$\frac{1}{p} px = \int_{-\infty}^{t} \frac{dx}{dt} \, dt = x(t) - x(-\infty)$$

Hence

$$\frac{1}{p} px \neq x \qquad (2.2n)$$

$$y = \frac{1}{D(p)} x \qquad\qquad (2.4a)$$

means

$$D(p)y = x \qquad\qquad (2.4b)$$

and the equation

$$y = \frac{N(p)}{D(p)} x$$

means

$$D(p)y = N(p)x$$

Thus the operational equation

$$y = \frac{p + 3}{p^2 + 2p + 5} x$$

means

$$(p^2 + 2p + 5)y = (p + 3)x$$

or

$$\frac{d^2 y}{dt^2} + 2 \frac{dy}{dt} + 5y = \frac{dx}{dt} + 3x$$

Oftentimes the algebraic laws can be successfully applied to these operational equations. Consider, for example, the quantity $(p + 3)(p + 2)y$. We have

$$(p + 3)(p + 2)y = \left(\frac{d}{dt} + 3\right)\left(\frac{dy}{dt} + 2y\right)$$

$$= \frac{d}{dt}\left(\frac{dy}{dt} + 2y\right) + 3\left(\frac{dy}{dt} + 2y\right)$$

$$= \frac{d^2 y}{dt^2} + 5 \frac{dy}{dt} + 6y$$

$$= (p^2 + 5p + 6)y$$

Thus

$$(p + 3)(p + 2)y = (p^2 + 5p + 6)y$$

Hence we conclude that

$$(p + 3)(p + 2) = p^2 + 5p + 6$$

Simple extension of this process shows that the operators which are polynomials in p can be multiplied as though they were algebraic quantities. Conversely, polynomials in p can be factorized like an algebraic expression.

We may not, however, take for granted that every algebraic law is valid for operational equations. For example, cancellation of a common factor on two sides of an operational equation is not permissible in general. Consider the equation

$$px = py$$

Here we cannot cancel the common factor p because this equation represents

$$\frac{dx}{dt} = \frac{dy}{dt}$$

Integration of both sides yields

$$x = y + c$$

where c is an arbitrary constant. Hence, in general

$$x \neq y$$

Simple extension of this result leads to the conclusion that if

$$(p + a)N(p)x = (p + a)y \tag{2.5a}$$

then in general

$$N(p)x \neq y \tag{2.5b}$$

Also from Eqs. 2.5, it follows that if

$$x = \frac{p + a}{(p + a)N(p)} y \tag{2.6a}$$

then in general

$$x \neq \frac{1}{N(p)} y \tag{2.6b}$$

Thus, in general, cancellation of a common factor is not permitted. Under certain conditions, however, the common factors may be cancelled. It is shown in Appendix A that if $D(p)$ is a polynomial in p, then

$$D(p)\left[\frac{1}{D(p)} x\right] = x \tag{2.7a}$$

However,

$$\frac{1}{D(p)}[D(p)x] = x + \phi(t) \tag{2.7b}$$

where $\phi(t)$ is a function determined by $D(p)$. Thus a common factor may be cancelled only if the sequence of operations is as shown in Eq. 2.7a. This can also be seen from Eqs. 2.2 and 2.2n:

$$p\frac{1}{p}x = x$$

But

$$\frac{1}{p} px = x + c$$

$$\neq x$$

In general, when the sequence of operations is not known, it is safe not to cancel common factors. Such is the case in Eq. 2.6a.

It is now evident that we should be very cautious in applying algebraic rules to operational equations without prior justification.

Let us now turn our attention to the solution of simultaneous differential equations. Consider the network in Fig. 2.2. The loop equations for this network

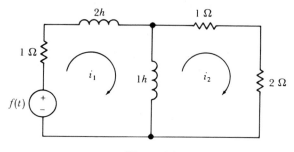

Figure 2.2

are

$$(3p + 1)i_1 - pi_2 = f(t) \qquad (2.8a)$$

$$-pi_1 + (p + 3)i_2 = 0 \qquad (2.8b)$$

These are two simultaneous differential equations in two unknowns i_1 and i_2.

In order to solve these equations we must first eliminate one variable from each equation, so that we have two differential equations, each in only one un-known. How shall we proceed? There is a great temptation to treat these equations as if they were algebraic equations for the purpose of eliminating variables. But is such a procedure valid? Fortunately yes! To show this, we shall employ step by step the elimination method used in simultaneous algebraic equations, and show that each step as applied to Eqs. 2.8 is also valid. If Eqs. 2.8 were algebraic, we could eliminate i_2 from Eq. 2.8a by multiplying Eq. 2.8a by $p + 3$, and Eq. 2.8b by p and then adding the resulting equations. This yields

$$(p + 3)(3p + 1)i_1 - p(p + 3)i_2 = (p + 3)f(t)$$
$$-p^2 i_1 + p(p + 3)i_2 = 0$$

Addition of these equations yields

$$[(p + 3)(3p + 1) - p^2]i_1 = (p + 3)f(t)$$

or

$$(2p^2 + 10p + 3)i_1 = (p + 3)f(t) \tag{2.9}$$

This is a differential equation in variable i_1 alone.

Let us now consider each step and see if it is justified for differential equations.

In step 1 we multiplied Eq. 2.8a by $(p + 3)$. Is such a multiplication valid? Yes indeed! Multiplication of any equation by $(p + 3)$ means (i) multiply the equation by p (that is, take the derivative of the equation) and (ii) add 3 times the original equation. There is nothing wrong with this operation. Thus multiplication of any equation by a polynomial in p is valid. The next step of addition of the two differential equations is obviously valid. Hence the entire process of elimination by treating these equations as algebraic equations is valid. We can therefore write these equations in the matrix form and use Cramer's rule directly. Equations 2.8 can be expressed in matrix form as

$$\begin{bmatrix} 3p + 1 & -p \\ -p & p + 3 \end{bmatrix} \begin{bmatrix} i_1 \\ i_2 \end{bmatrix} = \begin{bmatrix} f(t) \\ 0 \end{bmatrix} \tag{2.10}$$

Application of Cramer's rule to this equation yields

$$i_1 = \frac{p + 3}{2p^2 + 10p + 3} f(t) \tag{2.11a}$$

or

$$(2p^2 + 10p + 3)i_1 = (p + 3)f(t) \tag{2.11b}$$

which is the same result as obtained earlier. Similarly, we can obtain

$$i_2 = \frac{p}{2p^2 + 10p + 3} f(t) \tag{2.12a}$$

or

$$(2p^2 + 10p + 3)i_2 = pf(t) \tag{2.12b}$$

We shall once again remind the reader that Eqs. 2.11b and 2.12b are differential equations given by

$$2\frac{d^2i_1}{dt^2} + 10\frac{di_1}{dt} + 3i_1 = \frac{df}{dt} + 3f \tag{2.13a}$$

and

$$2\frac{d^2i_2}{dt^2} + 10\frac{di_2}{dt} + 3i_2 = \frac{df}{dt} \tag{2.13b}$$

These are two differential equations, each in one variable only. The elimination procedure discussed here can be extended to n simultaneous differential equations in n variables.

2.2   TRANSFER OPERATOR

Before discussing the techniques of solving differential equations such as Eqs. 2.13, we shall introduce a useful concept of *transfer operator* (also known as *system operator*). We have observed that, in general, the relationship between some response variable y and the input f of a system are related by a differential equation of the form†

$$D(p)y(t) = N(p)f(t) \tag{2.14}$$

where $D(p)$ and $N(p)$ are polynomials in p. This differential equation can also be expressed as

$$y(t) = \frac{N(p)}{D(p)} f(t) \tag{2.15}$$

We shall now define a transfer operator $H(p)$ which relates the response y to the input f as

$$H(p) = \frac{N(p)}{D(p)} \tag{2.16}$$

From Eqs. 2.15 and 2.16 it follows that

$$y(t) = H(p)f(t) \tag{2.17}$$

For the network in Fig. 2.2, the transfer operators $H_1(p)$ and $H_2(p)$ relating the response variables i_1 and i_2 to the input $f(t)$ are given by Eqs. 2.11a and 2.12a:

$$H_1(p) = \frac{p + 3}{2p^2 + 10p + 3} \tag{2.18a}$$

$$H_2(p) = \frac{p}{2p^2 + 10p + 3} \tag{2.18b}$$

and

$$i_1(t) = H_1(p)f(t) \tag{2.19a}$$

$$i_2(t) = H_2(p)f(t) \tag{2.19b}$$

It must be stressed here again that these are not algebraic equations but are differential equations in operational notation. It is important that the reader has a proper image of the transfer operator $H(p)$. Consider again the equation

$$y(t) = H(p)f(t)$$

Here $H(p)$ is not a factor multiplying $f(t)$ but is an operator which operates on $f(t)$ to transform it into $y(t)$. Thus the operator p transforms $f(t)$ into its derivative, the operator $(p + \alpha)$ transforms $f(t)$ into its derivative plus α times $f(t)$. The

---

†This is true for linear, time-invariant systems.

operator $1/p$ transforms $f(t)$ into its integral

$$\frac{1}{p} f(t) = \int_{-\infty}^{t} f(\tau)\, d\tau$$

It will also be shown in Sec. 2.6 that the operator $1/(p - \lambda)$ represents an integral transformation

$$\frac{1}{p - \lambda} f(t) = \int_{-\infty}^{t} e^{\lambda(t-\tau)} f(\tau)\, d\tau$$

Thus $H(p)$ represents some operation.

For a given system there are several response variables, and there is one transfer operator which relates each response variable to the input. Hence for a given system there will be many transfer operators. If however we are interested in one and only one response variable, say y, then the system is completely specified or characterized for this purpose by only one transfer operator $H(p)$ which relates y to the input f. Symbolically, we may represent this fact as shown in Fig. 2.3.

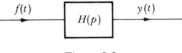

Figure 2.3

Example 2.1.   Find the transfer operator $H(p)$ which relates the response $i(t)$ to the input voltage $f(t)$ in the circuit shown in Fig. 2.4.

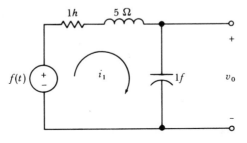

Figure 2.4

The loop equation for this circuit is given by

$$\left(p + 5 + \frac{6}{p}\right)i = f(t) \tag{2.20a}$$

Multiplying this equation by p (differentiating throughout) yields

$$(p^2 + 5p + 6)i = pf(t)$$

and

$$i = \frac{p}{p^2 + 5p + 6} f(t) \tag{2.20b}$$

$$= H(p)f(t)$$

Therefore

$$H(p) = \frac{p}{p^2 + 5p + 6} \tag{2.21}$$

Example 2.2. Find the transfer operators $H_1(p)$ and $H_2(p)$ which relates the response currents i_1 and i_2 to the input voltage $f(t)$ in the network shown in Fig. 2.1. The loop equations for this network are already found in Eqs. 2.3. Note that these are simultaneous integrodifferential equations rather than simultaneous differential equations.

The elimination procedure using Cramer's rule has been shown to be valid only for simultaneous differential equations. In general, it cannot be applied to simultaneous integrodifferential equations directly.† This problem can be handled either by (i) converting integrodifferential equation(s) into differential equation(s) through differentiation (multiplication by p) or (ii) by using charge variables instead of current variables. The second alternative has certain advantages over the first. A charge variable q is an integral of a current variable i. Thus

$$q = \int_{-\infty}^{t} i \, dt = \frac{1}{p} i$$

Thus instead of using i_1 and i_2, if we use $(1/p)i_1$ and $(1/p)i_2$ as variables, Eq. 2.3 can be converted into differential equations. We can rewrite Eqs. 2.3 as

$$\begin{bmatrix} p^2 + p + 1 & -1 \\ \\ -1 & 2p^2 + p + 1 \end{bmatrix} \begin{bmatrix} \frac{1}{p} i_1 \\ \\ \frac{1}{p} i_2 \end{bmatrix} = \begin{bmatrix} f(t) \\ \\ 0 \end{bmatrix} \tag{2.22}$$

These are simultaneous differential equations in variables $(1/p)i_1$ and $(1/p)i_2$. Note that these variables represent circulating charges q_1 and q_2 in the two loops. Thus by suitably choosing the variables, we can convert a set of integrodifferential equations into a set of differential equations. Application of Cramer's rule to Eq. 2.22 yields

$$\frac{1}{p} i_1 = \frac{2p^2 + p + 1}{p(2p^3 + 3p^2 + 4p + 2)} f(t)$$

---

†This is because the operators p and $1/p$ do not commute—that is,

$$p \frac{1}{p} x \neq \frac{1}{p} px$$

See Eqs. 2.2 and 2.2n.

$$\frac{1}{p}\, i_2 = \frac{1}{p(2p^3 + 3p^2 + 4p + 2)}\, f(t)$$

Multiplying both these equations by p and remembering that $p(1/p) = 1$, we obtain

$$i_1 = \frac{2p^2 + p + 1}{2p^3 + 3p^2 + 4p + 2}\, f(t) \qquad (2.23a)$$

$$i_2 = \frac{1}{2p^3 + 3p^2 + 4p + 2}\, f(t) \qquad (2.23b)$$

If $H_1(p)$ and $H_2(p)$ are the transfer operators relating the response variables i_1 and i_2 to the input $f(t)$, then from Eq. 2.23, it is obvious that

$$H_1(p) = \frac{2p^2 + p + 1}{2p^3 + 3p^2 + 4p + 2} \qquad (2.24a)$$

$$H_2(p) = \frac{1}{2p^3 + 3p^2 + 4p + 2} \qquad (2.24b)$$

If we had applied Cramer's rule directly to integrodifferential Eqs. 2.3, we would have still obtained the correct results in Eqs. 2.24. Then why so much fuss about converting integrodifferential equations into differential equations? The answer is provided in the next example.

Example 2.3.   Find the transfer operator $H(p)$ relating the loop current i_1 to the input voltage $f(t)$ for the network in Fig. 2.5.

The loop equations of this network are

$$\begin{bmatrix} 3p & -p & -p \\ -p & p+1 & -1 \\ -p & -1 & p+1+\dfrac{1}{p} \end{bmatrix} \begin{bmatrix} i_1 \\ i_2 \\ i_3 \end{bmatrix} = \begin{bmatrix} 0 \\ f(t) \\ 0 \end{bmatrix} \qquad (2.25)$$

These are simultaneous integrodifferential equations. In order to eliminate variables to find the appropriate transfer operator, we must convert these equations into a set of differential equations. We note that in the set 2.25, integration is associated with only i_3. Hence, these equations can be converted into differential equations if we choose i_1, i_2, and $(1/p)i_3$ as variables. We can rewrite Eq. 2.25 as

$$\begin{bmatrix} 3p & -p & -p^2 \\ -p & p+1 & -p \\ -p & -1 & p^2+p+1 \end{bmatrix} \begin{bmatrix} i_1 \\ i_2 \\ (1/p)i_3 \end{bmatrix} = \begin{bmatrix} 0 \\ f(t) \\ 0 \end{bmatrix} \qquad (2.26)$$

This is a set of simultaneous differential equations in variables i_1, i_2, and $(1/p)i_3$. Application of Cramer's rule to Eq. 2.26 yields

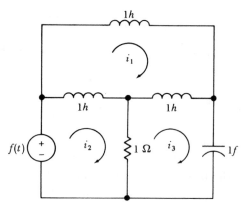

Figure 2.5

$$i_1 = \frac{p(p^2 + 2p + 1)}{p(p^3 + 2p^2 + 2p + 3)} f(t) \qquad (2.27a)$$

Hence

$$H(p) = \frac{p(p^2 + 2p + 1)}{p(p^3 + 2p^2 + 2p + 3)} \qquad (2.27b)$$

Note that we cannot cancel the common factor p in the right-hand side of this equation (see Eq. 2.6).

Now let us see what happens if we had applied Cramer's rule directly to integrodifferential Eqs. 2.25. The reader can easily verify that this yields

$$i_1 = \frac{p^2 + 2p + 1}{p^3 + 2p^2 + 2p + 3} f(t) \qquad (2.28)$$

Comparison of Eq. 2.28 with Eq. 2.27a shows that the common factor p is lost in Eq. 2.28. It will be seen in a later section that this will cause a loss of a constant term in the final solution (in the zero-input response).† It is therefore obvious that for correct results, one should convert integrodifferential equations into differential equations by choosing appropriate variables.

Example 2.4.   Find transfer operators $H_1(p)$ and $H_2(p)$ relating voltages v_1 and v_2 to the input current $f(t)$ in the network shown in Fig. 2.6.

---

†It can be shown that if the zero-input response of a system has no constant term, then the direct application of Cramer's rule to integrodifferential equations also yields correct results. However, to guard against such rare possibilities where the zero-input response may contain a constant term (or a polynomial in t), we need to convert integrodifferential equations into differential equations. In electrical networks such a situation arises when there are all inductor loop(s) and/or all capacitor cut set(s).

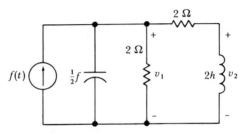

Figure 2.6

The node equations are

$$\left(\frac{p}{2} + 1\right)v_1 - \frac{1}{2}v_2 = f(t) \tag{2.29a}$$

$$-\frac{1}{2}v_1 + \left(\frac{1}{2p} + \frac{1}{2}\right)v_2 = 0 \tag{2.29b}$$

This is a set of integrodifferential equations in variables v_1 and v_2. We observe that integration is associated only with the variable v_2. Hence this set can be converted into a set of differential equations by choosing variables v_1 and $(1/p)v_2$. We can therefore rewrite these equations as

$$\begin{bmatrix} (p/2) + 1 & -(p/2) \\ -(1/2) & (1/2) + (p/2) \end{bmatrix} \begin{bmatrix} v_1 \\ (1/p)v_2 \end{bmatrix} = \begin{bmatrix} f(t) \\ 0 \end{bmatrix}$$

These are differential equations in variables v_1 and $(1/p)v_2$.   Application of Cramer's rule to this set of equations yields

$$v_1 = \frac{2(p + 1)}{p^2 + 2p + 2} f(t) \tag{2.29c}$$

and

$$\frac{1}{p} v_2 = \frac{2}{p^2 + 2p + 2} f(t)$$

or

$$v_2 = \frac{2p}{p^2 + 2p + 2} f(t) \tag{2.29d}$$

If $H_1(p)$ and $H_2(p)$ are the transfer operators relating the response voltages v_1 and v_2 to the input $f(t)$, then

$$v_1 = H_1(p)f(t)$$
$$v_2 = H_2(p)f(t)$$

where

$$H_1(p) = \frac{2(p + 1)}{p^2 + 2p + 2} \tag{2.29e}$$

$$H_2(p) = \frac{2p}{p^2 + 2p + 2} \tag{2.29f}$$

2.3   SOLUTION OF LINEAR DIFFERENTIAL EQUATIONS

We now come to the final step in our analysis problem. In the previous section the procedure to eliminate variables from simultaneous differential (or integrodifferential) equations was developed. This resulted in differential equations each in only one variable and each of the form

$$y(t) = H(p)f(t) = \frac{N(p)}{D(p)} f(t)$$

or

$$D(p)y(t) = N(p)f(t) \tag{2.30}$$

where y is some response variable, $f(t)$ is the input, and $D(p)$ and $N(p)$ are polynomials in p representing some differential operators. The general form of Eq. 2.30 is given by

$$(p^n + a_{n-1}p^{n-1} + \cdots + a_1 p + a_0)y(t)$$
$$= (b_m p^m + b_{m-1}p^{m-1} + \cdots + b_1 p + b_0)f(t) \tag{2.31a}$$

We remind the reader once again that this equation is not an algebraic equation but a differential equation given by

$$\frac{d^n y}{dt^n} + a_{n-1}\frac{d^{n-1}y}{dt^{n-1}} + \cdots + a_1\frac{dy}{dt} + a_0 y = b_m\frac{d^m f}{dt^m} + b_{m-1}\frac{d^{m-1}f}{dt^{m-1}}$$
$$+ \cdots + b_1\frac{df}{dt} + b_0 f(t) \tag{2.31b}$$

This is a linear equation. We can easily verify that the relationship between $y(t)$ and $f(t)$ is linear. If $y_1(t)$ is the solution of this equation corresponding to the input $f_1(t)$, then

$$(p^n + a_{n-1}p^{n-1} + \cdots + a_0)y_1(t) = (b_m p^m + \cdots + b_0)f_1(t)$$

and if $y_2(t)$ is the solution corresponding to $f_2(t)$, then

$$(p^n + a_{n-1}p^{n-1} + \cdots + a_0)y_2(t) = (b_m p^m + \cdots + b_0)f_2(t)$$

Multiplying these two equations by an arbitrary constant k and adding, we obtain

$$(p^n + \cdots + a_0)k[y_1(t) + y_2(t)] = (b_m p^m + \cdots + b_0)k[f_1(t) + f_2(t)]$$

It is obvious from this equation that $k[y_1(t) + y_2(t)]$ is the solution corresponding to $f(t) = k[f_1(t) + f_2(t)]$. This is the linearity condition. Hence Eq. 2.31 is a linear differential equation.   The coefficients a_i's and b_k's may be constants or functions of t.   For time-invariant systems these coefficients are constants and the corresponding equation (Eq. 2.31) is then a linear differential equation with constant coefficients.   For linear systems with time-varying parameters, the coefficients will in general be functions of t, and the corresponding equation (Eq. 2.31) is then a linear differential equation with time-varying coefficients.   In this book we shall mainly be concerned with the former type.

In Eq. 2.31 the variable $y(t)$ represents some response variable for the input $f(t)$.   As discussed in Chapter 1, the response y consists of two components (i) zero-input response y_x (ii) zero-state response y_f (see Eq. 1.10).   The zero-input response is the response that exists due to initial conditions (or initial state) alone with the input $f(t) = 0$.   The zero-state response is the response that exists when the input is $f(t)$ and the system is initially in zero state.   Thus the response $y(t)$ is given by

$$y(t) = y_x(t) + y_f(t) \tag{2.32}$$

Usually we are given the input $f(t)$ for $t \geq 0$, and the initial conditions (initial state) are given at the instant just before the input is applied (at $t = 0$).   Hence the initial conditions are given at $t = 0^-$ (just before $t = 0$).   As seen in Chapter 1, this information is necessary to obtain the response for $t \geq 0$.   It must also be remembered that the input $f(t)$ may or may not be known for $t < 0$.   The effect of the input for $t < 0$ is all available in the initial conditions (for the purpose of finding the response for $t \geq 0$).   Hence we do not care about the knowledge of $f(t)$ for $t < 0$.   In the discussion here we are assuming the initial moment quite arbitrarily to be $t = 0$.   It can be easily generalized for any value of t.

We shall now show that the solution of the differential equation 2.31 is given by Eq. 2.32:

$$y(t) = y_x(t) + y_f(t)$$

where $y_x(t)$ is the zero-input response satisfying the equation

$$(p^n + a_{n-1}p^{n-1} + \cdots + a_0)y_x = 0 \tag{2.33a}$$

and $y_f(t)$ is the solution of†

$$(p^n + a_{n-1}p^{n-1} + \cdots + a_0)y_f = (b_m p^m + b_{m-1} p^{m-1} + \cdots + b_0)f(t) \tag{2.33b}$$

This can be shown by adding Eqs. 2.33a and 2.33b.   This yields

$$(p^n + a_{n-1}p^{n-1} + \cdots + a_0)(y_x + y_f) = (b_m p^m + b_{m-1}p^{m-1} + \cdots + b_0)f(t)$$

Comparison of this equation with Eq. 2.31a shows that the desired solution is

$$y(t) = y_x(t) + y_f(t)$$

Hence the solution of Eq. 2.31 has a decomposition property—that is, the response can be expressed as a sum of zero-input component and zero-state component.

---

†Subject to zero initial conditions—that is, $y(0) = \dot{y}(0) = \cdots = y^{(n-1)} = 0$.

2.4   ZERO-INPUT RESPONSE

The zero-input response is the response when the input is zero.   If Eq. 2.31 relates the response variable y to input $f(t)$, then the zero-input response is the solution of the equation

$$(p^n + a_{n-1}p^{n-1} + \cdots + a_1 p + a_0)y(t) = 0 \qquad (2.34)$$

To find the solution of this differential equation, we first consider a simple case

$$(p - \lambda)y = 0 \qquad (2.35)$$

This equation is

$$\frac{dy}{dt} - \lambda y = 0$$

or

$$\frac{dy}{y} = \lambda \, dt$$

Therefore

$$\int \frac{dy}{y} = \lambda \int dt$$

and

$$\ln y = \lambda t + k$$

where k is a constant of integration.   Therefore

$$y = ce^{\lambda t} \qquad (2.36)$$

where $c = e^k$ is an arbitrary constant.

Let us now consider the differential equation

$$(p^2 + 4p + 3)y = 0 \qquad (2.37a)$$

or

$$(p + 1)(p + 3)y = 0 \qquad (2.37b)$$

This equation is not an algebraic equation but a differential equation.   In the past we have observed many situations where we could treat it as an algebraic equation. But we cannot take this for granted without verifying.   Here again we shall consider for the time being that we can treat this equation as an algebraic equation (we shall soon verify that this is legitimate).   On this assumption Eq. 2.37b will be satisfied in two ways as follows:

$$(p + 1)y = 0 \qquad (2.38a)$$

or

$$(p + 3)y = 0 \tag{2.38b}$$

These are two first-order differential equations of the form Eq. 2.35 and the solutions are given by (see Eq. 2.36), $y = c_1 e^{-t}$ and $y = c_2 e^{-3t}$ respectively, where c_1 and c_2 are arbitrary constants. Obviously if each of these solutions satisfies Eq. 2.37b, the general solution of Eq. 2.37b will be the sum of these two solutions. Hence the general solution of Eq. 2.37b is given by

$$y(t) = c_1 e^{-t} + c_2 e^{-3t} \tag{2.39}$$

This will be the desired solution provided our assumption that Eq. 2.38 follow from Eq. 2.37 were true. We shall now show that this is indeed true. Equation 2.37a can be written in two ways as follows:

$$p(p + 3)y + 1(p + 3)y = 0 \tag{2.40}$$

or

$$p(p + 1)y + 3(p + 1)y = 0 \tag{2.41}$$

Obviously Eq. 2.40 will be satisfied if

$$(p + 3)y = 0$$

and Eq. 2.41 will be satisfied by

$$(p + 1)y = 0$$

But both Eqs. 2.40 and 2.41 represent the same equation (Eq. 2.37). Hence Eq. 2.37 is satisfied if $(p + 1)y = 0$ or $(p + 3)y = 0$, and our assumption earlier is justified.

Thus the solution of Eq. 2.37 is given by

$$y(t) = c_1 e^{-t} + c_2 e^{-3t}$$

where c_1 and c_2 are arbitrary constants. Any arbitrary values for these constants will satisfy Eq. 2.37. This can be easily verified as follows:

$$y = c_1 e^{-t} + c_2 e^{-3t} \tag{2.42a}$$

$$\frac{dy}{dt} = -c_1 e^{-t} - 3c_2 e^{-3t} \tag{2.42b}$$

$$\frac{d^2 y}{dt^2} = c_1 e^{-t} + 9c_2 e^{-3t} \tag{2.42c}$$

and

$$\frac{d^2 y}{dt^2} + 4 \frac{dy}{dt} + 3y = 0 \tag{2.42d}$$

Thus solution in Eq. 2.42a satisfies Eq. 2.37 regardless of the values of constants c_1 and c_2. In any given problem, however, we have some additional constraints such as initial conditions (or initial state). These constraints determine the specific values of these constants. For example in the solution of Eq. 2.37, if the initial conditions (initial states) are given as

$$y(0) = 1 \qquad \text{and} \qquad \frac{dy}{dt}(0) = 0 \tag{2.43}$$

Then substitution of these conditions in Eqs. 2.42a and 2.42b when $t = 0$, yields unique solution for c_1 and c_2.

$$1 = c_1 + c_2$$
$$0 = -c_1 - 3c_2$$

This yields

$$c_1 = \tfrac{3}{2} \qquad \text{and} \qquad c_2 = -\tfrac{1}{2}$$

Thus the solution of Eq. 2.42d, subject to initial conditions (initial state) in Eq. 2.43, is given by

$$y(t) = \tfrac{3}{2} e^{-t} - \tfrac{1}{2}e^{-3t} \tag{2.44}$$

We shall now generalize our results for the nth-order differential equation. Let us factorize the polynomial $D(p)$ in Eq. 2.34 as

$$(p - \lambda_1)(p - \lambda_2)\cdots(p - \lambda_n)y = 0 \tag{2.45}$$

We observe that Eq. 2.45 will be satisfied if

$$(p - \lambda_1)y = 0$$

or

$$(p - \lambda_2)y = 0$$
$$\vdots$$
$$(p - \lambda_n)y = 0$$

These are n first-order differential equations of the form in Eq. 2.35. The solutions of these equations are given by (see Eq. 2.36), $c_1 e^{\lambda_1 t}$, $c_2 e^{\lambda_2 t}$, ..., $c_n e^{\lambda_n t}$ respectively where $c_1, c_2, \ldots, c_n$ are arbitrary constants. Obviously the general solution of Eq. 2.45 is given by the sum of these n solutions. Hence the general solution of Eq. 2.34 (or Eq. 2.45) is given by

$$y(t) = c_1 e^{\lambda_1 t} + c_2 e^{\lambda_2 t} + \cdots + c_n e^{\lambda_n t} \tag{2.46}$$

where $c_1, c_2, \ldots, c_n$ are arbitrary constants which can be determined from additional n pieces of information usually given in the form of initial conditions (the initial state).

If the initial conditions are given to be $y(0)$, $y'(0)$, $y''(0)$, ..., $y^{(n-1)}(0)$, then substitution of these conditions in Eq. 2.46 and its $n - 1$ derivatives at $t = 0$ yields

$$y(0) = c_1 + c_2 + \cdots + c_n$$
$$y'(0) = \lambda_1 c_1 + \lambda_2 c_2 + \cdots + \lambda_n c_n$$
$$y''(0) = \lambda_1^2 c_1 + \lambda_2^2 c_2 + \cdots + \lambda_n^2 c_n \qquad (2.47a)$$
$$\vdots$$
$$y^{(n-1)}(0) = \lambda_1^{(n-1)} c_1 + \lambda_2^{(n-1)} c_2 + \cdots + \lambda_n^{(n-1)} c_n$$

The constants $c_1, c_2, \ldots, c_n$ can be determined by solving these simultaneous equations. One may use matrix method (Appendix D) to solve these equations if desired. In the matrix form, Eqs. 2.47a can be expressed as

$$
\begin{bmatrix} y(0) \\ y'(0) \\ y''(0) \\ \vdots \\ y^{(n-1)}(0) \end{bmatrix}
=
\begin{bmatrix}
1 & 1 & 1 & & 1 \\
\lambda_1 & \lambda_2 & \lambda_3 & \cdots & \lambda_n \\
\lambda_1^2 & \lambda_2^2 & \lambda_3^2 & \cdots & \lambda_n^2 \\
\vdots & \vdots & \vdots & & \vdots \\
\lambda_1^{n-1} & \lambda_2^{n-1} & \lambda_3^{n-1} & \cdots & \lambda_n^{n-1}
\end{bmatrix}
\begin{bmatrix} c_1 \\ c_2 \\ c_3 \\ \vdots \\ c_n \end{bmatrix}
\qquad (2.47b)
$$

The matrix formed by λ's in Eq. 2.47b is known as *Vandermonde matrix*. Inversion of this equation yields†

$$
\begin{bmatrix} c_1 \\ c_2 \\ \vdots \\ c_n \end{bmatrix}
=
\begin{bmatrix}
1 & 1 & 1 & \cdots & 1 \\
\lambda_1 & \lambda_2 & \cdot & \cdots & \lambda_n \\
\vdots & \vdots & \vdots & \vdots & \vdots \\
\lambda_1^{n-1} & \lambda_2^{n-1} & \cdot & \cdots & \lambda_n^{n-1}
\end{bmatrix}^{-1}
\begin{bmatrix} y(0) \\ y'(0) \\ \vdots \\ y^{(n-1)}(0) \end{bmatrix}
\qquad (2.47c)
$$

The inverse of a matrix can be found by using a formula developed in Appendix D (Eq. D.32). For a 2×2 matrix the inverse is given by a simple formula:

$$
\begin{bmatrix} a & b \\ c & d \end{bmatrix}^{-1}
= \frac{1}{ad - bc}
\begin{bmatrix} d & -b \\ -c & a \end{bmatrix}
\qquad (2.47d)
$$

---

†In deriving $\mathbf{V}^{-1}$ the inverse of Vandermonde matrix, we need $|\mathbf{V}|$, the determinant of $\mathbf{V}$. It can be shown that for Vandermonde matrix $\mathbf{V}$ in Eq. 2.47b, $|\mathbf{V}|$ is given by

$$|\mathbf{V}| = (\lambda_2 - \lambda_1)(\lambda_3 - \lambda_1)(\lambda_4 - \lambda_1)\cdots(\lambda_n - \lambda_1)$$
$$(\lambda_3 - \lambda_2)(\lambda_4 - \lambda_2)\cdots(\lambda_n - \lambda_2)$$
$$(\lambda_4 - \lambda_3)\cdots(\lambda_n - \lambda_3)$$
$$\cdot \quad \cdot \quad \cdot \quad \cdot \quad \cdot \quad \cdot$$
$$(\lambda_n - \lambda_{n-1})$$

Thus for the 3×3 case

$$|\mathbf{V}| = (\lambda_2 - \lambda_1)(\lambda_3 - \lambda_1)(\lambda_3 - \lambda_2)$$

Thus from the knowledge of the roots $\lambda_1, \lambda_2, \ldots, \lambda_n$ of the polynomial $D(p)$ and the initial conditions $y(0), y'(0), \ldots, y^{(n-1)}(0)$, we can compute the arbitrary constants $c_1, c_2, \ldots, c_n$ as shown in Eq. 2.47c.

ZERO-INPUT RESPONSE FROM THE TRANSFER OPERATOR

Consider the differential equation

$$D(p)y(t) = N(p)f(t) \tag{2.48a}$$

For zero-input response, the relevant differential equation is

$$D(p)y(t) = 0$$

where $D(p)$ is nth-degree polynomial in p and is factorized as

$$(p - \lambda_1)(p - \lambda_2)\cdots(p - \lambda_n)y = 0 \tag{2.48b}$$

Note that at $p = \lambda_1, \lambda_2, \ldots, \lambda_n$, $D(p) = 0$. For this reason these values of p are called *zeros* of $D(p)$. The zero-input solution is given by

$$y(t) = c_1 e^{\lambda_1 t} + c_2 e^{\lambda_2 t} + \cdots + c_n e^{\lambda_n t} \tag{2.48c}$$

Let us write Eq. 2.48a in the form

$$y = \frac{N(p)}{D(p)} f(t) = H(p)f(t)$$

Where $H(p)$ is the transfer function relating the response y to the input $f(t)$. The values of p which make $D(p) = 0$ are the values which make $H(p) = \infty$. These values are called *poles* of $H(p)$. Obviously, $\lambda_1, \lambda_2, \ldots, \lambda_n$ are poles of $H(p)$. The zero-input response can be immediately written from the knowledge of poles of the transfer operator. The poles of a function are easily located in its denominator. In conclusion, the zero-input response of any variable can be written from the knowledge of the poles of the transfer operator relating the variable to the input. For the network in Fig. 2.1, for example, the transfer operator relating the response variable i_1 to the input $f(t)$ is given by (see Eq. 2.24a)

$$H_1(p) = \frac{2p^2 + p + 1}{2p^3 + 3p^2 + 4p + 2}$$

$$= \frac{2p^2 + p + 1}{2(p + 0.7)(p + 0.4 - j1.13)(p + 0.4 + j1.13)}$$

Hence the poles of $H_1(p)$ are -0.7, $-0.4 + j1.13$, and $-0.4 - j1.13$, and the zero-input response i_1 is given by

$$i_1(t) = c_1 e^{-0.7t} + c_2 e^{(-0.4+j1.13)t} + c_3 e^{(-0.4-j1.13)t} \tag{2.48d}$$

where the arbitrary constants c_1, c_2, and c_3 are determined by initial conditions (see Eq. 2.47c). The terms $c_2 e^{(-0.4+j1.13)t}$ and $c_3 e^{(-0.4-j1.13)t}$ are complex. However $i_1(t)$ is a real function of time. This is possible only if the two complex terms on the right-hand side of Eq. 2.48d are conjugate of each other. This means c_2 and c_3 are

conjugate of each other. This result is true in general (see Appendix B). It therefore follows that in Eq. 2.48d

$$c_3 = c_2*$$

Let

$$c_2 = |\, c_2\,|\, e^{j\theta}$$

Then

$$c_3 = |c_2|\, e^{-j\theta}$$

and

$$c_2 e^{(-0.4+j1.13)t} + c_3 e^{(-0.4-j1.13)t} = |c_2|\, e^{-0.4t}[e^{j(1.13t+\theta)} + e^{-j(1.13t+\theta)}]$$
$$= 2\,|c_2|\, e^{-0.4t}\cos(1.13t + \theta)$$

and Eq. 2.48d becomes

$$i_1(t) = c_1 e^{-0.7t} + 2\,|c_2|\, e^{-0.4t}\cos(1.13t + \theta) \qquad (2.48e)$$

It is evident from this discussion that the zero-input response contains monoton-
ically decaying or growing exponentials of the form $e^{\lambda_j t}$ for real poles λ_j of $H(p)$ and
exponentially decaying or growing sinusoids corresponding to complex poles of
$H(p)$. The complex poles $\lambda_j = \alpha \pm j\beta$, it will give rise to the term $ce^{\alpha t}\cos(\beta t + \theta)$.
It can be seen that if the complex pole λ_j has negative real part it will cause an
exponentially decaying sinusoid. Similarly if it has positive real part it will give rise
to exponentially growing sinusoid in the zero-input response. From this discussion
it can be seen that the nature of the zero-input response is completely determined
by the poles of the transfer operator $H(p)$.

Example 2.5.   Find the response $i(t)$ in the network shown in Fig. 2.4 when the
source $f(t) = 0$ and the initial conditions (initial state) are given as $i(0) = 1$ and
$i'(0) = 2$. The loop equation for this network is (see Eq. 2.20a)

$$\left(5 + p + \frac{6}{p}\right)i = f(t) \qquad (2.49)$$

Since we are interested in finding zero-input response (response due to initial
states), we let $f(t) = 0$ in Eq. 2.49:

$$\left(5 + p + \frac{6}{p}\right)i = 0$$

or

$$(p^2 + 5p + 6)i = 0$$

or

$$(p + 2)(p + 3)i = 0$$

The solution of this equation is given by

$$i(t) = c_1 e^{-2t} + c_2 e^{-3t} \qquad (2.50)$$

Note that the transfer operator relating i to $f(t)$ is given by (Eq. 2.21)

$$H(p) = \frac{p}{p^2 + 5p + 6} = \frac{p}{(p+2)(p+3)}$$

It is obvious that the poles of $H(p)$ are -2 and -3. Hence the zero-input response can be immediately written as shown in Eq. 2.50. The arbitrary constants c_1 and c_2 in Eq. 2.50 can be determined from the initial conditions $i(0) = 1$, and $i'(0) = 2$. Substituting $t = 0$ in Eq. 2.50 and its derivative, we obtain

and
$$\left. \begin{array}{l} i(0) = 1 = c_1 + c_2 \\[2em] i'(0) = 2 = -2c_1 - 3c_2 \end{array} \right\} \implies \begin{array}{l} c_1 = 5 \\[1em] c_2 = -4 \end{array}$$

or we may use Eq. 2.47c directly. From this equation we have

$$\begin{bmatrix} c_1 \\ c_2 \end{bmatrix} = \begin{bmatrix} 1 & 1 \\ -2 & -3 \end{bmatrix}^{-1} \begin{bmatrix} 1 \\ 2 \end{bmatrix} = \begin{bmatrix} 3 & 1 \\ -2 & -1 \end{bmatrix} \begin{bmatrix} 1 \\ 2 \end{bmatrix} = \begin{bmatrix} 5 \\ -4 \end{bmatrix} \tag{2.51}$$

and

$$i(t) = 5e^{-2t} - 4e^{-3t}$$

Example 2.6.   In the network shown in Fig. 2.4 find the zero-input response $i(t)$ if the initial conditions (initial state) are given as

$$i(0) = 1 \qquad \text{and} \qquad v_0(0) = 10$$

This example is identical to Example 2.5 except that the initial state is given in a form which cannot be directly used in the solution of $i(t)$ in Eq. 2.50. In order to solve for constants c_1 and c_2 in Eq. 2.50, we need $i(0)$ and $i'(0)$. The former is given and the latter must be derived from the given initial condition $i(0) = 1$ and $v_0(0) = 10$. This can be easily done by observing that the loop KVL equation is given by

$$5i(t) + \frac{di}{dt} + v_0(t) = f(t)$$

Letting $t = 0$ and realizing that we are given $f(t) = 0$, we have

$$5i(0) + \frac{di}{dt}(0) + v_0(0) = 0$$

$$5 + \frac{di}{dt}(0) + 10 = 0$$

and

$$\frac{di}{dt}(0) = -15$$

Thus the new initial state (derived from the given initial state) is $i(0) = 1$ and

$i'(0) = -15$. Substitution of these values in Eq. 2.51 we obtain

$$\begin{bmatrix} c_1 \\ c_2 \end{bmatrix} = \begin{bmatrix} 3 & 1 \\ -2 & -1 \end{bmatrix} \begin{bmatrix} 1 \\ -15 \end{bmatrix} = \begin{bmatrix} -12 \\ 13 \end{bmatrix}$$

and

$$i(t) = -12e^{-2t} + 13e^{-3t}$$

Example 2.7.   Find the zero-input response voltages v_1 and v_2 for the network shown in Fig. 2.6 if the initial conditions (the initial state) are as follows: $v_1(0) = 1$ and $v_1'(0) = 2$.   The transfer operators which relate the voltages v_1 and v_2 to the input $f(t)$ were found in Eqs. 2.29c and 2.29d, respectively.   For zero-input case $f(t) = 0$.   Cross multiplication of these equations and setting $f(t) = 0$ yields

$$(p^2 + 2p + 2)v_1 = 0$$
$$(p^2 + 2p + 2)v_2 = 0$$

Note that both v_1 and v_2 satisfy identical equations.   But this does not mean that v_1 and v_2 are identical.   As seen before, the solution of these equations contains arbitrary constants.   These constants will be different for v_1 and v_2.   The form of the solution, however, is the same.   This is true in general.   For a given system the zero-input response of all variables has the same form.   This topic will be discussed in Chapter 6 in more detail.   Let us consider the equation for v_1.

$$(p^2 + 2p + 2)v_1 = 0$$

or

$$[p + (1 - j1)][p + (1 + j1)]v_1 = 0$$

Thus

$$\lambda_1 = -1 + j1 \text{ and } \lambda_2 = -1 - j1$$

Hence the solution of this equation is given by

$$v_1(t) = c_1 e^{(-1+j1)t} + c_2 e^{(-1-j1)t} \tag{2.52a}$$

Alternatively, we observe that the transfer function relating v_1 to $f(t)$ is given by (Eq. 2.29e)

$$H_1(p) = \frac{2(p + 1)}{p^2 + 2p + 2} = \frac{2(p + 1)}{(p + 1 - j1)(p + 1 + j1)}$$

The poles of $H_1(p)$ are $-1 + j1$ and $-1 - j1$.   Hence the zero-input solution for v_1 is given by Eq. 2.52a.

To find the arbitrary constants c_1 and c_2, we let $t = 0$ in Eq. 2.52a and its derivative and use the fact that $v_1(0) = 1$ and $v_1'(0) = 2$.   Thus

$$v_1(0) = 1 = c_1 + c_2$$
$$v_1'(0) = 2 = (-1 + j1)c_1 + (-1 - j1)c_2$$

Solution of these two simultaneous equations yields

$$c_1 = \frac{1 - 3j}{2} \quad \text{and} \quad c_2 = \frac{1 + 3j}{2}$$

Alternatively, we can use Eq. 2.47c to obtain the arbitrary constants c_1 and c_2. Thus

$$\begin{bmatrix} c_1 \\ c_2 \end{bmatrix} = \begin{bmatrix} 1 & 1 \\ -1+j1 & -1-j1 \end{bmatrix}^{-1} \begin{bmatrix} 1 \\ 2 \end{bmatrix} = \begin{bmatrix} \dfrac{1-j1}{2} & -\dfrac{j}{2} \\ \dfrac{1+j1}{2} & \dfrac{j}{2} \end{bmatrix} \begin{bmatrix} 1 \\ 2 \end{bmatrix} = \begin{bmatrix} \dfrac{1-3j}{2} \\ \dfrac{1+3j}{2} \end{bmatrix} \quad (2.52b)$$

Hence

$$\begin{aligned} v_1 &= \frac{1 - 3j}{2} e^{(-1+j1)t} + \frac{1 + 3j}{2} e^{(-1-j1)t} \\ &= \frac{\sqrt{10}}{2} e^{-j(71.5°)} e^{(-1+j1)t} + \frac{\sqrt{10}}{2} e^{j(71.5°)} e^{(-1-j1)t} \\ &= \frac{\sqrt{10}}{2} e^{-t} [e^{j(t-71.5°)} + e^{-j(t-71.5°)}] \\ &= \sqrt{10}\, e^{-t} \cos(t - 71.5°) \end{aligned}$$

As seen before the zero-input response v_2 has the same form as that of v_1. Hence

$$v_2(t) = a_1 e^{(-1+j1)t} + a_2 e^{(-1-j1)t} \qquad (2.53)$$

To determine the arbitrary constants a_1 and a_2, we need the initial conditions (initial state) $v_2(0)$ and $v_2'(0)$. The given initial conditions are $v_1(0) = 1$ and $v_1'(0) = 2$ [or $pv_1(0) = 2$]. The required initial conditions $v_2(0)$ and $v_2'(0)$ or $pv_2(0)$ must be obtained from the given initial condition. This is most conveniently done by using the system equations found in Eqs. 2.29a and b. These equations are good for all values of t. Let us consider these equations for $t = 0$. In addition, in the present situation $f(t) = 0$. Hence these equations become

$$\left(\frac{p}{2} + 1\right)v_1(0) - \frac{1}{2}v_2(0) = 0$$

$$-\frac{1}{2}v_1(0) + \left(\frac{1}{2p} + \frac{1}{2}\right)v_2(0) = 0$$

Rearranging these equations (multiply the second equation by $2p$), we obtain

$$\tfrac{1}{2} pv_1(0) + v_1(0) - \tfrac{1}{2} v_2(0) = 0 \qquad (2.54a)$$

$$-pv_1(0) + (1 + p)v_2(0) = 0 \qquad (2.54b)$$

We are given that $v_1(0) = 1$ and $pv_1(0) = 2$. Substitution of these values in Eq. 2.54a yields

$$v_2(0) = 4$$

Substituting $v_2(0) = 4$, $pv_1(0) = 2$ in Eq. 2.54b, we obtain

$$pv_2(0) = -2$$

Hence the desired initial conditions are

$$v_2(0) = 4 \quad \text{and} \quad v_2'(0) = -2.$$

The arbitrary constants a_1 and a_2 in Eq. 2.53 are obtained by using Eq. 2.47c.

$$\begin{bmatrix} a_1 \\ a_2 \end{bmatrix} = \begin{bmatrix} 1 & 1 \\ -1+j1 & -1-j1 \end{bmatrix}^{-1} \begin{bmatrix} 4 \\ -2 \end{bmatrix} = \begin{bmatrix} 2-j1 \\ 2+j1 \end{bmatrix}$$

and

$$\begin{aligned} v_2 &= (2-j1)e^{(-1+j1)t} + (2+j1)e^{(-1-j1)t} \\ &= \sqrt{5}\, e^{j(-26.6°)}e^{(-1+j1)t} + \sqrt{5}\, e^{j(26.6°)}e^{(-1-j1)t.} \\ &= \sqrt{5}e^{-t}[e^{j(t-26.6°)} + e^{-j(t-26.6°)}] \\ &= 2\sqrt{5}\, e^{-t}\cos(t - 26.6°) \end{aligned} \tag{2.55}$$

DERIVED INITIAL CONDITIONS

It is evident from the earlier discussion that the solution of an nth order differential equation has n arbitrary constants and there must be an additional n independent pieces of information to find these constants. These n pieces of information can be given in various ways. This information may be given in the form of initial values of some response variables and their derivatives. These are the initial conditions (or the initial state). Since there are infinite derivatives of each variable, there are virtually infinite possibilities of specifying initial conditions or the initial state. This bears out our observation in Chapter 1 that there are infinite possible ways of specifying initial state of a system. The initial conditions (initial state) may be given in a suitable form so that they may be used directly to evaluate the arbitrary constants, or they may be given in a form which cannot be used directly. This is clearly shown in Examples 2.6 and 2.7. In Example 2.6 we are given initial conditions $i(0)$ and $v_0(0)$, whereas we need the conditions $i(0)$ and $i'(0)$. In this case we must derive $i'(0)$ from the knowledge of $i(0)$ and $v_0(0)$. In Example 2.7 we are given $v_1(0)$ and $v_1'(0)$. These conditions can be used directly in evaluating the response $v_1(t)$. However to obtain the response $v_2(t)$, we need to know $v_2(0)$ and $v_2'(0)$. These conditions are obtained from the given conditions $v_1(0)$ and $v_1'(0)$. In general any desired initial conditions can be obtained from the given initial conditions by using various system equations.

MULTIPLE ROOTS

The solution of Eq. 2.48b is given in Eq. 2.48c. In writting this solution, we have implicity assumed that $\lambda_1, \lambda_2, \ldots, \lambda_n$, the roots of $D(p) = 0$ are distinct. If some of the roots coincide, then the form of solution is slightly modified.

The solution of differential equation

$$(p - \lambda)^2 y = 0$$

is given by†

$$y = c_1 e^{\lambda t} + c_2 t e^{\lambda t}$$

In a similar way it can be shown that the solution of

$$(p - \lambda)^r y = 0$$

is given by

$$y(t) = (c_0 + c_1 t + \cdots + c_{r-1} t^{r-1}) e^{\lambda t} \qquad (2.56a)$$

It now follows that the solution of differential equation

$$(p - \lambda_1)^r (p - \lambda_{r+1}) \cdots (p - \lambda_n) y = 0$$

is given by

$$y(t) = (c_0 + c_1 t + \cdots + c_{r-1} t^{r-1}) e^{\lambda_1 t} + c_{r+1} e^{\lambda_{r+1} t} + \cdots + c_n e^{\lambda_n t} \qquad (2.56b)$$

Example 2.8.   Find the zero-input response $y(t)$ of a system whose transfer operator $H(p)$, relating $y(t)$ to the input $f(t)$ is given as

$$H(p) = \frac{(2p^2 + 8p + 3)}{(p + 1)(p + 3)^2}$$

and the initial conditions (initial state) is given by $y(0) = 2$, $y'(0) = 1$ and $y''(0) = 0$.

---

†This may be proved as follows:
Let the solution be given by

$$y = v(t) e^{\lambda_1 t}$$

where $v(t)$ is unknown; then

$$py = v'(t) e^{\lambda_1 t} + \lambda_1 v(t) e^{\lambda_1 t}$$

and

$$p^2 y = v''(t) e^{\lambda_1 t} + 2\lambda_1 v'(t) e^{\lambda_1 t} + \lambda_1^2 v(t) e^{\lambda_1 t}$$

and

$$(p - \lambda)^2 y = p^2 y - 2p\lambda y + \lambda^2 y = 0$$

Substituting for y, py, and $p^2 y$ in this equation, we obtain

$$v''(t) e^{\lambda t} = 0$$

or

$$v''(t) = 0 \qquad \text{and} \qquad v(t) = c_1 + c_2 t$$

Hence

$$y(t) = v(t) e^{\lambda t} = (c_1 + c_2 t) e^{\lambda t}$$

In this case $D(p)$ has two coincident roots at -3.   Hence according to Eq. 2.56b, the solution is given by

$$y(t) = c_1 e^{-t} + (c_2 + c_3 t)e^{-3t} \tag{2.57a}$$

To determine arbitrary constants c_1, c_2, and c_3, we substitute the initial conditions in Eq. 2.57a and its derivatives.   This yields

$$\left. \begin{aligned} y(0) &= 2 = c_1 + c_2 \\ y'(0) &= 1 = -c_1 - 3c_2 + c_3 \\ y''(0) &= 0 = c_1 + 9c_2 - 6c_3 \end{aligned} \right\} \implies \begin{aligned} c_1 &= 6 \\ c_2 &= -4 \\ c_3 &= -5 \end{aligned}$$

and

$$y(t) = 6e^{-t} - 4e^{-3t} - 5te^{-3t} \tag{2.57b}$$

2.5   SOME INSIGHT INTO ZERO-INPUT BEHAVIOR OF A SYSTEM

The zero-input component of the output is caused not by the *external input* (driving functions) but by *internal conditions*.   It is very illuminating to understand this phenomenon.   If a system is disturbed momentarily from its rest position and the distrubance is then removed, the system will not come to rest, but, in general, will have a response after the disturbance is removed.†   For example, a pendulum set in motion will eventually return to its rest position.   Similarly a spring, mass, and a dashpot system when disturbed momentarily from the equilibrium position, will eventually return to the original position.   The momentary disturbance implies some external input applied for a short time.   After this input (disturbance) is removed, the subsequent motion must be sustained by the system on its own, without the benefit of external input.   In other words, the system is restricted to only such type of response as will be sustained by the system without any input.   As seen in Sec. 2.4, every system can sustain some form of response without the need of any input (zero-input response).

As an example, consider an R-L circuit driven by a voltage source as shown in Fig. 2.7.   If we consider the mesh current $i(t)$ as response, then the transfer oper-

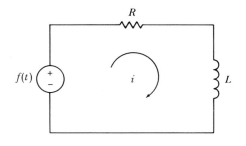

Figure 2.7

---

†There are exceptions.

ator, $H(p)$, relating $i(t)$ to the input $f(t)$ is given by

$$H(p) = \frac{1}{R + Lp}$$

The transfer operator $H(p)$ has only one pole at $-R/L$. Hence the zero-input response is given by $Ae^{-Rt/L}$. From the above discussion it follows that this current can be sustained by this system without a driving function [input voltage $f(t) = 0$]. This is indeed true and can be easily verified by finding the voltage $f(t)$ required to sustain this current. From Fig. 2.7, we have

$$f(t) = Ri(t) + L\frac{di}{dt}$$

If

$$i(t) = Ae^{-Rt/L}$$

then

$$f(t) = RAe^{-Rt/L} + LA(-R/L)e^{-Rt/L} = 0$$

It is evident that the current $Ae^{-Rt/L}$ can be sustained by the system without any input (driving function). Every system has some form of output (response) which it can sustain without any input (driving function). This is by definition the zero-input response. This response is also called the natural response† of the system because it is the response executed by the system on its own without any external driving function. It is an inherent or a natural attribute of the system.

Let us discuss the earlier example of a pendulum and a spring, mass, and dashpot system. If a pendulum is disturbed from its equilibrium position it exhibits an oscillatory response of decaying amplitude. This is evidently the natural response (zero-input response) of the pendulum. Similarly, when a spring, mass, and dashpot system is disturbed from the rest position and then left to itself, it will return to its rest position through its natural response [zero-input response which may be oscillatory (with exponentially decaying amplitude) or nonoscillatory (monotonically decaying exponential)].

If the transfer operator $H(p)$ has n poles $\lambda_1, \lambda_2, \ldots, \lambda_n$, the natural response (the zero-input response) will contain n exponential terms as

$$y_x(t) = c_1 e^{\lambda_1 t} + c_2 e^{\lambda_2 t} + \cdots + c_n e^{\lambda_n t} \tag{2.58}$$

Each of the exponentials in $y_x(t)$ is also known as a *natural mode* of the system. If the transfer function has n poles, then the natural response has n natural modes.

It must be remembered that although the natural response is executed without an external force (external input), it does not mean that the natural response dissipates no energy. Actually the initial disturbance in a system is equivalent to

---

†The term *natural response* has broader implications than *zero-input response*. It connotes a class of signals. Any expression of the form of Eq. 2.58 is a natural response. Zero-input response is a member of this class. See Sec. 2.17 for further discussion of natural response.

storing some energy in the system. This amounts to saying that the initial disturbance creates nonzero initial conditions (initial state) in the system. This stored energy is dissipated in the natural response (zero-input) executed by the system.

RESONANCE PHENOMENA

We have discussed above the natural behavior of a system. We observe that the natural response can be sustained by a system on its own, without any need for external driving function (input). In other words, the system offers no resistance or obstacle to the signals of natural mode. Imagine then what will happen if we actually drive the system with the input of the form of the natural mode. This is like pouring gasoline on a fire in a dry forest. The response of the system to such inputs will naturally be very high (see Prob. 2.36). This phenomenon is known as *resonance*. Thus the resonance will occur for any of the n-input waveforms $e^{\lambda_1 t}$, $e^{\lambda_2 t}, \ldots, e^{\lambda_n t}$. This phenomenon is not particularly noticeable for the modes for which λ_k is real, but it can be very pronounced for the modes corresponding to complex λ_k [complex poles of $H(p)$]. Since the poles occur in pairs of complex conjugates (see Appendix B), if $\lambda_k = \alpha + j\beta$ there must be a pole $\lambda_k^* = \alpha - j\beta$. The natural modes corresponding to such a pair of poles was found (Eq. 2.48e) to be of the form $re^{\alpha t} \cos(\beta t + \theta)$. Hence the natural mode is a sinusoidal signal with angular frequency β, and whose amplitude is varying exponentially as $e^{\alpha t}$. If we apply the input of this form, the response will be high. If α is small, then the natural mode is practically a sinusoidal signal with angular frequency β. It follows that the response of the system to sinusoidal input of angular frequency β (input of the form of natural mode) will be very high. It can be shown that if $\alpha = 0$, the response of the system to sinusoidal input of angular frequency β is infinite, and is finite but very high to inputs of angular frequencies in the immediate vicinity of β (see Prob. 2.36). The resonance behavior is shown in Fig. 2.8 for four different

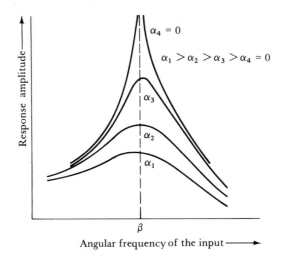

Figure 2.8

values of α. It can be seen from this figure that as $\alpha \rightarrow 0$, the resonance becomes more pronounced and the system becomes more selective. Electrical filters which are required to allow certain band of frequencies to pass, and to suppress the remaining frequency components are designed on the phenomenon of resonance. The poles of the transfer functions are chosen in the vicinity of the frequencies of the passband signals.

2.6   ZERO-STATE RESPONSE

As discussed in Chapter 1 and earlier in the present chapter, response of a system has two components: (i) the response component due to initial conditions or initial state (zero-input response) and (ii) the response component due to sources or inputs (zero-state response). We have already developed a method of *obtaining* zero-input response. We shall now consider techniques of *evaluating* zero-state response. It shall be remembered that this component depends not on initial state but entirely on the inputs.

To begin with, let us consider a first-order differential equation

$$y = \frac{1}{p - \lambda} f(t) \tag{2.59a}$$

or

$$(p - \lambda)y = f(t) \tag{2.59b}$$

where y is a response variable and $f(t)$ is the input. The solution for y consists of y_x, the zero-input component and y_f the zero-state component. The zero-input component y_x can be immediately written from Eq. 2.58 as

$$y_x = ce^{\lambda t} \tag{2.60a}$$

At $t = 0^-$, the entire y is given by y_x (because the input is not yet applied and so $y_f = 0$). Hence $y_x(0) = y(0)$ and letting $t = 0$ in Eq. 2.60a, we obtain

$$y(0) = c$$

and

$$y_x = y(0)e^{\lambda t} \tag{2.60b}$$

To find the complete solution of Eq. 2.59b, we now express it in differential form:

$$\frac{dy}{dt} - \lambda y = f(t) \tag{2.61}$$

Multiplying both sides by $e^{-\lambda t}$, we obtain

$$e^{-\lambda t} \frac{dy}{dt} - \lambda e^{-\lambda t} y(t) = e^{-\lambda t} f(t) \tag{2.62}$$

We now recognize the left-hand side of Eq. 2.62 to be $(d/dt)\,[e^{-\lambda t}y(t)]$. Hence

$$\frac{d}{dt}\,[e^{-\lambda t}y(t)] = e^{-\lambda t}f(t)$$

Integrating both sides of this equation, from 0 to t, we obtain

$$e^{-\lambda t}y(t)\,\Big|_{0}^{t} = \int_{0}^{t} e^{-\lambda \tau}f(\tau)\,d\tau$$

or

$$e^{-\lambda t}y(t) - y(0) = \int_{0}^{t} e^{-\lambda \tau}f(\tau)\,d\tau$$

and

$$e^{-\lambda t}y(t) = y(0) + \int_{0}^{t} e^{-\lambda \tau}f(\tau)\,d\tau$$

Hence

$$y(t) = y(0)e^{\lambda t} + e^{\lambda t}\int_{0}^{t} e^{-\lambda \tau}f(\tau)\,d\tau \tag{2.63a}$$

$$= \underbrace{y(0)e^{\lambda t}}_{\substack{\text{zero-input} \\ \text{response } y_x}} + \underbrace{\int_{0}^{t} e^{\lambda(t-\tau)}f(\tau)\,d\tau}_{\text{zero-state response } y_f} \tag{2.63b}$$

The first term on the right-hand side is immediately recognized as the zero-input response y_x (see Eq. 2.60b). Hence the second term must be y_f, the zero-state component. This conclusion is also evident from the fact that for $f(t) = 0$, this term goes to zero and it does not depend upon initial conditions.

The solution in Eq. 2.63b can be generalized for any initial instant t_0 instead of 0. It is left as an exercise for the reader to show that the solution of Eq. 2.61 can also be expressed as

$$y(t) = \underbrace{y(t_0)e^{\lambda(t-t_0)}}_{\text{zero-input response } y_x} + \underbrace{\int_{t_0}^{t} e^{\lambda(t-\tau)}f(\tau)\,d\tau}_{\text{zero-state response } y_f} \tag{2.63c}$$

Example 2.9.   Find the current $i(t)$ in the circuit in Fig. 2.7, if $R = 2$, $L = 1$, and the input $f(t)$ is a unit step function $u(t)$ shown in Fig. 2.9 and the initial condition $i(0) = 2$. The unit step function $u(t)$ shown in Fig. 2.10 has a value of 1 for $t > 0$ and is zero for $t < 0$.

The loop equation of the circuit is given by

$$i(t) = \frac{1}{p + 2}\,f(t)$$

Figure 2.9

The solution of this equation can be written immediately from Eq. 2.63b as

$$i(t) = i(0)e^{-2t} + \int_0^t e^{-2(t-\tau)}u(\tau)\, d\tau$$

$$= 2e^{-2t} + \int_0^t e^{-2(t-\tau)}u(\tau)\, d\tau$$

since $u(\tau) = 1$ for $\tau > 0$,

$$i(t) = 2e^{-2t} + e^{-2t}\int_0^t e^{2\tau}\, d\tau$$

$$= 2e^{-2t} + \tfrac{1}{2}(1 - e^{-2t})$$

$$= \tfrac{1}{2} + \tfrac{3}{2}e^{-2t}$$

HIGHER-ORDER CASE

Let us recapitulate our results. For a differential equation,

$$\frac{dy}{dt} - \lambda y = f(t)$$

or

$$y(t) = \frac{1}{p - \lambda}f(t) \tag{2.64a}$$

The zero-state response $y_f(t)$ is given by

$$y_f(t) = \int_0^t e^{\lambda(t-\tau)}f(\tau)\, d\tau \tag{2.64b}$$

It is easy to see that if

$$y(t) = \frac{k}{p - \lambda}f(t) \tag{2.64c}$$

then

$$y_f(t) = \int_0^t ke^{\lambda(t-\tau)} f(\tau) \, d\tau \qquad (2.64d)$$

Let us now consider the second-order equation

$$(p^2 + 4p + 3)y(t) = (3p + 5)f(t) \qquad (2.65a)$$

or

$$y(t) = \frac{3p + 5}{p^2 + 4p + 3} f(t)$$

$$= \frac{3p + 5}{(p + 1)(p + 3)} f(t) \qquad (2.65b)$$

If this were an algebraic equation, we could have expanded the right-hand side of Eq. 2.65b by partial-fraction expansion as (see Appendix B)

$$y(t) = \left(\frac{1}{p + 1} + \frac{2}{p + 3} \right) f(t) \qquad (2.65c)$$

$$= \frac{1}{p + 1} f(t) + \frac{2}{p + 3} f(t) \qquad (2.65d)$$

y now is represented as a sum of two terms. With an eye on Eqs. 2.64c and 2.64d, we could have immediately written the zero-state response as

$$y_f(t) = \int_0^t e^{-(t-\tau)} f(\tau) \, d\tau + \int_0^t 2e^{-3(t-\tau)} f(\tau) \, d\tau \qquad (2.66a)$$

$$= \int_0^t [e^{-(t-\tau)} + 2e^{-3(t-\tau)}] f(\tau) \, d\tau \qquad (2.66b)$$

However Eq. 2.65b is not an algebraic equation and we may not expand it by partial fractions unless it is proved that such an operation is permitted. Fortunately for us, such an expansion is valid and gives correct results. The proof of this statement is as follows. Let

$$z(t) = \left[\frac{1}{p + 1} + \frac{2}{p + 3} \right] f(t)$$

Multiplication of both sides of this equation by $(p + 1)(p + 3)$ yields

$$(p + 1)(p + 3)z(t)$$

$$= \left[(p + 3)(p + 1) \left(\frac{1}{p + 1} \right) + (p + 1)(p + 3) \left(\frac{2}{p + 3} \right) \right] f(t) \quad (2.67)$$

From Eq. 2.7a it follows that the common factors on the right-hand side of Eq. 2.67 can be canceled. Hence

$$(p + 1)(p + 3)z(t) = [(p + 3) + 2(p + 1)] f(t)$$

$$= (3p + 5)f(t)$$

Therefore

$$z(t) = \frac{3p + 5}{(p + 1)(p + 3)} f(t)$$

Comparison with Eq. 2.65b yields

$$y(t) = z(t)$$

or

$$\frac{3p + 5}{(p + 1)(p + 3)} f(t) = \left[\frac{1}{p + 1} + \frac{2}{p + 3} \right] f(t)$$

The validity of partial-fraction expansion has been proved in this case. This result can be extended to a general case

$$(p^n + a_{n-1}p^{n-1} + \cdots + a_1 p + a_0)y(t)$$
$$= (b_m p^m + b_{m-1}p^{m-1} + \cdots + b_1 p + b_0)f(t) \quad (2.68a)$$

or

$$y(t) = H(p)f(t) \tag{2.68b}$$

where

$$H(p) = \frac{b_m p^m + b_{m-1}p^{m-1} + \cdots + b_1 p + b_0}{p^n + a_{n-1}p^{n-1} + \cdots + a_1 p + a_0} \tag{2.69a}$$

$$= \frac{\mathcal{N}(p)}{(p - \lambda_1)(p - \lambda_2)\cdots(p - \lambda_n)} \tag{2.69b}$$

Expanding the right-hand side by partial fractions (Appendix B), we have

$$H(p) = \frac{k_1}{p - \lambda_1} + \frac{k_2}{p - \lambda_2} + \cdots + \frac{k_n}{p - \lambda_n} \tag{2.69c}$$

The solution of Eq. 2.68b can now be immediately written as

$$y(t) = y_x(t) + y_f(t) \tag{2.70a}$$

where

$$y_x(t) = c_1 e^{\lambda_1 t} + c_2 e^{\lambda_2 t} + \cdots + c_n e^{\lambda_n t} = \sum_{j=1}^{n} c_j e^{\lambda_j t} \tag{2.70b}$$

$$y_f(t) = \int_0^t [k_1 e^{\lambda_1 (t-\tau)} + k_2 e^{\lambda_2 (t-\tau)} + \cdots + k_n e^{\lambda_n (t-\tau)}] f(\tau) \, d\tau \tag{2.70c}$$

$$y(t) = \underbrace{\sum_{j=1}^{n} c_j e^{\lambda_j t}}_{\substack{\text{zero-input component} \\ y_x}} + \underbrace{\int_0^t h(t - \tau) f(\tau) \, d\tau}_{\substack{\text{zero-state component} \\ y_f}} \tag{2.71}$$

where the arbitrary constants c_j are determined from the initial state (see Eq. 2.47c) and we define $h(t)$ as

$$h(t) = k_1 e^{\lambda_1 t} + k_2 e^{\lambda_2 t} + \cdots + k_n e^{\lambda_n t} \tag{2.72}$$

Note that constants $k_1, k_2, \ldots, k_n$ are not arbitrary but are obtained from the partial fractions of $H(p)$ as shown in Eq. 2.69c. It is evident that the solution of Eq. 2.68b can be written immediately from the knowledge of $H(p)$, the transfer operator relating the response variable $y(t)$ to the input $f(t)$. Let us recapitulate the result here. The complete solution of the differential equation

$$\begin{aligned} y(t) &= H(p)f(t) \\ &= \frac{N(p)}{(p - \lambda_1)(p - \lambda_2)\cdots(p - \lambda_n)} f(t) \end{aligned}$$

is given by

$$y(t) = \underbrace{\sum_{j=1}^{n} c_j e^{\lambda_j t}}_{\text{zero-input response}} + \underbrace{\int_0^t h(t - \tau)f(\tau)\, d\tau}_{\text{zero-state response}} \tag{2.73}$$

where $c_1, c_2, \ldots, c_n$ are arbitrary constants to be determined from the initial state (initial condition) and $h(t)$ is obtained by expanding $H(p)$ by partial fractions.

$$H(p) = \frac{k_1}{p - \lambda_1} + \frac{k_2}{p - \lambda_2} + \cdots + \frac{k_n}{p - \lambda_n} \tag{2.74a}$$

and

$$h(t) = k_1 e^{\lambda_1 t} + k_2 e^{\lambda_2 t} + \cdots + k_n e^{\lambda_n t} \tag{2.74b}$$

Table 2.1 shows various forms of $H(p)$ and the corresponding $h(t)$. If $H(p)$ has multiple roots, the partial-fraction expansion will include terms of the form $k/(p - \lambda)r$ (see Appendix B). The function $h(t)$ corresponding to such term is shown† in pair 4. Note that the results in pair 2 follows from pair 1, when $\lambda = \alpha \pm j\beta$. Pair 3 gives alternate form of pair 2 (the polar form of coefficients).

Consider for example

$$\begin{aligned} H(p) &= \frac{p^3 + 9p^2 + 24p + 18}{(p + 1)(p^2 + 2p + 2)(p + 2)^2} \\ &= \frac{p^3 + 9p^2 + 24p + 18}{(p + 1)(p + 1 + j1)(p + 1 - j1)(p + 2)^2} \end{aligned}$$

The partial fractions of $H(p)$ are given by (see Appendix B)

$$H(p) = \frac{k_1}{p + 1} + \frac{k_2}{p + 1 - j1} + \frac{k_2^*}{p + 1 + j1} + \frac{a_0}{(p + 2)^2} + \frac{a_1}{p + 2}$$

---

†This can be easily shown by using the concept of convolution integral developed in Sec. 2.8.

TABLE 2.1

No.	Transfer Operator $H(p)$	$h(t)$
1	$\dfrac{k}{p - \lambda}$	$ke^{\lambda t}$
2	$\dfrac{c_1 + jc_2}{p - (\alpha + j\beta)} + \dfrac{c_1 - jc_2}{p - (\alpha - j\beta)}$	$2e^{\alpha t}(c_1 \cos \beta t - c_2 \sin \beta t)$
3	$\dfrac{re^{j\theta}}{p - (\alpha + j\beta)} + \dfrac{re^{-j\theta}}{p - (\alpha - j\beta)}$	$2re^{\alpha t}\cos(\beta t + \theta)$
4	$\dfrac{k}{(p - \lambda)^r}$, r integral	$\dfrac{k}{(r - 1)!}\, t^{r-1}e^{\lambda t}$

Using the results in Appendix B, we have

$$k_1 = (p + 1)H(p)\Big|_{p=-1} = \frac{p^3 + 9p^2 + 24p + 18}{(p^2 + 2p + 2)(p + 2)^2}\Big|_{p=-1} = 2$$

$$k_2 = (p + 1 - j1)H(p)\Big|_{p=-1+j1}$$

$$= \frac{p^3 + 9p^2 + 24p + 18}{(p + 1)(p + 1 + j1)(p + 2)^2}\Big|_{p=-1+j1} = -(2 + j1) = \sqrt{5}e^{-j153°}$$

$$a_0 = (p + 2)^2 H(p)\Big|_{p=-2} = \frac{p^3 + 9p^2 + 24p + 18}{(p + 1)(p^2 + 2p + 2)}\Big|_{p=-2} = 1$$

and

$$a_1 = \frac{d}{dp}\left[(p + 2)^2 H(p)\right]\Big|_{p=-2} = 2$$

Hence

$$H(p) = \frac{2}{p + 1} - \frac{\sqrt{5}e^{-j153°}}{p + 1 - j1} + \frac{\sqrt{5}e^{j153°}}{p + 1 + j1} + \frac{2}{p + 2} + \frac{1}{(p + 2)^2}$$

Using Table 2.1, we can now immediately write $h(t)$ as

$$h(t) = 2e^{-t} - 2\sqrt{5}\, e^{-t}(\cos t - 153°) + 2e^{-2t} + te^{-2t}$$

Example 2.10.   Find the current $i(t)$ in the network shown in Fig. 2.4 when $f(t) = u(t)$ and the initial conditions are given as $i(0) = 1$ and $i'(0) = 2$.

From Eq. 2.20b, we have

$$i(t) = \frac{p}{(p + 2)(p + 3)}\, f(t)$$

Partial-fraction expansion (Appendix B) yields

$$i(t) = \left(\frac{-2}{p + 2} + \frac{3}{p + 3}\right) f(t)$$

Hence (see Table 2.1)

$$h(t) = -2e^{-2t} + 3e^{-3t}$$

and

$$i(t) = c_1 e^{-2t} + c_2 e^{-3t} + \int_0^t h(t - \tau)f(\tau)\, d\tau, \qquad t \geq 0 \qquad (2.75)$$

We shall first find the arbitrary constants c_1 and c_2. At $t = 0$ the integral in Eq. 2.75 is zero, hence

$$i(0) = c_1 + c_2$$

Taking the derivative of Eq. 2.75 and letting $t = 0$, we have

$$i'(0) = -2c_1 - 3c_2$$

Thus from the given values of $i(0)$ and $i'(0)$, we have†

$$\left.\begin{array}{l} 1 = c_1 + c_2 \\ 2 = -2c_1 - 3c_2 \end{array}\right\} \implies \begin{array}{l} c_1 = 5 \\ c_2 = -4 \end{array}$$

Hence

$$i(t) = 5e^{-2t} - 4e^{-3t} + \int_{0^-}^t h(t - \tau)f(\tau)\, d\tau \qquad (2.76)$$

$$= 5e^{-2t} - 4e^{-3t} + \int_0^t [-2e^{-2(t-\tau)} + 3e^{-3(t-\tau)}]\, u(\tau)\, d\tau$$

$$= 5e^{-2t} - 4e^{-3t} + (-2e^{-2t}) \int_0^t e^{2\tau}\, d\tau + 3e^{-3t} \int_0^t e^{3\tau}\, d\tau$$

$$= \underbrace{(5e^{-2t} - 4e^{-3t})}_{i_x(t)} + \underbrace{e^{-2t} - e^{-3t}}_{i_f(t)}$$

$$= 6e^{-2t} - 5e^{-3t} \qquad (2.77)$$

IMPORTANT NOTE ON EVALUATING ARBITRARY CONSTANTS

The total response y is given by $y_x + y_f$. As seen from Eq. 2.71, the component $y_f(t) = 0$ at $t = 0$. This is because the lower limit of the integral representing y_f is $t = 0$. The initial conditions, on the other hand, are given for $t = 0^-$, when only y_x, the zero-input component, is present. Therefore when applying the initial conditions at $t = 0$, $y = y_x$ and hence the conditions must be applied only to the component y_x. For this reason in Example 2.10, we used the conditions

$$i(0) = i_x(0) = c_1 + c_2$$

---

†Alternately we may use Vandermonde matrix (Eq. 2.47c) to obtain the arbitrary constants c_1 and c_2.

and

$$i'(0) = i'_x(0) = -2c_1 - 3c_2$$

It is obvious that we can use the Vandermonde matrix (Eq. 2.47c) to derive the arbitrary constants $c_1, c_2, \ldots, c_n$ in the zero-input component of the total response in Eq. 2.71.

In general we shall assume that initial conditions are given at $t = 0^-$, the instant immediately preceding $t = 0$, when the input is applied. The zero-state response will start appearing from the moment $t = 0$. Consequently the lower limit of integration in Eq. 2.71 should be 0^- (just before the input is applied†). Hence

$$y_f(t) = \int_{0^-}^{t} h(t - \tau)f(\tau)\, d\tau \qquad (2.78)$$

COMMON FACTORS IN $H(p)$

If $H(p)$ has a common factor $(p - \lambda_i)$, in the numerator and the denominator, the coefficient k_i of the partial fraction corresponding to the pole $(p - \lambda_i)$ vanishes. This can be seen from the following example.

$$\frac{2(p + 3)}{(p + 1)(p + 3)} = \frac{2}{p + 1} + \frac{0}{p + 3}$$

$$= \frac{2}{p + 1}$$

It can be easily seen that this result is true in general. We therefore conclude that, in determining the zero-state response, common factors in the numerator and the denominator of $H(p)$ should be canceled. In determining zero-input response, however, the common factors may not be canceled.

2.7   UNIT IMPULSE FUNCTION

The unit impulse function was originally defined by P. A. M. Dirac by the equations

$$\int_{-\infty}^{\infty} \delta(t)\, dt = 1$$

$$\delta(t) = 0, t \neq 0 \qquad (2.79)$$

Thus the impulse function $\delta(t)$ is zero every where except at the origin and it has a unit area under it. Intuitively this function may be conceived as a narrow pulse (width ϵ) with a large height ($h = 1/\epsilon$) and unit area ($h\epsilon = 1$). In the limit as the width $\epsilon \to 0$, its height $1/\epsilon$ increases but the area always remains unity (Fig. 2.10). Symbolically, the impulse function is represented by a spike as shown in Fig. 2.11.

---

†This is done to account for the presence of an impulse function and its derivatives at $t = 0$, in the input.

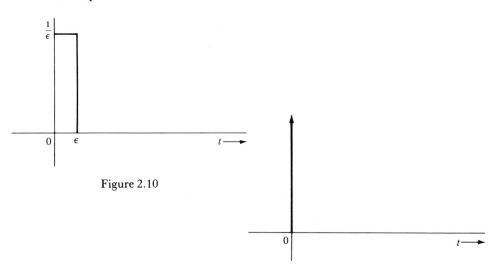

Figure 2.10

Figure 2.11

Since $\delta(t) = 0$ for all values of t, except at the origin, the product $\delta(t)\phi(t)$ will also be zero for all t except at $t = 0$. At $t = 0$, $\phi(t) = \phi(0)$. Hence $\delta(t)\phi(t) = \delta(t)\phi(0)$ and

$$\int_{-\infty}^{\infty} \delta(t)\phi(t)\,dt = \int_{-\infty}^{\infty} \delta(t)\phi(0)\,dt$$

$$= \phi(0) \int_{-\infty}^{\infty} \delta(t)\,dt$$

$$= \phi(0) \tag{2.80}$$

Thus Eq. 2.80 follows as a consequence of definition 2.79 of the impulse function. By a similar argument, we can extend this result to show that

$$\int_{-\infty}^{\infty} \phi(t)\delta(t - t_0)\,dt = \phi(t_0) \tag{2.81}$$

This fact is also obvious from Fig. 2.12. This property of impulse function to sample or to sift the value of a function $\phi(t)$ at the instant where the impulse exists is known as the *sampling property*, or sifting property of an impulse function. In conclusion, if we multiply a certain function by a unit impulse and integrate the product, the result is the value of the function at the instant where the impulse is located. Thus if we multiply $\phi(t)$ by $\delta(t - t_0)$ and integrate, the result is $\phi(t_0)$, the value of $\phi(t)$ at $t = t_0$ where the impulse is located. Using this result we can easily show that

$$\int_{-\infty}^{\infty} \phi(x - t)\delta(t)\,dt = \phi(x) \tag{2.82}$$

It should be noted that if we multiply a function by an impulse $\delta(t)$, the product

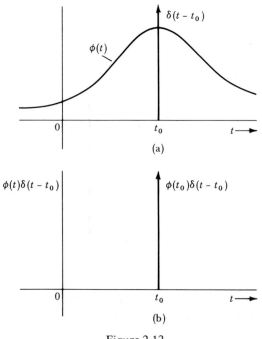

Figure 2.12

exists only at $t = 0$ and is zero everywhere else. Obviously the limits of integration need not be $-\infty$ to ∞ but (a, b) as long as the interval (a, b) include the location of the impulse. Thus Eq. 2.82 may be expressed as

$$\int_a^b \phi(t)\delta(t)\,dt = \phi(0) \tag{2.83a}$$

$$\text{for}\quad a < 0 < b$$

$$\int_a^b \phi(x - t)\delta(t)\,dt = \phi(x) \tag{2.83b}$$

On the other hand, if the interval (a, b) excludes the impulse location ($t = 0$ in Eqs. 2.83), then the integral is zero. Thus

$$\int_{-\infty}^t \delta(\tau)\,d\tau = \begin{cases} 1 & t > 0 \\ 0 & t < 0 \end{cases}$$

Hence

$$\int_{-\infty}^t \delta(\tau)\,d\tau = u(t)$$

and it follows that

$$\frac{d}{dt}u(t) = \delta(t)$$

The impulse function is obviously not a function in the ordinary mathematical sense where a function is defined for every value of t. The impulse function is actually a more general kind of function known in literature as a *generalized function* or *distribution*.† The concept of a generalized function is of relatively recent origin. The crude definition of the impulse function in Eq. 2.79, although not edifying mathematically, proves adequate for many problems. Generalized functions are defined by their integral property. The impulse function in this rigorous approach is defined as a function $\delta(t)$ which has a property‡

$$\int_{-\infty}^{\infty} \delta(t)\phi(t)\, dt \;=\; \phi(0)$$

Thus we do not say anything about the values of $\delta(t)$ at various values of t as we do in case of ordinary functions. We define $\delta(t)$ by its effect on other function $\phi(t)$ through an integral operation. Thus in the rigorous approach the impulse function is defined by the sampling property (Eq. 2.80) whereas in our heuristic approach we define it as in Eq. 2.79 and the sampling property follows as a consequence of this definition.

2.8 IMPULSE RESPONSE OF A SYSTEM

We shall now find the zero-state response of a system to unit impulse input. It will be shown that the zero-state response to unit impulse input is given by $h(t)$ defined earlier.

The zero-state response $y_f(t)$ to the input $\delta(t)$ is given (Eq. 2.78) by

$$y_f(t) \;=\; \int_{0^-}^{t} h(t - \tau)\delta(\tau)\, d\tau \tag{2.84}$$

Using the sampling property of the impulse function (Eq. 2.83b) the integral in Eq. 2.84 is immediately seen to be $h(t)$. Hence

$$y_f(t) \;=\; h(t) \tag{2.85}$$

This is a very important result, and we shall recapituate it here.

If a response variable y is related to the input $f(t)$ by the differential equation

$$y = H(p)f(t)$$

$$= \left[\frac{k_1}{p - \lambda_1} + \frac{k_2}{p - \lambda_2} + \cdots + \frac{k_n}{p - \lambda_n} \right] f(t)$$

Then y_f, the zero-state response to unit impulse input, is given by

$$y_f(t) = h(t) = k_1 e^{\lambda_1 t} + k_2 e^{\lambda_2 t} + \cdots + k_n e^{\lambda_n t} \tag{2.86}$$

---

†L. Schwartz, *Théorie des Distributions*, Vols. I and II (Paris: Hermann et Cie., 1950, 1951); M. J. Lighthill, *Fourier Analysis and Generalized Functions* (London: Cambridge University Press, 1955).

‡It is assumed that $\phi(t)$ is a function which has derivatives of any order at the origin and it approaches 0 as $t \to \infty$.

In the literature $h(t)$ is often referred to as the *unit impulse response* of the system. But it must be clearly understood that $h(t)$ is actually a unit impulse response for zero-state condition. The unit impulse response $h(t)$ of a system corresponding to various transfer operators is listed in Table 2.1.

The observant reader will not fail to notice the similarity in the form of $y_x(t)$, the zero-input response of the system and $h(t)$, the zero-state unit impulse response of the same system. From Eq. 2.70b,

$$y_x(t) = c_1 e^{\lambda_1 t} + c_2 e^{\lambda_2 t} + \cdots + c_n e^{\lambda_n t} \tag{2.87}$$

and the zero-state unit impulse response $h(t)$ is given by (see Eq. 2.72)

$$h(t) = k_1 e^{\lambda_1 t} + k_2 e^{\lambda_2 t} + \cdots + k_n e^{\lambda_n t} \tag{2.88}$$

In $y_x(t)$, the constants $c_1, c_2, \ldots, c_n$ are arbitrary depending on the initial state, whereas in $h(t)$, the constants $k_1, k_2, \ldots, k_n$ are fixed and are obtained from the transfer operator.†

The reason for similarity is not difficult to understand. The component $y_x(t)$ is the zero-input component—that is, it is the response of the system to initial energy storages (initial state) and when the external input is zero. Next consider the input $\delta(t)$. This input exists only at $t = 0$ and for $t > 0$, there is no input. The application of impulse creates energy storages. Hence at $t = 0^+$, we have no input but some energy storages created by the impulse. Obviously the response of the system for $t > 0$ will be similar to zero-input response in form (natural response).

2.9   THE CONVOLUTION INTEGRAL

The zero-state response $y_f(t)$ due to input $f(t)$ is given by an integral

$$y_f(t) = \int_0^t h(t - \tau)f(\tau)\, d\tau \tag{2.89a}$$

This form of integral commonly occurs in areas of physical sciences and mathematics and is given a special name the *convolution integral*.

Given two functions $f_1(t)$ and $f_2(t)$, we form the integral

$$q(t) = \int_{-\infty}^{\infty} f_1(\tau)f_2(t - \tau)\, d\tau \tag{2.89b}$$

This integral defines the convolution of function $f_1(t)$ and $f_2(t)$. The convolution integral is also expressed symbolically as

$$q(t) = f_1(t) * f_2(t) \tag{2.89c}$$

---

†In some cases where $H(p)$ contains a common factor in the numerator and the denominator, the exponential corresponding to the common factor will be missing in $h(t)$ because we can cancel common factor for the purpose of evaluating zero-state response (see p. 60). This exponential may or may not be missing in the zero-input component. In addition, if in $H(p)$ the numerator polynomial is of an order equal to or greater than the denominator polynomial, then $h(t)$ will contain the impulse function and its derivatives which do not appear in the zero-input response.

Let us consider the convolution of $f(t)$ with $h(t)$:

$$f(t) * h(t) = \int_{-\infty}^{\infty} f(\tau)h(t - \tau)\, d\tau \qquad (2.90a)$$

We are considering the case where the input $f(t)$ starts at $t = 0$. Hence the function $f(\tau)$ starts at $\tau = 0$ and is zero for $\tau < 0$. Therefore the integrand in Eq. 2.90a is zero for $\tau < 0$. Hence the lower limit of the integral in this case may be changed to 0 (actually 0^-). Similarly, we recognize that $h(t)$ is the response of the system (in zero-state) to an unit impulse input $\delta(t)$. Since $\delta(t)$ starts at $t = 0$, $h(t)$ cannot possibly exist for $t < 0$. Hence $h(t) = 0$ for $t < 0$. Thus the function $h(\cdot)$ is zero for negative values of the argument. Obviously the function $h(t - \tau)$ is zero when the argument $t - \tau < 0$ or when $\tau > t$. This means in Eq. 2.90a, the integrand $f(\tau)h(t - \tau) = 0$ for $\tau > t$. Consequently the upper limit of the integral can be replaced by t. Hence

$$f(t) * h(t) = \int_{-\infty}^{\infty} f(\tau)h(t - \tau)\, d\tau \qquad (2.90b)$$

$$= \int_{0}^{t} f(\tau)h(t - \tau)\, d\tau \qquad \text{when both } f(t) \text{ and} \qquad (2.90c)$$
$$h(t) = 0 \text{ for } t < 0$$

Note that the lower limit is actually 0^- (slightly before $t = 0$). Thus $y_f(t)$ the zero-state response to the input $f(t)$ can be expressed as

$$y_f(t) = f(t) * h(t) \qquad (2.91a)$$

where $h(t)$ is the unit impulse response of the system (in zero state).

In conclusion, the zero-state response of a system is given by the convolution of the input $f(t)$, with $h(t)$, the unit impulse response of the system. A short list of convolutions is given in Table 2.2. From the knowledge of $f(t)$ and $h(t)$, the zero-state response can be found from this table directly without actually performing the integration. It should be observed that convolution of any function $f(t)$ with unit impulse function $\delta(t)$ results in the function $f(t)$ itself. This is the result of the sampling property of the impulse function

$$f(t) * \delta(t) = \int_{-\infty}^{\infty} f(\tau)\delta(t - \tau)\, d\tau \qquad (2.91b)$$

From Eq. 2.81, it follows that the integral on the right-hand side is $f(t)$. Thus

$$f(t) * \delta(t) = f(t) \qquad (2.91c)$$

Therefore convolution of a function with unit impulse yields the function itself. This is a very useful result.

The table of convolution integral (Table 2.2) is very convenient in evaluating y_f, the zero-state response of linear systems, since

$$y_f = f(t) * h(t)$$

TABLE 2.2
Convolution Table

	$f_1(t)$	$f_2(t)$	$f_1(t) * f_2(t) = f_2(t) * f_1(t)$
1	$f(t)$	$\delta(t)$	$f(t)$
2	$e^{\lambda t}u(t)$	$u(t)$	$\dfrac{-1}{\lambda}(1 - e^{\lambda t})u(t)$
3	$u(t)$	$u(t)$	$tu(t)$
4	$e^{\lambda_1 t}u(t)$	$e^{\lambda_2 t}u(t)$	$\dfrac{1}{\lambda_2 - \lambda_1}[e^{\lambda_2 t} - e^{\lambda_1 t}]u(t) \qquad \lambda_1 \neq \lambda_2$
5	$e^{\lambda t}u(t)$	$e^{\lambda t}u(t)$	$te^{\lambda t}u(t)$
6	$t^n u(t)$	$e^{\lambda t}u(t)$	$\dfrac{n!}{\lambda^{n+1}}e^{\lambda t}u(t) - \displaystyle\sum_{j=0}^{n}\dfrac{n!}{\lambda^{j+1}(n-j)!}t^{n-j}u(t)$
7	$t^m u(t)$	$t^n u(t)$	$\dfrac{m!\,n!}{(m+n+1)!}t^{m+n+1}u(t)$
8	$t^m e^{\lambda_1 t}u(t)$	$t^n e^{\lambda_2 t}u(t)$ $\lambda_1 \neq \lambda_2$	$\displaystyle\sum_{j=0}^{m}\dfrac{(-1)^j m!\,(n+j)!}{j!(m-j)!(\lambda_1 - \lambda_2)^{n+j+1}}t^{m-j}e^{\lambda_1 t}u(t)$ $+ \displaystyle\sum_{k=0}^{n}\dfrac{(-1)^k n!\,(m+k)!}{k!(n-k)!(\lambda_2 - \lambda_1)^{m+k+1}}t^{n-k}e^{\lambda_2 t}u(t)$
9	$e^{-\alpha t}\cos(\beta t + \theta)u(t)$	$e^{\lambda t}u(t)$	$\left[\dfrac{\cos(\theta - \phi)}{\sqrt{(\alpha + \lambda)^2 + \beta^2}}e^{\lambda t} - \dfrac{e^{-\alpha t}\cos(\beta t + \theta - \phi)}{\sqrt{(\alpha + \lambda)^2 + \beta^2}}\right]u(t),$ $\phi = \tan^{-1}\dfrac{-\beta}{\alpha + \lambda}$

In Sec. 2.10, it will be shown that

$$f_1(t) * f_2(t) = f_2(t) * f_1(t)$$

Hence

$$y_f(t) = f(t) * h(t) = h(t) * f(t)$$

when $f(t)$ and $h(t)$ are known, the zero-state response y_f can be immediately obtained from Table 2.2. As an example consider the problem in Example 2.10. In this case $h(t)$ is given by

$$h(t) = -2e^{-2t} + 3e^{-3t}$$

Note that $h(t)$ is the response of the system to unit impulse input applied at $t = 0$. Hence $h(t) = 0$ for $t < 0$. Therefore it is appropriate to write

$$h(t) = (-2e^{-2t} + 3e^{-3t})u(t)$$

and

$$f(t) = u(t)$$

Hence

$$y_f(t) = h(t) * u(t)$$
$$= (-2e^{-2t} + 3e^{-3t})u(t) * u(t)$$
$$= -2e^{-2t}u(t) * u(t) + 3e^{-3t}u(t) * u(t)$$

These convolutions are found from Table 2.2 (pair 2). This gives

$$y_f(t) = -2[\tfrac{1}{2}(1 - e^{-2t})u(t)] + 3[\tfrac{1}{3}(1 - e^{-3t})u(t)]$$
$$= (e^{-2t} - e^{-3t})u(t)$$

which is the same result as obtained earlier. Note that since we are talking about the response for $t > 0$, $u(t) = 1$, and one may ignore $u(t)$ that appears on the right hand side of $y_f(t)$.

2.10   SOME CONVOLUTION RELATIONSHIPS

Convolution represents some kind of operation and the symbolic representation in Eq. 2.89c suggests that convolution is a special kind of multiplication.

It is indeed true that some of the algebraic rules applicable to ordinary multiplication also apply to convolution relationship.

COMMUTATIVE LAW

$$f_1(t) * f_2(t) = f_2(t) * f_1(t) \tag{2.92a}$$

This relationship is easily proved as follows

$$f_1(t) * f_2(t) = \int_{-\infty}^{\infty} f_1(\tau)f_2(t - \tau)\, d\tau$$

changing the variable τ to $t - x$, we obtain

$$f_1(t) * f_2(t) = \int_{-\infty}^{\infty} f_2(x)f_1(t - x)\, dx$$
$$= f_2(t) * f_1(t)$$

This implies that

$$y_f(t) = f(t) * h(t) = h(t) * f(t)$$
$$= \int_{0}^{t} f(\tau)h(t - \tau)\, d\tau = \int_{0}^{t} h(\tau)f(t - \tau)\, d\tau \tag{2.92b}$$

DISTRIBUTIVE LAW

$$f_1(t) * [f_2(t) + f_3(t)] = f_1(t) * f_2(t) + f_1(t) * f_3(t) \tag{2.92c}$$

The proof is trivial.

ASSOCIATIVE LAW

$$f_1(t) * [f_2(t) * f_3(t)] = [f_1(t) * f_2(t)] * f_3(t) \tag{2.92d}$$

Each term in this equation represents a double integral in two dummy variables τ_1 and τ_2. For reasonably well-behaved functions the order of integration can be changed. This proves Eq. 2.92d. The verification is left as an exercise for the reader.

2.11   CONVOLUTION INTEGRAL AS A SUPERPOSITION INTEGRAL

In Chapter 1 it was indicated that the analysis of linear systems can be considerably simplified by exploiting the principle of superposition. The input may be separated or decomposed into several simpler functions. This permits the response to each component to be found with less difficulty. The desired response is given by the sum of the responses to all the components of the input function. An arbitrary input function may be represented by sum of functions such as exponential, impulse, step, ramp and so on. The case of exponential function decomposition will be considered in Chapters 3 and 4. We shall now show that the convolution integral representing $y_f(t)$ in our response can be interpreted as a superposition integral representing the response of the system to various impulse components of the input $f(t)$.

A function $f(t)$ may be represented as a sum of narrow pulses of width Δt. The height of nth pulse located at $t = n\Delta t$ is $f(n\Delta t)$ as shown in Fig. 2.13a. In the limit as $\Delta t \to 0$, a pulse can be considered as an impulse of the strength equal to the area of the pulse. The response due to $f(t)$ is the sum of the responses due to all the impulse components of $f(t)$. We shall now find the response at $t = t_1$, due to pulse located at $t = n\Delta t$ (shaded pulse in Fig. 2.13a). The response due to unit impulse input $\delta(t)$ is

$$h(t)$$

The shaded pulse in Fig. 2.13a has a strength $f(n\Delta t)\Delta t$, and is shifted from the origin by $n\Delta t$. Hence the response due to this pulse is (see Fig. 2.13b)

$$\lim_{\Delta t \to 0} [f(n\Delta t)\Delta t] h(t - n\Delta t)$$

Therefore the response at $t = t_1$ due to the shaded pulse is (see Fig. 2.13b)

$$\lim_{\Delta t \to 0} [f(n\Delta t)\Delta t] h(t_1 - n\Delta t) \tag{2.93}$$

The response due to all the impulse components is the sum of the responses of the form in Eq. 2.93. Thus the total response at $t = t_1$ is $y_f(t_1)$, given by (see Fig. 2.13c)

$$y_f(t_1) = \lim_{\Delta t \to 0} \Sigma [f(n\Delta t)\Delta t] h(t_1 - n\Delta t) \tag{2.94a}$$

Note that the pulses lying beyond $t = t_1$ (Fig. 2.13a) have no effect on the response at $t = t_1$. Hence the summation in Eq. 2.94a should be performed over the limits $n\Delta t = 0$, to $n\Delta t = t_1$. Therefore

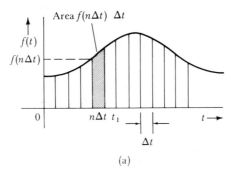

(a)

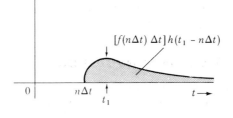

(b)

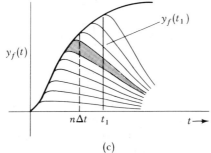

(c)

Figure 2.13

$$y_f(t_1) = \lim_{\Delta t \to 0} \sum_{n\Delta t = 0}^{n\Delta t = t_1} f(n\Delta t)h(t_1 - n\Delta t)\, \Delta t \qquad (2.94b)$$

In the limit as $\Delta t \to 0$, the discrete sum becomes an integral (continuous sum)

$$y_f(t_1) = \int_0^{t_1} f(t)h(t_1 - t)\, dt$$

In the integral on the right-hand side, t is a dummy variable and can be changed to another variable τ. Thus

$$y_f(t_1) = \int_0^{t_1} f(\tau)h(t_1 - \tau)\, d\tau$$

The response at any instant t is therefore given by

$$y_f(t) = \int_0^t f(\tau)h(t - \tau)\,d\tau \qquad (2.95)$$

This is exactly the same result as obtained earlier by the direct method (Eq. 2.71). The lower limit of integration in Eq. 2.95 should be taken as 0^- in general.† Therefore

$$y_f(t) = \int_{0^-}^t f(\tau)h(t - \tau)\,d\tau \qquad (2.96)$$

The result expressed in Eq. 2.96 is precisely the zero-state response obtained earlier. Thus the convolution integral in the zero-state response actually is a super-position integral representing the continuous sum (integral) of responses to the continuum of impulses into which the input $f(t)$ was decomposed.

2.12   GRAPHICAL VIEW OF CONVOLUTION

The graphical interpretation of convolution is very useful in system analysis as well as in communication theory (signal analysis). It permits one to grasp visually the results of many abstract relationships.‡ In linear systems, graph-ical convolution is very helpful in analysis if $f(t)$ and $h(t)$ are known only graph-ically. To illustrate, consider $f(t)$ and $h(t)$ shown in Fig. 2.14a. We shall now find the convolution $f(t) * h(t)$ graphically. By definition,

$$y_f(t) = f(t) * h(t) = \int_{-\infty}^{\infty} f(\tau)h(t - \tau)\,d\tau \qquad (2.97a)$$

The variable of integration is τ. The functions $f(\tau)$ and $h(-\tau)$ are shown in Fig. 2.14b. Note that $h(-\tau)$ is the mirror image of $h(\tau)$ about the vertical axis. If we conceive $h(\tau)$ as a rigid wire frame then $h(-\tau)$ is obtained by rotating this frame about the vertical axis. The term $h(t - \tau)$ represents the function $h(-\tau)$ shifted by an amount t along the positive τ axis. Figure 2.14c shows $h(t_1 - \tau)$. The value of the convolution integral at $t = t_1$ is $y_f(t_1)$ and is given by (see Eq. 2.97a)

$$y_f(t_1) = \int_{-\infty}^{\infty} f(\tau)h(t_1 - \tau)\,d\tau \qquad (2.97b)$$

Thus the value of the convolution integral at $t = t_1$ is $y_f(t_1)$ and is given by Eq. 2.97b, being the area under the product of $f(\tau)$ and $h(t_1 - \tau)$. The product of $f(\tau)$ and $h(t_1 - \tau)$ and corresponding area A_1 under this product curve is shown in Fig. 2.14c. Thus $y_f(t_1) = A_1$. This area A_1 represents the value of the convolu-tion function $f(t) * h(t)$ at $t = t_1$ and is plotted as shown in Fig. 2.14e. We repeat

---

†See comments on p. 72 for the discussion on the lower limit of integration.
‡See, for example, B. P. Lathi, *Communication Systems* (New York: Wiley, 1968), Chapters 1, 3, 5.

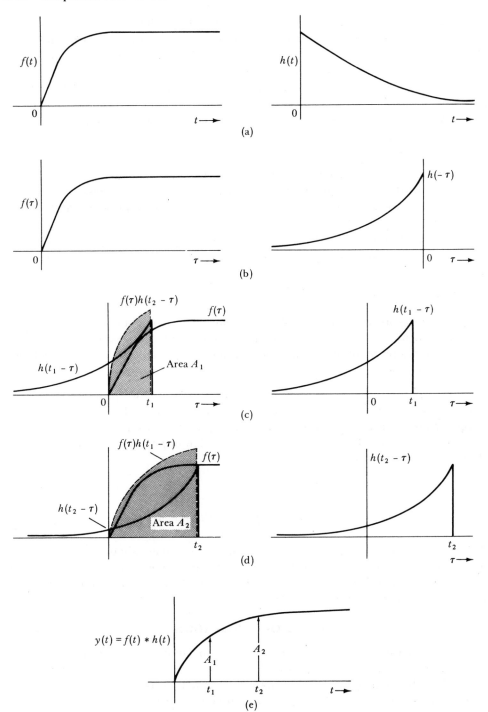

Figure 2.14

the procedure for $t = t_2$ as shown in Fig. 2.14d. This is done by shifting $h(-\tau)$ by amount t_2, multiplying the shifted function by $f(\tau)$ and finding the area A_2 under the product curve. Thus $y_f(t_2) = A_2$. In this way we choose different values of t, shift the function $h(-\tau)$ accordingly, and find the area under the new product curve. These areas represent values of the convolution function $f(t) * h(t)$ at the respective values of t. The plot of the area under the product curve as a function of the amount of shift t, represents the desired convolution function $f(t) * h(t)$ as shown in Fig. 2.14e.

To summarize the steps in convolution:

1. Fold or rotate the function $h(\tau)$ about the vertical axis to obtain $h(-\tau)$.
2. Consider the folded curve as a rigid wire frame and progress it along the τ-axis by an amount say t_0. The rigid frame now represents the function $h(t_0 - \tau)$.
3. The product of the fixed function $f(\tau)$ and the function represented by the displaced frame is $f(\tau)h(t_0 - \tau)$, and the area under this curve is given by

$$\int_{-\infty}^{\infty} f(\tau)h(t_0 - \tau)\, d\tau = f(t) * h(t)\, |_{t=t_0}$$

This represents the value of the convolution function at $t = t_0$.

4. Repeat this procedure for different values of t by successively progressing the frame by different amount and finding the value of $f(t) * h(t)$ at those values of t.

To find the value of $f(t) * h(t)$ for positive values of t, we progress the frame along the positive τ-axis (to the right), whereas for negative values of t the frame is progressed in the negative direction (toward the left of the origin).

It was shown in Eq. 2.92a that the convolution of $f(t)$ with $h(t)$ is identical to the convolution of $h(t)$ with $f(t)$. That is,

$$f(t) * h(t) = h(t) * f(t)$$

$$= \int_{-\infty}^{\infty} h(\tau)f(t - \tau)\, d\tau$$

Therefore we may keep $h(\tau)$ fixed and take the mirror image about the vertical axis of $f(\tau)$ and vary this rigid frame by various amount. We obtain the same result either ways. It is left as an exercise for the reader to verify this graphically using the functions $f(t)$ and $h(t)$ in Fig. 2.14a.

LIMITS OF INTEGRATION

The general limits of integration for the convolution integral are from $-\infty$ to ∞. It was shown earlier that if both the functions to be convolved are zero for $t < 0$, then the limits of integration may be changed to from 0 to t. This fact is quite obvious from Fig. 2.14. The product function exists from 0 to t only. The function $f(\tau)$ is zero for $\tau < 0$, and $h(t - \tau)$ is zero for $\tau > t$. Hence the product is nonzero only for $0 < \tau < t$, and the limits can be replaced by 0 to t. The lower limit 0 here is actually 0^-.   In general the caution of using 0^- for the lower limit is necessary only if the integrand contains an impulse function or its derivatives at the origin.   In such cases if the lower limit of 0^+ is used, the impulse and/or its derivatives (which exist only at the origin) will be ignored in the process of integration, giving erroneous results.

2.13   NUMERICAL TECHNIQUES OF CONVOLUTION

We shall now give a numerical method of computing convolution function. Let us consider

$$y_f(t) = f(t) * h(t)$$

where $f(t)$ and $h(t)$ are functions as shown in Fig. 2.15.   For the purpose of convolution, we take $f(\tau)$ and fold $h(\tau)$ about the vertical axis to obtain $h(-\tau)$ as

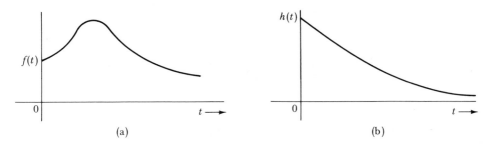

Figure 2.15

shown in Fig. 2.16a.   Next we approximate both the functions by staircase functions† as shown in Fig. 2.16a.   The accuracy in this approximation can be improved by reducing the increment T.

We now shift $h(-\tau)$ by T seconds and multiply the two functions and find the area under the product curve as shown in Fig. 2.16b.   The area is obviously $Tf_0 h_1$.   Hence

$$y_f(T) = Tf_0 h_1$$

To obtain $y_f(2T)$, we shift $h(-\tau)$ by $2T$ to the right.   This can be conveniently done by using two strips shown in Fig. 2.16c.   The sequence of values of the function $f(t)$ for $t = 0, T, 2T, \ldots$, etc. are written on the upper strip.   Let $f(nT) = f_n$. Similarly, the sequence of values of $h(t)$ for $f = 0, T, 2T$, etc., are given on the

---

†Note that we have approximated $f(t)$ by a staircase function in such a way that the approximated function lies to the right of the original function.   This appears as a delay of $T/2$.   Thus this approximation adds a delay factor of $T/2$.   To compensate for this delay we approximate $h(t)$ such that the approximated function lies to the left of the original function giving the negative delay $-T/2$.   That is why the first step in $f(t)$ staircase is f_0 whereas the first step in $h(t)$ staircase is h_1.

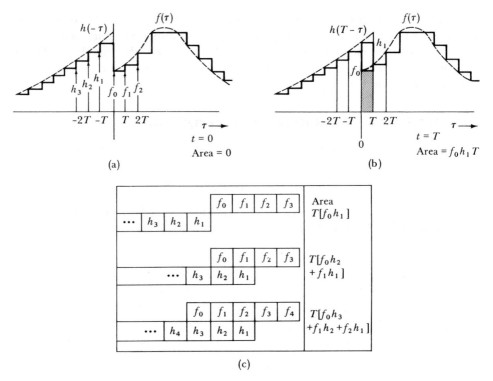

Figure 2.16

lower strip in reverse order (Fig. 2.16c) with the notation $h(nT) = h_n$. Shifting the function $h(-\tau)$ by nT seconds is equivalent to shifting the lower strip to the right by the same amount. The area under the product curve is given by T times the sum of the products of overlapping numbers. Thus to obtain $y_f(nT)$, we shift the lower strip by n notches to the right, multiply all the overlapping numbers and add the products. T times this sum is $y_f(nT)$. It can be seen from Fig. 2.16c that

$$y_f(0) = 0$$
$$y_f(T) = Tf_0h_1$$
$$y_f(2T) = T[f_0h_2 + f_1h_1]$$
$$y_f(3T) = T[f_0h_3 + f_1h_2 + f_2h_1]$$

In general,

$$y_f(nT) = T\sum_{m=1}^{n} h_m f_{n-m} = T\sum_{m=1}^{n} f_m h_{n-m} \qquad (2.98)$$

This equation holds true when both $f(t)$ and $h(t)$ are causal, that is, both are zero for $t < 0$. For a general case, the summation should be carried out over $m = -\infty$ to ∞.

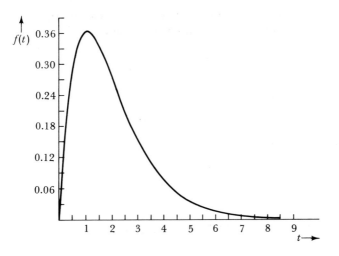

Figure 2.17

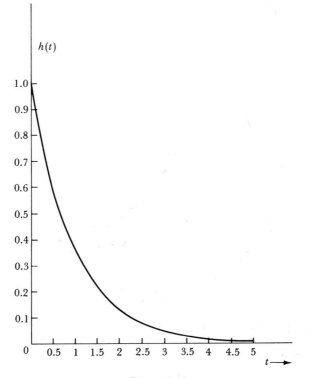

Figure 2.18

As an example consider voltage $f(t)$ whose waveform appears in Fig. 2.17 to be applied to the input of a system whose transfer operator is given by

$$H(p) = \frac{1}{p + 1} \tag{2.99}$$

and

$$h(t) = e^{-t}u(t)$$

Hence

$$y_f(t) = f(t) * e^{-t}u(t) \tag{2.100}$$

and

$$y_f(nT) = T \sum_{m=1}^{n} f_m h_{n-m}$$

where

$$f_m = f(mT) \qquad \text{and} \qquad h_{n-m} = h[(n - m)T]$$

The values of f_m and h_{n-m} are read from the data (Figs. 2.17 and 2.18). The computational work can be conveniently performed on a digital computer. The result of such a computation is shown in Fig. 2.19 for $T = 0.05$. The exact response is also shown in the same figure.

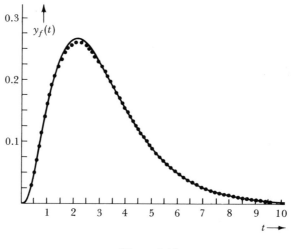

Figure 2.19

2.14   INITIAL-CONDITION GENERATORS

The response of a system arises because of (1) initial conditions, and (2) inputs or sources. Thus the initial conditions act as some kind of sources. One wonders if it is possible to replace initial conditions in a system with equivalent sources. If this can be done we will be able to convert a system with initial conditions and sources into an equivalent system with zero initial conditions and sources only.   Such a system will have no zero-input component.   The total response will be given entirely by the zero-state response due to all the sources.

Initial conditions can indeed be so replaced. In electrical networks the initial conditions are often specified by initial voltages of capacitors, and initial currents in inductors. It can be shown that a capacitor with initial voltage $v(0)$ can be represented by the same capacitor (with zero initial voltage) in series with a voltage source of magnitude $v(0)$, or in parallel with a current source $Cv(0)\delta(t)$. These equivalent arrangements are shown in Fig. P-2.32a. Similarly, an inductor L with initial current $i(0)$ can be represented by the inductor L (with zero initial current) in series with a voltage source $Li(0)\delta(t)$, or a current source $i(0)$ in parallel, as shown in Fig. P-2.32b. A similar source substitution for initial conditions can be effected in mechanical systems.

In general it can be shown that in a linear system any set of initial conditions can always be represented by equivalent sources.

2.15   MULTIPLE INPUTS

When a linear system contains more than one input, any of the response variables can be obtained by using the principle of superposition. If $f_1(t)$, $f_2(t)$, $\ldots, f_m(t)$ are the m inputs acting on a system, then the effect of each input can be evaluated independently of the remaining inputs (principle of superposition). The desired response is the sum of m such components (zero-state response), and the zero-input component. The differential equation that relates the output variable y to these inputs is of the form

$$y(t) = H_1(p)f_1(t) + H_2(p)f_2(t) + \cdots + H_m(p)f_m(t) \qquad (2.101a)$$

where $H_k(p)$ is the transfer operator relating y to $f_k(t)$. If $\lambda_1, \lambda_2, \ldots, \lambda_n$ are numbers such that each one is a pole of at least one of the m transfer functions, then the zero-input response is given by $\Sigma c_j e^{\lambda_j t}$. If $h_k(t)$ is the unit impulse response associated with the transfer operator $H_k(p)$, then the total response is given by

$$y(t) = \sum_{j=1}^{n} c_j e^{\lambda_j t} + \sum_{k=1}^{m} f_k(t) * h_k(t) \qquad (2.101b)$$

The arbitrary constants c_j can be evaluated from the initial conditions (Eq. 2.47c).

Example 2.11.   Find the transfer operators which relate the outputs i_1 and i_2 to the inputs $f_1(t)$ and $f_2(t)$ in Fig. 2.20.

The mesh equations are given by

$$\begin{bmatrix} \dfrac{1}{p} + \dfrac{1}{5} & \dfrac{1}{5} \\[2ex] \dfrac{1}{5} & \dfrac{6}{5} + \dfrac{p}{2} \end{bmatrix} \begin{bmatrix} i_1 \\[2ex] i_2 \end{bmatrix} = \begin{bmatrix} f_1(t) \\[2ex] f_2(t) \end{bmatrix}$$

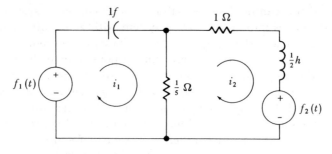

Figure 2.20

Application of Cramer's rule to this equation yields

$$i_1 = \frac{p(5p + 12)f_1(t) - 2pf_2(t)}{p^2 + 7p + 12}$$

$$= \frac{p(5p + 12)}{p^2 + 7p + 12} f_1(t) + \frac{-2p}{p^2 + 7p + 12} f_2(t) \qquad (2.102a)$$

$$= H_{11}(p)f_1(t) + H_{12}(p)f_2(t) \qquad (2.102b)$$

where $H_{11}(p)$ and $H_{12}(p)$, and the transfer operators relating i_1 to inputs $f_1(t)$ and $f_2(t)$ respectively, are given by

$$H_{11}(p) = \frac{p(5p + 12)}{p^2 + 7p + 12} \qquad \text{and} \qquad H_{12}(p) = \frac{-2p}{p^2 + 7p + 12}$$

Similarly, we obtain

$$i_2 = \frac{-2p}{p^2 + 7p + 12} f_1(t) + \frac{2(p + 5)}{p^2 + 7p + 12} f_2(t) \qquad (2.102c)$$

$$= H_{21}(p)f_1(t) + H_{22}(p)f_2(p) \qquad (2.102d)$$

where $H_{21}(p)$ and $H_{22}(p)$, and the transfer operators relating i_2 to inputs $f_1(t)$ and $f_2(t)$ respectively, are given by

$$H_{21}(p) = \frac{-2p}{p^2 + 7p + 12} \qquad \text{and} \qquad H_{22}(p) = \frac{2(p + 5)}{p^2 + 7p + 12}$$

Let us consider $i_2(t)$. With partial fractions of $H_{21}(p)$, and $H_{22}(p)$, Eq. 2.102c can be written as

$$i_2(t) = \left(\frac{6}{p + 3} - \frac{8}{p + 4}\right) f_1(t) + \left(\frac{4}{p + 3} - \frac{2}{p + 4}\right) f_2(t)$$

Hence

$$i_2(t) = c_1 e^{-3t} + c_2 e^{-4t} + (6e^{-3t} - 8e^{-4t}) * f_1(t) + (4e^{-3t} - 2e^{-4t}) * f_2(t)$$

2.16   RESPONSE OF A SYSTEM TO EXPONENTIAL INPUT SIGNAL

The exponential signal plays an important role in the analysis of linear system. The frequency-domain techniques of analysis of linear systems discussed in the following chapter are based on resolving any given input in terms of exponential components and then finding the response to these components and adding the result to obtain the desired response. More on this topic will be said later. Here we shall find the response of a system to the exponential input $f(t) = e^{st}$.

Let the response variable y be related to the input $f(t)$ by

$$y = H(p)f(t) \tag{2.103}$$

where

$$H(p) = \frac{N(p)}{(p - \lambda_1)(p - \lambda_2)\cdots(p - \lambda_n)} \tag{2.104a}$$

$$= \frac{k_1}{p - \lambda_1} + \frac{k_2}{p - \lambda_2} + \cdots + \frac{k_n}{p - \lambda_n} \tag{2.104b}$$

$$= \sum_{j=1}^{n} \frac{k_j}{p - \lambda_j} \tag{2.104c}$$

Hence

$$h(t) = \sum_{j=1}^{n} k_j e^{\lambda_j t} \tag{2.105}$$

The solution for y can now be written immediately as

$$y(t) = \sum_{j=1}^{n} c_j e^{\lambda_j t} + \sum_{j=1}^{n} k_j [f(t) * e^{\lambda_j t}]$$

where the c_j's are arbitrary constants determined by initial conditions (using Eq. 2.47c). When the input $f(t) = e^{st}$, this equation becomes

$$y(t) = \sum_{j=1}^{n} c_j e^{\lambda_j t} + \sum_{j=1}^{n} k_j [e^{st} * e^{\lambda_j t}]$$

Use of convolution Table 2.2 yields

$$y(t) = \sum_{j=1}^{n} c_j e^{\lambda_j t} + \sum_{j=1}^{n} \frac{k_j}{s - \lambda_j} (e^{st} - e^{\lambda_j t})$$

$$= \underbrace{\sum_{j=1}^{n} c_j e^{\lambda_j t}}_{y_x} + \underbrace{\left[-\sum_{j=1}^{n} \frac{k_j}{s - \lambda_j} e^{\lambda_j t} + e^{st} \sum_{j=1}^{n} \frac{k_j}{s - \lambda_j} \right]}_{y_f(t)} \tag{2.106a}$$

From Eq. 2.104c we observe that the last summation in Eq. 2.106a is $H(s)$. If we let

$$\frac{-k_j}{s - \lambda_j} = b_j$$

then Eq. 2.106a can be expressed as

$$y(t) = \underbrace{\sum_{j=1}^{n} c_j e^{\lambda_j t}}_{y_x} + \underbrace{\sum_{j=1}^{n} b_j e^{\lambda_j t} + H(s)e^{st}}_{y_f} \qquad (2.106b)$$

If we let

$$a_j = c_j + b_j \qquad (2.106c)$$

then

$$y(t) = \sum_{j=1}^{n} a_j e^{\lambda_j t} + H(s)e^{st} \qquad (2.107)$$

This is the result we seek. It shows that the response of a linear time-invariant system to the input e^{st} contains a component of the same form (viz., e^{st}) and has a magnitude $H(s)$. This may be considered as an alternative definition of the function $H(s)$, which is commonly known as the *transfer function* (or system function).

2.17   NATURAL RESPONSE AND FORCED RESPONSE

So far the response was expressed as a sum of two components; one due to the initial state (zero-input component) and the other due to the input (zero-state component). The two components y_x and y_f are clearly shown in Eq. 2.106b. The zero-input component y_x consists of sum of natural modes $c_j e^{\lambda_j t}$. We notice that the zero-state component y_f also has natural-mode terms $b_j e^{\lambda_j t}$. When all the natural-mode terms are combined together in one term, the response appears as shown in Eq. 2.107. The component formed by all natural mode terms is known as the *natural response* (see Eq. 2.58). The remaining part of the response is called the *forced response*. Thus we have (see Eq. 2.107)

$$y(t) = \underbrace{\sum_{j=1}^{n} a_j e^{\lambda_j t}}_{\text{natural response}} + \underbrace{H(s)e^{st}}_{\text{forced response}} \qquad (2.108)$$

The natural response is not the zero-input response, although they both have the same form. This fact is clear from Eqs. 2.106b and 2.108.

Although our discussion here on the natural and the forced response was with reference to Eq. 2.106, where the input was exponential signal e^{st}, it is equally valid for any other input. In general the zero-state component contains natural-mode terms. When we combine all the natural mode terms of the response, the resulting component is the natural response and the remainder is the forced response.   We shall show this by the example below.

Example 2.12.   Find the natural and the forced components of the output $y(t)$ for the system whose transfer operator $H(p)$ is given by

$$H(p) = \frac{1}{p + 1}$$

and

$$f(t) = tu(t); \, y(0) = 2$$

The unit impulse response $h(t)$ is given by

$$h(t) = e^{-t}u(t)$$

Therefore

$$y(t) = ce^{-t} + f(t) * h(t)$$

$$= ce^{-t} + \int_0^t \tau e^{-(t-\tau)} \, d\tau \tag{2.109}$$

The value of the arbitrary constant c is obtained by letting $t = 0$ in Eq. 2.109 and using $y(0) = 2$.   Thus

$$y(0) = 2 = c + 0 \Longrightarrow c = 2$$

Hence

$$y(t) = 2e^{-t} + tu(t) * e^{-t}u(t)$$

The convolution term is obtained from the table of convolution (Table 2.2).   Hence

$$y(t) = \underbrace{2e^{-t}}_{\text{zero-input}} + \underbrace{e^{-t} + t - 1}_{\text{zero-state}}$$

$$= \underbrace{3e^{-t}}_{\text{natural}} + \underbrace{(t - 1)}_{\text{forced}}$$

Note that the zero-input response is $2e^{-t}$ whereas the natural response is $3e^{-t}$.

2.18   TRANSIENT AND STEADY-STATE RESPONSE

In many instances, the response of a system consists of a part which decays with time and eventually vanishes and the remaining part which persists indefinitely.   These components are called the transient component and the steady-state component respectively.   In Example 2.12, the response is $3e^{-t} + t - 1$. The term $3e^{-t}$ dies out whereas the component $t - 1$ persists indefinitely.   It is therefore evident that $3e^{-t}$ is a transient component and $t - 1$ is the steady-state component.   In this particular example, the natural response is also the transient response and the forced response is also the steady-state response.   This is not true in general, however.   Consider for example the case of exponential input signal e^{st}.

The response to the input e^{st} is given in Eq. 2.107. If Re $s \geq 0$, the component $H(s)e^{st}$ will persist indefinitely and the component $\Sigma a_j e^{\lambda_j t}$ will decay with time for stable systems.† In this case the natural response and the transient response are identical. Similarly the forced response and the steady-state response are the same. But consider the case when Re $s < 0$. Here the term $H(s)e^{st}$ also decays with time and eventually vanishes. Thus the entire response in this case is transient. In unstable systems where one of the poles λ_j is such that Re $\lambda_j \geq 0$, then the term $a_j e^{\lambda_j t}$ persists indefinitely whereas $H(s)e^{st}$ will vanish if Re $s < 0$. In these cases the transient and the natural response are not the same.

The transient and steady-state components are particularly meaningful when the input is a sinusoidal signal. In this case the transient and the natural response are identical. Similarly the steady-state and the forced components are the same (sinusoids).

2.19   STEADY-STATE RESPONSE TO SINUSOIDAL INPUTS

When sinusoidal signals are applied to a linear system, the response consists of a transient component and a steady-state component which is also a sinusoidal signal of the same frequency as the input. For a stable system the transient component decays with time, whereas the sinusoidal signal maintains the constant amplitude. As a result, after a sufficient time interval only the steady-state (sinusoidal) component is observed.

Sinusoidal signals can be expressed in terms of complex exponentials, and the steady-state response to sinusoidal signals $\cos \omega t$ and $\sin \omega t$ can be readily found in terms of the steady-state response to $e^{j\omega t}$.

Let $H(s)$ be the transfer function relating the response variable y to the input $f(t)$. For input $e^{j\omega t}$, the steady-state response will be $H(j\omega)e^{j\omega t}$ (see Eq. 2.107, with $s = j\omega$), and for the input $e^{-j\omega t}$ the steady-state response will be $H(-j\omega)e^{-j\omega t}$. Since

$$\cos \omega t = \tfrac{1}{2}[e^{j\omega t} + e^{-j\omega t}]$$

we can use principle of superposition to find the steady state response. Thus $y_{ss}(t)$, the steady-state response to $\cos \omega t$ is given by

$$y_{ss}(t) = \tfrac{1}{2}[H(j\omega)e^{j\omega t} + H(-j\omega)e^{-j\omega t}]$$

Since $H(j\omega)e^{j\omega t}$ is complex conjugate of $H(-j\omega)e^{-j\omega t}$, we have

$$y_{ss}(t) = \text{Re}\,[H(j\omega)e^{j\omega t}] \tag{2.110a}$$

In a similar way, we can show that $y_{ss}(t)$, the steady-state response of the system to the input $\sin \omega t$ is given by

$$y_{ss}(t) = \text{Im}\,[H(j\omega)e^{j\omega t}] \tag{2.110b}$$

If we use an arrow to indicate the steady-state response for a certain input, then

---

†Stable systems by definition have a natural response which decays with time. See Chapter 4 for further discussion.

we may express these results symbolically as

$$e^{j\omega t} \rightarrow H(j\omega)e^{j\omega t} \tag{2.111a}$$

$$\text{Re } e^{j\omega t} = \cos \omega t \rightarrow \text{Re } [H(j\omega)e^{j\omega t}] \tag{2.111b}$$

$$\text{Im } e^{j\omega t} = \sin \omega t \rightarrow \text{Im } [H(j\omega)e^{j\omega t}] \tag{2.111c}$$

$H(j\omega)$ is complex in general and can be expressed in polar form as

$$H(j\omega) = |H(j\omega)| e^{j\theta(\omega)}$$

and $\qquad\qquad\qquad\qquad\qquad\qquad\qquad\qquad\qquad\qquad\qquad\qquad$ (2.112)

$$H(j\omega)e^{j\omega t} = |H(j\omega)| e^{j(\omega t + \theta)}$$

Therefore

$$\text{Re } [H(j\omega)e^{j\omega t}] = |H(j\omega)| \cos(\omega t + \theta)$$

and $\qquad\qquad\qquad\qquad\qquad\qquad\qquad\qquad\qquad\qquad\qquad\qquad$ (2.113)

$$\text{Im } [H(j\omega)e^{j\omega t}] = |H(j\omega)| \sin(\omega t + \theta)$$

As a result, the steady-state response of the system to a driving function $\cos \omega t$ is given by $|H(j\omega)| \cos(\omega t + \theta)$, and the steady-state response to a driving function $\sin \omega t$ is given by $|H(j\omega)| \sin(\omega t + \theta)$. Symbolically,

$$\cos \omega t \rightarrow |H(j\omega)| \cos(\omega t + \theta)$$
$$\sin \omega t \rightarrow |H(j\omega)| \sin(\omega t + \theta) \tag{2.114}$$

In general,

$$\cos(\omega t + \phi) \rightarrow |H(j\omega)| \cos(\omega t + \phi + \theta)$$

The term $|H(j\omega)|$ therefore represents the magnitude of the steady-state response to a unit sunusoidal input, and it is evident from Eq. 2.114 that the response function is shifted in phase by angle θ from the input function. As can be seen from Eq. 2.112, the angle θ is the phase angle of $H(j\omega)$. That is,

$$\angle H(j\omega) = \theta(\omega)$$

The knowledge of $H(j\omega)$ is therefore sufficient to evaluate the steady-state response of a linear system to a sinusoidal driving function. Very often it is stated that the response of a linear system to a unit sinusoidal signal is $H(j\omega)$. This statement is meaningless unless it is understood in the proper context. The response of a system to any real function must be real, and hence $H(j\omega)$ cannot be a response of a real function. What we actually mean by the response being $H(j\omega)$ is that the magnitude of the response is $|H(j\omega)|$ and is shifted in phase from the driving function by $\angle H(j\omega)$ (see Fig. 2.22).

As an example, consider a system with transfer function

$$H(s) = \frac{1}{s + 1}$$

$$H(j\omega) \;=\; \frac{1}{j\omega + 1} \;=\; \frac{1}{\sqrt{\omega^2 + 1}}\, e^{-j\,tan^{-1}\omega}$$

Thus

$$|H(j\omega)| \;=\; \frac{1}{\sqrt{\omega^2 + 1}}$$

$$\theta(\omega) \;=\; -\tan^{-1}\omega$$

Both the magnitude $|H(j\omega)|$, and the phase $\theta(\omega)$ are plotted in Fig. 2.21. These

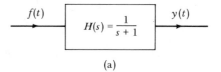

(a)

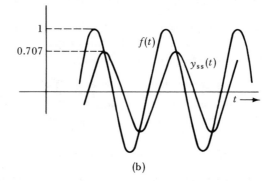

(b)

Figure 2.21

plots give the information about the steady-state behavior of the system over the entire frequency range. Thus when the input signal is

$$f(t) \;=\; \sin t$$

the value of $\omega = 1$. From Fig. 2.21 we read $|H(j\omega)|$ and $\theta(\omega)$ for $\omega = 1$ as 0.707 and $-45°$ respectively. Hence the steady-state response will be given by

$$y_{ss}(t) \;=\; 0.707 \sin(t - 45°)$$

This is shown in Fig. 2.22.
    If the input is

$$f(t) \;=\; \sin 3t$$

we read from Fig. 2.21 $|H(j\omega)|$ and $\theta(\omega)$ for $\omega = 3$ to be 0.316 and $-71°$ respectively. Hence

$$y_{ss}(t) \;=\; 0.316 \sin(3t - 71°)$$

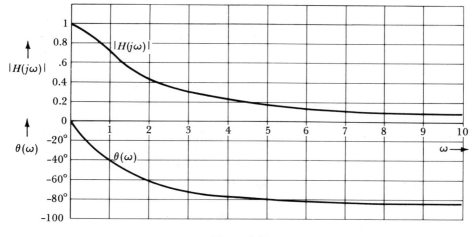

Figure 2.22

The magnitude and phase plots prove useful in designing feedback systems and we shall meet them again in Chapter 5.

2.20   BODE PLOTS

The plotting of the magnitude and phase functions (Fig. 2.22) is remarkably facilitated by using logarithmic scales. Instead of plotting $|H(j\omega)|$, we plot $20 \log_{10} |H(j\omega)|$ as a function of ω. The logarithmic magnitude thus defined

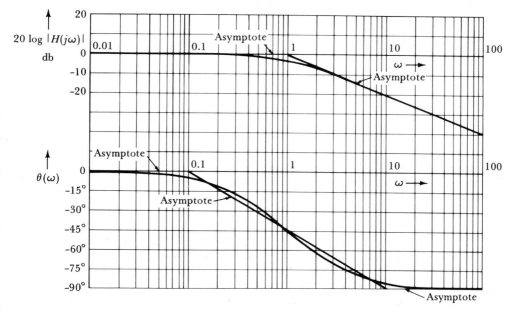

Figure 2.23

has a unit of decibel (db). The log-magnitude and phase plots as a function of ω (also on logarithmic scale) are known as *Bode plots*. These plots can be sketched with remarkable ease even for a complex transfer function by using the asymptotic behavior of the log magnitude and phase functions. The magnitude and phase plots in Fig. 2.22 when sketched as log-magnitude and phase plots appear as shown in Fig. 2.23. Note that the frequency ω is also on logarithmic scale. The log-magnitude is $20 \log |H(j\omega)|$ db. Thus for $\omega = 1$, $|H(j\omega)| = 0.707$. Hence the log magnitude is $20 \log_{10} 0.707 = -3$ db. Observe the asymptotic behavior of log-magnitude as well as phase function. As mentioned earlier, Bode plots can be sketched with considerable ease even for a complex transfer function, using asymptotic techniques. These techniques are fully developed in Appendix C for those who may be unfamiliar with this remarkable tool.

PROBLEMS

2.1.   Following differential equations are expressed in terms of operational motation $p = d/dt$. Write them in the form of differential equations.

(a) $(p + 2)x = pf(t)$

(b) $(p + 1)x = (p + 1)f(t)$

(c) $x = \dfrac{p + 1}{2p + 3} f(t)$

(d) $(p^2 + 2p + 3)x = (p + 1)f(t) + (2p + 1)g(t)$

(e) $(p + 1)(p + 2)x = p(p + 3)f(t)$

2.2.   You are given the following simultaneous differential equations. By using the process of elimination obtain differential equations each in one variable only.

(a) $(p + 2)x_1 - x_2 = f(t)$
    $-x_1 + (p + 2)x_2 = 0$

(b) $(p + 2)x_1 - (p + 1)x_2 = 0$
    $-(p + 1)x_1 + (2p + 1)x_2 = f(t)$

(c) $(p - 3)x_1 - 6x_2 = (p + 1)f(t)$
    $px_1 + (p - 3)x_2 = 0$

(d) $px_1 + (3p + 1)x_2 = 0$
    $-x_1 + (p - 1)x_2 = f(t)$

2.3.   In each of the four sets of simultaneous differential equations in Prob. 2.2, find the transfer operators relating x_1 and x_2 to the input $f(t)$.

2.4.   For each of the networks shown in Fig. P-2.4 find the transfer operator which relates

(a) i_1 to $f(t)$ (b) i_2 to $f(t)$ (c) v_0 to $f(t)$.

2.5.   Water flows in a tank (Fig. P-2.5) at a rate of q_i units/sec and flows out through the outflow valve at a rate of q_0 units/sec. Determine the transfer operator relating the outflow q_0 to the input q_i. The outflow rate is proportional to the head h. Thus $q_0 = Rh$, where R is the valve resistance. Determine also the transfer operator relating h to q_i. (*Hint:* The net inflow of water in time Δt is $(q_i - q_0)\Delta t$. This must be equal to $A\Delta h$, where A is the cross section of the tank.)

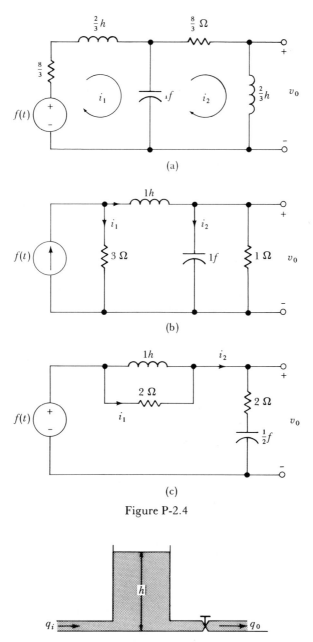

(a)

(b)

(c)

Figure P-2.4

Figure P-2.5

2.6.   The attitude of an aircraft can be controlled by three sets of surfaces: elevators, rudder, and ailerons shown shaded in Fig. P-2.6. By manipulating these surfaces, one can set the aircraft on a desired flight path. The roll angle ϕ can be controlled by deflection, in opposite direction, of the two aileron surfaces as shown in Fig. P-2.6. Assuming

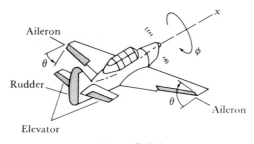

Figure P-2.6

only rolling motion, determine the transfer operator relating the roll angle ϕ to the input (aileron deflection) θ. The moment of inertia of the plane about the x-axis is $\mathcal{J}$. The torque generated about the x-axis due to aileron deflection θ is $k\theta$. The viscous friction due to air generates an opposing torque proportional to the roll velocity, and is given by $D\dot{\phi}$.

2.7.   A hydraulic controller is commonly used in steering rocket vehicle, or aircraft attitude control surfaces (elevators, rudder, ailerons shown in Fig. P-2.6). In this arrangement (Fig. P-2.7), small force can control heavy loads. A small valve (motion x) admits oil under high pressure. This oil moves a large piston (motion y). Here the input is x, and the output is y, or θ. Determine the transfer operator relating y to x, and θ to x. (*Hint:* In time Δt, the oil flow in the large cylinder is $kx\Delta t$, and this oil moves the large piston by an amount Δy. Hence $kx\Delta t = A\Delta y$, where A is the piston cross section.)

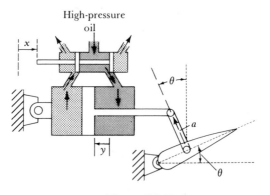

Figure P-2.7

2.8.   Mounting a machine such as an air compressor or a diesel engine creates a problem of vibration. It is necessary to minimize the effect of these undesirable vibrations on the mount (or the ground). This is usually done by a vibration absorber consisting of a spring and a dashpot. Two arrangements are shown in Fig. P-2.8. The force $f(t)$ is a periodic force generated by the machine. The force transmitted to the ground is considerably reduced by these arrangements (This is a mechanical filtering problem).

(a) For each of the arrangements, find the transfer operator relating x, the displacement of the mount to the input force $f(t)$.

(b) For each of the arrangements, find the transfer operator which relates the force transmitted to the floor to the input force $f(t)$.

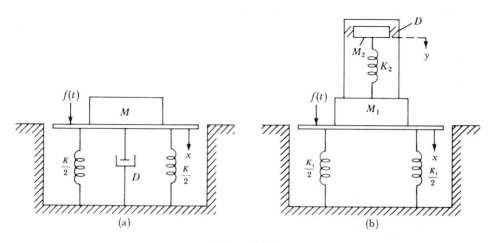

Figure P-2.8

2.9. An accelerometer to measure the acceleration of an aircraft or missile is shown in Fig. P-2.9. It consists of a test mass (seismic mass) M restrained by springs K and viscous damping D, and mounted on a platform P. The platform maintains a fixed orientation in inertial space (except for short transients) with the help of gyroscopes. Three such platforms are needed to measure the three components of acceleration in space. Figure P-2.9 shows a horizontal accelerometer. The platform motion is x. The motion of the mass M is y. In practice, we can measure only the relative motion z of the mass M with respect to the platform. Thus

$$z = y - x$$

The platform motion x acts as the input and the motion z is the output. Determine the transfer operator relating z to x.

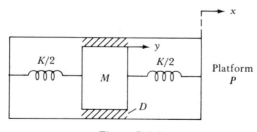

Figure P-2.9

2.10. For the armature controlled motor in Fig. 1.10, determine

(a) the transfer operator relating θ, the angular position of the load, to the input voltage $f(t)$.

(b) the transfer operator relating ω, the angular velocity of the load, to the input voltage $f(t)$.

2.11.   For the field controlled motor in Prob. 1.13, determine

(a) the transfer operator relating θ, the angular position of the load, to the input voltage $f(t)$.

(b) the transfer operator relating ω, the angular velocity of the load to the input voltage $f(t)$.

2.12.   A response variable y and the input $f(t)$ are related by

(a) $y(t) = \dfrac{p + 3}{p^2 + 3p + 2}\, f(t)$

(b) $y(t) = \dfrac{p + 1}{p^2 + 2p + 2}\, f(t)$

(c) $y(t) = \dfrac{p}{p^2 + 2p + 1}\, f(t)$

In each of the three cases the initial conditions are given as $y(0) = 1$, $y'(0) = 2$.   Find and roughly sketch the zero-input response $y(t)$ in each case.

2.13.   The transfer operator of a certain system is known to be

$$H(p) = \frac{p + 4}{p(p^2 + 3p + 2)}$$

Find and roughly sketch the zero-input response y if it is given that $y(0) = 0$, $y'(0) = 1$, $y''(0) = 0$.

2.14.   Repeat Prob. 2.13 if

$$H(p) = \frac{-(3p + 8)}{p(p^2 + 4p + 8)}$$

and $y(0) = 1$, $y'(0) = 0$ and $y''(0) = 0$

2.15.   Repeat Prob. 2.13 if

$$H(p) = \frac{3p + 1}{p(p + 1)^2}$$

if $y(0) = y'(0) = 0$ and $y''(0) = 1$

2.16.   In each of the four cases in Prob. 2.2, the initial conditions are given as $x_1(0) = 2$, and $x_1'(0) = 1$.   Find the zero-input solution for $x_1(t)$, and $x_2(t)$.   [*Hint:* To find $x_2(t)$, derive the initial conditions $x_2(0)$ and $x_2'(0)$ from the given initial conditions and the set of differential equations when $f(t) = 0$.]

2.17.   In each of the four cases in Prob. 2.2, the initial conditions are given as $x_1(0) = 1$ and $x_2(0) = 2$.   Find the zero-input solution for $x_1(t)$ and $x_2(t)$.

2.18.   A coasting freight car is brought to a stop by a bumper (Fig. P-2.18) which has a spring-and-damper action (in parallel).   The car strikes the bumper at $t = 0$ with a velocity v_0.   Find and sketch the motion (x and v) of the car for $t \geq 0$ if the parameters are (in MKS units),

(a) $M = 10,000$, $K = 160,000$,   and $D = 100,000$, $v_0 = 1$
(b) $M = 10,000$, $K = 100,000$,   and $D = 40,000$, $v_0 = 1$

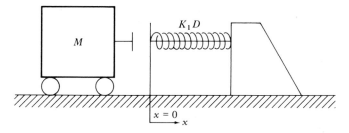

Figure P-2.18

[*Hint:* The car is not attached to the bumper. Therefore as soon as *x* becomes zero, the car will continue to move in opposite direction. Hence the solution obtained here will be correct until *x* changes sign. After that the car will move with constant velocity in opposite direction (assuming zero friction).]

2.19.   Find and roughly sketch the form of the zero-input response of the vibration absorber shown in Fig. P-2.8a if (in MKS units)

(a) $M = 1000$, $D = 40,000$,   and $K = 100,000$
(b) $M = 1000$, $D = 12,000$,   and $K = 100,000$
(c) $M = 1000$, $D = 0$,   and $K = 100,000$
(d) Can you explain the purpose of the dashpot *D* in the vibration absorber?

2.20.   Find and sketch roughly the complete response *y* if the transfer operator relating the output *y* to the input $f(t)$ is given by

$$H(p) = \frac{p + 3}{p^2 + 3p + 2}$$

and if

(a) $f(t) = u(t), y(0) = 1$ and $y'(0) = 2$
(b) $f(t) = e^{-3t}u(t), y(0) = 1$ and $y'(0) = 2$

2.21.   Repeat Prob. 2.20 if

$$H(p) = \frac{p + 1}{p^2 + 2p + 1}$$

and $f(t) = u(t), y(0) = 1$, and $y'(0) = 2$.
2.22.   Repeat Prob. 2.20 if

$$H(p) = \frac{p}{p^2 + 2p + 1}$$

and if

(a) $f(t) = u(t), y(0) = 1$ and $y'(0) = 2$
(b) $f(t) = e^{-t}u(t), y(0) = 1$ and $y'(0) = 2$

2.23.   Repeat Prob. 2.20 if

$$H(p) = \frac{p + 4}{p(p^2 + 3p + 2)}$$

and if

(a) $f(t) = e^{-3t}u(t)$, $y(0) = y'(0) = y''(0) = 0$
(b) $f(t) = u(t)$, $y(0) = y'(0) = y''(0) = 0$

2.24.   Repeat Prob. 2.20 if

$$H(p) = \frac{(3p + 8)}{p(p^2 + 4p + 8)}$$

and $f(t) = u(t), y(0) = y'(0) = y''(0) = 0$.

2.25.   Repeat Prob. 2.20 if

$$H(p) = \frac{3p + 1}{p(p + 1)^2}$$

and $f(t) = u(t), y(0) = y'(0) = y''(0) = 0$.

2.26.   Repeat Prob. 2.20 if

$$H(p) = \frac{p^2 + 4p + 5}{p^2 + 3p + 2}$$

and $f(t) = u(t), y(0) = y'(0) = 0$. *Hint:* Here $H(p)$ is an improper fraction and can be expressed as

$$H(p) = 1 + \frac{p + 3}{p^2 + 3p + 2}$$

Hence

$$y(t) = \left(1 + \frac{p + 3}{p^2 + 3p + 2}\right) f(t) = f(t) + \frac{p + 3}{p^2 + 3p + 2} f(t)$$

2.27.   Using the sampling property of the impulse function, evaluate the following integrals.

$$\int_{-\infty}^{\infty} \delta(t - 2) \sin t \, dt$$

$$\int_{-\infty}^{\infty} \delta(t + 3) e^{-t} \, dt$$

$$\int_{-\infty}^{\infty} (t^3 + 4) \delta(1 - t) \, dt$$

2.28.   Evaluate graphically $f_1(t) * f_2(t)$ and $f_2(t) * f_1(t)$ for functions shown in Fig. P-2.28.

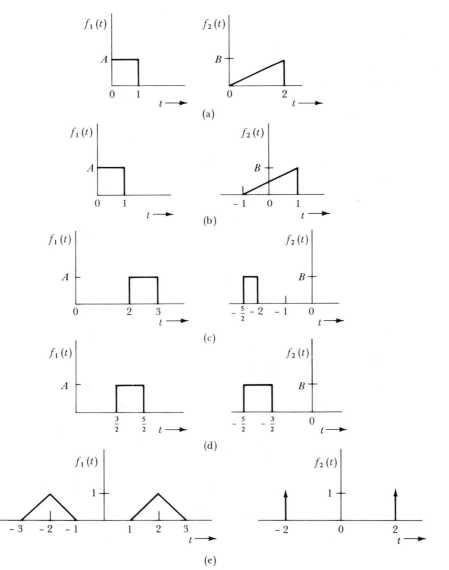

(a)

(b)

(c)

(d)

(e)

Figure P-2.28

2.29. Find the zero-state response when

$$H(p) = \frac{1}{p+1}, \qquad f(t) = e^{2t}u(-t)$$

Give a rough sketch of the input and the output.

2.30. Find the zero-state response when

$$H(p) = \frac{1}{p+1}, \qquad f(t) = e^{-2t}u(t) + e^{2t}u(-t).$$

Give a rough sketch of the input and the output.

2.31.   Find the zero-state response when

$$H(p) \; = \; \frac{1}{p \, + \, 1}, \qquad f(t) \; = \; (3e^{2t} \, - \, 2e^{t})u(-t)$$

Give a rough sketch of the input and the output.

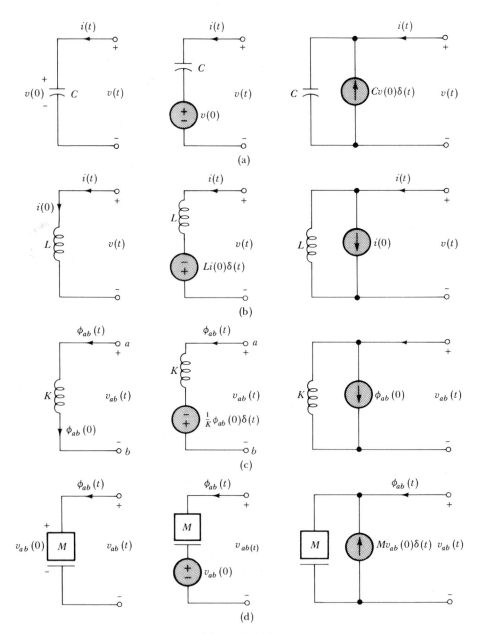

Figure P-2.35

2.32.   Find the zero-state response when

$$H(p) = \frac{1 - p}{1 + p}, \qquad f(t) = e^t u(-t)$$

Give a rough sketch of the input and the output.

2.33.   For a linear-time invariant system described by Eq. 2.31 if $y(t)$, is the zero-state response to the input $f(t)$, then show that $y'(t)$ is the zero-state response to the input $f'(t)$, and $\int_{-\infty}^{t} y(\tau)\,d\tau$ is the zero-state response to the input $\int_{-\infty}^{t} f(\tau)\,d\tau$. Hence show that zero-state response of a system to unit step function is $\int_{0}^{t} h(\tau)\,d\tau$.

2.34.   If $g(t)$ is zero-state response of a system to unit step input, then show that the zero-state response $y_f(t)$ of the system to the input $f(t)$ is given by

$$y_f(t) = g(t) * f'(t) = f'(t) * g(t)$$

2.35.   Initial conditions in a system can always be represented by equivalent sources. In electrical systems initially charged capacitor can be represented by an uncharged capacitor and a source as shown in Fig. P-2.35a. (See page 94.)   Show that the terminal relations of all three arrangements in Fig. P-2.35a are identical and are given by

$$v(t) = v(0) + \frac{1}{C} \int_{0}^{t} i(\tau)\,d\tau$$

Similarly, show that the terminal equations of all three arrangements in Fig. P-2.35b are identical, and are given by

$$i(t) = i(0) + \frac{1}{L} \int_{0}^{t} v(\tau)\,d\tau$$

2.36.   In Prob. 2.20 find the complete response and determine (i) zero-input and zero-state components, (ii) natural and forced components, and (iii) transient and steady-state components.

2.37.   Repeat Prob. 2.36 for the case in Prob. 2.22.

2.38.   A response y of a system is related to the input $f(t)$ by a linear differential equation (of the form Eq. 2.31) as follows:

$$D(p)y(t) = N(p)f(t)$$

If $y_n(t)$ and $y_\phi(t)$ are the natural and the forced components of $y(t)$, then show that

$$D(p)y_n(t) = 0$$
$$D(p)y_\phi(t) = N(p)f(t)$$

2.39.   For the vibration absorber shown in Fig. P-2.7a, the parameters are given as (in MKS units), $M = 1000$, $D = 1000$, and $K = 4000$. The input force $f(t)$ is a sinusoidal signal of magnitude 1000, and frequency $\omega = 100$.

(a) Find the steady-state response x (displacement of the platform).
(b) Sketch roughly the amplitude of the steady-state response x as the angular frequency ω of the input varies from 0 to ∞.   Pay careful attention to the system behavior in the vicinity of $\omega = 2$.   Comment.
(c) Repeat part (b) if $D = 0$ in Fig. P-2.8a.

3

Frequency-Domain Analysis

3.1 INTRODUCTION

The system analysis technique used in Chapter 2 is labeled as time-domain technique because we use the variable t (time) as the independent variable.

It was shown that zero-state response given by the convolution integral in this method can be viewed as a superposition integral. The input signal is expressed as a continuous sum (integral) of impulse components. The response is given by the continuous sum (integral) of the responses to the impulse components of the input. This procedure is permitted for the linear systems because the superposition principle applies to such systems.

This principle opens many avenues for analyzing these systems. Instead of using impulse components to represent the input $f(t)$, we may use other kinds of functions, such as exponential functions, unit step functions, ramp functions, and other function of higher powers of t. It can be shown that all these possibilities except the exponential functions can be used with a slight modification of the convolution techniques.† The use of exponential signals is considered in this chapter. This approach is called *frequency-domain analysis* for reasons which will soon be apparent.

3.2 THE EXPONENTIAL SIGNAL IN LINEAR SYSTEMS

The exponential function is perhaps the most important function to engineers, physicists, and mathematicians alike. This is because the exponential function has the almost magic property that its derivative and the integral yield a function proportional to itself. This can be seen from the fact that

$$\frac{d}{dt} e^{st} = se^{st} \quad \text{and} \quad \int e^{st} dt = \frac{1}{s} e^{st} \tag{3.1}$$

This function has been liberally used by engineers and physicists starting

---

†For further discussion on this see B. P. Lathi, *Signals, Systems, and Communications* Chap. 10. (New York: Wiley, 1965).

96

from such simple applications as phasors in analysis of steady-state networks to more sophisticated application in connection with the Schrödinger equation in quantum mechanics. Many of the phenomena observed in nature can be described by exponential functions. It is therefore natural that we choose exponential functions for the purpose of analyzing linear systems by the principle of superposition. We must, however, justify the choice on firmer grounds. It turns out that the main reason for using exponential functions in analyzing linear systems has to do with certain important properties of that function. They are as follows.

1. Every function or waveform encountered in practice can always be expressed as a sum (discrete or continuous) of various exponential functions.
2. The response of a linear time invariant system to an exponential input e^{st} is also an exponential function† $H(s)e^{st}$.

By exponential function we mean a function

$$Ae^{st} \qquad \text{for } -\infty < t < \infty \qquad (3.2)$$

where s is complex in general and the function is eternal, that is, it starts at $t = -\infty$. The complex quantity s, the index of the exponential, is known as the *complex frequency*. The reason for this designation will become clear later. Note that since s is complex in general, the following functions can also be categorized as exponential functions:

(a) A constant, $k = ke^{0t}$
(b) A monotonic exponential, e^{at}
(c) A sinusoidal function, $\sin \omega t = (1/2j)(e^{j\omega t} - e^{-j\omega t})$
(d) An exponentially varying sinusoid,

$$e^{-at} \sin \omega t = \frac{1}{2j} [e^{(-a+j\omega)t} - e^{(-a-j\omega)t}]$$

Exponential functions, therefore, cover a variety of important waveforms encountered in practice. In general, the exponent s is complex:

$$s = \sigma + j\omega \qquad (3.3)$$

Therefore

$$e^{st} = e^{(\sigma+j\omega)t}$$
$$= e^{\sigma t} e^{j\omega t}$$

The waveforms of e^{st} depend upon the nature of s. If s is real (that is, if $\omega = 0$), then e^{st} is given by $e^{\sigma t}$ and can be represented by monotonically increasing or decreasing function of time (Fig. 3.1a), depending upon whether $\sigma > 0$ or $\sigma < 0$. On the other hand, if s is imaginary (that is, $\sigma = 0$), then e^{st} is given by $e^{j\omega t}$. The function $e^{j\omega t}$ is complex and has real and imaginary parts:

---

†There is a certain limitation on this statement. It is true only if the natural component of the system decays faster than the magnitude of the function e^{st}.

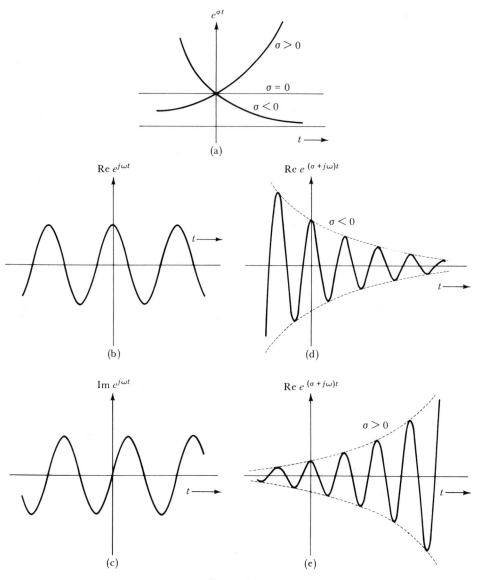

Figure 3.1

$$e^{j\omega t} = \cos \omega t + j \sin \omega t$$

Thus

$$\text{Re}\,(e^{j\omega t}) = \cos \omega t$$

and

$$\text{Im}\,(e^{j\omega t}) = \sin \omega t$$

The function $e^{j\omega t}$ can therefore be represented graphically by a real part and an imaginary part. Each of these functions is a sinusoidal function of frequency ω and has a constant amplitude (Figs. 3.1b and 3.1c).

We can extend this discussion to the case where s is complex—that is both σ and ω are nonzero, as follows:

$$e^{st} = e^{\sigma t} e^{j\omega t}$$

$$= e^{\sigma t} (\cos \omega t + j \sin \omega t)$$

$$= e^{\sigma t} \cos \omega t + j e^{\sigma t} \sin \omega t$$

Therefore the function e^{st} has real and imaginary parts when s is complex. Here, both functions $e^{\sigma t} \cos \omega t$ and $e^{\sigma t} \sin \omega t$ represent functions oscillating at angular frequency ω, with the amplitude increasing or decreasing exponentially depending upon whether σ is positive or negative (Figs. 3.1d and 3.1e).

If s is purely imaginary, then e^{st} is represented by $e^{j\omega t}$ where ω connotes the frequency of the signal. This same concept is extended to the case when s is complex. We call the variable s in the function e^{st} the *complex frequency*. This designation, however, is slightly misleading because we always associate the term *frequency* with a periodic function. Now when the variable s is real (that is, $\omega = 0$), then e^{st} is represented by $e^{\sigma t}$ which increases or decreases monotonically (Fig. 3.1a). Yet, according to this designation, the signal $e^{\sigma t}$ has a complex frequency σ. It will be more appropriate to say that the frequency of the signal e^{st} is given by the imaginary part of the variable† s. We shall, however, follow the common convention used by engineers to designate s as the complex frequency of signal e^{st}. The complex frequency s can be conveniently represented on a complex frequency plane (s-plane) as shown in Fig. 3.2. The horizontal axis is the real axis (σ-axis) and the vertical axis is the imaginary axis (ω-axis). Thus the imaginary part (ω) of s represents the frequency of oscillation of e^{st} and the real part (σ) gives the information about the rate of change (increase or decrease) of the amplitude of e^{st}. Along the real axis (σ-axis), $\omega = 0$, and the frequency of oscillation is zero, implying that the signals associated with the σ-axis are exponential signals with monotonically increasing or decreasing amplitudes (Fig. 3.1a). Along the imaginary axis (ω-axis), $\sigma = 0$ and $e^{\sigma t} = 1$, implying that the signals associated with the $j\omega$-axis are oscillating signals (sinusoidal) with constant amplitude (Figs. 3.1b and 3.1c). For the signals shown in Figs. 3.1d and 3.1e, the frequency s is complex and does not lie on either axis. For Fig. 3.1d, σ is negative and s lies to the left of the imaginary axis. On the other hand, for Fig. 3.1e, σ is positive and s lies to the right of the imaginary axis. Note that for $\sigma > 0$, the amplitudes of the signals increases exponentially and for $\sigma < 0$, the amplitude decays exponentially. Thus the s-plane is distinguished into two parts: the left half-plane (LHP), which represents exponentially decaying signals, and the right half-plane (RHP), which represents exponentially growing signals.

---

†It will be more appropriate to use a complex variable $\lambda = \omega + j\sigma$ for the complex frequency. Thus the complex frequency of the signal $e^{j\lambda t}$ is $\lambda = \omega + j\sigma$. Mathematicians often use this notation.

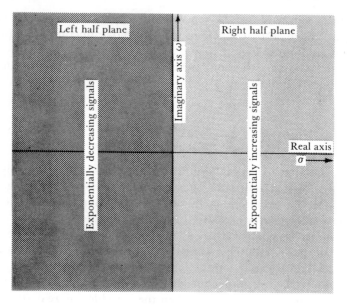

Figure 3.2

Thus each point on the complex frequency plane corresponds to a certain mode of the exponential function. At this point one may wonder about the frequencies lying along the $-j\omega$-axis. The frequencies of the signals represented along the $-j\omega$-axis appear to be negative according to our designation of the frequency. What is a negative frequency? By its very definition, the frequency is an inherently positive quantity. The number of times that a function passes through a fixed point, say zero, in 1 sec is always positive. How shall we then interpret a negative frequency? The confusion arises because we are defining frequency not as number of cycles per second of a particular waveform but as an index of the exponential function. Hence the negative frequencies are associated with negative exponent:

$$e^{-j\omega t} = \cos \omega t - j \sin \omega t$$

Thus the complex frequency $s = -j\omega$ represents signals oscillating at angular frequency ω. It should be noted that a signal of negative frequencies $(s = -j\omega)$ and a signal of positive frequency $(s = j\omega)$ can be combined to obtain a real signal:

$$e^{j\omega t} + e^{-j\omega t} = 2 \cos \omega t \tag{3.4}$$

Similarly, signals of two complex conjugate frequencies $(\sigma + j\omega)$ and $(\sigma - j\omega)$ can form real signals:

$$e^{(\sigma+j\omega)t} + e^{(\sigma-j\omega)t} = 2e^{\sigma t} \cos \omega t \tag{3.5}$$

This result is quite significant. It will be seen in a later chapter that any real function of time encountered in practice can always be expressed as a discrete or a continuous sum of exponential functions which occur in pairs with complex conjugate frequencies.† It will also be shown that any periodic function can be expressed as sum of discrete exponential functions:

$$f(t) = C_1 e^{s_1 t} + C_2 e^{s_2 t} + \cdots + C_n e^{s_n t} + \cdots + \qquad (-\infty < t < \infty)$$

$$= \sum_r C_r e^{s_r t} \qquad\qquad\qquad (3.6)$$

Here C_r represents the amplitude of the rth exponential of frequency s_r. Any nonperiodic function can be expressed as a continuous sum of eternal exponential functions as

$$f(t) = \int_{s_A}^{s_B} C(s) e^{st} \, ds \qquad (-\infty < t < \infty) \qquad (3.7)$$

Here $C(s)$ expresses the amplitude distribution of various exponentials. Note that Eq. 3.7 is an extension of Eq. 3.6. In Eq. 3.6, $f(t)$ is expressed as the sum of discrete exponential components of frequencies $s_1, s_2, \ldots, s_r$, etc., whereas in Eq. 3.7 $f(t)$ is expressed as a continuous sum of the exponentials of all the frequencies lying along the path from s_A to s_B in the complex frequency plane (s-plane). The amplitude distribution of various exponentials along this path is given by $C(s)$.

In all of this discussion it is important to realize that the exponentials that we are talking about are eternal—that is, they start at $t = -\infty$. Any function $f(t)$ can be expressed as a sum (discrete or continuous) of these eternal exponentials. Let us consider a function $f(t)$ which exists only for $t \geq 0$ and is identically zero for $t < 0$. One might wonder whether it will be possible to express such a function in terms of a sum of eternal functions, which exist over the entire interval $(-\infty < t < \infty)$. The answer is yes! Such functions can always be expressed as a sum of eternal exponential functions. These exponential functions add in such a way as to cancel one another for $t < 0$ and yield the desired function for $t \geq 0$.

3.3 RESPONSE OF LINEAR SYSTEMS TO EXPONENTIAL INPUTS

Let us consider a system with zero initial conditions (zero-state system). The response of such a system to exponential input e^{st} was shown to be (see Eq. 2.106b)

$$y_f(t) = \underbrace{\sum_{j=1}^{n} b_j e^{\lambda_j t}}_{\text{natural}} + \underbrace{H(s) e^{st}}_{\text{forced}} \qquad (3.8)$$

where $\lambda_1, \lambda_2, \ldots, \lambda_n$ are the poles of $H(s)$. The first term on the right hand side of Eq. 3.8 is the natural component and the second term is the forced component of the response (see Eq. 2.106b).

---

†It can be shown by using the theory of complex variables that although a real function of time can be represented by a continuous sum of complex exponential functions occurring in pairs of complex conjugate frequencies, it is not the only way of representation.

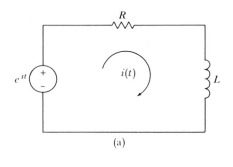

(a)

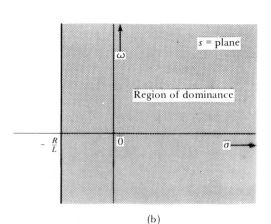

(b)

Figure 3.3

Consider for example the circuit shown in Fig. 3.3a, with the input e^{st} and the response variable $i(t)$. The transfer function $H(s)$ is given by

$$H(s) = \frac{1}{R + Ls}$$

Hence the transfer function has one pole at $-R/L$. The response $i(t)$ can now be written as

$$i(t) = Ae^{-Rt/L} + H(s)e^{st}$$

$$= Ae^{-Rt/L} + \frac{1}{R + Ls} e^{st} \tag{3.9}$$

Notice that the response $i(t)$ has two components: $Ae^{-Rt/L}$, the natural component whose nature is characteristic of the network alone, and the forced component $H(s)e^{st}$, which has the same frequency as that of the driving function. The complex frequency of the natural component is $-R/L$ and hence this component decays with time. The forced component $H(s)e^{st}$ may or may not decay with time, depending upon the value of s. If it is assumed that the natural component decays faster than the forced component, then after a long time the natural component will become negligible compared to the forced component. In other words, the forced component will dominate the natural component.†

---

†This discussion is not restricted to stable systems where the natural component decays with time but applies as well to unstable systems where the natural component grows with time. In such cases we choose the exponential function e^{st} which also grows with time at a faster rate than the natural component. The forced component thus will dominate the natural component. In such cases s must lie in the right half of the complex frequency plane.

In the previous example, if

$$\mathrm{Re}\ s = \sigma > - \frac{R}{L} \qquad (3.10)$$

then it is evident that after a long time, the magnitude of $e^{-Rt/L}$ will become negligible compared to the magnitude of e^{st}. Under these conditions the forced component will dominate the natural component, and hence after a sufficiently long time the response of the system to the exponential function e^{st} will consist entirely of the forced component $H(s)e^{st}$.

If the magnitude of the exponential function e^{st} decays slower† than the natural component of the system, then the signal e^{st} is said to satisfy the *dominance condition* for the system. For the example of the *RL* circuit considered above, Eq. 3.10 gives the dominance condition (Fig. 3.3b).

Now suppose the function e^{st} were applied at $t = -\infty$ instead of $t = 0$ and, further, suppose that the natural component decays faster than the function e^{st}; that is, if the dominance condition is satisfied, then at any finite time t, the natural component will have vanished and the response will consist entirely of the forced component $H(s)e^{st}$, which is also an exponential function of frequency s. We therefore reach a very important conclusion: the zero-state response of a linear system to an eternal exponential function e^{st} $(-\infty < t < \infty)$ consists entirely of an exponential function of the same frequency. Of course, it is understood that the statement is true only for those values of s which will satisfy the dominance condition.

We can demonstrate this easily for the case of *R-L* circuit in Fig. 3.3a. Let this circuit be switched on at $t = T$ and assume that the network at this time is in zero-state, that is $i(T) = 0$. From Eq. 3.9 we have

$$i(T) = Ae^{-RT/L} + \frac{e^{sT}}{R + Ls} = 0 \qquad (3.11)$$

and

$$A = \frac{-e^{(s+R/L)T}}{R + Ls} \qquad (3.12a)$$

If the circuit is switched on at $T = -\infty$, then from the above equation, we have $A = 0$, provided Re $(s + R/L) > 0$, or Re $s > -R/L$ (that is, if the dominance condition is satisfied). This yields

$$i(t) = H(s)e^{st} \qquad (3.12b)$$

Thus the entire response to the input of eternal exponential e^{st} is given by the forced component $H(s)e^{st}$ alone for values of s lying in the shaded region in Fig. 3.3b

---

†For unstable systems, the natural component grows with time and the dominance condition is satisfied when the magnitude of e^{st} grows faster than the natural component.

(the region of dominance). The values of s in this region satisfy the dominance condition ($\text{Re } s > -R/L$) for this network.

Note that if $\text{Re } s < -R/L$ (that is if the dominance condition is not satisfied), then from Eq. 3.12a, we have $A = -\infty$ and the response $i(t)$ to an eternal exponential input is infinitely large. In this case the natural component is infinite and dominates the finite forced component.

It must be clearly understood that if a driving function e^{st} is applied to a system at any finite time, say $t = 0$, then the response of the system will consist of a natural component as well as the forced component $H(s)e^{st}$. If, however, the exponential signal e^{st} is applied at $t = -\infty$ (that is, the eternal exponential signal), then at any finite time the response will be given entirely by $H(s)e^{st}$ (provided, of course, that the dominance condition is satisfied). This distinction should be properly understood. We shall repeat it once again. The response of a linear time-invariant system to an eternal exponential signal e^{st} is $H(s)^{st}$. If, however, the exponential signal e^{st} is applied at some finite time, then the response will consist of the natural component and the forced component $H(s)e^{st}$. The forced component of response, therefore, becomes the entire response when the signal is applied at $t = -\infty$, and provided that the dominance condition is satisfied.

THE DOMINANCE CONDITION

For the dominance condition to be satisfied the signal e^{st} must outlive the natural component.† If $\lambda_1, \lambda_2, \ldots, \lambda_n$ are the poles of $H(s)$, then the natural component consists of sum of the exponentials $e^{\lambda_j t}$ (Eq. 3.8), where the λ_j's are complex in general. The signal $H(s)e^{st}$ can outlive this component only if

$$\text{Re } s > \text{Re } \lambda_j \qquad (j = 1, 2, \ldots, n) \tag{3.13}$$

This implies that values of s which lie to the right of all poles of $H(s)$ satisfy the dominance condition. For the circuit in Fig. 3.3a, $H(s)$ has only one pole at $-R/L$. Hence for the dominance condition to be satisfied, we must choose these values of s for which $\text{Re } s > -R/L$ (Fig. 3.3b).

3.4   FOUNDATIONS OF FREQUENCY-DOMAIN ANALYSIS

We are now in a position to understand the frequency-domain approach which uses the exponential basis signals to represent the input and exploits the principle of superposition in analysis of linear systems.

As shown in Sec. 3.3, the response of the system to an eternal exponential input e^{st} is given by $H(s)e^{st}$. We now wish to find the response of this system to some

---

†For unstable systems, where the natural component grows with time, this implies that the signal e^{st} grows faster than the natural component. This is true only if

$$\text{Re } s > \text{Re } \lambda_j \qquad (j = 1, 2, \ldots, n)$$

Hence Eq. 3.13 represents the criterion for the dominance condition for stable as well as unstable systems.

input $f(t)$. It will be shown later in this chapter that an arbitrary function $f(t)$ can be expressed as a sum (either discrete or continuous) of eternal exponential functions. If $f(t)$ is a periodic function over the entire interval $(-\infty < t < \infty)$, then it can be expressed as a discrete sum:

$$f(t) = C_1 e^{s_1 t} + C_2 e^{s_2 t} + \cdots + C_n e^{s_n t} + \cdots$$

$$= \sum_r C_r e^{s_r t} \qquad (-\infty < t < \infty) \tag{3.14a}$$

According to the principle of superposition, the response of the system to $f(t)$ will be given by the sum of the responses of the system to individual exponential components. The zero-state response $y_f(t)$ is therefore given by

$$y_f(t) = C_1 H(s_1) e^{s_1 t} + C_2 H(s_2) e^{s_2 t} + \cdots + C_r H(s_r) e^{s_r t} + \cdots$$

$$= \sum_r C_r H(s_r) e^{s_r t} \qquad (-\infty < t < \infty) \tag{3.14b}$$

If, however, $f(t)$ is a nonperiodic function, then it can be expressed as a continuous sum of eternal exponential functions, that is

$$f(t) = \int_{s_A}^{s_B} C(s) e^{st} \, ds \tag{3.15a}$$

Here $f(t)$ is expressed as a continuous sum of exponentials of all frequencies lying along the path from s_A to s_B in the complex frequency plane (s-plane) and $C(s)$ represents the amplitude distribution of various frequency components along this path. The response $y_f(t)$ will also be given by a continuous sum of the responses of the system to individual exponential components:

$$y_f(t) = \int_{s_A}^{s_B} H(s) C(s) e^{st} \, ds \qquad (-\infty < t < \infty) \tag{3.15b}$$

Note that it is tacitly assumed in this discussion that the variable s in the above equations satisfies dominance condition. That is, $\mathrm{Re}\ s > \mathrm{Re}\ \lambda_j (j = 1, 2, \ldots, n)$, where $\lambda_1, \lambda_2, \ldots, \lambda_n$ are the poles of the transfer function $H(s)$. This means we must express the input in terms of eternal exponentials e^{st} whose complex frequencies lie to the right of all poles of $H(s)$.

SYSTEM SPECIFICATION

We have outlined a method of determining the zero-state response of a system from the knowledge of the transfer function $H(s)$. The transfer function carries the information required to determine the zero-state response of a system. A system is often specified by its transfer function $H(s)$ rather than by its transfer operator $H(p)$. The difference between $H(s)$ and $H(p)$ should be noted carefully. Although $H(s)$ is obtained by substituting s for p in $H(p)$, there are conceptual differences between $H(s)$ and $H(p)$. The entity $H(p)$ is an operator and p is not a variable

On the other hand, $H(s)$ is a function of a variable s. Consequently, common factors in the numerator and the denominator of $H(s)$ can be canceled. As long as there are no common factors in the numerator and the denominator of $H(p)$, the transfer function $H(s)$ carries all the information contained in the transfer operator $H(p)$. Hence $H(s)$ contains all the information required to determine the zero-input as well as zero-state response of the system. On the other hand, if there are common factors in the numerator and the denominator of $H(p)$, they will be canceled in $H(s)$. In this case $H(s)$ will have complete information required to determine the zero-state response, but only a partial information required to determine the zero-input response.

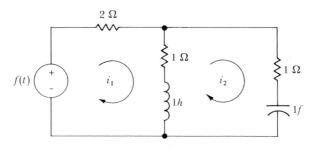

Figure 3.4

As an example, consider the network in Fig. 3.4. The loop equations are

$$(p + 3)i_1 - (p + 1)i_2 = f(t)$$

$$-(p + 1)i_1 + \left(p + 2 + \frac{1}{p}\right)i_2 = 0$$

These integrodifferential equations can be converted into differential equations by using variables i_1 and $(1/p)i_2$. In matrix form the above loop equations can be expressed as

$$\begin{bmatrix} (p + 3) & -p(p + 1) \\ -(p + 1) & p^2 + 2p + 1 \end{bmatrix} \begin{bmatrix} i_1 \\ (1/p)i_2 \end{bmatrix} = \begin{bmatrix} f(t) \\ 0 \end{bmatrix}$$

Cramer's rule yields

$$i_1 = \frac{p^2 + 2p + 1}{3(p^2 + 2p + 1)} f(t)$$

$$= \frac{(p + 1)^2}{3(p + 1)^2} f(t)$$

Thus the transfer operator $H(p)$ relating i_1 to $f(t)$ is given by

$$H(p) = \frac{(p + 1)^2}{3(p + 1)^2}$$

But the transfer function $H(s)$ relating i_1 to $f(t)$ is given by

$$H(s) = \tfrac{1}{3}$$

Note that $H(p)$ has the information required to determine the zero-input as well zero-state response. However, $H(s)$ has lost the information about the zero-input response.

We shall now devise techniques of representing inputs in terms of eternal exponential signals. Once this is done, the response to any input can be found as a sum of the responses to its exponential components in a manner discussed in this section. To begin with, we shall consider the case of periodic inputs.

A vector can be represented in terms of its components. There are an infinite number of possible ways of representing a vector in terms of its components. The reader is familiar with representing a vector in terms of orthogonal components (basis vectors). A signal can also be represented in terms of basis signals and there is a perfect analogy between vectors and signals.† There are an infinite number of possible ways of expressing a signal in terms of component signals. Here we shall restrict ourselves to signal representation in terms of sinusoidal or exponential signal components. We shall begin with representing periodic functions by such components and then extend the results to general signals.

3.5   EXPONENTIAL REPRESENTATION OF PERIODIC INPUTS: THE FOURIER SERIES

A periodic signal has one basic waveform which repeats itself periodically over $-\infty < t < \infty$. An example of a periodic signal is shown in Fig. 3.5. It is obvious that such function remains unchanged by a shift of integral number of the period. We therefore define a function $f(t)$ to be periodic if

$$f(t) = f(t \pm nT) \qquad (n \text{ integral}) \tag{3.16}$$

The smallest value of T that satisfies Eq. 3.16 is called the *period*.

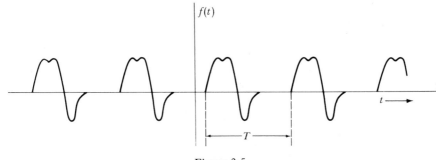

Figure 3.5

---

†For a complete discussion of such analogy see B. P. Lathi, *Signals, Systems, and Communication,* (New York: Wiley, 1965), Chap. 3.

Let us consider a signal $f(t)$ formed by adding sinusoidal signals of angular frequencies $0, \omega_0, 2\omega_0, \ldots, k\omega_0$.

$$f(t) = a_0 + a_1 \cos \omega_0 t + a_2 \cos 2\omega_0 t + \cdots + a_k \cos k\omega_0 t$$
$$\qquad + b_1 \sin \omega_0 t + b_2 \sin 2\omega_0 t + \cdots + b_k \sin k\omega_0 t$$

$$= a_0 + \sum_{n=1}^{k} a_n \cos n\omega_0 t + b_n \sin n\omega_0 t$$

It can be easily seen that this signal is periodic with period $T = 2\pi/\omega_0$. This follows from the fact that

$$f(t + T) = a_0 + \sum_{n=1}^{k} a_n \cos n\omega_0 (t + T) + b_n \sin n\omega_0 (t + T)$$

$$= a_0 + \sum_{n=1}^{k} a_n \cos (n\omega_0 t + 2n\pi) + b_n \sin (n\omega_0 t + 2n\pi)$$

$$= a_0 + \sum_{n=1}^{k} a_n \cos n\omega_0 t + b_n \sin n\omega_0 t \qquad (3.17)$$

$$= f(t)$$

Hence the summation of sinusoidal signals of angular frequencies $0, \omega_0, 2\omega_0, \ldots,$ $k\omega_0$ is a periodic signal with period T.

It is obvious that by changing the values of coefficients a_i's and b_i's in Eq. 3.17, we can construct a variety of periodic signals.

The converse of this result is also true. Any periodic signal $f(t)$ of period T can be expressed† as a sum of sinusoidal signals of angular frequencies $0, \omega_0, \ldots,$ $n\omega_0, \ldots,$ where $\omega_0 = 2\pi/T$:

$$f(t) = a_0 + a_1 \cos \omega_0 t + a_2 \cos 2\omega_0 t + \cdots + a_n \cos n\omega_0 t$$
$$\qquad + \cdots + b_1 \sin \omega_0 t + b_2 \sin 2\omega_0 t + \cdots + b_n \sin n\omega_0 t + \cdots$$

$$= a_0 + \sum_{n=1}^{\infty} a_n \cos n\omega_0 t + b_n \sin n\omega_0 t; \qquad \omega_0 = \frac{2\pi}{T} \qquad (3.18)$$

This equation shows that a periodic signal consists of sinusoidal components of (angular) frequencies $0, \omega_0, 2\omega_0, \ldots, n\omega_0, \ldots,$ and so on. Note that zero angular frequency represents a dc component. The amplitudes a_i's and b_i's of these sinusoidal components depend upon the nature of the waveform $f(t)$. To evaluate the amplitudes of these sinusoidal components, we shall first integrate both sides of Eq. 3.18 over one period $(t_0, t_0 + T)$ where t_0 is arbitrary. Thus

$$\int_{t_0}^{t_0+T} f(t) \, dt = a_0 \int_{t_0}^{t_0+T} dt + \int_{t_0}^{t_0+T} \sum_{n=1}^{\infty} (a_n \cos n\omega_0 t + b_n \sin n\omega_0 t) \, dt$$

---

†See B. P. Lathi, *ibid.*, Chap. 3.

Interchanging the order of summation and integration in the second term on the right-hand side, we obtain

$$\int_{t_0}^{t_0+T} f(t)\, dt = a_0 T + \sum_{n=1}^{\infty} a_n \int_{t_0}^{t_0+T} \cos n\omega_0 t\, dt + b_n \int_{t_0}^{t_0+T} \sin n\omega_0 t\, dt \qquad (3.19)$$

Note that the sinusoidal signals $\cos n\omega_0 t$ and $\sin n\omega_0 t$ both have a period $2\pi/n\omega_0 = T/n$. Hence there are n full cycles of these signals over the integration interval of T seconds in Eq. 3.19. The integrals of right-hand side of Eq. 3.19 represent the areas under n complete cycles of sinusoidal signals. This is obviously zero. Hence

$$\int_{t_0}^{t_0+T} f(t)\, dt = a_0 T$$

or

$$a_0 = \frac{1}{T} \int_{t_0}^{t_0+T} f(t)\, dt \qquad (3.20)$$

To evaluate the amplitude a_k of the kth cosine component we multiply Eq. 3.18 by $\cos k\omega_0 t$ and integrate over one period $(t_0, t_0 + T)$. This yields (after interchanging the order of summation and integration)

$$\int_{t_0}^{t_0+T} f(t) \cos k\omega_0 t\, dt = a_0 \int_{t_0}^{t_0+T} \cos k\omega_0 t\, dt + \sum_{n=1}^{\infty} a_n \int_{t_0}^{t_0+T} \cos n\omega_0 t \cos k\omega_0 t\, dt$$

$$+ \sum_{n=1}^{\infty} b_n \int_{t_0}^{t_0+T} \sin n\omega_0 t \cos k\omega_0 t\, dt \qquad (3.21)$$

In evaluating these integrals we use the following results:

$$\int_{t_0}^{t_0+T} \cos n\omega_0 t \cos k\omega_0 t\, dt = \begin{cases} 0 & k \neq n \\ T/2 & k = n \end{cases} \qquad (3.22a)$$

$$\int_{t_0}^{t_0+T} \sin n\omega_0 t \sin k\omega_0 t\, dt = \begin{cases} 0 & k \neq n \\ T/2 & k = n \end{cases} \qquad (3.22b)$$

$$\int_{t_0}^{t_0+T} \sin n\omega_0 t \cos k\omega_0 t\, dt = 0 \qquad (3.22c)$$

The first integral on the right-hand side of Eq. 3.21 is zero because it represents the area of k complete cycles of a sinusoidal signal. Also from Eqs. 3.22a and 3.22b it follows that on the right-hand side of Eq. 3.21 all the remaining terms except one (cosine integral when $n = k$) are zero. From Eq. 3.22a, this one non-zero term is seen to be $a_k T/2$. Hence

$$\int_{t_0}^{t_0+T} f(t) \, \cos k\omega_0 t \, dt = \frac{a_k T}{2}$$

and

$$a_k = \frac{2}{T} \int_{t_0}^{t_0+T} f(t) \, \cos k\omega_0 t \, dt \qquad\qquad k \neq 0 \qquad (3.23a)$$

Similarly, we can show that b_k, the amplitude of the sin $k\omega_0 t$ component of $f(t)$, is given by

$$b_k = \frac{2}{T} \int_{t_0}^{t_0+T} f(t) \, \sin k\omega_0 t \, dt \qquad\qquad\qquad\qquad (3.23b)$$

The series on the right-hand side of Eq. 3.18 is known as a *trigonometric Fourier series*.† Thus a periodic signal can be represented by a trigonometric Fourier series. Note that since a periodic function by definition exists for all time, the sinusoidal signals in the trigonometric Fourier series are eternal (start at $t = -\infty$).

A trigonometric Fourier series in Eq. 3.18 can be expressed in alternate form as

$$f(t) = c_0 + \sum_{n=0}^{\infty} c_n \, \cos(n\omega_0 t + \theta_n) \qquad\qquad\qquad (3.24a)$$

where by trigonometric identity we have

$$c_0 = a_0, \qquad c_n = \sqrt{a_n^2 + b_n^2}, \qquad \text{and} \qquad \theta_n = -\tan^{-1}\left(\frac{b_n}{a_n}\right) \qquad (3.24b)$$

Note that the angular frequency ω_0 has the same period (T) as that of $f(t)$ and is therefore called the *fundamental* (angular) frequency, and $n\omega_0$ is called the *nth harmonic* frequency of $f(t)$. It is evident from Eq. 3.24b that the amplitude of the nth harmonic of $f(t)$ is c_n, and it has a phase angle of θ_n. It can be seen that the nth harmonic executes n complete cycles in one period T because its frequency is n times the fundamental frequency.

As an example, consider the periodic signal in Fig. 3.6. This is a rectified sine wave. Let us find the Fourier series for this signal when $T = 1$. Since $T = 1$, $\omega_0 = 2\pi$, and the Fourier series consists of components of angular frequencies $0, 2\pi, 4\pi, 6\pi, \ldots$, and can be expressed as

$$f(t) = a_0 + \sum_{n=1}^{\infty} a_n \, \cos 2n\pi t + b_n \, \sin 2n\pi t$$

---

†It can be seen from Eqs. 3.23 that the existence of the Fourier coefficients a_i's and b_i's is guaranteed if $f(t)$ is absolutely integrable over one period, that is,

$$\int_{t_0}^{t_0+T} |f(t)| \, dt < \infty$$

In addition, for the uniform convergence of the series $f(t)$ must be finite and have finite number of discontinuities in one period. These are known as *Dirichlet conditions* for the Fourier series expansion. These conditions are sufficient but not strictly necessary.

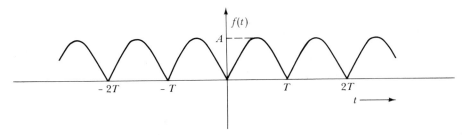

Figure 3.6

where

$$a_0 = \frac{1}{T} \int_{t_0}^{t_0 + T} f(t)\, dt$$

where t_0 is arbitrary.   Let us choose $t_0 = 0$.   Also $f(t) = A \sin \pi t$ over the interval $(0, 1)$. Hence

$$a_0 = \int_0^1 A \sin \pi t\, dt = \frac{A}{\pi} (-\cos \pi t)\Big|_0^1 = \frac{2A}{\pi}$$

and

$$a_n = 2A \int_0^1 \sin \pi t \cos 2n\pi t\, dt = \frac{-4A}{\pi (4n^2 - 1)}$$

and

$$b_n = 2A \int_0^1 \sin \pi t \sin 2n\pi t\, dt = 0$$

and

$$f(t) = \frac{2A}{\pi} - \frac{4A}{\pi} \sum_{n=1}^{\infty} \frac{1}{4n^2 - 1} \cos 2n\pi t$$

For this particular case all the sine terms are zero.   The trigonometric Fourier series can be written as

$$f(t) = \frac{2A}{\pi} - \frac{4A}{\pi} \left(\frac{1}{3} \cos 2\pi t + \frac{1}{15} \cos 4\pi t + \frac{1}{35} \cos 6\pi t + \cdots \right)$$

$$= \frac{2A}{\pi} - \frac{4A}{\pi} \sum_{n=1}^{\infty} \frac{1}{4n^2 - 1} \cos 2\pi nt$$

Since a sinusoidal signal of angular frequency $n\omega_0$ can be expressed in terms of exponential signals $e^{jn\omega_0 t}$ and $e^{-jn\omega_0 t}$, it follows that a periodic signal of period T can also be represented by a series consisting of eternal exponential components as

$$f(t) = F_0 + F_1 e^{j\omega_0 t} + F_2 e^{j2\omega_0 t} + \cdots + F_n e^{jn\omega_0 t} + \cdots$$
$$\cdot + F_{-1} e^{-j\omega_0 t} + F_{-2} e^{-j2\omega_0 t} + \cdots + F_{-n} e^{-jn\omega_0 t} + \cdots \quad (3.25a)$$

$$= \sum_{n=-\infty}^{\infty} F_n e^{jn\omega_0 t} \qquad \omega_0 = \left(\frac{2\pi}{T}\right) \tag{3.25b}$$

The amplitudes F_k in this series can be obtained by multiplying both sides of Eq. 3.25 by $e^{-jk\omega_0 t}$ and integrating over one cycle $(t_0, t_0 + T)$, for any value of t_0. This yields

$$\int_{t_0}^{t_0+T} f(t)e^{-jk\omega_0 t} \, dt = \int_{t_0}^{t_0+T} \left(e^{-jk\omega_0 t} \sum_{n=-\infty}^{\infty} F_n e^{jn\omega_0 t} \right) dt \tag{3.26a}$$

Interchanging the order of integration and summation on the right-hand side, we have

$$\int_{t_0}^{t_0+T} f(t)e^{-jk\omega_0 t} \, dt = \sum_{n=-\infty}^{\infty} F_n \int_{t_0}^{t_0+T} e^{j(n-k)\omega_0 t} \, dt \tag{3.26b}$$

Now consider the integral on the right-hand side:

$$I = \int_{t_0}^{t_0+T} e^{j(n-k)\omega_0 t} \, dt = \int_{t_0}^{t_0+T} dt = T \qquad (n = k)$$

and

$$I = \frac{1}{j(n-k)\omega_0} e^{j(n-k)\omega_0 t} \Big|_{t_0}^{t_0+T} \qquad (n \neq k)$$

$$= \frac{1}{j(n-k)\omega_0} \left[e^{j(n-k)\omega_0 (t_0+T)} - e^{j(n-k)\omega_0 t_0} \right]$$

$$= \frac{e^{j(n-k)\omega_0 t_0}}{j(n-k)\omega_0} \left[e^{j(n-k)\omega_0 T} - 1 \right] = 0 \qquad (n \neq k)$$

since $\omega_0 T = 2\pi$ and $e^{j(n-k)\omega_0 T} = e^{j2\pi(n-k)} = 1$.

Hence all terms on the right-hand side of Eq. 3.26b are zero except for one when $n = k$. Thus

$$\int_{t_0}^{t_0+T} f(t)e^{-jk\omega_0 t} \, dt = F_k T$$

and

$$F_k = \frac{1}{T} \int_{t_0}^{t_0+T} f(t)e^{-jk\omega_0 t} \, dt \tag{3.27}$$

The representation in Eq. 3.25 is called *exponential Fourier series* representation of a periodic function.

In conclusion, a periodic waveform $f(t)$ with a period T can be expressed by an exponential Fourier series†:

†To be more precise, the right hand side of Eq. 3.28a converges to $f(t)$ in the mean, that is, the mean square of the error (averaged over one period) approaches 0, where the error ϵ is defined as

$$f(t) = \sum_{n=-\infty}^{\infty} F_n e^{jn\omega_0 t} \qquad \left(\omega_0 = \frac{2\pi}{T}\right) \qquad (3.28a)$$

The amplitudes F_n are complex in general and are given by

$$F_n = \frac{1}{T} \int_{t_0}^{t_0+T} f(t)e^{-jn\omega_0 t}\, dt \qquad (3.28b)$$

for any value of t_0.

It is evident that a periodic waveform $f(t)$ can be expressed by eternal exponentials as shown in Eq. 3.28a. The representation is valid for $-\infty < t < \infty$.

In Eq. 3.28b, t_0 can have any value. Oftentimes it proves convenient in integration to choose a certain value of t_0. For $t_0 = 0$ and $-T/2$, the limits of integration in Eq. 3.28b are $(0, T)$ and $(-T/2, T/2)$ respectively.

Example 3.1.   Find the exponential Fourier series for the rectified sine wave in Fig. 3.6 when $T = 1$. In this case $\omega_0 = 2\pi/T = 2\pi$ and $f(t)$ can be expressed as

$$f(t) = \sum_{n=-\infty}^{\infty} F_n e^{j2\pi nt}$$

and

$$F_n = \frac{1}{T} \int_0^T f(t)e^{-j2\pi nt}\, dt$$

$$= \int_0^1 A \sin \pi t e^{-j2\pi nt}\, dt$$

$$= \frac{-2A}{\pi(4n^2 - 1)}$$

Hence

$$f(t) = \frac{2A}{\pi} - \frac{2A}{\pi}\left(\frac{1}{3} e^{j2\pi t} + \frac{1}{15} e^{j4\pi t} + \frac{1}{35} e^{j6\pi t} + \cdots + \right)$$

$$- \frac{2A}{\pi}\left(\frac{1}{3} e^{-j2\pi t} + \frac{1}{15} e^{-j4\pi t} + \frac{1}{35} e^{-j6\pi t} + \cdots + \right) \qquad (3.29a)$$

$$= -\frac{2A}{\pi} \sum_{n=-\infty}^{\infty} \frac{1}{(4n^2 - 1)} e^{j2\pi nt} \qquad (3.29b)$$

$$\epsilon = f(t) - \sum_{n=-\infty}^{\infty} F_n e^{jn\omega_0 t}$$

See B. P. Lathi, *Signals, Systems, and Communication* (New York: Wiley, 1965), Chapter 3.

THE FOURIER SPECTRUM

A Fourier series expansion of a periodic function is really equivalent to resolving the function in terms of its components of various frequencies. In the exponential Fourier series, a periodic function is expressed as a sum of exponential functions of complex frequencies, 0, $\pm j\omega_0$, $\pm 2j\omega_0, \ldots$, etc. The significance of negative frequencies has already been pointed out in Sec. 3.2. Both the signals $e^{j\omega t}$ and $e^{-j\omega t}$ oscillate at frequency ω. They may be looked upon as two phasors rotating in opposite direction which when added yield a real function of time oscillating at angular frequency ω:

$$e^{j\omega t} + e^{-j\omega t} = 2 \cos \omega t$$

For a periodic signal $f(t)$ of period T, the exponential Fourier series is given by Eq. 3.25. It is evident that the function $f(t)$ contains components of various frequencies and thus possesses its spectrum of frequencies. If we specify $f(t)$ we can find its spectrum. Conversely, if the spectrum is known one can find the corresponding periodic function $f(t)$. We therefore have two ways of specifying a periodic function $f(t)$: the time-domain representation, where $f(t)$ is expressed as a function of time, and the frequency-domain representation, where the spectrum (that is, the amplitudes of various exponential components) is specified. Note that the spectrum exists only at discrete frequencies. Thus the spectrum is not continuous but exists only at some discrete values of ω. It is therefore a *discrete spectrum,* sometimes referred to as a *line spectrum.*

The frequency spectrum can be specified by plotting coefficients F_n versus ω. In general, the amplitudes F_n are complex, and hence can be described by magnitude and phase. Therefore, in general, we need two line spectra: the *magnitude spectrum* and the *phase spectrum* for the frequency-domain representation of a periodic function. In many of the cases, however, the amplitudes of frequency components are either real or imaginary, and thus it is possible to describe the function by only one spectrum.

Consider the periodic function in Example 3.1 (Fig. 3.6). This is a rectified sine wave, and the exponential Fourier series was found to be

$$f(t) = \frac{2A}{\pi} - \frac{2A}{3\pi} e^{j2\pi t} - \frac{2A}{15\pi} e^{j4\pi t} - \frac{2A}{35\pi} e^{j6\pi t}$$

$$- \cdots - \frac{2A}{3\pi} e^{-j2\pi t} - \frac{2A}{15\pi} e^{-j4\pi t} - \frac{2A}{35\pi} e^{-j6\pi t} - \cdots$$

The spectrum exists at $\omega = 0$, $\pm 2\pi$, $\pm 4\pi$, $\pm 6\pi, \ldots$, etc., and the corresponding magnitudes are $2A/\pi$, $-2A/3\pi$, $-2A/15\pi$, $-2A/35\pi, \ldots$, etc. Note that all the amplitudes are real and consequently it is necessary to plot only one spectrum. The required frequency spectrum is shown in Fig. 3.7. It is evident from this figure that the spectrum is symmetrical about the vertical axis passing through the origin. This is not a coincidence. We shall presently show that the magnitude spectrum of every periodic function is *symmetrical* about a vertical axis passing through the

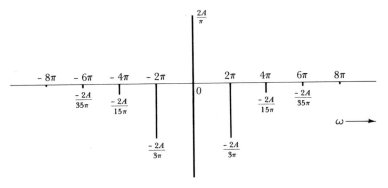

Figure 3.7

origin. This can be easily demonstrated. The coefficient F_n is given by

$$F_n = \frac{1}{T} \int_{t_0}^{t_0+T} f(t)e^{-jn\omega_0 t} \, dt \tag{3.30a}$$

and

$$F_{-n} = \frac{1}{T} \int_{t_0}^{t_0+T} f(t)e^{jn\omega_0 t} \, dt \tag{3.30b}$$

It is evident from these equations that the coefficients F_n and F_{-n} are complex conjugates of each other, when $f(t)$ is a real function of t. Thus

$$F_n^* = F_{-n}; \qquad \text{hence} \qquad |F_n| = |F_{-n}|$$

It therefore follows that the magnitude spectrum is symmetrical about the vertical axis passing through the origin, and hence is an even function of ω.

If F_n is real, then F_{-n} is also real and F_n is equal to F_{-n}. If F_n is complex, let

$$F_n = |F_n| e^{j\theta_n} \tag{3.31a}$$

then

$$F_{-n} = |F_n| e^{-j\theta_n} \tag{3.31b}$$

The phase of F_n is θ_n; however, the phase of F_{-n} is $-\theta_n$. Hence it is obvious that the phase spectrum is *antisymmetrical* (an odd function) and the magnitude spectrum is *symmetrical* (an even function) about the vertical axis passing through the origin.

INTERPRETATION OF THE MAGNITUDE AND PHASE SPECTRUM

The exponential Fourier series (Eq. 3.25a) can be written as

$$f(t) = F_0 + \sum_{n=1}^{\infty} (F_n e^{jn\omega_0 t} + F_{-n} e^{-jn\omega_0 t})$$

For a real signal $f(t)$,

$$F_n = |F_n| e^{j\theta_n} \quad \text{and} \quad F_{-n} = |F_n| e^{-j\theta_n}$$

Hence

$$f(t) = F_0 + \sum_{n=1}^{\infty} |F_n| e^{j(n\omega_0 t + \theta_n)} + |F_n| e^{-j(n\omega_0 t + \theta_n)}$$

$$= F_0 + \sum_{n=1}^{\infty} 2 |F_n| \cos(n\omega_0 t + \theta_n)$$

Comparison of this equation with Eq. 3.24a shows that the magnitude $|F_n|$ of the nth coefficient in the exponential Fourier series represents half the magnitude of the sinusoidal component of frequency $n\omega_0$ in the waveform $f(t)$, and hence the magnitude spectrum (Fig. 3.7) represents half the magnitude of sinusoidal components of respective frequencies (with exception of dc, where it represents the full dc component). Therefore the magnitude spectrum of exponential Fourier series directly gives us the picture of magnitudes of sinusoidal components of various frequencies in the waveform $f(t)$. The phase spectrum directly tells us the phases of the various frequency components in $f(t)$.

3.6   RESPONSE OF LINEAR SYSTEMS TO PERIODIC INPUTS

We are now in a position to find the response of a system to a periodic input. The periodic input $f(t)$ is expressed in terms of eternal exponentials of the form $e^{jn\omega_0 t}$. The complex frequencies of these exponentials are of the form $s = \pm jn\omega_0 t$. In the s-plane these are all located along $j\omega$ axis ($\sigma = 0$). The response to an eternal exponential $e^{jn\omega_0 t}$ is $H(jn\omega_0)e^{jn\omega_0 t}$ provided the frequency $jn\omega_0$ satisfies the dominance condition. This means the frequencies $jn\omega_0$ must lie to the right of all the poles $\lambda_1, \lambda_2, \ldots, \lambda_k$ of the transfer function $H(s)$. The stable system, by definition, has a zero-input component that decays with time.† This means all the terms of the form $ce^{\lambda_j t}$ decay with time implying that Re $\lambda_j < 0$. In other words, all the poles of $H(s)$ lie in the LHP. Obviously then the value of $s = jn\omega_0$ satisfies the dominance condition (Re $s = 0 >$ Re λ_j). Hence, if

$$f(t) = \sum_{n=-\infty}^{\infty} F_n e^{jn\omega_0 t}$$

then $y(t)$, the zero-state response, is given by

$$y(t) = \sum_{n=-\infty}^{\infty} F_n H(jn\omega_0)e^{jn\omega_0 t} \tag{3.32}$$

If some of the poles of $H(s)$ lie in the RHP, then the natural component grows with time. Such systems are (by definition) called unstable systems. Since the exponentials along the $j\omega$-axis cannot dominate growing exponentials, the domi-

---

†See Chapter 6.

nance condition cannot be satisfied. Consequently this analysis is not valid for unstable systems.†

The passive, lossless systems (such as *L-C* networks or M-K systems), have poles of $H(s)$ along the $j\omega$-axis and the dominance condition is not satisfied. As a result Eq. 3.32 is not valid.

Example 3.2.   A rectifier is used to obtain dc signal from ac signals. It consists of a device to rectify the ac signal, followed by a low-pass filter (Fig. 3.8a).   A simple low-pass filter is shown in Fig. 3.8a. Find the output $y(t)$ when the input is $A \sin \pi t$.

The rectified signal $f(t)$ in this case is a periodic signal shown in Fig. 3.8b.

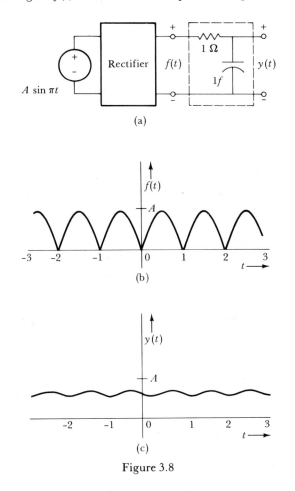

Figure 3.8

---

†It can be shown that the response obtained by Eq. 3.32 represents only a forced component for unstable systems. The natural component must be computed separately.

The Fourier series for $f(t)$ is already found in Eq. 3.29b. The transfer function of the low-pass filter is readily found as

$$H(s) = \frac{1}{s + 1}$$

and from Eq. 3.29b,

$$f(t) = \frac{-2A}{\pi} \sum_{n=-\infty}^{\infty} \frac{1}{(4n^2 - 1)} e^{j2\pi nt}$$

Note that $\omega_0 = 2\pi$. From Eq. 3.32 it now follows that

$$y(t) = \frac{-2A}{\pi} \sum_{n=-\infty}^{\infty} \frac{1}{4n^2 - 1} \left(\frac{1}{j2\pi n + 1} \right) e^{j2\pi nt}$$

$$= \frac{-2A}{\pi} \sum_{n=-\infty}^{\infty} \frac{1 - j2\pi n}{(4n^2 - 1)(4\pi^2 n^2 + 1)} e^{j2\pi nt}$$

The response is also given by a Fourier series. Hence the zero-state response of a linear system to a periodic input (which begins in the remote past $t = -\infty$) is also periodic as shown in Fig. 3.8c.

We have found the zero-state component. What about the zero-input component? By definition, the periodic input begins at $t = -\infty$. Hence the zero-input as well as zero-state component begin at $t = -\infty$. But since this (Fourier) analysis is valid only for stable systems for which the zero-input response decays with time exponentially, at any finite time the zero-input component will have vanished. We are therefore left with only the zero-state response. Hence at any finite time t the total response is given by the zero-state term alone.

3.7   LIMITATIONS ON FOURIER SERIES METHOD OF ANALYSIS

We have developed a method of representing periodic signals by exponentials whose frequencies lie along the $j\omega$-axis. This representation (Fourier series) is valuable in itself. However, as a tool in analyzing linear systems it has serious limitations and consequently has limited utility for the following reasons:

1. It can be used only for those inputs which are periodic. Most of the inputs in practice are nonperiodic.
2. The method applies only to stable systems, that is, the systems whose natural response decays with time.

The first limitation can be overcome if we can represent a nonperiodic input $f(t)$ in terms of exponential components. This can indeed be done by what is known as Fourier transform representation of $f(t)$.

3.8   EXPONENTIAL REPRESENTATION OF NONPERIODIC INPUTS

Representation of nonperiodic signals by eternal exponential signals can be accomplished by a simple limiting process, and we shall show that non-periodic signals, in general, can be expressed as a continuous sum (integral) of eternal exponential signals.

Consider a nonperiodic signal $f(t)$ shown in Fig. 3.9a. We desire to represent this function by eternal exponential functions. For this purpose we shall construct

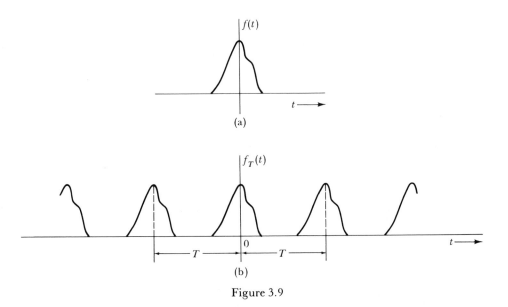

(a)

(b)

Figure 3.9

a new periodic function $f_T(t)$ with period T, where the function $f(t)$ repeats itself every T seconds as shown in Fig. 3.9b. The period T is made large enough so that there is no overlap between the pulses of the shape of $f(t)$. This new function $f_T(t)$ is a periodic function, and consequently can be represented with an exponential Fourier series. In the limit if we let T become infinite, then the pulses in the periodic function repeat after an infinite interval. Hence in the limit as $T \to \infty$, $f_T(t)$ and $f(t)$ are identical. That is,

$$\lim_{T\to\infty} f_T(t) = f(t)$$

Thus the Fourier series representing $f_T(t)$ will also represent $f(t)$, if we let $T = \infty$ in this series.

The exponential Fourier series for $f_T(t)$ can be represented as

$$f_T(t) = \sum_{n=-\infty}^{\infty} F_n e^{jn\omega_0 t} \qquad (3.33a)$$

where

$$\omega_0 = \frac{2\pi}{T}$$

and

$$F_n = \frac{1}{T} \int_{-T/2}^{T/2} f_T(t) e^{-jn\omega_0 t} \, dt \tag{3.33b}$$

The term F_n represents the amplitude of the component of frequency $n\omega_0$. We shall now let T become very large. As T becomes larger, ω_0 (the fundamental frequency) becomes smaller and the spectrum becomes denser. As seen from Eq. 3.33b, the amplitudes of individual components become smaller too. The shape of the frequency spectrum, however, is unaltered. In the limit when $T = \infty$, the magnitude of each component becomes infinitesimally small, but now there are also an infinite number of frequency components ($\omega_0 \to 0$). The spectrum exists for every value of ω and is no longer a discrete but a continuous function of ω. To illustrate this point, let us make a slight change in notation. Since in the limit as $T \to \infty$, $\omega_0 \to 0$, the quantity ω_0 is an infinitesimal and may be denoted by $\Delta\omega$, and

$$T = \frac{2\pi}{\omega_0} = \frac{2\pi}{\Delta\omega}$$

From Eq. 3.33b,

$$TF_n = \int_{-T/2}^{T/2} f_T(t) e^{-jn\Delta\omega t} \, dt \tag{3.34a}$$

The quantity TF_n is a function of $jn\Delta\omega$. So let

$$TF_n = F(jn\Delta\omega) \tag{3.34b}$$

Then from Eqs. 3.33a and 3.34b we have

$$f_T(t) = \sum_{n=-\infty}^{\infty} \frac{F(jn\Delta\omega)}{T} e^{(jn\Delta\omega)t}$$

$$= \sum_{n=-\infty}^{\infty} \left[\frac{F(jn\Delta\omega)}{2\pi} \Delta\omega \right] e^{(jn\Delta\omega)t} \tag{3.35a}$$

Equation 3.35a shows that $f_T(t)$ can be expressed as a sum of eternal exponentials of frequencies 0, $\pm j\Delta\omega$, $\pm j2\Delta\omega$, $\pm 3j\Delta\omega$, ..., etc. The amplitude of the component of frequency $jn\Delta\omega$ is $F(jn\Delta\omega)\,\Delta\omega/2\pi$. In the limit as $T \to \infty$, $\Delta\omega \to 0$, $f_T(t) \to f(t)$,

$$f(t) = \lim_{T \to \infty} f_T(t) = \lim_{\Delta\omega \to 0} \frac{1}{2\pi} \sum_{n=-\infty}^{\infty} F(jn\Delta\omega) e^{(jn\Delta\omega)t} \, \Delta\omega \tag{3.35b}$$

The right-hand side is by definition the integral

$$f(t) = \frac{1}{2\pi} \int_{-\infty}^{\infty} F(j\omega)e^{j\omega t}\,d\omega \qquad (3.36a)$$

where the variable $n\Delta\omega$ in the discrete summation now becomes a continuous variable ω. The function $F(j\omega)$ is $F(jn\Delta\omega)$ in Eq. 3.34b. Thus using Eq. 3.33b,

$$F(j\omega) = \lim_{\Delta\omega \to 0} F(jn\Delta\omega)$$

$$= \lim_{T \to \infty} \int_{-T/2}^{T/2} f_T(t)e^{-jn\Delta\omega t}\,dt$$

$$= \int_{-\infty}^{\infty} f(t)e^{-j\omega t}\,dt \qquad (3.36b)$$

We have thus succeeded in representing a nonperiodic function $f(t)$ in terms of exponential functions. Equation 3.36a represents $f(t)$ as a continuous sum of exponential functions with frequencies lying in the interval $(-\infty < \omega < \infty)$. The amplitude of the component of any frequency ω is proportional to $F(j\omega)$. Therefore $F(j\omega)$ represents the frequency spectrum of $f(t)$ and is called the *spectral density* function. Note, however, that the frequency spectrum now is continuous and exists at all values of ω. The spectral density function $F(j\omega)$ can be evaluated from Eq. 3.36b.

Equations 3.36a and 3.36b are usually referred to as the *Fourier transform pair*. Equation 3.36b is known as the *direct Fourier transform* of $f(t)$, and Eq. 3.36a is known as the *inverse Fourier transform* of $F(j\omega)$. Symbolically, these transforms are also written as

$$F(j\omega) = \mathcal{F}[f(t)] \qquad \text{and} \qquad f(t) = \mathcal{F}^{-1}[F(j\omega)]$$

Thus $F(j\omega)$ is the direct Fourier transform of $f(t)$, and $f(t)$ is the inverse Fourier transform of $F(j\omega)$, where

$$F(j\omega) = \mathcal{F}[f(t)] = \int_{-\infty}^{\infty} f(t)e^{-j\omega t}\,dt \qquad (3.37a)$$

and†

$$f(t) = \mathcal{F}^{-1}[F(j\omega)] = \frac{1}{2\pi} \int_{-\infty}^{\infty} F(j\omega)e^{j\omega t}\,d\omega \qquad (3.37b)$$

---

†To be more precise the inverse Fourier transform converges to $f(t)$ in the mean, that is the mean square error approaches zero, where the error ϵ is defined as

$$\epsilon = f(t) - \frac{1}{2\pi} \int_{-\infty}^{\infty} F(\omega)e^{j\omega t}\,d\omega$$

See B. P. Lathi, *Signals, Systems, and Communications* (New York: Wiley, 1965), Chap. 3.

SOME REMARKS ABOUT THE CONTINUOUS SPECTRUM FUNCTION

We have expressed a nonperiodic function $f(t)$ as a continuous sum of exponential functions with frequencies in the interval $-\infty$ to ∞. The amplitude of a component of any frequency ω is infinitesimal, but is proportional to $F(j\omega)$, where $F(j\omega)$ is the spectral density.

The concept of a continuous spectrum is sometimes confusing because we generally picture the spectrum as existing at discrete frequencies and with finite amplitudes in the manner of a periodic function. The continuous spectrum concept can be appreciated by considering the analogous concrete phenomenon. One familiar example of a continuous distribution is the loading of a beam. Consider a beam loaded with weights F_1, F_2, $F_3, \ldots, F_7$ units at uniformly spaced points $x_1, x_2, \ldots, x_7$, as shown in Fig. 3.10a. The beam is loaded at seven discrete points,

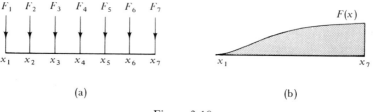

(a)   (b)

Figure 3.10

and the total weight on the beam is given by the sum of these loads at seven discrete points:

$$W_T = \sum_{r=1}^{7} F_r$$

Next, consider the case of a continuously loaded beam as shown in Fig. 3.10b. The loading density is a function of x, and let it be given by $F(x)$ in kg per meter. The total weight on the beam is now given by a continuous sum of the weight—that is, the integral of $F(x)$ over the entire length:

$$W_T = \int_{x_1}^{x_7} F(x)\, dx$$

In the former case of discrete loading, the weight existed only at discrete points. At other points there was no loading. On the other hand, in the continuously distributed case, the loading exists at every point but at any one point the loading is zero. However, the loading in a small distance dx is given by $F(x)\, dx$. Therefore, $F(x)$ represents the relative magnitude of loading at a point x. An exactly analogous situation exists in the case of signals and their frequency spectrum. A periodic signal can be represented by a sum of discrete exponentials with finite amplitudes:

$$f(t) = \sum_{n=-\infty}^{\infty} F_n e^{j\omega_n t} \qquad (\omega_n = n\omega_0)$$

For a nonperiodic function, the distribution of exponentials becomes continuous — that is, the spectrum function exists at every value of ω. At any one frequency ω the amplitude of that frequency component is zero.   The total contribution in an infinitesimal interval $d\omega$ is given by $(1/2\pi)F(j\omega)\,d\omega$ and the function $f(t)$ can be expressed in terms of the continuous sum of such infinitesimal components:

$$f(t) = \frac{1}{2\pi} \int_{-\infty}^{\infty} F(j\omega)e^{j\omega t}\,d\omega$$

The factor 2π in this equation may be removed, if desired, by performing the integration with respect to variable f instead of ω. We have

$$\omega = 2\pi f$$

and

$$d\omega = 2\pi df$$

The Fourier transform pair (Eq. 3.37) can now be expressed as

$$F(j2\pi f) = \int_{-\infty}^{\infty} f(t)e^{-j2\pi f t}\,dt$$

and

$$f(t) = \int_{-\infty}^{\infty} F(j2\pi f)e^{j2\pi f t}\,df$$

in our discussion, however, we shall continue to use variable ω instead of f.

3.9   TIME-DOMAIN AND FREQUENCY-DOMAIN REPRESENTATION OF A SIGNAL

The Fourier transform is a tool that resolves a given signal into its exponential components. The function $F(j\omega)$ is the direct Fourier transform of $f(t)$ and represents relative amplitudes of various frequency components.   Therefore $F(j\omega)$ is the frequency-domain representation of $f(t)$. Time-domain representation specifies a function at each instant of time, whereas frequency-domain representation specifies the relative amplitudes of the frequency components of the function. Either representation uniquely specifies the function.[†]   However, the function $F(j\omega)$ is complex in general, and needs two plots for its graphical representation.

$$F(j\omega) = |F(j\omega)|\,e^{j\theta(\omega)}$$

[†]This is true at all instants except those where there is a jump discontinuity. In the vicinity of the points of jump discontinuities of $f(t)$, the integral 3.37b does not converge to $f(t)$. This is called Gibb's phenomenon. See, for example, A. Papoulis, *The Fourier Integral and Its Applications* (New York: McGraw-Hill, 1962).

Thus $F(j\omega)$ may be represented by a magnitude plot $|F(j\omega)|$ and a phase plot $\theta(\omega)$. In many cases, however, $F(j\omega)$ is either real or imaginary and only one plot is necessary. $F(j\omega)$, however, is in general a complex function of ω, and we shall now show that for a real function $f(t)$,

$$F(-j\omega) = F^*(j\omega)$$

we have

$$F(j\omega) = \int_{-\infty}^{\infty} f(t)e^{-j\omega t}\, dt$$

Hence

$$F(-j\omega) = \int_{-\infty}^{\infty} f(t)e^{j\omega t}\, dt$$

If $f(t)$ is a real function of time, then it is evident from the two integrals above that $F(j\omega)$ and $F(-j\omega)$ are conjugates. Hence

$$F(-j\omega) = F^*(j\omega) \tag{3.38}$$

Hence if

$$F(j\omega) = |F(j\omega)|\, e^{j\theta(\omega)}$$

then

$$F(-j\omega) = |F(j\omega)|\, e^{-j\theta(\omega)}$$

It is evident from these equations that the *magnitude spectrum* $|F(j\omega)|$ *is an even function* of ω and the *phase spectrum* $\theta(\omega)$ is an *odd function* of ω.

EXISTENCE OF THE FOURIER TRANSFORM

For the Fourier transform of $f(t)$ to exist, the integral $\int_{-\infty}^{\infty} f(t)e^{-j\omega t}\, dt$ must be finite. Since $e^{j\omega t}$ has a unit magnitude, a sufficient condition for the existence of $F(j\omega)$ is that $f(t)$ be absolutely integrable, that is,

$$\int_{-\infty}^{\infty} |f(t)|\, dt < \infty$$

This condition, however, is not necessary for the existence of the Fourier transform. Functions such as $\sin \omega_0 t$, $u(t)$, etc., are not absolutely integrable yet do have Fourier transforms.

Any waveform that can be generated in a laboratory satisfies all the above conditions and has a Fourier transform. A function $e^{at}u(t)$ for $a > 0$ does not have Fourier transform, but then strictly speaking we cannot generate such a waveform in the laboratory (for $0 < t < \infty$).

3.10   FOURIER TRANSFORMS OF SOME FUNCTIONS

SINGLE-SIDED EXPONENTIAL $e^{-at}u(t)$

$$f(t) = e^{-at}u(t)$$

and

$$F(j\omega) = \int_0^\infty e^{-at}e^{-j\omega t}\,dt$$

$$= \int_0^\infty e^{-(a+j\omega)t}\,dt$$

$$= \frac{-1}{a + j\omega}\,e^{-(a+j\omega)t}\,\Big|_0^\infty \tag{3.39a}$$

$$= \frac{1}{a + j\omega}\qquad \text{for}\quad a > 0 \tag{3.39b}$$

Observe that if a is negative in Eq. 3.39a, the expression goes to infinity, and the Fourier transform does not exist. This is also seen from the fact that $e^{-at}u(t)$ does not satisfy the condition of absolute integrability if a is negative; in which case the function represents a growing exponential. The spectral density function of $e^{-at}u(t)$ is therefore given by $1/(a + j\omega)$. Note that this is a complex function and can be expressed as

$$\frac{1}{a + j\omega} = \frac{1}{\sqrt{a^2 + \omega^2}}\,e^{-j[\tan^{-1}(\omega/a)]}$$

The magnitude function $1/\sqrt{a^2 + \omega^2}$ and the phase function $\theta(\omega) = -\tan^{-1}(\omega/a)$ are shown in Fig. 3.11.

It is evident that the magnitude function is an even function, whereas the phase function is an odd function of ω.

We have thus expressed the signal $e^{-at}u(t)$ in terms of a continuous sum of

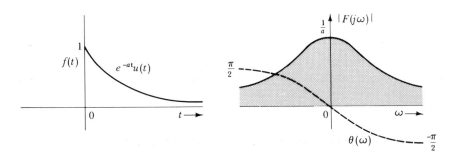

Figure 3.11

TABLE 3.1
Table of Fourier Transforms

$f(t)$	$F(j\omega)$
1. $e^{-at}u(t)$	$1/(a + j\omega)$
2. $te^{-at}u(t)$	$1/(a + j\omega)^2$
3. $|t|$	$-2/\omega^2$
4. $\delta(t)$	1
5. 1	$2\pi\delta(\omega)$
6. $u(t)$	$\pi\delta(\omega) + 1/j\omega$
7. $\cos\omega_0 t\, u(t)$	$\dfrac{\pi}{2}[\delta(\omega - \omega_0) + \delta(\omega + \omega_0)] + \dfrac{j\omega}{\omega_0^2 - \omega^2}$
8. $\sin\omega_0 t\, u(t)$	$\dfrac{\pi}{2j}[\delta(\omega - \omega_0) - \delta(\omega + \omega_0)] + \dfrac{\omega_0}{\omega_0^2 - \omega^2}$
9. $\cos\omega_0 t$	$\pi[\delta(\omega - \omega_0) + \delta(\omega + \omega_0)]$
10. $\sin\omega_0 t$	$j\pi[\delta(\omega + \omega_0) - \delta(\omega - \omega_0)]$
11. $e^{-at}\cos\omega_0 t\, u(t)$	$\dfrac{j\omega + a}{(j\omega + a)^2 + \omega_0^2}$
12. $\dfrac{1}{\omega_n\sqrt{1 - \zeta^2}}e^{-\zeta\omega_n t}\sin(\omega_n\sqrt{1 - \zeta^2})t$	$\dfrac{1}{(j\omega)^2 + 2\zeta\omega_n(j\omega) + \omega_n^2}$
13. $\dfrac{\omega_0}{2\pi}\text{Sa}\dfrac{(\omega_0 t)}{2}$	$G_{\omega_0}(\omega)$
14. $G_\tau(t)$	$\tau\text{Sa}\left(\dfrac{\omega\tau}{2}\right)$
15. $1 - \dfrac{|t|}{\tau}\ \cdots\ |t| < \tau$ $\quad 0 \qquad\qquad \cdots\ |t| > \tau$	$\tau\left[\text{Sa}\left(\dfrac{\omega\tau}{2}\right)\right]^2$
16. $e^{-a|t|}$	$\dfrac{2a}{a^2 + \omega^2}$
17. $e^{-(t/2\tau)^2}$	$\tau\sqrt{2\pi}\,e^{-\tau^2\omega^2/2}$
18. $\displaystyle\sum_{n=-\infty}^{\infty}\delta(t - nT)$	$\displaystyle\omega_0\sum_{n=-\infty}^{\infty}\delta(\omega - n\omega_0)$

eternal exponential functions. These exponential components add in such a way as to yield zero value for $t < 0$ and add up to e^{-at} for $t > 0$.

Table 3.1 gives a table of Fourier transforms of several functions.

TRANSFORM OF A GATE FUNCTION

A gate function $G_\tau(t)$ is a rectangular pulse, as shown in Fig. 3.12a and is defined by

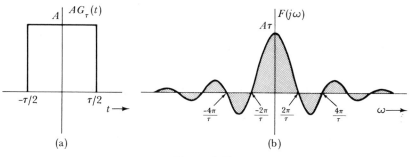

Figure 3.12

$$G_\tau(t) = \begin{cases} 1 & |t| < \tau/2 \\ 0 & |t| > \tau/2 \end{cases}$$

The Fourier transform of this function is given by

$$\begin{aligned} F(j\omega) &= \int_{-\tau/2}^{\tau/2} Ae^{-j\omega t}\, dt \\ &= \frac{A}{j\omega}\left(e^{j\omega\tau/2} - e^{-j\omega\tau/2}\right) \\ &= A\tau\,\frac{\sin(\omega\tau/2)}{\omega\tau/2} \end{aligned}$$

The transform $F(j\omega)$ contains a function of the form $\sin x/x$. This function is very useful in both systems theory and signal theory and is given a special name *sampling function* and is denoted by $\mathrm{Sa}(x)$.

$$\mathrm{Sa}(x) = \frac{\sin x}{x} \tag{3.40}$$

This function plotted in Fig. 3.13. It can be seen that this function oscillates with amplitudes decaying as $1/x$, and has zero at multiples of π. Using this notation, we can express the Fourier transform of a gate function as

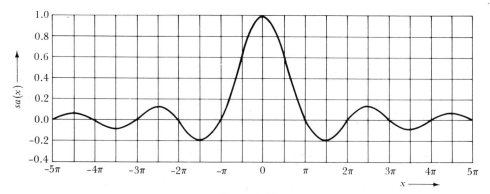

Figure 3.13

$$F(j\omega) = A\tau \text{Sa}\left(\frac{\omega\tau}{2}\right)$$

The spectrum $F(j\omega)$ of the gate function is shown in Fig. 3.12b.

TRANSFORM OF UNIT IMPULSE FUNCTION

The Fourier transform of a unit impulse is given by

$$\mathcal{F}[\delta(t)] = \int_{-\infty}^{\infty} \delta(t) e^{-j\omega t} dt$$

Using the sampling property of impulse function (Eq. 2.80), $F(j\omega)$ can be written immediately as

$$\mathcal{F}[\delta(t)] = 1 \tag{3.41}$$

It is evident that unit impulse function has a uniform spectral density over the entire frequency interval. In other words, an impulse function contains all frequency components in the same relative amount (Fig. 3.14).

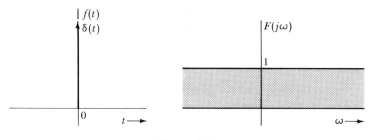

Figure 3.14

From Eq. 3.41 we have

$$\mathcal{F}^{-1}[1] = \delta(t)$$

or

$$\frac{1}{2\pi} \int_{-\infty}^{\infty} e^{j\omega t} d\omega = \delta(t)$$

or, in general,

$$\int_{-\infty}^{\infty} e^{jxy} dx = 2\pi\delta(y)$$

Changing the variable $x = -z$, we obtain

$$\int_{-\infty}^{\infty} e^{-jzy} dz = 2\pi\delta(y)$$

Hence it follows that

$$\int_{-\infty}^{\infty} e^{\pm jxy} dx = 2\pi\delta(y) \tag{3.42}$$

This is an important result.† We can use it to find Fourier transforms of several other functions.

TRANSFORM OF A CONSTANT $f(t) = 1$

By definition,

$$\mathcal{F}[1] = \int_{-\infty}^{\infty} e^{-j\omega t}\, dt$$

From Eq. 3.42 it immediately follows that

$$\mathcal{F}[1] = 2\pi\delta(\omega) \tag{3.43}$$

Thus the frequency spectrum of $f(t) = 1$ is an impulse located at $\omega = 0$ (Fig. 3.15). This is an expected result, since $f(t) = 1$ is a dc signal and hence the entire frequency spectrum of $f(t)$ is concentrated at the dc frequency ($\omega = 0$).

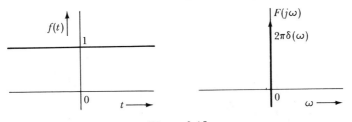

Figure 3.15

TRANSFORM OF ETERNAL EXPONENTIAL $e^{j\omega_0 t}$

By definition,

$$\mathcal{F}[e^{j\omega_0 t}] = \int_{-\infty}^{\infty} e^{j\omega_0 t} e^{-j\omega t}\, dt$$

$$= \int_{-\infty}^{\infty} e^{-j(\omega - \omega_0)t}\, dt$$

From Eq. 3.42 it immediately follows that

$$\mathcal{F}[e^{j\omega_0 t}] = 2\pi\delta(\omega - \omega_0) \tag{3.44}$$

Thus the frequency spectrum of eternal exponential $e^{j\omega_0 t}$ is an impulse located at $\omega = \omega_0$. This is also an expected result since eternal exponential $e^{j\omega_0 t}$ is a signal with complex frequency ω_0. Hence, the entire frequency spectrum is concentrated at $\omega = \omega_0$.

---

†This result is meaningful only when interpreted in the framework of generalized functions mentioned in Sec. 2.7. The integral on the left-hand side of Eq. 3.42 does not exist in the ordinary sense.

TRANSFORM OF COS $\omega_0 t$ AND SIN $\omega_0 t$

Since

$$\cos \omega_0 t = \tfrac{1}{2}[e^{j\omega_0 t} + e^{-j\omega_0 t}]$$

From Eq. 3.44 we can immediately write†

$$\mathfrak{F}[\cos \omega_0 t] = \pi[\delta(\omega + \omega_0) + \delta(\omega - \omega_0)] \qquad (3.45a)$$

The spectrum (Fig. 3.16) consists of two impulses, each of strength π and located

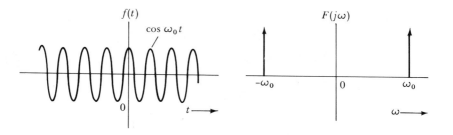

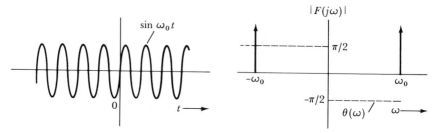

Figure 3.16

at $\pm\omega_0$. In a similar way, we can show that

$$\mathfrak{F}[\sin \omega_0 t] = j\pi[\delta(\omega + \omega_0) - \delta(\omega - \omega_0)]$$

The transform of $\sin \omega_0 t$ can be expressed alternately by noting that $\pm j = e^{\pm j\pi/2}$. Hence

$$\mathfrak{F}[\sin \omega_0 t] = \pi[\delta(\omega + \omega_0)e^{j\pi/2} + \delta(\omega - \omega_0)e^{-j\pi/2}] \qquad (3.45b)$$

The spectrum consists of two impulses each of strength π and located at $\pm\omega_0$. Note that the factor $e^{\pm j\pi/2}$ represents phase function $\theta(\omega)$ indicating a phase shift of $-\pi/2$ for $\sin \omega_0 t$ in relation to $\cos \omega_0 t$. This is also an expected result.

---

†Here we are implicitly using the linearity property (see Eq. 3.52)

$$\mathfrak{F}[a_1 f_1(t) + a_2 f_2(t)] = a_1 \mathfrak{F}[f_1(t)] + a_2 \mathfrak{F}[f_2(t)]$$

The proof is trivial.

TRANSFORM OF A PERIODIC SIGNAL

A general periodic signal $f(t)$ with period T can be expressed by a Fourier series as

$$f(t) = \sum_{n=-\infty}^{\infty} F_n e^{jn\omega_0 t}$$

where $\omega_0 = 2\pi/T$. Hence

$$\mathcal{F}[f(t)] = \mathcal{F} \sum_{n=-\infty}^{\infty} F_n e^{jn\omega_0 t}$$

$$= \sum_{n=-\infty}^{\infty} F_n \mathcal{F}[e^{jn\omega_0 t}]$$

From Eq. 3.44 it follows that

$$\mathcal{F}[f(t)] = 2\pi \sum_{n=-\infty}^{\infty} F_n \delta(\omega - n\omega_0) \tag{3.46}$$

This result shows that the frequency spectrum of a periodic signal is not continuous but is discrete and is concentrated at fundamental frequency ω_0 and all its harmonics. It is also an expected result and hardly needs any elaboration.

TRANSFORM OF A UNIFORM TRAIN OF IMPULSES

A uniform train of impulses consists of unit impulses spaced T seconds apart (Fig. 3.17a) and is denoted by a symbol $\delta_T(t)$. This is a periodic signal and its

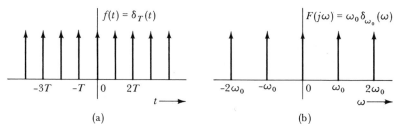

Figure 3.17

Fourier transform can be written as (see Eq. 3.46)

$$\mathcal{F}[\delta_T(t)] = 2\pi \sum_{n=-\infty}^{\infty} F_n \delta(\omega - n\omega_0)$$

where F_n is the coefficient of the nth harmonic in Fourier series expansion of $\delta_T(t)$. Since F_n is given by (see Eq. 3.28b),

$$F_n = \frac{1}{T} \int_{-T/2}^{T/2} \delta_T(t) e^{-jn\omega_0 t} dt$$

Note that $\delta_T(t) = \delta(t)$ in the interval $|t| < T/2$ (Fig. 3.17a). Hence

$$F_n = \frac{1}{T} \int_{-T/2}^{T/2} \delta(t) e^{-jn\omega_0 t} \, dt$$

From sampling property (Eq. 2.83b) it follows that $F_n = 1/T$, and

$$\mathcal{F}[\delta_T(t)] = \frac{2\pi}{T} \sum_{n=-\infty}^{\infty} \delta(\omega - n\omega_0) = \omega_0 \delta_{\omega_0}(\omega) \qquad (3.47)$$

This shows that the frequency spectrum of impulse train $\delta_T(t)$ is also a train of impulses spaced ω_0 ($= 2\pi/T$) rps apart (Fig. 3.17b).

3.11   SYSTEM ANALYSIS USING EXPONENTIALS ALONG $j\omega$-AXIS

We have seen that the complete response of a linear system to eternal exponential e^{st} is $H(s)e^{st}$, provided s satisfies the dominance condition.   For stable systems, all the poles of $H(s)$ lie in LHP ($\sigma < 0$). Hence the values of s for which $\sigma \geq 0$ will satisfy the dominance condition for any stable system.   The eternal exponential $e^{j\omega t}$ obviously satisfies this condition (since $s = j\omega$ implies $\sigma = 0$). Hence the response of a stable linear system to eternal exponential $e^{j\omega t}$ is $H(j\omega)e^{j\omega t}$. Symbolically we may represent this fact as

$$e^{j\omega t} \rightarrow H(j\omega)e^{j\omega t} \qquad (3.48a)$$

Fourier transform is a tool which allows us to represent a general function $f(t)$ as a continuous sum of eternal exponentials of the form $e^{j\omega t}$.   This can be seen from the equation

$$f(t) = \frac{1}{2\pi} \int_{-\infty}^{\infty} F(j\omega)e^{j\omega t} \, d\omega$$

The nature of this (continuous) sum becomes clear by expressing it as a discrete sum (see Eq. 3.35b)

$$f(t) = \lim_{\Delta\omega \to 0} \sum_{n=-\infty}^{\infty} \frac{F(jn\Delta\omega)\Delta\omega}{2\pi} e^{(jn\Delta\omega)t} \qquad (3.48b)$$

From this equation we observe that $f(t)$ is composed of exponentials of the form $e^{j(n\Delta\omega)t}$ and having a magnitude $F(jn\Delta\omega)\Delta\omega/2\pi$. From Eq. 3.48a we can find the response to this component as

$$\frac{F(jn\Delta\omega)\Delta\omega}{2\pi} e^{j(n\Delta\omega)t} \rightarrow \frac{H(jn\Delta\omega)F(jn\Delta\omega)}{2\pi} e^{j(n\Delta\omega)t}\Delta\omega \qquad (3.48c)$$

The zero-state response $y(t)$ to $f(t)$ will be given by the sum of responses to all the exponentials $e^{j(n\Delta\omega)t}$ in Eq. 3.48c over the range $-\infty < n < \infty$. This is given by

$$y(t) = \lim_{\Delta\omega \to 0} \sum_{n=-\infty}^{\infty} \frac{H(jn\Delta\omega)F(jn\Delta\omega)}{2\pi} e^{j(n\Delta\omega)t}\Delta\omega \qquad (3.49a)$$

$$= \frac{1}{2\pi} \int_{-\infty}^{\infty} H(j\omega)F(j\omega)e^{j\omega t}\, d\omega \qquad (3.49b)$$

$$= \mathcal{F}^{-1}[H(j\omega)F(j\omega)] \qquad (3.49c)$$

This is a very significant result. The zero-state response of a linear system to input $f(t)$ is given by $\mathcal{F}^{-1}[H(j\omega)F(j\omega)]$. To recapitulate, if a response variable $y(t)$ is related to the input $f(t)$ by a transfer function $H(s)$, then the zero-state response $y(t)$ is given by

$$y(t) = \mathcal{F}^{-1}[H(j\omega)F(j\omega)] \qquad (3.49d)$$

and

$$\mathcal{F}[y(t)] = H(j\omega)F(j\omega) \qquad (3.49e)$$

Thus the Fourier transform or the spectral density function of the (zero-state) response is given by $H(j\omega)F(j\omega)$. We may represent the system in Frequency domain as shown in Fig. 3.18.

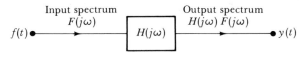

Figure 3.18

It must be remembered that the response thus obtained (Eq. 3.49d) is the response due to the input only and hence is the zero-state response. If $\lambda_1, \lambda_2, \ldots, \lambda_n$ are the poles† of $H(s)$, then the zero input response is $\sum_j c_j e^{\lambda_j t}$, where c_j are arbitrary constants. Thus the total response is given by

$$y(t) = \underbrace{\sum_{j=1}^{n} c_j e^{\lambda_j t}}_{\substack{\text{zero-input}\\\text{component}}} + \underbrace{\frac{1}{2\pi} \int_{-\infty}^{\infty} H(j\omega)F(j\omega)e^{j\omega t}\, d\omega}_{\text{zero-state component}} \qquad (3.50a)$$

$$= \sum_{j-1}^{n} c_j e^{\lambda_j t} + \mathcal{F}^{-1}[H(j\omega)F(j\omega)] \qquad (3.50b)$$

---

†Strictly speaking, these are the poles of the transfer operator $H(p)$. If there are any common factors in the numerator and the denominator of $H(p)$, they will be lost by way of cancellation in $H(s)$. We shall assume here that either $H(p)$ has no common factors or if the common factors are present, they are not canceled for computing the zero-input response. It should be remembered that cancellation of common factors does not affect the zero-state response. In short, the transfer function $H(s)$ has information required to determine the zero-state response but may not have the information required to determine the zero-input response if the common factors in $H(p)$ are canceled.

The arbitrary constants c_j can be found from the initial conditions (Eq. 2.47c). The reader should compare Eq. 3.50a with Eq. 2.73, the time-domain solution, and observe the similarity between two expressions. Both consist of same zero-input components. The zero-state component in each expression is given by an integral. In Eq. 2.73 (time-domain analysis) the input is expressed as a continuous sum of impulse components. The zero-state response is found by adding the response to these impulse components of the input. In Eq. 3.50a (frequency-domain analysis) on the other hand, the input is expressed as a continuous sum of eternal exponential components. The zero-state response is found by adding the responses to these eternal exponential components of the input. In both cases the integration required to obtain zero-state component can be avoided by the use of tables (convolution table and Fourier transform table).

Example 3.3.   Find the response $y(t)$ for the system (Fig. 3.19) when the input $f(t) = e^{-t}u(t)$ and the initial conditions (initial state) are zero. The transfer func-

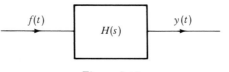

Figure 3.19

tion $H(s)$ which relates $y(t)$ to $f(t)$ is given by

$$H(s) = \frac{s}{s^2 + 5s + 6}$$

$$= \frac{s}{(s + 2)(s + 3)}$$

and

$$H(j\omega) = \frac{j\omega}{(j\omega + 2)(j\omega + 3)}$$

$$\mathfrak{F}[f(t)] = \mathfrak{F}[e^{-t}u(t)] = \frac{1}{j\omega + 1}$$

Hence

$$F(j\omega)H(j\omega) = \frac{j\omega}{(j\omega + 1)(j\omega + 2)(j\omega + 3)}$$

Since the initial conditions are zero, $y(t)$ is given by the zero-state component. From Eq. 3.50b we have

$$y(t) = \mathfrak{F}^{-1}\left[\frac{j\omega}{(j\omega + 1)(j\omega + 2)(j\omega + 3)}\right]$$

Using partial-fraction expansion, (Appendix B), we obtain

$$y(t) = \mathfrak{F}^{-1}\left[\frac{-1/2}{j\omega + 1} + \frac{2}{j\omega + 2} - \frac{3/2}{j\omega + 3}\right]$$

From Table 3.1 the inverse Fourier transform can be immediately written as

$$y(t) = \left(-\tfrac{1}{2} e^{-t} + 2e^{-2t} - \tfrac{3}{2} e^{-3t}\right)u(t)$$

Example 3.4. In Example 3.3, find $y(t)$ if the initial conditions are not zero but given by $y(0) = 2$ and $y'(0) = 1$.

The response in this case will have a zero-input component, as well as zero-state component. The zero state component $y_f(t)$ is already found in Example 3.3.

$$y_f(t) = \left(-\tfrac{1}{2} e^{-t} + 2e^{-2t} - \tfrac{3}{2} e^{-3t}\right)u(t)$$

The poles of the transfer function $H(s)$ are -2 and -3. Hence the zero-input component $y_x(t)$ is given by (see Eq. 2.46)

$$y_x(t) = c_1 e^{-2t} + c_2 e^{-3t}$$

where the constants c_1 and c_2 are given by (see Eq. 2.47c)

$$\begin{bmatrix} c_1 \\ c_2 \end{bmatrix} = \begin{bmatrix} 1 & 1 \\ -2 & -3 \end{bmatrix}^{-1} \begin{bmatrix} 2 \\ 1 \end{bmatrix}$$

$$= \begin{bmatrix} 3 & 1 \\ -2 & -1 \end{bmatrix} \begin{bmatrix} 2 \\ 1 \end{bmatrix} = \begin{bmatrix} 7 \\ -5 \end{bmatrix}$$

Hence

$$y_x(t) = 7e^{-2t} - 5e^{-3t}$$

Therefore

$$y(t) = y_x(t) + y_f(t)$$
$$= -\tfrac{1}{2} e^{-t} + 9e^{-2t} - \tfrac{13}{2} e^{-3t}$$

3.12 THE UNIT IMPULSE RESPONSE

The zero-state response of a system to input $f(t)$ was found to be

$$y(t) = \mathfrak{F}^{-1}[F(j\omega)H(j\omega)]$$

In the input $f(t) = \delta(t)$, then by definition $y(t) = h(t)$ and from Eq. 3.41, $F(j\omega) = \mathfrak{F}[\delta(t)] = 1$. Thus

$$h(t) = \mathfrak{F}^{-1}[H(j\omega)] \qquad (3.51a)$$

and

$$\mathfrak{F}[h(t)] = H(j\omega)$$

Thus

$$H(j\omega) = \int_{-\infty}^{\infty} h(t)e^{-j\omega t}\,dt \qquad (3.51b)$$

and

$$h(t) = \frac{1}{2\pi} \int_{-\infty}^{\infty} H(j\omega)e^{j\omega t}\,d\omega \qquad (3.51c)$$

This is a very significant result. The transfer function $H(j\omega)$ is the Fourier transform of the unit impulse response $h(t)$.

3.13   SOME PROPERTIES OF THE FOURIER TRANSFORM

We shall now discuss some useful properties of the Fourier transform. For convenience the relationship between $f(t)$ and its Fourier transform $F(j\omega)$ will be denoted by a double arrow:

$$f(t) \leftrightarrow F(j\omega)$$

implying that $F(j\omega)$ is the Fourier transform of $f(t)$ and $f(t)$ is the inverse Fourier transform of $F(j\omega)$.

1. LINEARITY PROPERTY

If

$$f_1(t) \leftrightarrow F_1(j\omega)$$
$$f_2(t) \leftrightarrow F_2(j\omega)$$

then for any arbitrary constants a_1 and a_2,

$$a_1 f_1(t) + a_2 f_2(t) \leftrightarrow a_1 F_1(j\omega) + a_2 F_2(j\omega) \qquad (3.52)$$

The proof is trivial. The above is also valid for finite sums:

$$a_1 f_1(t) + a_2 f_2(t) + \cdots + a_n f_n(t) \leftrightarrow a_1 F_1(j\omega) + a_2 F_2(j\omega) + \cdots + a_n F_n(j\omega)$$

2. SCALING PROPERTY

If

$$f(t) \leftrightarrow F(j\omega)$$

then for a real constant a,

$$f(at) \leftrightarrow \frac{1}{|a|} F\left(\frac{j\omega}{a}\right) \qquad (3.53)$$

Proof.   For a positive real constant a,

$$\mathfrak{F}[f(at)] = \int_{-\infty}^{\infty} f(at)e^{-j\omega t}\,dt$$

Let $x = at$. Then for positive constant a,

$$\mathcal{F}[f(at)] = \frac{1}{a} \int_{-\infty}^{\infty} f(x)e^{(-j\omega/a)x}\,dx$$

$$= \frac{1}{a} F\left(\frac{j\omega}{a}\right)$$

Hence

$$f(at) \leftrightarrow \frac{1}{a} F\left(\frac{j\omega}{a}\right)$$

Similarly, it can be shown that if $a < 0$,

$$f(at) \leftrightarrow -\frac{1}{a} F\left(\frac{j\omega}{a}\right)$$

Consequently it follows that

$$f(at) \leftrightarrow \frac{1}{|a|} F\left(\frac{j\omega}{a}\right)$$

Significance of the Scaling Property

The function $f(at)$ represents the function $f(t)$ compressed in the time scale by a factor of a. Similarly, a function $F(j\omega/a)$ represents a function $F(j\omega)$ expanded in the frequency scale by the same factor a. The scaling property therefore states that compression in the time domain is equivalent to expansion in the frequency domain, and vice versa. This result is also obvious intuitively, since compression in

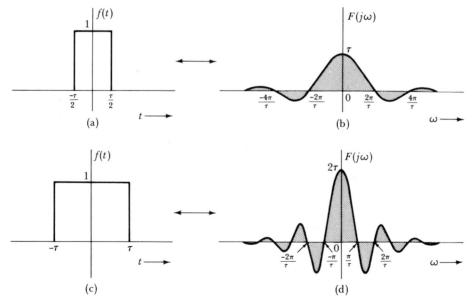

Figure 3.20. Compression in the time domain is equivalent to expansion in the frequency domain.

the time scale by a factor a means that the function is varying rapidly by the same factor, and hence the frequencies of its components will be increased by the factor a. We therefore expect its frequency spectrum to be expanded by the factor a in the frequency scale. Similarly, if a function is expanded in the time scale, it varies slowly, and hence the frequencies of its components are lowered. Thus the frequency spectrum is compressed. As an example, consider the signal $\cos \omega_0 t$. This signal has frequency components at $\pm \omega_0$. The signal $\cos 2\omega_0 t$ represents compression of $\cos \omega_0 t$ by a factor of 2, and its frequency components lie at $\pm 2\omega_0$. It is therefore evident that the frequency spectrum has been expanded by a factor of 2. The effect of scaling is demonstrated in Fig. 3.20.

3. TIME-SHIFTING PROPERTY

If

$$f(t) \leftrightarrow F(j\omega)$$

then

$$f(t - t_0) \leftrightarrow F(j\omega)e^{-j\omega t_0} \tag{3.54}$$

Proof.

$$\mathscr{F}[f(t - t_0)] = \int_{-\infty}^{\infty} f(t - t_0)e^{-j\omega t}\, dt$$

Let

$$t - t_0 = x$$

then

$$\mathscr{F}[f(t - t_0)] = \int_{-\infty}^{\infty} f(x)e^{-j\omega(x + t_0)}\, dx$$

$$= F(j\omega)e^{-j\omega t_0}$$

This theorem states that if a function is shifted in the time domain by t_0 seconds, then its magnitude spectrum $|F(j\omega)|$ remains unchanged, but the phase spectrum is changed by an amount $-\omega t_0$. This result is also obvious intuitively, since the shifting of a function in the time domain really does not change the frequency components of the signal but each component is shifted by an amount t_0. A shift of time t_0 for a component frequency ω is equivalent to a phase shift of ωt_0. This can be seen from the equation

$$\cos \omega(t - t_0) = \cos [\omega t - \omega t_0]$$

It is obvious that a time shift of t_0 is equivalent to a phase shift (lag) of ωt_0 for a sinusoidal signal of frequency ω.

We may state that a shift of t_0 in the time domain is equivalent to multiplication by $e^{-j\omega t_0}$ in the frequency domain.

Example 3.5. Find the Fourier transform of a rectangular pulse shown in Fig. 3.21. The function in Fig. 3.21 is really a gate function $G_\tau(t)$ shifted by $\tau/2$

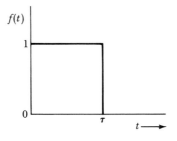

Figure 3.21

seconds. Hence it may be represented as $G_\tau(t - \tau/2)$. From Table 3.1,

$$G_\tau(t) \leftrightarrow \tau\mathrm{Sa}(\omega\tau/2)$$

Therefore

$$G_\tau(t - \tau/2) \leftrightarrow \tau\mathrm{Sa}(\omega\tau/2)e^{-j\omega\tau/2}$$

4. FREQUENCY-SHIFTING PROPERTY

If

$$f(t) \leftrightarrow F(j\omega)$$

then

$$f(t)e^{j\omega_0 t} \leftrightarrow F[j(\omega - \omega_0)] \qquad (3.55)$$

Proof.

$$\mathcal{F}[f(t)e^{j\omega_0 t}] = \int_{-\infty}^{\infty} f(t)e^{j\omega_0 t}e^{-j\omega t}\, dt$$

$$= \int_{-\infty}^{\infty} f(t)e^{-j(\omega - \omega_0)t}\, dt$$

$$= F[j(\omega - \omega_0)]$$

The theorem states that a shift of ω_0 in the frequency domain is equivalent to multiplication by $e^{j\omega_0 t}$ in the time domain. Note the dual nature between the time-shift and the frequency-shift theorems. It is evident that multiplication by a factor $e^{j\omega_0 t}$ translates the whole frequency spectrum $F(j\omega)$ by an amount ω_0. Hence this theorem is also known as the *frequency-translation* theorem.

In carrier-type control systems (ac control systems) and in long distance transmission of signals over a channel, it is often desirable to translate the frequency spectrum of the signal $f(t)$. This is usually accomplished by multiplying the signal $f(t)$ by a sinusoidal signal. This process is known as *modulation*. Since a sinusoidal signal of frequency ω_0 can be expressed as the sum of exponentials, it is evident that multiplication of a signal $f(t)$ by a sinusoidal signal (modulation) will translate the whole frequency spectrum. This can be easily shown by observing the identity

$$f(t) \cos \omega_0 t = \tfrac{1}{2} [f(t)e^{j\omega_0 t} + f(t)e^{-j\omega_0 t}]$$

Using the frequency-shifting property it follows that if

$$f(t) \leftrightarrow F(j\omega)$$

then

$$f(t) \cos \omega_0 t \leftrightarrow \tfrac{1}{2} \{F[j(\omega + \omega_0)] + F[j(\omega - \omega_0)]\} \qquad (3.56)$$

Similarly it can be shown that

$$f(t) \sin \omega_0 t \leftrightarrow \frac{j}{2} \{F[j(\omega + \omega_0)] - F[j(\omega - \omega_0)]\} \qquad (3.57)$$

Thus the process of modulation translates the frequency spectrum by the amount $\pm\omega_0$. This is a very useful result.

An example of frequency translation caused by modulation is shown in Fig. 3.22. This result is also known as the *modulation theorem*.

The signal $\cos \omega_0 t$ is known as the *carrier*. Multiplication of $\cos \omega_0 t$ by $f(t)$ is

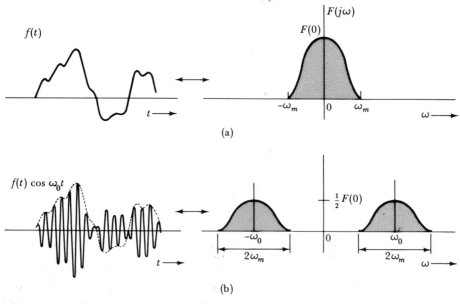

(a)

(b)

Figure 3.22

really equivalent to multiplying the amplitude of the carrier $\cos \omega_0 t$ in proportion to the modulating signal $f(t)$. Hence this type of modulation is called *amplitude modulation* (AM).

Applications of Modulation

The type of modulation described above is useful in many applications where it is important to shift the signal spectrum. This occurs in the following situations.

1. For message communication, whether by telephone line or by radio or TV

broadcast, it is not possible to transmit directly more than one signal at a time because of cross-interference. By modulating different messages by different carrier frequencies one can shift the spectra of various signals so that they do not overlap. This allows transmission of several long-distance conversations simultaneously on one telephone channel. It also permits broadcasting from several radio and TV stations in the same area without interference. In all these cases the modulated signals at the receiver are demodulated (spectrum shifted to original position through another modulation) to obtain original signals.

2. From electromagnetic theory it can be shown that for effective radiation of energy the antenna size must be of the order of the wavelength of the signal. Audio signal frequencies are so low (large wavelength) as to necessitate impracticably large antenna for radiation. Here shifting the audio spectrum to higher frequencies (shorter wavelength) through modulation solves the problem.

3. Amplification of very low-frequency signals poses a problem because it requires impracticably large coupling capacitors between stages. In this case the low-frequency signal to be amplified is first modulated to shift its spectrum to a higher frequency where it can be easily amplified. The amplified signal is then demodulated. The so-called chopper amplifiers use this principle.

Demodulation

The process of recovering original signal $f(t)$ from the modulated signal is known as *demodulation*. Demodulation can be accomplished by another modulation (multiplication by $\cos \omega_0 t$) of the modulated signal. This shifts the spectrum of the modulated signal by $\pm \omega_0$ (and multiplies by a factor of $\frac{1}{2}$). It can be easily seen that such a shift of the spectrum in Fig. 3.22b yields the new spectrum as shown in Fig. 3.23a. The original signal $f(t)$ can be recovered by passing the spectrum in Fig. 3.23a through a low-pass filter (shown dotted in Fig. 3.23a) that suppresses the spectrum centered at $\pm 2\omega_0$ and allows passage to the spectrum centered at the origin. The schematic of a demodulator is shown in Fig. 3.23b. The results derived here can also be obtained as follows. The modulation of the modulated signal $f(t) \cos \omega_0 t$ yields $f(t) \cos^2 \omega_0 t$. Also,

$$f(t) \cos^2 \omega_0 t = \tfrac{1}{2} f(t) [1 + \cos 2\omega_0 t]$$
$$= \tfrac{1}{2} [f(t) + f(t) \cos 2\omega_0 t]$$

Hence this signal contains the original signal component $\frac{1}{2} f(t)$, whose spectrum is centered at the origin and the modulated signal $\frac{1}{2} f(t) \cos 2\omega_0 t$ whose spectrum is centered at $\pm 2\omega_0$. This is precisely what we found in Fig. 3.23a. The low-pass filter in Fig. 3.23b removes the unwanted component $f(t) \cos 2\omega_0 t$.

The demodulation process described here necessitates generation of a carrier $\cos \omega_0 t$ which must be identical in frequency and phase to the carrier at the modulator. Generation of such a carrier requires elaborate and expensive equipment. The need for local carrier at the demodulator can be avoided if a large amount of free carrier is added to the modulated signal. When a free carrier is added, the resulting signal is $f(t) \cos \omega_0 t + A \cos \omega_0 t$. This can be expressed as $[A + f(t)] \cos \omega_0 t$. This signal is shown in Fig. 3.23c. Note that if A is large enough so that $A + f(t) > 0$ for all t, then the envelope of the modulated signal in Fig. 3.23c

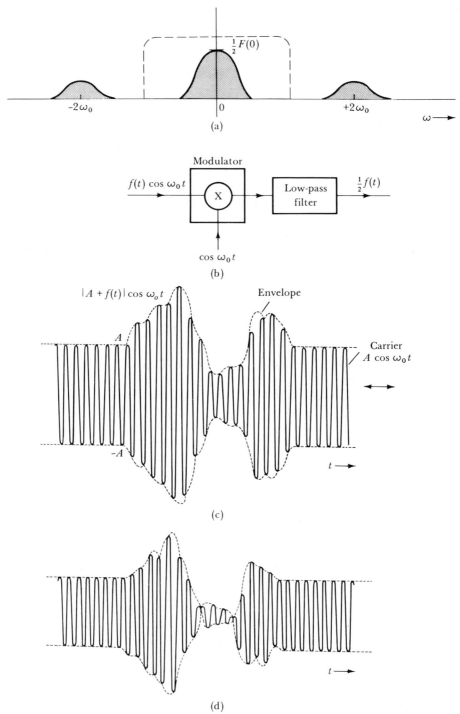

Figure 3.23

is $A + f(t)$. A simple envelope detector which consists of diodes and R-C elements†
can now be used to obtain the envelope $[A + f(t)]$ from the signal in Fig. 3.23c.
Another scheme to obtain the envelope is discussed in Prob. 3.27 (Fig: P-3.27).
Note, however, if A is not large enough to make $A + f(t) > 0$ for all t, then the
envelope of $[A + f(t)] \cos\omega_0 t$ is no longer $A + f(t)$ but will have rectified segment
when $A + f(t) < 0$ as shown in Fig. 3.23d. The signal $f(t)$ can be recovered
from $A + f(t)$ by supressing the dc component A by using a large capacitor.
Note that the demodulation in this case can be carried out by using simple envelope
detector and without the need for generating a local carrier. This scheme is commonly used in public radio broadcast where it is necessary to reduce the cost of
receivers. The price is paid in terms of expensive transmitter which must now
transmit the additional power in the carrier $A \cos\omega_0 t$. But this is justified in view
of the fact that there is one transmitter for multitude of receivers. This scheme is
known as amplitude modulation (AM). For point to point communication where
there is one receiver for each transmitter, the earlier scheme without the carrier
(suppressed carrier, AM) is commonly used.

In amplitude modulation the amplitude of the carrier $\cos\omega_0 t$ is varied in proportion to the message $f(t)$. We can transmit information by modulating the frequency or the phase of the carrier in proportion to the message $f(t)$. These
schemes also shift the frequency spectrum of $f(t)$ but in a rather nonlinear way.
The shifted spectrum no longer resembles the original spectrum. More information
on these forms of modulation, frequency modulation (FM) and phase modulation
(PM) may be found in the literature.‡

5. TIME DIFFERENTIATION AND INTEGRATION

If

$$f(t) \leftrightarrow F(j\omega)$$

then*

$$\frac{df}{dt} \leftrightarrow (j\omega) F(j\omega) \tag{3.58}$$

and**

$$\int_{-\infty}^{t} f(\tau)\, d\tau \leftrightarrow \frac{1}{j\omega} F(j\omega) \tag{3.59}$$

Proof,

$$f(t) = \frac{1}{2\pi} \int_{-\infty}^{\infty} F(j\omega) e^{j\omega t}\, d\omega$$

---

†See, for example, B. P. Lathi, *Communication Systems*, (New York: 1968) Wiley.
‡Lathi, *ibid.*
*Equation 3.58 does not guarantee the existence of the transform of df/dt. It merely
says that if the transform exists, it is given by $j\omega F(\omega)$.

**This is true provided that $F(0) = 0$, implying that $\int_{-\infty}^{\infty} f(t)\, dt = 0$. For a general

case Eq. 3.59 becomes

$$\int_{-\infty}^{t} f(\tau)\, d\tau \leftrightarrow F(j\omega)\delta(\omega) + \frac{1}{j\omega} F(j\omega)$$

Therefore

$$\frac{df}{dt} = \frac{1}{2\pi} \frac{d}{dt} \int_{-\infty}^{\infty} F(j\omega)\, e^{j\omega t} d\omega$$

Changing the order of differentiation and integration, we get

$$\frac{df}{dt} = \frac{1}{2\pi} \int_{-\infty}^{\infty} j\omega F(j\omega)\, e^{j\omega t} d\omega$$

It is now evident from the above equation that

$$\frac{df}{dt} \leftrightarrow j\omega F(j\omega)$$

In a similar way the result can be extended to the nth derivative.

$$\frac{d^n f}{dt^n} \leftrightarrow (j\omega)^n F(j\omega)$$

Proof of Eq. 3.59 can be easily obtained from Eq. 3.58 and is left as an exercise for the reader.

The time differentiation property can be used to solve linear differential equations. Consider the equation

$$y(t) = H(p)\, f(t)$$
$$= \frac{b_m p^m + b_{m-1} p^{m-1} + \cdots + b_1 p + b_0}{p^n + a_{n-1} p^{n-1} + \cdots + a_1 p + a_0}\, f(t)$$

or

$$(p^n + a_{n-1} p^{n-1} + \cdots + a_1 p + a_0)\, y(t) = (b_m p^m + \cdots + b_0)\, f(t)$$

$$(3.60)$$

Let

$$y(t) \leftrightarrow Y(j\omega)$$

and

$$f(t) \leftrightarrow F(j\omega)$$

Then from the time-differentiation property,

$$p^k y(t) \leftrightarrow (j\omega)^k Y(j\omega)$$
$$p^k f(t) \leftrightarrow (j\omega)^k F(j\omega)$$

$$(3.61)$$

Taking the Fourier transform of both sides of Eq. 3.60 and substituting the results in Eq. 3.61, we obtain

$$[(j\omega)^n + a_{n-1}(j\omega)^{n-1} + \cdots + a_0]\, Y(j\omega) = [b_m(j\omega)^m + \cdots + b_0]\, F(j\omega)$$

or

$$Y(j\omega) = \frac{b_m(j\omega)^m + \cdots + b_0}{(j\omega)^n + a_{n-1}(j\omega)^{n-1} + \cdots + a_0}\, F(j\omega)$$
$$= H(j\omega)\, F(j\omega)$$

and

$$y(t) = \mathcal{F}^{-1}[H(j\omega)F(j\omega)]$$

Which is precisely the result obtained earlier. Note, however, that this is only a zero-state component.

6. TIME CONVOLUTION

If

$$f_1(t) \leftrightarrow F_1(j\omega)$$

and

$$f_2(t) \leftrightarrow F_2(j\omega)$$

then

$$\int_{-\infty}^{\infty} f_1(\tau) f_2(t - \tau) \, d\tau \leftrightarrow F_1(j\omega) F_2(j\omega) \tag{3.62a}$$

that is,

$$f_1(t) * f_2(t) \leftrightarrow F_1(j\omega) F_2(j\omega) \tag{3.62b}$$

Proof

$$\mathcal{F}[f_1(t) * f_2(t)] = \int_{-\infty}^{\infty} e^{-j\omega t} \left[\int_{-\infty}^{\infty} f_1(\tau) f_2(t - \tau) \, d\tau \right] dt$$

$$= \int_{-\infty}^{\infty} f_1(\tau) \left[\int_{-\infty}^{\infty} e^{-j\omega t} f_2(t - \tau) \, dt \right] d\tau$$

From the time-shifting theorem (Eq. 3.54), it is evident that the integral inside the bracket, on the right-hand side, is equal to $F_2(j\omega)e^{-j\omega\tau}$. Hence

$$\mathcal{F}\left[f_1(t) * f_2(t) \right] = \int_{-\infty}^{\infty} f_1(\tau) \, e^{-j\omega\tau} F_2(j\omega) \, d\tau$$

$$= F_1(j\omega) F_2(j\omega)$$

We had anticipated this result earlier. From time-domain in analysis, we have (for the zero-state case)

$$y(t) = h(t) * f(t)$$

and from frequency-domain analysis, we have (for the zero-state case)

$$y(t) = \mathcal{F}^{-1}[H(j\omega)F(j\omega)]$$

Hence

$$h(t) * f(t) \leftrightarrow H(j\omega)F(j\omega)$$

This is precisely the result of time convolution property in Eq. 3.62.

7. FREQUENCY CONVOLUTION

If

$$f_1(t) \leftrightarrow F_1(j\omega)$$

and

$$f_2(t) \leftrightarrow F_2(j\omega)$$

then

$$f_1(t)f_2(t) \leftrightarrow \frac{1}{2\pi} F_1(j\omega) * F_2(j\omega) \tag{3.63}$$

This theorem can be proved in exactly the same way as the time-convolution theorem because of the symmetry in the direct and inverse Fourier transform.

We therefore conclude that the convolution of two functions in the time domain is equivalent to multiplication of their spectra in the frequency domain, and that multiplication of two functions in the time domain is equivalent to convolution of their spectra in the frequency domain.

3.14   DISTORTIONLESS TRANSMISSION

Consider a system (Fig. 3.24a) with transfer function $H(s)$. The input spectrum $F(j\omega)$ and the output spectrum $Y(j\omega)$ are related by

$$Y(j\omega) = H(j\omega)F(j\omega) \tag{3.64}$$

The system processes the input spectrum and modifies it by a factor $H(j\omega)$.

Thus a component of some frequency ω_0 undergoes a magnitude change by a factor $|H(j\omega_0)|$ and a phase shift by an amount $\angle H(j\omega_0)$. The system acts as a

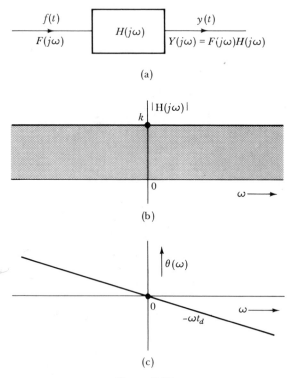

Figure 3.24

filter which may boost certain frequency components and attenuate others depending upon the nature of $H(j\omega)$. It also changes the relative phases of various frequency components. As a result, the output may not resemble the input at all. The output is a modified or distorted version of the input. In many applications the systems are deliberately designed to modify (or distort in a prescribed way) the input signal. However, in certain applications such as amplifiers it is very desirable that the output resemble the input closely and all forms of distortions be minimized. We shall now determine the conditions required for a distortionless transmission.

Distortion can occur in linear systems in two ways: (i) The amplitudes of various frequency components may undergo different amplification (amplitude distortion). This arises if $|H(j\omega)|$ is not constant for all values of ω. (ii) The relative phases of various components may be disturbed (phase distortion). This occurs if $\angle H(j\omega)$ does not satisfy certain conditions given below.

For the input $f(t)$, if the output is $kf(t - t_d)$, where k and t_d are constants, then we say that the transmission is *distortionless*. In this case the output is an amplified (by a factor k) and delayed (by t_d seconds) version of $f(t)$. The waveform of the output retains the form of the input. Consequently the transmission is distortionless. Hence for distortionless transmission the output $y(t)$ must be of the form

$$y(t) = kf(t - t_d) \tag{3.65}$$

From the time-shifting property (Eq. 3.54) we have

$$Y(j\omega) = kF(j\omega)e^{-j\omega t_d}$$

But

$$Y(j\omega) = F(j\omega)H(j\omega)$$

Hence

$$H(j\omega) = ke^{-j\omega t_d} \tag{3.66}$$

Thus for a distortionless transmission, the transfer function $H(j\omega)$ must have a form in Eq. 3.66. It follows that

$$|H(j\omega)| = k \tag{3.67a}$$

$$\angle H(j\omega) = \theta(\omega) = -\omega t_d \tag{3.67b}$$

The magnitude function $|H(j\omega)|$ must be constant for all frequencies. The phase spectrum $\theta(\omega)$ must be a linear function of ω as shown in Fig. 3.24. Note that the transmission delay is given by the slope of the phase function.

$$t_d = -\frac{d\theta}{d\omega}$$

It can be seen that for distortionless transmission, the phase shift introduced by the system in a component of frequency ω should be proportional to ω. Thus for distortionless transmission, higher frequencies should undergo proportionately higher phase shifts. It is not difficult to see why this must be so. Consider a signal consisting of two frequency components

$$f(t) = \cos \omega_1 t + \cos \omega_2 t$$

The delayed signal $f(t - t_d)$ is given by

$$f(t - t_d) = \cos \omega_1 (t - t_d) + \cos \omega_2 (t - t_d)$$
$$= \cos (\omega_1 t - \omega_1 t_d) + \cos (\omega_2 t - \omega_2 t_d)$$

It is obvious that delay t_d implies a phase shift $\omega_1 t_d$ for the component of frequency ω_1 and a phase shift $\omega_2 t_d$ for the component of frequency ω_2. Thus for a given delay the phase shift is proportional to the frequency.

BANDWIDTH OF TRANSMISSION

In practice, it is impossible to design an ideal distortionless system. This is because the response of all practical systems diminishes at higher frequencies. We therefore design systems to have distortionless characteristics over a certain frequency range of interest. The range of frequencies over which the magnitude $|H(j\omega)|$ remains within a ratio of $\sqrt{2}$ (3 db) is called the *bandwidth* of transmission (Fig. 3.25). This is somewhat arbitrary but widely accepted criterion of measuring a system bandwidth.

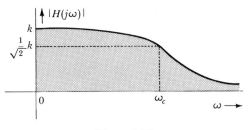

Figure 3.25

3.15   IDEAL FILTERS

By definition, an ideal low-pass filter allows distortionless transmission of all frequency components of frequencies below cutoff frequency ω_c and attenuate completely all frequency components above ω_c. The magnitude and phase characteristics of an ideal low pass filter are shown in Fig. 3.26a. Note that the phase function $\theta(\omega) = -\omega t_d$ implies a delay of t_d seconds in the transmission (see Eq. 3.67). The unit impulse response of this filter is given by $\mathfrak{F}^{-1}[H(j\omega)]$ and is found from Table 3.1, pair 13 (with delay of t_d) to be

$$h(t) = \frac{\omega_c}{\pi} \operatorname{Sa} \omega_c (t - t_d) \tag{3.68}$$

The unit impulse response $h(t)$ is shown in Fig. 3.26b. Observe that the impulse response $h(t)$ is nonzero for negative values of t. This is very strange in view of the fact that $h(t)$ is the response of the system to unit impulse which is applied at $t = 0$. How can a response exist even before the input is applied? Can the system anticipate the input? Unfortunately, no real-life system can exhibit such a foresight!

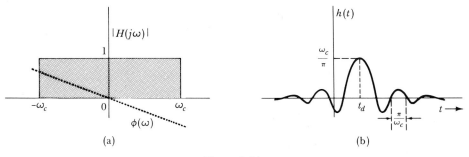

Figure 3.26

Hence we must conclude that the ideal low-pass filter in Fig. 3.26 is physically un-realizable. Using similar argument, one can show that other ideal filters (such as high-pass or bandpass filters) are physically unrealizable.

The time-domain criterion of physical realizability is that a response cannot occur before the input is applied. This is the *causality* condition. Thus the unit impulse, response $h(t)$ of a physically realizable system must be causal—that is, $h(t) = 0$ for $t < 0$.

In the frequency domain it can be shown that† the causality condition implies that the magnitude function $|H(j\omega)|$ must satisfy

$$\int_{-\infty}^{\infty} \frac{|\ln|H(j\omega)||}{1 + \omega^2} \, d\omega < \infty \qquad (3.69)$$

where the function $H(j\omega)$ is such that the area under $|H(j\omega)|^2$ over $\omega = -\infty$ to ∞ is finite. This is known as the *Paley-Wiener criterion*. Note that if $|H(j\omega)| = 0$, over a finite bandwidth, the integral in Eq. 3.69 becomes infinite because $|\ln|H(j\omega)|| = \infty$. Hence for a physically realizable system, $|H(j\omega)|$, may be zero at some discrete frequencies, but cannot be zero over a finite band-width. It is therefore obvious that ideal filter characteristics such as shown in Fig. 3.26a (or ideal high-pass, or bandpass characteristics) are physically unrealizable.

Figure 3.27a shows a low-pass filter realized by a simple *RLC* network. The transfer function $H(s)$, of this filter is readily found as

$$H(s) = \frac{V_0(s)}{V_i(s)} = \frac{1 / \left(\frac{1}{R} + Cs \right)}{Ls + 1 / \left(\frac{1}{R} + Cs \right)}$$

$$= \frac{1}{1 + LCs^2 + Ls/R}$$

Choosing $R = \sqrt{L/C}$ and defining $\omega_c = 1/\sqrt{LC}$, we have

---

†R. E. A. C. Paley and N. Wiener, *Fourier Transform in the Complex Domain* (New York: 1934). American Mathematical Society Colloquium Publication 19.

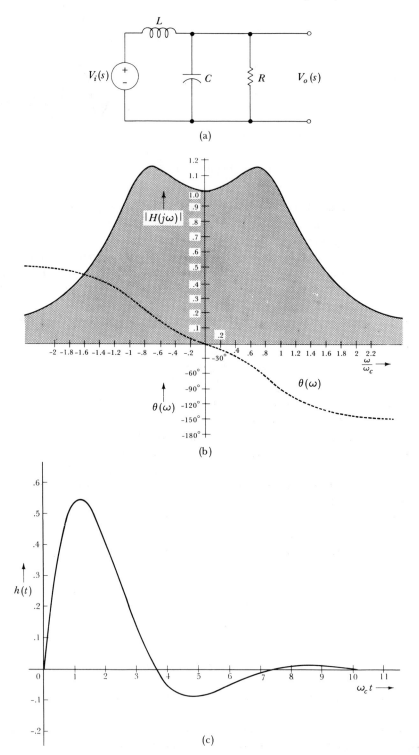

Figure 3.27

$$H(s) = \frac{\omega_c^2}{s^2 + \omega_c s + \omega_c^2}$$

Hence

$$H(j\omega) = \frac{\omega_c^2}{(j\omega)^2 + \omega_c(j\omega) + \omega_c^2}$$

The unit impulse response $h(t)$, is the inverse Fourier transform of $H(j\omega)$. From Table 3.1 (pair 12), we find

$$h(t) = \frac{2}{\sqrt{3}} \, \omega_c e^{-\omega_c t/2} \sin\left(\frac{\sqrt{3}}{2} \, \omega_c t\right)$$

The magnitude $|H(j\omega)|$, the phase $\theta(\omega)$, and the unit impulse response $h(t)$ are all shown in Fig. 3.27. Compare these functions with those for the ideal low-pass filter in Fig. 3.26.

In practice we can achieve filter characteristics close to ideal. The design of the Butterworth filters, for example, is obtained through a transfer function whose magnitude characteristic is given by

$$|H(j\omega)| = \frac{1}{\sqrt{1 + \left(\dfrac{\omega}{\omega_c}\right)^{2n}}} \tag{3.70}$$

These characteristics are shown in Fig. 3.28 for various values of n. Note that the

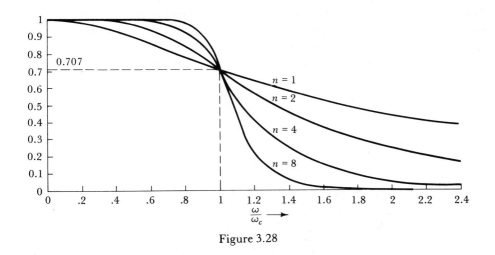

Figure 3.28

characteristics approach the ideal low-pass behavior as n increases. The bandwidth of these filters is ω_c regardless of the value of n.

For a unit bandwidth, ($\omega_c = 1$) the magnitude function in Eq. 3.70 can be realized by following transfer functions†:

---

†For more information on filters, see F. Kuo, *Network Analysis and Synthesis*, (New York: Wiley, 1966); L. Weinberg, *Network Analysis and Synthesis* (New York: McGraw-Hill, 1962).

$$n = 1, \qquad H(s) = \frac{1}{s + 1}$$

$$n = 2, \qquad H(s) = \frac{1}{s^2 + \sqrt{2}s + 1}$$

$$n = 4, \qquad H(s) = \frac{1}{(s^2 + 0.76536s + 1)(s^2 + 1.84776s + 1)} \qquad (3.71)$$

and so on.

3.16   BANDWIDTH AND RISE-TIME RELATIONSHIP

A step function is a signal of much practical significance. It occurs frequently in control systems. In many communication systems the messages consist of a sequence of rectangular pulses which can be represented as sum of step functions. Hence amplification and/or transmission of step function with reasonable fidelity is a problem of much practical significance. We have seen that the distortion in signal transmission depends upon the bandwidth of the system transmitting the signal. The larger the bandwidth, the smaller is the distortion. It is not difficult to see how a bandwidth of a system will affect transmission of a step input. In step input there is a jump discontinuity which means rapid variation in signal. This implies presence of very-high-frequency components in the step input. If these high-frequency components are suppressed because of a small bandwidth, the output will no longer show a jump discontinuity but will exhibit a gradual rise in the output. The output will rise faster (smaller rise time) for larger bandwidths. Thus some kind of inverse relationship exists between the rise time and the bandwidth. We shall now demonstrate that this is indeed true. The rise time is inversely proportional to the bandwidth of transmission.

Let us apply a unit step input to the low-pass filter in Fig. 3.26a. The output $y(t)$ is given by

$$y(t) = h(t) * u(t)$$

$$= \frac{\omega_c}{\pi} \mathrm{Sa}[\omega_c(t - t_d)] * u(t)$$

$$= \frac{\omega_c}{\pi} \int_{-\infty}^{\infty} \mathrm{Sa}[\omega_c(\tau - t_d)] u(t - \tau) \, d\tau$$

Note that $u(\cdot)$ is zero for negative arguments. Hence $u(t - \tau) = 0$ for $\tau > t$ and the upper limit of the integral may be replaced by t. Moreover $u(t - \tau) = 1$ for the remaining range of τ. Hence

$$y(t) = \frac{\omega_c}{\pi} \int_{-\infty}^{t} \mathrm{Sa}[\omega_c(\tau - t_d)] \, d\tau \qquad (3.72)$$

Note that the unit step response $y(t)$ in Eq. 3.72 is the integral of the unit impulse response $h(t)$ (Eq. 3.68). This is an expected result (see Prob. 2.30).

Letting $x = \omega_c(\tau - t_d)$ in Eq. 3.72, we obtain

$$y(t) = \frac{1}{\pi} \int_{-\infty}^{\omega_c(t - t_d)} \mathrm{Sa}(x) \, dx \qquad (3.73)$$

Unfortunately this integral cannot be evaluated in a close form. It has been computed and widely tabulated in mathematical tables.† Using this data, the step response $y(t)$ is plotted as a function of $\omega_c t$ in Fig. 3.29.   The rise time t_r is defined as the time required for the response to rise from 10 to 90% of the final value.‡ From Fig. 3.29 we read $\omega_c t_r = 2.8$. Hence

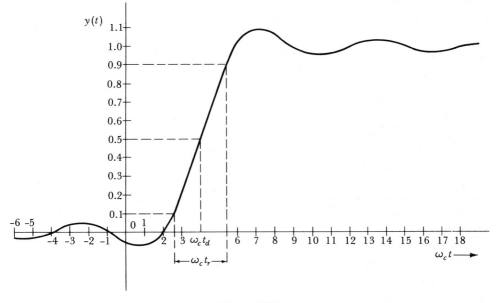

Figure 3.29

<hr />

†These tables tabulate a sine integral function $\text{Si}(t)$ defined as

$$\text{Si}(t) = \int_0^t \text{Sa}(x)\, dx$$

The step response $y(t)$ in Eq. 3.73 can be expressed in terms of $\text{Si}(t)$ as

$$y(t) = \frac{1}{\pi} [- \text{Si}(-\infty) + \text{Si}[\omega_c (t - t_d)]]$$

from integral tables, it is found that

$$\text{Si}(-\infty) = - \frac{\pi}{2}$$

$$y(t) = \frac{1}{2} + \frac{1}{\pi} \text{Si}[\omega_c (t - t_d)]$$

For tabulation of sine integral function, see E. Jahnke and F. Emde, *Table of Functions* (New York: Dover, 1945).

‡Several other definitions of rise time are used in the literature. For the case under consideration it can be shown that the rise time is inversely proportional to the bandwidth regardless of the definition of the rise time.

$$t_r = \frac{2.8}{\omega_c} \qquad\qquad (3.74)$$

Thus the rise time is inversely proportional to the bandwidth.

3.17 THE SAMPLING THEOREM

The sampling theorem has deep significance in signal theory. It states the following.

A band-limited signal which has no spectral components above a frequency f_m Hz is uniquely determined by its values at uniform intervals less than or equal to $1/2f_m$ sec apart.

This theorem is known as the *uniform sampling theorem,* since it pertains to the specifications of a given signal by its samples at uniform intervals. This condition implies that if the Fourier transform of $f(t)$ is zero beyond a certain frequency ω_m rps (or f_m Hz), then the complete information about $f(t)$ is contained in its samples spaced uniformly at a distance less than $1/2f_m$ sec. The function $f(t)$ is sampled once every T sec ($T \le 1/2f_m$ sec) or at a rate greater than or equal to $2f_m$ samples per second (Fig. 3.30). The successive samples are labeled as f_0, f_1, f_2 ..., etc. The sampling theorem states that these samples contain the information about $f(t)$ for every value of t. The sampling rate, however, must be at least twice the highest frequency f_m Hz present in the spectrum of $f(t)$. Stated another way, the signal must be sampled at least twice during each period or cycle of its highest frequency component.

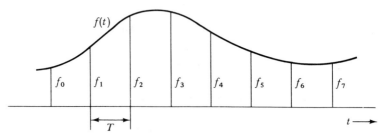

Figure 3.30

The sampling theorem can be easily proved with the help of the frequency-convolution theorem. Consider a band-limited signal $f(t)$ which has no spectral components above ω_m rps. This means $F(j\omega) = 0$ for $|\omega| > \omega_m$ as shown in Fig. 3.31b. We now multiply $f(t)$ by a periodic impulse train $\delta_T(t)$ (Fig. 3.31c). The product $f(t)\delta_T(t)$ is a sequence of impulses located at regular intervals of T seconds and having strengths equal to values of $f(t)$ at corresponding instants (Fig. 3.31e). The product signal in Fig. 3.31e represents the function $f(t)$ sampled every T sec. We shall denote this sampled signal by $f_s(t)$:

$$f_s(t) = f(t)\delta_T(t) \qquad\qquad (3.75)$$

(a)   (b)

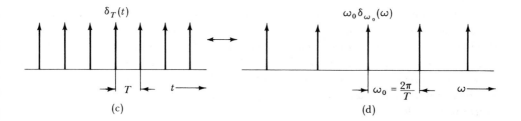

(c)   (d)

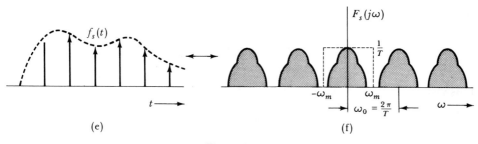

(e)   (f)

Figure 3.31

We have

$$f(t) \leftrightarrow F(j\omega) \tag{3.76}$$

Earlier we had shown (Eq. 3.47) that the Fourier transform of a uniform train of impulse function $\delta_T(t)$ is also a uniform train of impulse function $\omega_0 \delta_{\omega_0}(\omega)$ (Fig. 3.17b). The impulses have a strength of ω_0 each and are separated by a uniform interval $\omega_0 = 2\pi/T$.

$$\delta_T(t) \leftrightarrow \omega_0 \delta_{\omega_0}(\omega) \tag{3.77}$$

This is shown in Fig. 3.31c and d.

The Fourier transform of $f(t)\delta_T(t)$ will, according to frequency-convolution theorem (Eq. 3.63), be given by

$$f_s(t) = \frac{1}{2\pi}[F(j\omega) * \omega_0 \delta_{\omega_0}(\omega)]$$

Substituting $\omega_0 = 2\pi/T$, we obtain

$$f_s(t) \leftrightarrow \frac{1}{T}[F(j\omega) * \delta_{\omega_0}(\omega)] \qquad (3.78)$$

From Eq. 3.78 it is evident that the spectrum of the sampled signal $f_s(t)$ is given by the convolution of $F(j\omega)$ with a train of impulses. Functions $F(j\omega)$ and $\delta_{\omega_0}(\omega)$ can be convolved graphically by the procedure described in Chapter 2. In order to perform this operation, we fold back the function $\delta_{\omega_0}(\omega)$ about the vertical axis $\omega = 0$. Since $\delta_{\omega_0}(\omega)$ is an even function of ω, the folded function is same as the original function $\delta_{\omega_0}(\omega)$. We now superimpose the whole train of impulse on $F(j\omega)$ and progress it along the ω-axis. As each impulse passes by $F(j\omega)$, it reproduces $F(j\omega)$ by way of convolution (see Eq. 2.91c). Since the impulses are spaced at a distance ω_0, the operation of convolution yields $F(j\omega)$ repeating periodically every ω_0 rps as shown in Fig. 3.31f. The Fourier transform of $f_s(t)$ will be denoted by $F_s(j\omega)$. Note that $F_s(j\omega)$ consists of cycles of $F(j\omega)$ without overlapping as long as

$$\omega_0 \geq 2\omega_m \qquad (3.79a)$$

or

$$\frac{2\pi}{T} \geq 2(2\pi f_m)$$

This yields

$$T \leq \frac{1}{2f_m} \qquad (3.79b)$$

Therefore as long as $f(t)$ is sampled at a rate higher than $2f_m$ samples/second, $F_s(j\omega)$ will contain all the information about $F(j\omega)$. This means the sampled signal $f_s(t)$ contains all the information about $f(t)$. To recover $f(t)$ from $f_s(t)$, we allow the sampled signal $f_s(t)$ to pass through a low-pass filter which permits the transmission of all the components of frequencies below ω_m rps and attenuates all the frequency components above ω_m rps. The ideal filter characteristics required to recover $f(t)$ from $f_s(t)$ is shown dotted in Fig. 3.31f. The minimum rate of sampling is $2f_m$ samples/second and is known as the *Nyquist sampling rate* of a signal. The maximum interval of sampling is $T = 1/2 f_m$ and is known as the *Nyquist interval*.

If the samples are taken at the minimum rate (Nyquist rate), the low-pass filter must have a very sharp cutoff characteristic. This is undesirable from a practical view point. For this reason the samples are often taken at a rate much higher than the Nyquist rate. This separates the successive cycles of $F_s(j\omega)$ by a wider margin. The signal $f(t)$ can now be recovered from $f_s(t)$, using a low-pass filter with a gradual cutoff characteristic.

ALIASING ERROR

Strictly speaking, a band-limited signal does not exist in reality. It can be shown that if a signal is time-limited—that is, it exists over only a finite time interval, it contains the components of all frequencies.† However, for all practical signals, the magnitude spectrum $|F(j\omega)|$ reduces at higher frequencies. Most of the energy is carried by components lying within a certain frequency interval. The error introduced by ignoring the high-frequency tail is negligible. Thus for all practical purposes a signal can be considered to be band-limited at some value ω_m whose choice depends upon the accuracy desired.

We shall now consider the error introduced by sampling a signal below the Nyquist rate.

We have observed that if T is greater than Nyquist interval, then the successive cycles in $F_s(j\omega)$ will overlap as shown in Fig. 3.32. Note that because of the overlapping tail, $F_s(j\omega)$ no longer has the complete information about $F(j\omega)$ and it is no longer possible to recover $f(t)$ from $f_s(t)$. If the sampled signal $f_s(t)$ is passed through the low-pass filter, we shall get a spectrum which is not $F(j\omega)$ but is a distorted version because of two separate causes: (1) loss of the tail beyond $|\omega| > \omega_0/2$ and (2) addition of inverted tail at higher frequency end. This error, known as *aliasing error,* is shown shaded in Fig. 3.32.

The aliasing error can be eliminated by cutting the tail of $F(j\omega)$ beyond $|\omega| > \omega_0/2$ before it is sampled. By so doing, the overlap of successive cycles in $F_s(j\omega)$ is avoided. The only error in the recovery of $f(t)$ is that due to the missing tail beyond $|\omega| > \omega_0/2$.

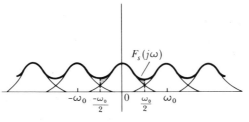

Figure 3.32

3.18   SOME APPLICATIONS OF THE SAMPLING THEOREM

The sampling theorem is an important concept, for it allows us to replace continuous band-limited signal by a discrete sequence of its samples without the loss of any information. The information content of a continuous band-

---

†This follows from the Paley-Wiener criterion (Eq. 3.69). If $F(j\omega)$ is bandlimited, $[|F(j\omega)| = 0$ for $|\omega| > \omega_m]$, then it violates Paley-Wiener condition and hence, its inverse transform $f(t)$ exists for all negative values of t. Thus a band-limited signal exists over an infinite time interval. Since all real-life signals exist over a finite time interval, they cannot be band-limited.

limited signal is thus equivalent to discrete pieces of information. Since the sampling principle specifies substantially the least number of discrete values necessary to reproduce a continuous signal, the problem of transmitting such a signal is reduced to that of transmitting a discrete number of values. Conversion of analog data into digital form (analog to digital conversion) is based on this principle. The sampling theorem opens the way for digital processing of analog signals. Sampled-data systems use samples of continuous-time signals for their operation. The sampling theorem proves most useful in the area of signal communication. Continuous-time messages are sampled, and the samples may now be transmitted instead of the continuous-time signal. While transmitting these samples the channel is occupied only over a fraction of the time; during idle periods we can transmit samples of other signals, and thus interweave samples of several signals, as shown in Fig. 3.33. In this way one can transmit several messages simultaneously over a

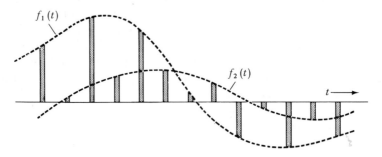

Figure 3.33. Time multiplexing of two signals.

channel. This technique of transmitting several messages simultaneously is known as *time-division multiplexing* (TDM) and is an alternative to *frequency-division multiplexing* (FDM), discussed earlier, where several messages are transmitted simultaneously through modulation (shifting frequency spectra in a nonoverlapping fashion). TDM has several practical advantages over FDM, including simplified equipment and relative immunity from interference within various messages (interchannel crosstalk). For these reasons TDM systems are being used more commonly in such applications as long-distance telephone communication.

Another interesting application is that of *pulse-code modulation* (PCM). The signal $f(t)$ to be transmitted is sampled and each sample is now approximated to the nearest allowable level, as shown in Fig. 3.34a. Here, there are 16 levels each separated by a value of 0.1. This process, known as *quantization,* reduces the possible number of values to be transmitted to a finite number (16 in Fig. 3.34a). Each of the 16 numbers can be represented by a certain pattern of pulses, as shown in Fig. 3.34b. Thus instead of transmitting the samples we transmit the corresponding pulse pattern (pulse code). The most important advantage of pulse-code modulation is its immunity to noise. To detect PCM all that is required is to distinguish between the presence and the absence of the pulse. PCM is therefore relatively immune to interference and distortion to unwanted noise over the

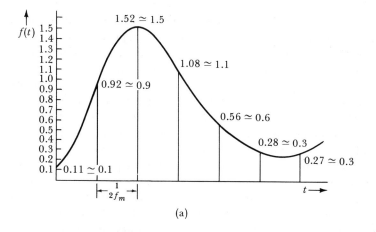

(a)

Digit	Binary equivalent	Pulse-code waveform
0	0000	
1	0001	
2	0010	
3	0011	
4	0100	
5	0101	
6	0110	
7	0111	
8	1000	
9	1001	
10	1010	
11	1011	
12	1100	
13	1101	
14	1110	
15	1111	

(b)

Figure 3.34

channel. This is a serious problem in other modes of communication, such as AM (discussed earlier). The channel noise distorts the received signal and there is no way of getting rid of this disturbance. If the noise is too large, the received signal may be worthless.

The second advantage of PCM lies in its ability to communicate over very long distances. As is well known, transmitted messages become weaker as they travel over a channel, and hence are periodically amplified at intermediate repeater stations. Unfortunately the noise also gets amplified, and is acquired over the entire length of the channel with cumulative effect. As a result of amplification, the noise level may become so high as to make the desired signal unintelligible.

This condition sets a limit to the distance over which continuous-time messages may be transmitted. For PCM, however, at a repeater station the pulses contaminated with noise can be detected and new clean pulses generated and transmitted to the next repeater station, where the same process is repeated. This cleansing and regeneration operation is not possible for other modes. For this reason data can be transmitted over a much larger distance with PCM.

The only error introduced in PCM is that due to quantization. This error can however be reduced as much as desired by increasing quantizing levels. The price for these advantages is paid in terms of increased bandwidth. This is because for each sample we are now transmitting more than one piece of data. For the code in Fig. 3.34b we are transmitting four pieces of information (four pulses) for each sample. The sampling theorem says that a channel of bandwidth f_m Hz can at most transmit $2f_m$ independent pieces of information. For the scheme in Fig. 3.34, we will need four times as much bandwidth for transmission as that needed to transmit the signal $f(t)$ directly.

SAMPLING THEOREM (FREQUENCY DOMAIN)

The sampling theorem in time domain has a dual, which states that a time-limited signal which is zero for $|t| > T$ is uniquely determined by the samples of its frequency spectrum at uniform intervals less than $1/2T$ Hz (or π/T rps) apart.

The proof of the theorem is left an exercise for the student. It is similar to that of the sampling theorem in time domain with the roles of $f(t)$ and $F(j\omega)$ reversed.

3.19   LIMITATIONS OF FOURIER ANALYSIS

Representation of signals by their frequency spectrum (Fourier transform) is a valuable tool, particularly in the analysis and processing of signals. In the area of systems analysis, however, it leaves something to be desired. The Fourier transform exists for a rather restricted class of signals. Admittedly this includes most of the signals of practical importance, but it also leaves out some important signals such as those whose amplitudes grow exponentially. Secondly, since Fourier transform expresses signals in terms of exponentials along the $j\omega$-axis, it can satisfy the dominance condition only for stable systems. For unstable systems, some of the transfer function poles lie in RHP. To satisfy dominance condition for unstable system, the signal, therefore, must be represented in terms of exponentials in RHP. In short, it is necessary to be able to express signal $f(t)$ in terms of generalized exponentials e^{st} where s is complex. This is done by generalization of complex frequency from $j\omega$ in the Fourier transform to $\sigma + j\omega$ in the Laplace transform. This is the subject of the next chapter.

PROBLEMS

3.1.   Locate the complex frequencies of the following signals in the s-plane (complex-frequency plane).

(a) e^{2t} (b) e^{-2t} (c) $\cos 2t$ (d) $\sin 2t$

(e) $e^{-t}\sin 5t$ (f) $e^{-t}\cos 5t$ (g) $e^{-t}\sin(-5t)$ (h) $e^{t}\cos(5t + \theta)$

3.2.   Write the general form of and sketch the signals whose complex frequencies are given by

(a) -1 (b) 2,   (c) $-1 + j2$ (d) $j3$ (e) $-j5$ (f) $1 - j2$

3.3.   Find the dominance condition for systems shown in Fig. 3.4.

3.4.   The transfer function of a system is given as

$$H(s) = \frac{1}{s + 5}$$

Find the response of this system to following eternal exponential inputs.

(a) e^{-4t} (b) e^{2t} (c) e^{-6t}

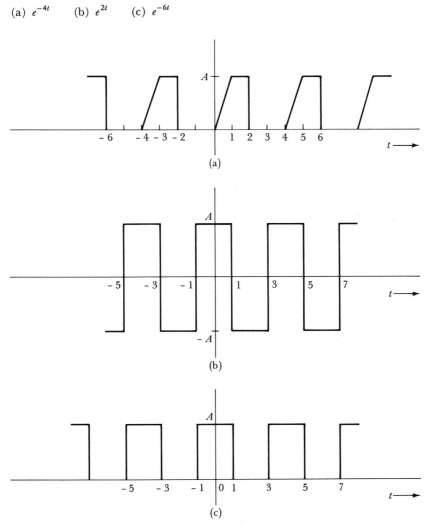

(a)

(b)

(c)

Figure P-3.6

3.5.   State with reasons if the following functions are periodic or nonperiodic. In case of periodic functions, find the period.

(a) $a \sin t + b \sin 3t$

(b) $a \sin 4t + b \cos 7t$

(c) $a \sin 3t + b \cos \pi t$

(d) $a \cos \pi t + b \sin 2\pi t$

(e) $a \sin \dfrac{5t}{2} + b \cos \dfrac{6t}{5} + c \sin \dfrac{t}{7}$

(f) $(a \sin 2t)^3$

(g) $(a \sin 2t + b \sin 5t)^2$

3.6.   Find the Fourier series and sketch the fourier spectrum of the periodic functions shown in Fig. P-3.6.

3.7.   A periodic input $f(t)$ shown in Fig. P-3.6b is applied to the input of a system whose transfer function is

$$H(s) = \frac{s}{s^2 + 2s + 3}$$

Find the response $y(t)$.

3.8.   If $F(j\omega)$ is the Fourier transform of $f(t)$, then show that when $f(t)$ is an even function of t,

$$F(j\omega) = 2 \int_0^\infty f(t) \cos \omega t \, dt$$

and when $f(t)$ is an odd function of t,

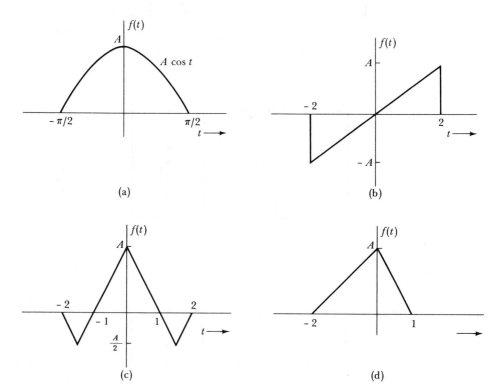

(a)

(b)

(c)

(d)

Figure P-3.9

$$F(j\omega) \; = \; -2j \int_0^{\infty} f(t) \sin \omega t \; dt$$

3.9.   Find Fourier transforms of functions shown in Fig. P-3.9.

3.10.   Find inverse Fourier transforms of functions shown in Fig. P-3.10.

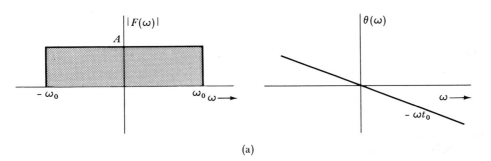

(a)

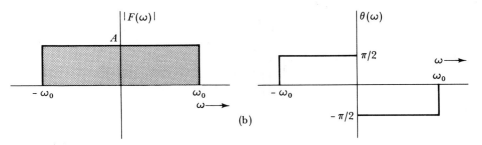

(b)

Figure P-3.10

3.11.   Determine the Fourier transform of $e^{at} u(-t)$ for $a > 0$.   (*Hint:* Use scaling property in Eq. 3.53 with $a = -1$.)

3.12.   Using the Fourier transform, determine the zero-state response of a system with transfer function

$$H(s) \; = \; \frac{s + 3}{s^2 + 3s + 2}$$

and the input $f(t)$ is given by

$$f(t) \; = \; e^{-4t} u(t)$$

3.13.   Repeat Prob. 3.12 if

$$f(t) \; = \; \cos 3t$$

3.14.   Repeat Prob. 3.12 for

$$f(t) \; = \; e^{2t} u(-t)$$

Give rough sketches of the input and the response.

3.15.   Repeat Prob. 3.12 for

$$f(t) = e^{-4t} u(t) + e^{2t} u(-t)$$

Give rough sketches of the input and the response.

3.16.   Repeat Prob. 3.12 for

$$H(s) = \frac{1 - s}{1 + s} \qquad f(t) = e^t u(-t)$$

Sketch the input and the output.

3.17.   Repeat Prob. 3.12 for

$$H(s) = \frac{1}{s + 1} \qquad f(t) = (3e^{2t} - 2e^{2t}) u(-t)$$

Give rough sketches of the input and the output.

3.18.   Repeat Prob. 3.12 for

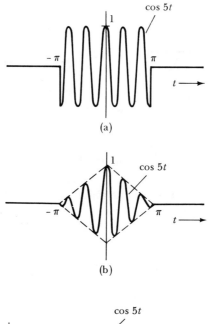

(a)

(b)

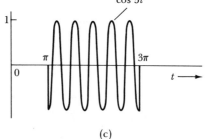

(c)

Figure P-3.22

$$H(s) = \frac{s^2 + 4s + 5}{s^2 + 3s + 2} \qquad f(t) = e^{-3t} u(t)$$

3.19. Repeat Prob. 3.12 for

$$H(s) = \frac{s + 3}{s^2 + 3s + 2} \qquad f(t) = u(t) - u(t - 1)$$

Sketch the input and the output.

3.20. Using the convolution property (Eq. 3.62) prove the results for pairs 1, 2, 4, and 5 in the convolution Table 2.2.

3.21. Derive the results of pair 11 in Table 3.1 from pair 1 (Table 3.1) and the modulation theorem.

3.22. Find and sketch Fourier transforms of signals shown in Fig. P-3.22.

3.23. A modulated signal $f(t) \cos \omega_0 t$ can be obtained by multiplying $f(t)$ by $\cos \omega_0 t$. It is, however, not necessary to have a pure sinusoidal signal. One can obtain a modulated signal $f(t) \cos \omega_0 t$ by multiplying $f(t)$ by any periodic signal of fundamental frequency ω_0. The wave shape is immaterial. Devise a scheme to accomplish this. (*Hint:* Express the periodic signal by a Fourier series.)

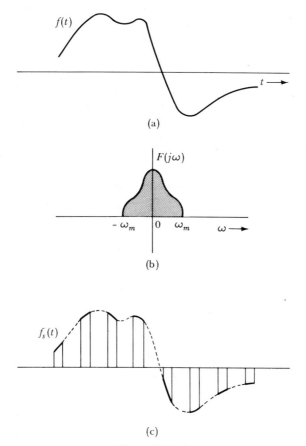

(a)

(b)

(c)

Figure P-3.26

3.24.   For the low-pass filter shown in Fig. 3.27a, the magnitude and phase characteristics are shown in Fig. 3.27b.

    (a) What is the nature of the signals that can be transmitted by this filter with reasonable fidelity?

    (b) For such signals, give the approximate value of the transmission delay t_d if $\omega_c = 1000$ in Fig. 3.27b. (*Hint:* $t_d = d\theta/d\omega$. Note that Fig. 3.27b gives θ, not as a function of ω, but as a function of ω/ω_c.)

3.25.   Repeat Prob. 24 for the low-pass filter whose magnitude and phase characteristics are shown in Fig. 2.21.

3.26.   In the text we considered an ideal sampler where the output was a sequence of impulses of strength equal to the corresponding sample values. In practice this is not possible. A sampler is a gate circuit which opens for a brief interval periodically. A function $f(t)$ and the sample signal $f_s(t)$ of a practical sampler are shown in Fig. P-3.26a and c. The Fourier transform of $f(t)$ is shown in Fig. P-3.26b.

    (a) Find the Fourier transform of the sampled signal $f_s(t)$. The sampling pulses are 0.1 μsec wide and the sampling rate is 10,000 samples/sec, and $\omega_m = 30{,}000$ rps.

    (b) Prove that the sampling theorem is valid even for the nonideal sampling scheme such as the one discussed in this problem. [*Hint for Part (a):* The sampled signal $f_s(t)$ may be considered as a product $f(t)p(t)$, where $p(t)$ is a rectangular pulse train of period 10^{-4} sec and pulse width 10^{-7} sec. Now use the frequency convolution theorem.]

3.27.   Show that the AM signal in Fig. 3.23c can be demodulated by the scheme shown in Fig. P-3.27. This arrangement yields the signal proportional to the envelope of the input signal. (*Hint:* A rectifier operation may be considered as multiplication of the input signal by a square wave of fundamental frequency ω_0. For full-wave rectifier, the square wave has a zero dc level and for a half-wave rectifier, the square wave has a dc level as shown in Fig. P-3.6c.)

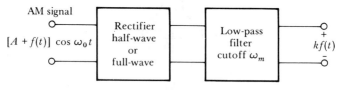

Figure P-3.27

3.28.   Fourier transform of a signal $f(t)$ is shown in Fig. P-3.28.

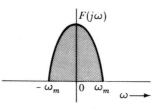

Figure P-3.28

(a) Give a crude sketch of the Fourier transform of $f^2(t)$ and $f^3(t)$. The sketch need not be very accurate. The important points are the spread of the spectra of $f^2(t)$ and $f^3(t)$.

(b) The input f and the output y of a nonlinear device can be described by a series

$$y = a_1 f + a_2 f^2 + a_3 f^3 + \cdots$$

Explain what happens to the signal spectrum when it passes through such a nonlinear device.

3.29.   A zero-order hold circuit shown in Fig. P-3.29 is often used to recover $f(t)$ from its samples. Find the unit impulse response of this circuit and from this information determine

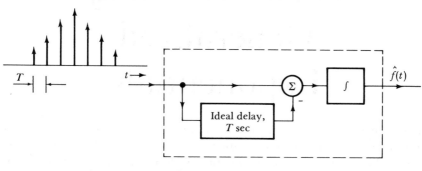

Figure P-3.29

(a) The output $f(t)$ when the input is a sampled signal (train of impulses) as shown in the figure.

(b) The magnitude spectrum $|H(j\omega)|$ for this circuit. Comment on the filter characteristics of this circuit.

3.30.   If a signal $f(t)$ is sampled at a Nyquist rate $(2f_m$ samples/sec), show that $f(t)$ can be expressed in terms of its samples as

$$f(t) = \sum_k f(kT) Sa(\omega_m t - k\pi)$$

where the sampling interval $T = 1/2f_m = \pi/\omega_m$. [*Hint:* $f(t)$ is obtained by passing the sampled signal through a low-pass filter of cutoff frequency ω_m. Hence $f(t)$ is obtained by convolving $f_s(t)$ with $h(t)$ of the low-pass filter.]

4

Frequency-Domain Analysis Using Generalized Exponentials

The Fourier transform is a tool which allows us to represent a function $f(t)$ by a continuous sum of exponential functions of the form $e^{j\omega t}$. The complex frequencies of these exponentials are restricted along the $j\omega$-axis in the complex-frequency plane. In general, however, it is much desirable to remove this restriction and to represent $f(t)$ in terms of exponentials of the form e^{st}, where $s = \sigma + j\omega$. This can be done by extending our earlier development for $s = j\omega$ to the case $s = \sigma + j\omega$. We will now show that such an extension leads us to representation of $f(t)$ as

$$f(t) = \frac{1}{2\pi j} \int_{\sigma-j\infty}^{\sigma+j\infty} F(s)\, e^{st} ds \tag{4.1}$$

where

$$F(s) = \int_{-\infty}^{\infty} f(t)\, e^{-st} dt \tag{4.2}$$

and

$$s = \sigma + j\omega$$

4.1 GENERALIZATION OF FREQUENCY: THE BILATERAL LAPLACE TRANSFORM

Let us begin with the Fourier transform pair

$$F(j\omega) = \int_{-\infty}^{\infty} f(t)\, e^{-j\omega t} dt \tag{4.3}$$

and

$$f(t) = \frac{1}{2\pi} \int_{-\infty}^{\infty} F(j\omega)\, e^{j\omega t} d\omega \tag{4.4}$$

168

Let a function $\phi(t)$ be defined as

$$\phi(t) = f(t) e^{-\sigma t} \tag{4.5}$$

where σ is real constant. Then

$$\mathcal{F}[\phi(t)] = \int_{-\infty}^{\infty} f(t) e^{-\sigma t} e^{-j\omega t} dt$$

$$= \int_{-\infty}^{\infty} f(t) e^{-(\sigma+j\omega)t} dt \tag{4.6}$$

It is evident from Eq. 4.3 that the above integral is $F(\sigma + j\omega)$. Thus

$$\mathcal{F}[\phi(t)] = F(\sigma + j\omega) \tag{4.7}$$

Hence

$$\phi(t) = \frac{1}{2\pi} \int_{-\infty}^{\infty} F(\sigma + j\omega) e^{j\omega t} d\omega \tag{4.8}$$

Substituting $\phi(t) = f(t) e^{-\sigma t}$ in Eq. 4.8, we obtain

$$f(t) e^{-\sigma t} = \frac{1}{2\pi} \int_{-\infty}^{\infty} F(\sigma + j\omega) e^{j\omega t} d\omega$$

Therefore

$$f(t) = \frac{1}{2\pi} \int_{-\infty}^{\infty} F(\sigma + j\omega) e^{(\sigma+j\omega)t} d\omega \tag{4.9}$$

The quantity $(\sigma + j\omega)$ is the complex frequency s and $d\omega = (1/j)ds$. The limits of integration for $\omega = -\infty$ to ∞ becomes $(\sigma - j\infty)$ to $(\sigma + j\infty)$ for variable s. Hence Eq. 4.9 becomes

$$f(t) = \frac{1}{2\pi j} \int_{\sigma-j\infty}^{\sigma+j\infty} F(s) e^{st} ds \tag{4.10a}$$

From Eqs. 4.6 and 4.7, we obtain

$$F(s) = \int_{-\infty}^{\infty} f(t) e^{-st} dt \tag{4.10b}$$

Equation 4.10a expresses $f(t)$ as a continuous sum of exponentials of complex frequency $s = \sigma + j\omega$. Note that the Fourier transform is just a special case of this, and can be obtained by letting $\sigma = 0$—that is, $s = j\omega$ in Eqs. 4.10a and 4.10b. The pair of equations 4.10a and 4.10b is called the *complex Fourier transform pair* or the *bilateral Laplace transform pair*.† It should be mentioned here that some authors use $\omega + j\sigma$ instead of $\sigma + j\omega$ as the generalized frequency variable.

The complex Fourier transform (or the bilateral Laplace transform) will be denoted symbolically as

$$F(s) = \mathcal{F}_c[f(t)]$$

---

†Also known as a *two-sided Laplace transform pair*.

and

$$f(t) = \mathfrak{F}_c^{-1}[F(s)]$$

Note that the complex transform $F(s)$ of $f(t)$ can be found by substituting s for $j\omega$ in its Fourier transform.† The Fourier transform discussed thus far $(s = j\omega)$ is also sometimes called an *ordinary Fourier transform* in contrast to a *complex Fourier transform* $(s = \sigma + j\omega)$.

As an example, consider the function $e^{-at}u(t)$. The Fourier transform of this function is given by (see Table 3.1)

$$\mathfrak{F}[e^{-at}u(t)] = \frac{1}{a + j\omega} \qquad (4.11)$$

The complex Fourier transform of $e^{-at}u(t)$ may be obtained directly from Eq. 4.10b, or may be found by replacing $j\omega$ by s in Eq. 4.11. Thus

$$\mathfrak{F}_c[e^{-at}u(t)] = \frac{1}{a + s}$$

4.2   CAUSAL FUNCTIONS AND UNILATERAL FREQUENCY TRANSFORMS

Thus far we have assumed that a general time function $f(t)$ exists in the entire interval $(-\infty < t < \infty)$. In practice, however, there is a large class of functions which start at some finite time, say $t = 0$, and have zero value for $t < 0$. Such functions are called *causal functions*. Henceforth we shall call a function causal if it is zero for negative values of t:

$$f(t) = 0 \qquad t < 0$$

For all of the physical systems, if the driving function is zero for the negative values of t, the response must also be zero for the negative values of t, since the system cannot anticipate the driving function. It therefore follows that for physical systems the response function of the causal excitation function must also be causal function.

Most of the analysis problems in practice involve causal functions. A system is generally excited at some finite time $t = 0$. We are therefore interested in the time interval $0 < t < \infty$ instead of the entire interval $-\infty < t < \infty$. The causal functions effect a great deal of simplification in the frequency transforms. When the frequency transforms are specialized for causal functions, they are called *unilateral transforms* (also known as *single-sided* or *right-handed transforms*). The bilateral Laplace transform, when specialized for causal functions, is known as the *unilateral Laplace transform*. It is also known as the *single-sided* or *right-handed Laplace transform*. Henceforth we shall refer to this transform simply as the *Laplace transform*.

---

†A note of caution is in order here. The complex Fourier transform can be found from an ordinary Fourier transform by replacing $j\omega$ by s only for absolutely integrable functions. This procedure cannot be applied for those functions whose Fourier transform exists in the limit (e.g., $u(t)$, $\sin \omega_0 t\, u(t)$, $\cos \omega_0 t\, u(t)$, etc.). For these functions we should evaluate the transform directly by using Eq. 4.10b.

For causal functions, the transform equations 4.10 reduce to

$$F(s) = \int_0^\infty f(t) e^{-st} dt \tag{4.12a}$$

$$f(t) = \frac{1}{2\pi j} \int_{\sigma-j\infty}^{\sigma+j\infty} F(s) e^{st} ds \tag{4.12b}$$

Equations 4.12a and 4.12b form a Laplace transform pair. Equation 4.12a represents the direct Laplace transform of $f(t)$, and Eq. 4.12b forms the inverse Laplace transform of $F(s)$. Symbolically, this pair is written as

$$F(s) = \mathcal{L}[f(t)]$$
$$f(t) = \mathcal{L}^{-1}[F(s)]$$

The lower limit of integration in Eq. 4.12a is 0. If $f(t)$ contains an impulse function or its derivatives at the origin, $f(t)$ is undefined at $t = 0$. To make sure that the impulse or its higher derivatives are included, we must choose the lower limit of 0^-. If the lower limit is chosen as 0^+, the impulse and its derivatives that are present, if any, at the origin will be ignored. Hence in general the lower limit should be taken as 0^-. Henceforth the lower limit in the integral of Eq. 4.12a will be tacitly assumed to be 0^-.

4.3   EXISTENCE OF THE LAPLACE TRANSFORM

For the Laplace transform to exist, the first condition is that the integral in Eq. 4.12a must exist. That is,

$$\int_0^\infty f(t) e^{-st} dt$$

must be finite. This integral will be finite, provided that

$$\int_0^\infty |f(t)| e^{-\sigma t} dt \quad < \infty \tag{4.13}$$

For a function $f(t)$, if real positive finite numbers M and α exist so that

$$|f(t)| < M e^{\alpha t} \tag{4.14}$$

for all positive t, $f(t)$ is said to be an exponential-order function or an E-function. The inequality 4.13 is always satisfied for an exponential-order function. This can be easily verified by substituting the inequality 4.14 in the inequality 4.13. The integral

$$\int_0^\infty |f(t)| e^{-\sigma t} dt < \int_0^\infty M e^{\alpha t} e^{-\sigma t} dt = \int_0^\infty M e^{(\alpha-\sigma)t} dt$$

$$= \frac{M}{\alpha - \sigma} e^{(\alpha-\sigma)t} \Big|_0^\infty$$

$$= \frac{M}{\sigma - \alpha} \qquad \sigma > \alpha$$

Thus by choosing $\sigma > \alpha$, the integral can be made absolutely convergent.

The exponential-order function may be defined alternately as follows: a function $f(t)$ is of exponential order if there exists a real positive number α such that

$$\lim_{t \to \infty} [e^{-\alpha t} f(t)] = 0 \qquad (4.15)$$

It can be easily shown that if $f(t)$ satisfies the condition of Eq. 4.15, then

$$\int_0^\infty |f(t)| e^{-\sigma t} dt$$

is finite for $\sigma > \alpha$. The region of convergence is therefore given by

$$\sigma > \alpha \qquad (4.16)$$

The Laplace transform converges for values of s given by

$$\mathrm{Re}\, s = \sigma > \alpha$$

where α satisfies the condition of Eq. 4.15.

The minimum value of α that satisfies the condition of Eq. 4.15 is known as the *abscissa of absolute convergence* for a given $f(t)$.

As an example, consider a unit step function $u(t)$. For this function

$$\lim_{t \to \infty} [u(t) e^{-\alpha t}] = 0$$

for any value of $\alpha > 0$. Hence the region of convergence of the Laplace transform of $u(t)$ is given by $\sigma > 0$. Similarly, it can be shown that the region of convergence for a function $e^{-at} u(t)$ is $\sigma > \alpha$.

The usefulness of the frequency-analysis technique depends upon whether or not all of the functions encountered in practice possess a Laplace transform. It can be stated here affirmatively that every physically possible signal is Laplace-transformable. Saying that a signal should be of exponential order is another way of saying that $f(t)$ should not grow more rapidly than $Me^{\alpha t}$ for any positive real M and α. There are some functions that do grow faster than $Me^{\alpha t}$; for example, e^{t^2} or t^t are functions that do not possess Laplace transform. But such functions are of no practical significance.

4.4 INTERPRETATION OF THE LAPLACE TRANSFORM

The ordinary Fourier transform is a tool for expressing a function $f(t)$ as a continuous sum of exponential functions of frequencies lying along the $j\omega$-axis:

$$f(t) = \frac{1}{2\pi} \int_{-\infty}^{\infty} F(j\omega) e^{j\omega t} d\omega$$

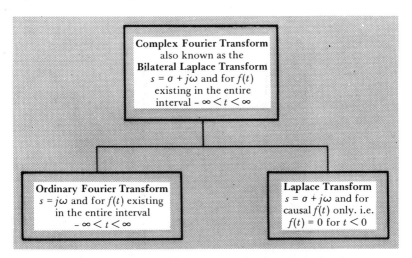

Figure 4.1 The ordinary Fourier transform and the Laplace transform as special cases of the complex Fourier transform.

Generalization of the ordinary Fourier transform leads to the bilateral Laplace transform by means of which it is possible to express a function $f(t)$ as a continuous sum of exponential functions of complex frequencies.

$$f(t) = \frac{1}{2\pi j} \int_{\sigma - j\infty}^{\sigma + j\infty} F(s)\, e^{st}\, ds$$

The above expression represents $f(t)$ as a continuous sum of exponential functions whose frequencies lie along the path $\sigma + j\omega$ for $\omega = -\infty$ to ∞. An ordinary Fourier transform is just a special case of the bilateral Laplace transform when $\sigma = 0$ or $s = j\omega$. The Laplace transform is really a bilateral Laplace transform, but it deals only with causal functions—those that have zero value for $t < 0$. Hence the Laplace transform is also a special case of the bilateral Laplace transform. The relationship between various transforms is shown in Fig. 4.1. The Laplace transform therefore expresses a causal function $f(t)$ as a continuous sum of exponentials of complex frequencies. This point will now be illustrated by an example.

Consider a function $e^{-at}u(t)$. For this function

$$F(s) = \int_{0}^{\infty} e^{-at}u(t)\, e^{-st}\, dt$$

$$= \frac{-1}{s + a} e^{-(s+a)t} \Big|_{0}^{\infty}$$

$$= \frac{-1}{s + a} e^{-(\sigma + j\omega + a)t} \Big|_{0}^{\infty}$$

$$= \frac{1}{s + a} \quad \begin{array}{l} \text{if } (\sigma + a) > 0 \\ \text{i.e., if } \sigma > -a \end{array} \tag{4.17}$$

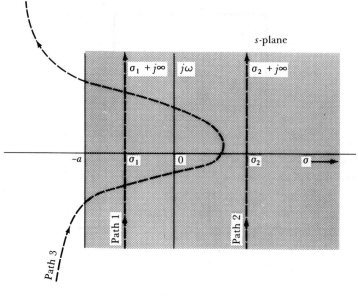

Figure 4.2

Note that if $(\sigma + a) < 0$, then the value of the above integral goes to infinity and the transform does not exist. Hence the region of convergence is given by Re $s > -a$. The region in the s plane where Re $s > -a$ is shown in Fig. 4.2. We can thus express the function $e^{-at}u(t)$ as a continuous sum of complex exponentials:

$$e^{-at}u(t) = \frac{1}{2\pi j} \int_{\sigma-j\infty}^{\sigma+j\infty} \frac{1}{s+a} e^{st}\, ds \qquad (4.18)$$

The function $1/(s + a)$ represents the relative magnitudes of the frequency components. The integral represents the continuous summation of exponential functions whose frequencies lie along the path $\sigma + j\omega$ from $\omega = -\infty$ to $\omega = \infty$. How shall we choose this path? From the above discussion it may appear that the path should lie in the region of convergence. This conclusion, however, needs some qualifications. It is sufficient to choose the entire path of integration in the region of convergence (Paths 1 and 2 in Fig. 4.2). However, it is not necessary for the entire path to lie in the region of convergence. It can be shown from the properties of analytic functions that the only requirement is that the path of integration lie to the right of the singularity points (poles) of $F(s)$. In the present case the singularity is at $s = -a$, and any path to the right of $s = -a$ (Path 3, for example) is permissible.

It is evident that there are infinite possible paths of integration of which three are shown in Fig. 4.2. The choice of Path 1 is equivalent to expressing $f(t)$ as a continuous sum of exponentially decaying sinusoids, since all the frequencies on this path lie in the LHP. On the other hand, Path 2 lies entirely in the RHP and hence this choice yields the representation of $f(t)$ as a continuous sum of exponentially growing sinusoids. Path 3 lies both in the LHP and RHP and hence corresponds to the representation of $f(t)$ as a continuous sum of exponentially growing and decaying sinusoids. The integral on the right-hand side of Eq. 4.18 yields the same result, namely $e^{-at}u(t)$, when integrated along any of these paths.

Note that we may just as well have chosen the path along the $j\omega$-axis. The choice of this path is equivalent to expressing $f(t)$ as a continuous sum of exponentials of the form $e^{j\omega t}$ (along the $j\omega$-axis). This is exactly what a Fourier transform accomplishes. Hence the special case when $s = j\omega$ reduces the Laplace transform to a Fourier transform. It is important to observe that there is a lower bound on the value of σ (in this case $-a$), but there is no upper bound. We can choose σ as high as we wish (as long as it is greater than the lower bound σ_m). Thus a signal can always be expressed as a continuous sum of exponentially growing sinusoids. Moreover, we may choose these sinusoids to grow as rapidly as we wish.

One more important point needs clearing up here. All of the frequency transforms discussed thus far express a function as a continuous sum of exponential functions which start at $t = -\infty$ (eternal exponentials). The Laplace transform therefore expresses a causal function $[f(t) = 0, t < 0]$ as a continuous sum of eternal exponential functions (starting at $t = -\infty$). This may appear rather surprising, since $f(t)$ is zero for negative values of t. It just so happens that all these eternal exponential components add in such a way as to yield a zero value for $t < 0$ and yield the desired function $f(t)$ for $t > 0$. To reiterate: the Laplace transform is a tool with which we can express a causal function as a continuous sum of eternal exponential functions of complex frequencies. The continuous sum is expressed by the integral

$$f(t) = \frac{1}{2\pi j} \int_{\sigma - j\infty}^{\sigma + j\infty} F(s) e^{st} ds$$

The function $F(s)$ represents the relative amplitude of the component of frequency s.

The inverse transform is obtained by evaluating the complex integral in Eq. 4.12b. The evaluation of such an integral requires a familiarity with the functions of complex variables in general and calculus of the residues in particular.† The integration is performed along a path from $(\sigma - j\infty)$ to $(\sigma + j\infty)$. Fortunately, however, the need for complex integration is obviated by using tables of the Laplace transforms. Standard tables of the Laplace transforms are available which list the direct and inverse transforms of a large number of functions encountered in practice. One column lists the function $f(t)$, and the other column gives the corresponding $F(s)$. The inverse transform of a function $F(s)$ can therefore be found directly from these tables, thus obviating the need for complex integration. Similarly, the direct transform $F(s)$ of a function $f(t)$ can be found from this table. A brief table is provided in Table 4.1.

---

†W. R. LePage, *Complex Variables and the Laplace Transform for Engineers* (New York: McGraw-Hill, 1961).

TABLE 4.1
Table of Laplace Transforms

$f(t) = \mathcal{L}^{-1}[F(s)]$	$F(s) = \mathcal{L}[f(t)]$
1. $u(t)$	$\dfrac{1}{s}$
2. $tu(t)$	$\dfrac{1}{s^2}$
3. $t^n u(t)$	$\dfrac{n!}{s^{n+1}}$
4. $e^{\lambda t} u(t)$	$\dfrac{1}{s - \lambda}$
5. $te^{\lambda t} u(t)$	$\dfrac{1}{(s - \lambda)^2}$
6. $t^n e^{\lambda t} u(t)$	$\dfrac{n!}{(s - \lambda)^{n+1}}$
7. $\sin \beta t\, u(t)$	$\dfrac{\beta}{s^2 + \beta^2}$
8. $\cos \beta t\, u(t)$	$\dfrac{s}{s^2 + \beta^2}$
9. $e^{\alpha t} \sin \beta t\, u(t)$	$\dfrac{\beta}{(s - \alpha)^2 + \beta^2}$
10. $e^{\alpha t} \cos \beta t\, u(t)$	$\dfrac{s - \alpha}{(s - \alpha)^2 + \beta^2}$
11. $2re^{\alpha t} \cos(\beta t + \phi)\, u(t)$	$\dfrac{re^{j\phi}}{s - \alpha - j\beta} + \dfrac{re^{-j\phi}}{s - \alpha + j\beta}$
12. $\dfrac{1}{\omega_n \sqrt{1 - \zeta^2}}\, e^{-\zeta \omega_n t} \sin(\omega_n \sqrt{1 - \zeta^2})t$	$\dfrac{1}{s^2 + 2\zeta\omega_n s + \omega_n^2}$
13. $\delta(t)$	1

In Table 4.1, note that every $f(t)$ is multiplied by $u(t)$. This is somewhat redundant for the Laplace transform since, by the very definition of this transform, it is understood that only causal $f(t)$ is used. Nevertheless this practice avoids certain unnecessary errors.

4.5   TRANSFORMS OF SOME USEFUL FUNCTIONS

Let us find Laplace transforms of some functions that are commonly encountered in practice. The exponential function starting at $t = 0$ is a very important one.

$$f(t) = e^{-at}u(t)$$

$$F(s) = \mathcal{L}\left[f(t)\right] = \int_0^\infty e^{-at}u(t)\,e^{-st}\,dt$$

$$= \int_0^\infty e^{-(s+a)t}\,dt$$

$$= \frac{1}{s+a}$$

Therefore

$$\mathcal{L}\left[e^{-at}u(t)\right] = \frac{1}{s+a} \tag{4.19a}$$

$$\mathcal{L}^{-1}\left[\frac{1}{s+a}\right] = e^{-at}u(t) \tag{4.19b}$$

Equation 4.19b also implies that

$$e^{-at}u(t) = \frac{1}{2\pi j}\int_{\sigma-j\infty}^{\sigma+j\infty} \frac{1}{s+a} e^{st}\,ds$$

A unit step function is a special case of $e^{-at}u(t)$ obtained by letting $a = 0$. From Eqs. 4.19 we can write

$$\mathcal{L}\left[u(t)\right] = \frac{1}{s}$$

$$\mathcal{L}^{-1}\left[\frac{1}{s}\right] = u(t)$$

A sinusoidal function can be written as a sum of two exponential functions:

$$\sin\beta t = \frac{e^{j\beta t} - e^{-j\beta t}}{2j}$$

$$\mathcal{L}\left[\sin\beta t\, u(t)\right] = \int_0^\infty \frac{e^{j\beta t} - e^{-j\beta t}}{2j} e^{-st}\,dt$$

$$= \int_0^\infty \frac{e^{-(s-j\beta)t} - e^{-(s+j\beta)t}}{2j}\,dt$$

$$= \frac{1}{2j}\left(\frac{1}{s-j\beta} - \frac{1}{s+j\beta}\right)$$

$$= \frac{\beta}{s^2 + \beta^2}$$

Similarly, it can be shown that

$$\mathcal{L}\left[\cos\beta t\, u(t)\right] = \frac{s}{s^2 + \beta^2}$$

4.6   SYSTEM ANALYSIS USING GENERALIZED EXPONENTIALS

Consider a system response $y(t)$ which is related to the input $f(t)$ by transfer function $H(s)$. The input $f(t)$ can be expressed as a continuous sum of exponentials as

$$f(t) = \frac{1}{2\pi j} \int_{\sigma-j\infty}^{\sigma+j\infty} F(s)\, e^{st}\, ds \tag{4.20}$$

Since the zero-state response of the system to input e^{st} is $H(s)e^{st}$, it follows that the (zero-state) response to input $f(t)$ will be given by the continuous sum of responses to all exponential components in Eq. 4.20. Hence

$$y(t) = \frac{1}{2\pi j} \int_{\sigma-j\infty}^{\sigma+j\infty} H(s)\, F(s)\, e^{st}\, ds$$

$$= \mathcal{L}^{-1}[F(s)\, H(s)]$$

If

$$\mathcal{L}[y(t)] = Y(s)$$

then

$$Y(s) = H(s)\, F(s) \tag{4.21}$$

This relationship is shown in Fig. 4.3a.

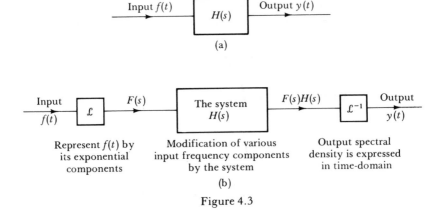

(a)

Represent $f(t)$ by its exponential components

Modification of various input frequency components by the system

Output spectral density is expressed in time-domain

(b)

Figure 4.3

At this point it is helpful to point out the difference between equations

$$y(t) = H(p)\, f(t) \tag{4.22a}$$

and

$$Y(s) = H(s)\, F(s) \tag{4.22b}$$

The first equation is a time-domain relationship between input and the output and in general represents a differential equation. We repeat this here again. Equation 4.22a is not an algebraic equation but a differential equation in general. On the other hand, Eq. 4.22b is an algebraic equation expressing the relationship between the Laplace transforms of the input and the output. It expresses the relationship between the input and the output in the frequency domain. The procedure for finding the zero-state response of a system to input $f(t)$ can be written down in following simple steps.

1. Evaluate $H(s)$, the transfer function relating the response variable $y(t)$ to the input $f(t)$.
2. Find the $F(s)$, the Laplace transform of the input $f(t)$ from the table of transforms.
3. The zero-state response $y(t)$ is given by the inverse Laplace transform of $H(s)F(s)$.

$$y(t) = \mathcal{L}^{-1}[H(s)F(s)]$$

From the table of transforms, find the inverse transform of $H(s)F(s)$. The procedure is illustrated in Fig. 4.3b. It must be remembered that $F(s)$ and $Y(s)$ are frequency domain representations of $f(t)$ and $y(t)$, respectively.

The response in Eq. 4.21 is a zero-state response—that is, the response when the initial conditions are zero. The complete response is given by

$$y(t) = \sum_{j=1}^{n} c_j e^{\lambda_j t} + \frac{1}{2\pi j} \int_{\sigma-j\infty}^{\sigma+j\infty} F(s)H(s)e^{st}ds \qquad (4.23a)$$

$$= \underbrace{\sum_{j=1}^{n} c_j e^{\lambda_j t}}_{\text{zero-input component}} + \underbrace{\mathcal{L}^{-1}[F(s)H(s)]}_{\text{zero-state component}} \qquad (4.23b)$$

The zero-input component can be evaluated in several ways: (i) using Eq. 2.47c, (ii) converting initial conditions into equivalent sources and finding the zero-state response to each source (see Prob. 2.35) and (iii) using the Laplace transform of the system differential equations to account for the initial conditions (discussed in Sec. 4.9).

Example 4.1. Find the output $y(t)$ of a system when the input $f(t) = 10\,u(t)$ and the transfer function $H(s)$ relating $y(t)$ is given by

$$H(s) = \frac{2s + 3}{s^2 + 2s + 5}$$

The initial conditions (initial state) are all zero. In this case

$$F(s) = \mathcal{L}[10u(t)] = \frac{10}{s}$$

and

$$Y(s) = F(s)H(s)$$

$$= \frac{10}{s}\frac{2s+3}{s^2+2s+5}$$

$$= \frac{10(2s+3)}{s(s+1-j2)(s+1+j2)}$$

The partial-fraction expansion of the right-hand side yields (see Appendix B)

$$Y(s) = \frac{6}{s} + \frac{4.59e^{j(229.5°)}}{s+1-j2} + \frac{4.59e^{-j(229.5°)}}{s+1+j2}$$

To find the inverse transform we use Table 4.1 (pairs 1 and 11):

$$y(t) = [6 + 9.18e^{-t}\cos(2t + 229.5°)]u(t)$$

Example 4.2.   Find the response $y(t)$ of a system when the input $f(t) = e^{-t}u(t)$ and the initial conditions are $y(0) = 2$, $y'(0) = 1$.   The transfer function relating $y(t)$ to the input $f(t)$ is given by

$$H(s) = \frac{s+5}{s^2+5s+6}$$

$$= \frac{s+5}{(s+2)(s+3)}$$

The poles of $H(s)$ are -2 and -3. Hence the response $y(t)$ is given by†

$$y(t) = \underbrace{c_1e^{-2t} + c_2e^{-3t}}_{y_x(t)} + \underbrace{\mathcal{L}^{-1}[F(s)H(s)]}_{y_f(t)}$$

The constants c_1, c_2 are given by (see Eq. 2.47c)

$$\begin{bmatrix} c_1 \\ c_2 \end{bmatrix} = \begin{bmatrix} 1 & 1 \\ -2 & -3 \end{bmatrix}^{-1} \begin{bmatrix} 2 \\ 1 \end{bmatrix}$$

$$= \begin{bmatrix} 3 & 1 \\ -2 & -1 \end{bmatrix} \begin{bmatrix} 2 \\ 1 \end{bmatrix} = \begin{bmatrix} 7 \\ -5 \end{bmatrix}$$

Hence

$$y_x(t) = 7e^{-2t} - 5e^{-3t} \qquad \text{for} \quad t \geq 0$$

$$y_f(t) = \mathcal{L}^{-1}[F(s)H(s)]$$

$$= \mathcal{L}^{-1}\left[\frac{s+5}{(s+1)(s+2)(s+3)}\right]$$

Expanding the right-hand side by partial fractions (Appendix B), we obtain

---

†Caveats of the footnote on p. 133 apply here also.

$$y_f(t) = \mathcal{L}^{-1}\left[\frac{2}{s+1} - \frac{3}{s+2} + \frac{1}{s+3}\right]$$

From Table 4.1 we obtain

$$y_f(t) = (2e^{-t} - 3e^{-2t} + e^{-3t})u(t)$$

Hence

$$y(t) = (2e^{-t} + 4e^{-2t} - 4e^{-3t})u(t)$$

THE DOMINANCE CONDITION

The result in Eq. 4.21 is based on the fact that the response of the system to eternal exponential e^{st} is $H(s)e^{st}$. This is true only if the exponential e^{st} dominates the exponentials $e^{\lambda_j t}$ where $\lambda_1, \lambda_2, \ldots, \lambda_n$ are the poles of $H(s)$. This is the dominance condition. The method of Laplace transform evidently holds true if a signal $f(t)$ is expressed as a continuous sum of exponentials e^{st} which dominate the exponentials $e^{\lambda_j t}$. In short, we must express $f(t)$ in terms of exponentials e^{st} such that s lies to the right of all poles of $H(s)$ in the complex frequency plane. This means

$$\mathrm{Re}\, s > \mathrm{Re}\, \lambda_j \qquad (j = 1, 2, \ldots, n) \qquad (4.24)$$

Thus the path of integration in Eq. 4.20 should be such that all values of s along this path will lie to the right of all poles of $H(s)$ in the s-plane. In Sec. 4.4 it was shown that the path of integration in Eq. 4.20 can be chosen in infinite ways. It was also shown that the path could be chosen to the right of some σ_m. We may choose the path as far right of this value σ_m as we please. Hence to satisfy the dominance condition, all we need do is shift the path of integration to the right of σ_m as well as all the poles† of $H(s)$. It is therefore evident that the dominance condition can always be satisfied by choosing the appropriate path. We therefore conclude that this method (Laplace transform) can be applied to all linear systems regardless of the location of the poles of the transfer function. We have buried the problem of dominance condition once and for all. Hence we shall dispense with the reference to dominance condition in all our future discussion, while using the Laplace transform (unilateral).

In the analysis of linear systems the advantages of the Laplace transform over the Fourier transform are now obvious. The Laplace transform can be used for all linear systems, whereas Fourier analysis can be applied only to stable systems. Moreover, the Fourier transform is capable of handling only a limited class of inputs. For these reasons the Laplace transform is almost universally adopted for

---

†It was mentioned in Sec. 4.4 that the path of integration in Eq. 4.20 should be such that it lies to the right of the poles of $F(s)$. Also from the discussion here it is obvious that in order to find the response $y(t)$ the path must be chosen to the right of the poles of $H(s)$. Hence the path of integration must be chosen to the right of poles of $F(s)$ as well as those of $H(s)$. This amounts to choosing a path to the right of poles of $F(s)H(s)$.

the analysis of linear systems.   For more general form of the Laplace transform (bilateral Laplace transform) the reader is referred to the author's earlier work†. For most of the applications, the input signals begin at some finite time and consequently the Laplace transform (unilateral) proves adequate.

UNIT IMPULSE RESPONSE

When the input to a system is a unit impulse, $f(t) = \delta(t)$ and $F(s) = 1$, and the unit impulse response $h(t)$ is given by

$$h(t) = \mathcal{L}^{-1}[F(s) H(s)]$$
$$= \mathcal{L}^{-1}[H(s)]$$

Hence

$$\mathcal{L}[h(t)] = H(s) \qquad (4.25)$$

Therefore the transfer function $H(s)$ is the Laplace transform of the unit impulse response $h(t)$.

4.7   COMMENTS ON THE TIME-DOMAIN AND THE FREQUENCY-DOMAIN ANALYSIS

We shall now view the time-domain and frequency-domain analysis of linear systems in proper perspective.   The more we reflect the more the similarities between the two methods become apparent.   In both cases the solution has similar form (Eqs. 2.73, 3.50a, or 4.23a). It consists of the same zero-input component and the zero-state component given by an integral (convolution integral in the time domain and the Fourier or the Laplace inversion integral in the case of the frequency domain).   The zero-state component represents the continuous sum (integral) of the responses to various elementary components (basis signals) into which the input is resolved.

In time-domain analysis we have observed that the zero-state response given by the convolution integral can be interpreted as the sum of responses of the system to impulse components of the input.   In the frequency-domain case the input is resolved into eternal exponential components and the response is obtained as a sum 'of the responses to these components. Hence the philosophy of the approach seems the same in both time-domain and frequency-domain analysis.   Only the basis signals used are different.   In time-domain we use impulse functions whereas in frequency-domain we use eternal exponential signals as the basis signals.   It is possible to use still another class of signals as basis signals.   That mysterious frequency-domain is in reality time-domain after all.   But she wears a glamorous disguise (or make up?)‡

---

†B. P. Lathi, *Signals, Systems, and Communication* (New York: Wiley, 1965).   See also W. R. LePage, *Complex Variables, and the Laplace Transforms for Engineers* (New York: McGraw-Hill, 1961).

‡The step function, the ramp function, and polynomials in t can be conveniently used as basis signals. See for example B. P. Lathi, *op. cit.*, Chap. 10.

It will also be realized that both the methods (time-domain and frequency-domain) of system analysis are identical to the very end. The variable p in one is replaced by a variable s in the other. In the end the solution is given by the convolution integral in time domain, and by the Laplace (or Fourier) inversion integral in frequency domain. In both cases the integration can be avoided by using the tables (convolution table and Laplace transform table).

There is one rather strange aspect of the frequency-domain method. In order to calculate the response at some instant t we need the knowledge about $f(t)$ for all time even beyond the instant t. This is because in order to find the response $y(t)$ we must first find $F(s)$, which is given by

$$F(s) = \int_0^\infty f(t)e^{-st}\,dt$$

Thus to compute $F(s)$ we need to know $f(t)$ from 0 to ∞. It therefore appears that to compute the response at some instant t we need to know the input for all instants, even those beyond t. This may give the wrong impression that the response at time t is influenced by the input occurring beyond t. For physical systems this is impossible, because the system cannot anticipate the input. How then do we explain this unreasonable behavior? The key to this problem lies in the fact that in the frequency-domain method we are expressing an input $f(t)$ over the entire interval of its existence by eternal exponentials. For such a decomposition we must observe $f(t)$ in its entirety from 0 to ∞ (or $-\infty$ to ∞ for the bilateral case). This is the reason why we need the knowledge about $f(t)$ over its entire life.

On the fundamental basis, however, this criticism of the frequency-domain method is completely unjustified. If we are interested in finding the response for $t < T$, we need observe $f(t)$ only over the interval (0 to T). We can define a finite time Laplace transform $F_T(s)$ as

$$F_T(s) = \int_0^T f(t)e^{-st}\,dt$$

and proceed as usual. This will give correct results for all $t < T$. Thus in order to find the response for $t < T$, we need to know $f(t)$ only for $t < T$.

4.8   SOME PROPERTIES OF THE LAPLACE TRANSFORM

Since the Laplace transform is merely a generalization of the Fourier transform, we expect it to possess properties similar to those of the Fourier transform with the variable s replacing $j\omega$. This indeed is the case. The reader may prove the following properties by methods similar to those used for the Fourier transform.

1.   LINEARITY PROPERTY

If

$$f_1(t) \leftrightarrow F_1(s) \qquad \text{and} \quad f_2(t) \leftrightarrow F_2(s)$$

then

$$a_1 f_1(t) + a_2 f_2(t) \leftrightarrow a_1 F_1(s) + a_2 F_2(s) \qquad (4.26)$$

2. SCALING PROPERTY

For $a > 0$,

$$f(at) \leftrightarrow \frac{1}{a} F\left(\frac{s}{a}\right) \qquad (4.27)$$

3. TIME-SHIFTING PROPERTY

For $t_0 > 0$,

$$f(t - t_0) \leftrightarrow F(s) e^{-st_0} \qquad (4.28a)$$

Note that $f(t)$ is zero for $t < 0$. Hence $f(t - t_0) = 0$ for $t < t_0$. Hence the function $f(t - t_0)$ starts at $t = t_0$ and is zero for $t < t_0$. This fact is sometimes inadvertently ignored, and may cause error. To avoid this, the time-shifting property is better expressed as

$$f(t - t_0) u(t - t_0) \leftrightarrow F(s) e^{-st_0} \qquad (4.28b)$$

Thus

$$\cos \omega_0 t \, u(t) \leftrightarrow \frac{s}{s^2 + \omega_0^2}$$

and

$$\cos \omega_0(t - t_0) u(t - t_0) \leftrightarrow \frac{s e^{-st_0}}{s^2 + \omega_0^2}$$

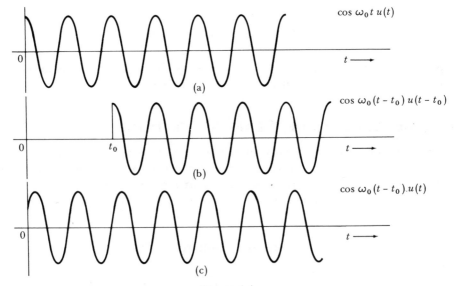

Figure 4.4

It is erroneous to write

$$\cos \omega_0 (t - t_0) \leftrightarrow \frac{s e^{-s t_0}}{s^2 + \omega_0^2}$$

The difference between functions $\cos \omega_0 (t - t_0) u(t - t_0)$ and $\cos \omega_0 (t - t_0) u(t)$ is shown in Fig. 4.4.

The time-shifting property proves very convenient in finding Laplace transforms of certain functions. Consider the function $f(t)$ shown in Fig. 4.5.   This

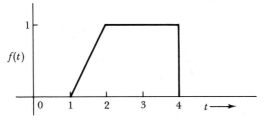

Figure 4.5

function can be expressed as

$$f(t) = (t - 1) u(t - 1) - (t - 2) u(t - 2) - u(t - 4)$$

From table of transforms and using the time-shifting property (Eq. 4.28b), we obtain

$$F(s) = \frac{1}{s^2} e^{-s} - \frac{1}{s^2} e^{-2s} - \frac{1}{s} e^{-4s} \qquad (4.29)$$

4. FREQUENCY-SHIFTING PROPERTY

$$f(t) e^{s_0 t} \leftrightarrow F(s - s_0) \qquad (4.30)$$

5. TIME-CONVOLUTION

$$f_1(t) * f_2(t) \leftrightarrow F_1(s) F_2(s) \qquad (4.31)$$

6. FREQUENCY CONVOLUTION

$$f_1(t) f_2(t) \leftrightarrow \frac{1}{2\pi j} [F_1(s) * F_2(s)] \qquad (4.32)$$

We had already anticipated the time-convolution property.   From time-domain analysis, the zero-state response $y(t)$ is given by

$$y(t) = f(t) * h(t)$$

From frequency-domain analysis, we have

$$y(t) = \mathcal{L}^{-1}[F(s) H(s)]$$

Hence

$$f(t) * h(t) \leftrightarrow F(s) H(s)$$

7. TIME-DIFFERENTIATION PROPERTY

If

$$f(t) \leftrightarrow F(s)$$

then

$$\frac{df}{dt} \leftrightarrow sF(s) - f(0^-) \qquad (4.33a)$$

This result can be easily extended to higher derivatives to yield

$$\frac{d^n f}{dt^n} \leftrightarrow s^n F(s) - s^{n-1} f(0^-) - s^{n-2} f'(0^-) - \cdots - f^{(n-1)}(0^-) \qquad (4.33b)$$

This property may be proved as follows. By definition,

$$\mathcal{L}\left[\frac{df}{dt}\right] = \int_0^\infty \frac{df}{dt} e^{-st} dt$$

Integration by parts of the right-hand side yields

$$\mathcal{L}\left[\frac{df}{dt}\right] = f(t)e^{-st} \Big|_{0^-}^\infty + s \int_{0^-}^\infty f(t)e^{-st} dt \qquad (4.34)$$

The quantity $f(t)e^{-st}$ goes to zero at $t = \infty$. This follows from the fact that s lies in the region of convergence where $f(t)e^{-st}$ goes to zero at $t = \infty$ (see Sec. 4.3, Eq. 4.15). Hence Eq. 4.34 becomes

$$\frac{df}{dt} = -f(0^-) + sF(s)$$

$$= sF(s) - f(0^-)$$

The extension of this result to the nth derivative yields Eq. 4.33b.
    Note that if the function $f(t)$ is causal, then

$$f(t) = 0, \qquad (t < 0)$$

Hence

$$f(0^-), f'(0^-), \ldots, f^{(n-1)}(0^-)$$

are all zero, and the time-differentiation property becomes

$$\frac{df}{dt} \leftrightarrow sF(s) \qquad (4.35a)$$

and

$$\frac{d^n f}{dt^n} \leftrightarrow s^n F(s) \tag{4.35b}$$

8.   TIME-INTEGRATION PROPERTY

If

$$f(t) \leftrightarrow F(s)$$

then

$$\int_0^t f(\tau)\, d\tau \leftrightarrow \frac{F(s)}{s} \tag{4.36}$$

Proof. By definition,

$$\mathcal{L}\left[\int_0^t f(\tau)\, d\tau\right] = \int_0^\infty \left(\int_0^t f(\tau)\, d\tau\right) e^{-st}\, dt$$

Integrating by parts, we get

$$\mathcal{L}\left[\int_0^t f(\tau)\, d\tau\right] = \left[\frac{-e^{-st}}{s} \int_0^t f(\tau)\, d\tau\right]_0^\infty + \frac{1}{s} \int_0^\infty f(t) e^{-st}\, dt \tag{4.37}$$

The term $e^{-st} \int_0^t f(\tau)\, d\tau$ at $t = \infty$ is zero for values of s lying in the region of convergence of $\int_0^t f(\tau)\, d\tau$. Also the integral term $\int_0^t f(\tau)\, d\tau$ is zero at $t = 0$. Hence the first term on the right-hand side of Eq. 4.37 is zero, and we get

$$\int_0^t f(\tau)\, d\tau \leftrightarrow \frac{F(s)}{s}$$

We can also show that

$$\int_{-\infty}^t f(\tau)\, d\tau \leftrightarrow \frac{F(s)}{s} + \frac{\displaystyle\int_{-\infty}^0 f(\tau)\, d\tau}{s} \tag{4.38}$$

Proof.

$$\int_{-\infty}^t f(\tau)\, d\tau = \int_{-\infty}^0 f(\tau)\, d\tau + \int_0^t f(\tau)\, d\tau$$

Taking Laplace transforms of both sides of this equation, and using Eq. 4.36, we have

$$\mathcal{L}\left[\int_{-\infty}^t f(\tau)\, d\tau\right] = \frac{\displaystyle\int_{-\infty}^0 f(\tau)\, d\tau}{s} + \frac{F(s)}{s}$$

Note that 0 here implies 0^- to be consistent with our convention of including the origin. The above result can be generalized to higher integrals. Note that for causal functions

$$f(t) = 0 \qquad \text{for} \qquad t < 0$$

and

$$\int_{-\infty}^{0} f(\tau)\, d\tau = 0$$

therefore

$$\int_{-\infty}^{t} f(\tau)\, d\tau \leftrightarrow \frac{F(s)}{s} \tag{4.39}$$

THE INITIAL AND FINAL VALUE

If the transform $F(s)$ of an unknown function $f(t)$ is given, then it is possible to determine the initial and the final value of $f(t)$—that is, the value of $f(t)$ at $t = 0^{+}$ and $t = \infty$. With the help of the initial and the final value theorem, this can be done without having to find the inverse transform of $F(s)$.

THE INITIAL-VALUE THEOREM

$$f(0^{+}) = \lim_{s \to \infty} sF(s) \tag{4.40}$$

Proof. Consider the transform of the derivative of $f(t)$. From Eq. 4.33a we have

$$sF(s) - f(0^{-}) = \int_{0-}^{\infty} \frac{df}{dt} e^{-st}\, dt$$

$$= \int_{0-}^{0+} \frac{df}{dt} e^{-st}\, dt + \int_{0+}^{\infty} \frac{df}{dt} e^{-st}\, dt$$

$$= f(t) \Big|_{0-}^{0+} + \int_{0+}^{\infty} \frac{df}{dt} e^{-st}\, dt$$

$$= f(0^{+}) - f(0^{-}) + \int_{0+}^{\infty} \frac{df}{dt} e^{-st}\, dt$$

Therefore

$$sF(s) = f(0^{+}) + \int_{0+}^{\infty} \frac{df}{dt} e^{-st}\, dt \tag{4.41}$$

Now if we let $s \to \infty$ in this equation, the integral on the right-hand side vanishes, and we have

$$\lim_{s \to \infty} sF(s) = f(0^{+})$$

Note that the left-hand side of the above equation may exist without the existence of $f(0^{+})$. Therefore this theorem should be applied only when $f(0^{+})$ exists.

THE FINAL-VALUE THEOREM

If $f(t)$ and its first derivative are Laplace-transformable, then

$$\lim_{s \to 0} sF(s) = f(\infty) \qquad (4.42)$$

Proof.   Let $s \to 0$ in Eq. 4.41. This yields

$$\lim_{s \to 0} sF(s) = f(0^+) + \int_{0^+}^{\infty} \frac{df}{dt} \, dt$$

$$= f(0^+) + f(t)\Big|_{0^+}^{\infty} = f(0^+) + f(\infty) - f(0^+)$$

$$= f(\infty)$$

Therefore

$$f(\infty) = \lim_{s \to 0} sF(s)$$

Here again the right-hand side of the above equation may exist without the existence of $f(\infty)$. Hence, it is important to know that $f(\infty)$ exists before applying this theorem.

We shall demonstrate the application of the initial and the final-value theorem by an example. Consider the response $Y(s)$ in Example 4.1.

$$Y(s) = \frac{10(2s + 3)}{s(s^2 + 2s + 5)}$$

TABLE 4.2
Some Properties of the Laplace Transform

Operation	Time domain, $f(t)$	Frequency domain, $F(s)$
Addition	$a_1 f_1(t) + a_2 f_2(t)$	$a_1 F_1(s) + a_2 F_2(s)$
Scaling $(a > 0)$	$f(at)$	$\dfrac{1}{a} F\left(\dfrac{s}{a}\right)$
Time shift $(t_0 > 0)$	$f(t - t_0)u(t - t_0)$	$F(s)e^{-st_0}$
Frequency shift	$f(t)e^{s_0 t}$	$F(s - s_0)$
Time differentiation	$\dfrac{df}{dt}$	$sF(s) - f(0^-)$
Time integration	$\displaystyle\int_{-\infty}^{t} f(t) \, dt$	$\dfrac{F(s)}{s} + \dfrac{\displaystyle\int_{-\infty}^{0} f(t)\, dt}{s}$
Time convolution	$f_1(t) * f_2(t)$	$F_1(s)F_2(s)$
Frequency convolution	$f_1(t)f_2(t)$	$\dfrac{1}{2\pi j}[F_1(s) * F_2(s)]$

We can calculate the initial value $y(0)$ and the final value $y(\infty)$ of the response using the above theorems:

$$y(0) = \lim_{s \to \infty} sY(s) = \lim_{s \to \infty} \frac{10(2s + 3)}{s^2 + 2s + 5} = 0$$

$$y(\infty) = \lim_{s \to 0} sY(s) = \lim_{s \to 0} \frac{10(2s + 3)}{s^2 + 2s + 5} = 6$$

The reader can easily verify these values. In Example 4.1, we found the response $y(t)$ to be

$$y(t) = 6 + 9.18e^{-t} \cos(2t + 229.5°)$$

This gives

$$y(0) = 0 \quad \text{and} \quad y(\infty) = 6$$

4.9   FREQUENCY-DOMAIN ANALYSIS: ANOTHER POINT OF VIEW

It is possible to analyze a linear system by using the Laplace transform without recourse to the concept of transfer function. The solution is obtained by transforming, directly, the integrodifferential equations of the system. This method leads to a slightly different point of view of the Laplace transform. The starting point in this method is the definition of the Laplace transform. For a given function $f(t)$ we define its Laplace transform $F(s)$ as

$$F(s) = \int_{0^-}^{\infty} f(t)e^{-st} \, dt$$

Starting with this definition, we prove the time-differentiation, and integration properties (Eqs. 4.33 and 4.38). We are now ready to solve integrodifferential equations.

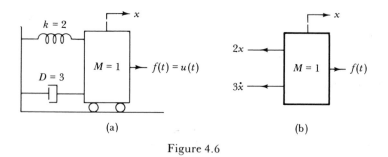

Figure 4.6

Let us consider the mechanical system in Fig. 4.6a. The free-body diagram is shown in Fig. 4.6b. The equation of motion is given by

$$\frac{d^2x}{dt^2} + 3\frac{dx}{dt} + 2x = u(t) \tag{4.43}$$

The initial conditions are given as

$$x(0) = 2 \quad \text{and} \quad x'(0) = 1$$

If we let

$$x(t) \leftrightarrow X(s) \tag{4.44a}$$

then

$$\frac{dx}{dt} \leftrightarrow sX(s) - x(0) = sX(s) - 2 \tag{4.44b}$$

and

$$\frac{d^2x}{dt^2} \leftrightarrow s[sX(s) - 2] - x'(0) = s^2X(s) - 2s - 1 \tag{4.44c}$$

also

$$u(t) \leftrightarrow \frac{1}{s} \tag{4.44d}$$

We now take the Laplace transform of Eq. 4.43. Using the results in Eqs. 4.44a, b, c, and d, we have

$$(s^2 + 3s + 2)X(s) = \frac{1}{s} + 2s + 7 \tag{4.45}$$

Observe that this is an algebraic equation. Thus the Laplace transform of a differential equation yields an algebraic equation. We can solve the algebraic equation to obtain $X(s)$ as

$$\begin{aligned}
X(s) &= \frac{2s^2 + 7s + 1}{s(s^2 + 3s + 2)} \\
&= \frac{2s^2 + 7s + 1}{s(s + 1)(s + 2)} \\
&= \frac{1/2}{s} + \frac{4}{s + 1} - \frac{5/2}{s + 2}
\end{aligned}$$

To obtain the desired solution $x(t)$, we take the inverse transform of $X(s)$:

$$x(t) = \mathcal{L}^{-1}[X(s)] = (\tfrac{1}{2} + 4e^{-t} - \tfrac{5}{2}e^{-2t})u(t)$$

In this method the integrodifferential equations in the t-domain are transformed into algebraic equations in the s-domain. We solve the algebraic equations and obtain the solution in the s-domain. To obtain the solution in the t-domain, we take the inverse transform of the s-domain solution. This method may be likened to use of logarithms in multiplication of two numbers. The operation of multiplication is inherently more difficult than that of addition, so we search for a method that will convert the operation of multiplication into that of addition. This is done by a kind of transformation (log transformation) where we transform each number into its logarithm (log-domain). We can now add the logarithms of two numbers. This gives us the answer in log-domain. To obtain the desired answer we take the inverse transform (antilog) of the answer in log-domain.

The second point of viewing the Laplace transform is similar to this. We shall illustrate this viewpoint by one more example.

Example 4.3. Let us analyze the network in Fig. 4.7 using this viewpoint. The mesh equations of the network are

$$\int_{-\infty}^{t} i_1(\tau)\, d\tau + \frac{1}{5}(i_1 - i_2) = f(t) \tag{4.46a}$$

$$-\frac{1}{5}i_1 + \frac{6}{5}i_2 + \frac{1}{2}\frac{di_2}{dt} = 0 \tag{4.46b}$$

We shall now take Laplace transforms of these equations. Let

$$i_1(t) \leftrightarrow I_1(s) \qquad \text{and} \qquad i_2(t) \leftrightarrow I_2(s) \tag{4.47a}$$

Then from Eq. 4.38

$$\int_{-\infty}^{t} i_1(\tau)\, d\tau \leftrightarrow \frac{I_1(s)}{s} + \frac{\int_{-\infty}^{0} i_1(\tau)\, d\tau}{s}$$

Note that the integral on the right-hand side of this equation represents the charge on the capacitor at $t = 0$ in Fig. 4.7. Since $C = 1$, this is also the voltage on the capacitor at $t = 0$ (initial voltage) which is given to be -5. Therefore

$$\int_{-\infty}^{t} i_1(\tau)\, d\tau \leftrightarrow \frac{I_1(s)}{s} - \frac{5}{s} \tag{4.47b}$$

Also since

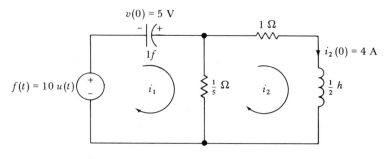

Figure 4.7

$$i_2(t) \leftrightarrow I_2(s)$$

then (see Eq. 4.33a)

$$\frac{di_2}{dt} \leftrightarrow sI_2(s) - i_2(0)$$

$$sI_2(s) - 4 \tag{4.47c}$$

Also

$$F(s) = \mathcal{L}[f(t)] = \mathcal{L}[10u(t)] = \frac{10}{s} \tag{4.47d}$$

We shall use Eqs. 4.47 b, c, and d in writing the Laplace transforms of system equations 4.46. These are

$$\frac{I_1(s)}{s} - \frac{5}{s} + \frac{1}{5}[I_1(s) - I_2(s)] = \frac{10}{s}$$

$$-\frac{1}{5}I_1(s) + \frac{6}{5}I_2(s) + \frac{s}{2}I_2(s) - 2 = 0$$

Rearranging these equations, we obtain

$$\begin{bmatrix} \dfrac{1}{s} + \dfrac{1}{5} & -\dfrac{1}{5} \\[2ex] -\dfrac{1}{5} & \dfrac{6}{5} + \dfrac{s}{2} \end{bmatrix} \begin{bmatrix} I_1(s) \\[2ex] I_2(s) \end{bmatrix} = \begin{bmatrix} \dfrac{15}{s} \\[2ex] 2 \end{bmatrix}$$

Application of Cramer's rule to this set of equation yields

$$I_1(s) = \frac{79s + 180}{s^2 + 7s + 12}$$

and

$$I_2(s) = \frac{2(2s + 25)}{s^2 + 7s + 12}$$

Hence

$$i_1(t) = \mathcal{L}^{-1} \frac{79s + 180}{s^2 + 7s + 12}$$

$$= \mathcal{L}^{-1} \frac{79s + 180}{(s + 3)(s + 4)}$$

$$= \mathcal{L}^{-1} \left(\frac{-57}{s + 3} + \frac{136}{s + 4} \right)$$

$$= (-57e^{-3t} + 136e^{-4t})u(t)$$

Similarly,

$$i_2(t) = \mathcal{L}^{-1} \frac{2(2s + 25)}{s^2 + 7s + 12}$$

$$= \mathcal{L}^{-1} \frac{2(2s + 25)}{(s + 3)(s + 4)}$$

$$= \mathcal{L}^{-1}\left(\frac{38}{s + 3} - \frac{34}{s + 4}\right)$$

$$= (38e^{-3t} - 34e^{-4t})u(t)$$

COMMENT ON TRANSFORM VIEWPOINT OF THE LAPLACE TRANSFORM

We have considered two interpretations of the Laplace transform and accordingly two ways of formulating the equations. The first approach, which is extensively treated in this book, conceives the Laplace transform as a tool for expressing the input in terms of eternal exponential components. The transfer function $H(s)$ represents the response of the system to an eternal exponential signal. The desired response is obtained by adding the responses to all exponential components (principle of superposition). This approach will be called the *frequency-analysis approach*. In this viewpoint, the transfer function plays major role.

On the other hand, the second viewpoint, discussed in the last few pages, conceives the Laplace transform as a tool for solving linear integrodifferential equations. As seen from time-differentiation and integration properties (Eqs. 4.35 and 4.38), the Laplace transform of a derivative or an integral of a function is an algebraic quantity. The Laplace transform of a differential equation in the t-domain converts it into an algebraic equation in the s-domain. The algebraic equations can be easily solved and the solution in the s-domain is readily obtained. The desired solution in the t-domain is given by the inverse Laplace transform of the solution in the s-domain. In this approach we do not concern ourselves with the transfer function. However, as a matter of convenience one may define the transfer function $H(s)$ as a ratio of the transform of the response to the transform of the input

$$H(s) = \frac{Y(s)}{F(s)} = \frac{\mathcal{L}[y(t)]}{\mathcal{L}[f(t)]} \tag{4.48}$$

where $y(t)$ is the zero-state response. The initial conditions are also automatically accounted for in this approach. Since this approach transforms the differential equations into algebraic equations it will be called the *transform viewpoint*.

The distinction between the two approaches (the frequency-analysis and the transform approach) should be clearly understood. The frequency-analysis approach is based upon classical dynamics which has been the basis of the system theory. A given driving function is expressed as a continuous sum of eternal exponential signals. The response to such an eternal exponential signal e^{st} is $H(s)e^{st}$. Hence the response of the system is given by a continuous sum of the responses to various exponential components. On the other hand, the transform approach views the Laplace transform as a machine which simplifies the solution of integrodifferential equations characterizing the system. There is no physical significance attached to any quantity. All one has to do is to feed these integro-differential equations to a Laplace machine which yields the products (algebraic equations in the s-domain) which are easier to handle. The solutions of these simplified equations are then fed back to the inverse Laplace machine to get the desired solutions. Hence the transform approach is a nonphysical abstract approach which gives little insight into the operation of the system. The frequency-analysis approach, on the other hand, gives physical viewpoint of the Laplace transform. The quantities $F(s)$ and $H(s)$ have meaningful interpretations. The Laplace transform of any function represents the relative amplitudes of various frequency components. The frequency-analysis viewpoint makes the frequency-domain analysis very meaningful to an engineer.

4.10   MULTIPLE INPUTS

The multiple-input case for a linear system can be handled by using the principle of superposition. Let $f_1(t), f_2(t), \ldots, f_m(t)$ be the inputs to a certain system and let $H_1(s), H_2(s), \ldots, H_m(s)$ represent the transfer functions relating a certain response variable $y(t)$ to these inputs respectively (Fig. 4.8). Then the response $y(t)$ can be obtained by considering one input at a time and assuming

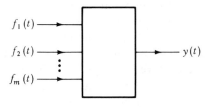

Figure 4.8

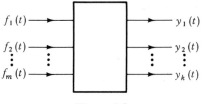

Figure 4.9

the remaining inputs to be zero. The sum of the m components thus obtained yields the total response. It can be easily seen that if $y(t) \leftrightarrow Y(s)$ and $f_k(t) \leftrightarrow F_k(s)$, $k = 1, 2, \ldots, m$, then (see Eq. 2.101a)

$$Y(s) = H_1(s)F_1(s) + H_2(s)F_2(s) + \cdots + H_m(s)F_m(s) \qquad (4.49)$$

where $Y(s)$ is the zero-state response. In general, a system has k outputs $y_1(t)$, $y_2(t), \ldots, y_k(t)$, and m inputs $f_1(t), f_2(t), \ldots, f_m(t)$ as shown in Fig. 4.9. If $H_{ij}(s)$

is the transfer function relating the output $y_i(t)$ to the input $f_j(t)$, then with usual notation, we can write

$$Y_1(s) = H_{11}(s)F_1(s) + H_{12}(s)F_2(s) + \cdots + H_{1m}(s)F_m(s)$$

$$Y_2(s) = H_{21}(s)F_1(s) + H_{22}(s)F_2(s) + \cdots + H_{2m}(s)F_m(s) \qquad (4.50a)$$

$$\cdots\cdots\cdots\cdots\cdots\cdots\cdots\cdots\cdots\cdots\cdots\cdots\cdots\cdots$$

$$Y_k(s) = H_{k1}(s)F_1(s) + H_{k2}(s)F_2(s) + \cdots + H_{km}(s)F_m(s)$$

or

$$\underbrace{\begin{bmatrix} Y_1(s) \\ Y_2(s) \\ \cdots \\ Y_k(s) \end{bmatrix}}_{\mathbf{Y}(s)} = \underbrace{\begin{bmatrix} H_{11}(s) & H_{12}(s) & \cdots & H_{1m}(s) \\ H_{21}(s) & H_{22}(s) & \cdots & H_{2m}(s) \\ \cdots\cdots\cdots\cdots\cdots\cdots\cdots\cdots \\ H_{k1}(s) & H_{k2}(s) & \cdots & H_{km}(s) \end{bmatrix}}_{\mathbf{H}(s)} \underbrace{\begin{bmatrix} F_1(s) \\ F_2(s) \\ \cdots \\ F_m(s) \end{bmatrix}}_{\mathbf{F}(s)} \qquad (4.50b)$$

or

$$\mathbf{Y}(s) = \mathbf{H}(s)\mathbf{F}(s) \qquad (4.50c)$$

4.11   BLOCK-DIAGRAM REPRESENTATION OF SYSTEMS

A system consists of number of components. A complex system may consist of a very large number of components and presents a problem so far as schematic representation is concerned. Fortunately, we are mostly concerned with the transmission characteristic (dynamic behavior, or the input-output relationship) of a system. Such a behavior can be specified by various transfer functions of the system. Therefore instead of showing the detailed schematic diagram of the interconnection of system elements, we may represent it with by its pertinent transfer functions. If in a system there is one input $f(t)$, and if we are interested in the response $y(t)$, then we may represent the system by a block diagram (Fig. 4.10a) specifying $H(s)$, the transfer function relating $y(t)$ to the input $f(t)$. The input(s) and output(s) may be represented as functions of time (as in Fig. 4.10a) or may be represented in the frequency domain (Laplace transforms as

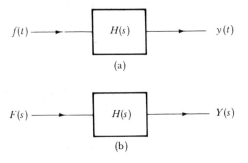

Figure 4.10

in Fig. 4.10b).   However, it is advisable to use frequency-domain description of signals (Fig. 4.10b) in order to avoid confusion and inadvertent errors caused by mixing time-domain and frequency-domain quantities (Fig. 4.10a) especially by the beginners.   With a little familiarity, however, one becomes adept enough to avoid these pitfalls.

We can use similar scheme to represent multiple-input, multiple-output system.   Let us consider a system with two inputs $f_1(t)$ and $f_2(t)$ and two outputs $y_1(t)$ and $y_2(t)$ (Fig. 4.11a).   For a linear system, the inputs and the outputs will be related as

$$
\begin{aligned}
Y_1(s) &= H_{11}(s)F_1(s) + H_{12}(s)F_2(s) \\
Y_2(s) &= H_{21}(s)F_1(s) + H_{22}(s)F_2(s)
\end{aligned}
\tag{4.51}
$$

This relationship can be represented by block diagrams using four decoupled systems as shown in Fig. 4.11b.   The summing device (shown by a circle with Σ inside) adds the incoming signals.   The reader can easily verify that the block diagram in Fig. 4.11b represents Eq. 4.51.

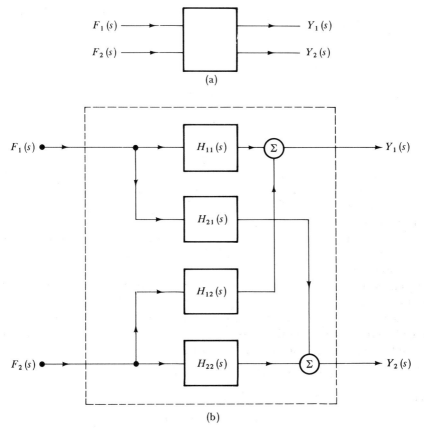

(a)

(b)

Figure 4.11

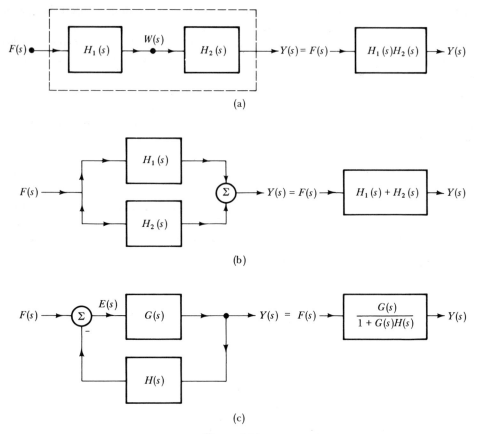

Figure 4.12

The block-diagram representation proves very convenient when we are interested only in certain response variables in a system. A system in general has several response variables. In most of the cases, however, we are interested only in few of response variables. The block diagram allows us to focus our attention only on the relationship relevant to these variables.

Complex systems in general can be represented by several subsystems interconnected in a certain manner. Each subsystem can be represented by a block diagram.

When two transfer functions appear in cascade as shown in Fig. 4.12a, the transfer function of the overall system is the product of the two transfer functions. This can be seen from the fact that in Fig. 4.12a

$$W(s) = H_1(s)F(s) \quad \text{and} \quad Y(s) = H_2(s)W(s)$$

Hence

$$Y(s) = H_1(s)H_2(s)F(s)$$

This result can be extended to any number of transfer functions in cascade.†

Similarly when two transfer functions $H_1(s)$ and $H_2(s)$ appear in parallel as shown in Fig. 4.12b, the transfer function is given by $H_1(s) + H_2(s)$ the sum of two transfer functions. The proof is trivial. This result can be extended to any number of systems in parallel.

When the output is fed back to the input as shown in Fig. 4.12c, the transfer function $Y(s)/F(s)$ can be computed as follows. The inputs to the summing device are $F(s)$ and $-H(s)Y(s)$. Therefore $E(s)$, the output of the summing device, is

$$E(s) = F(s) - H(s)Y(s)$$

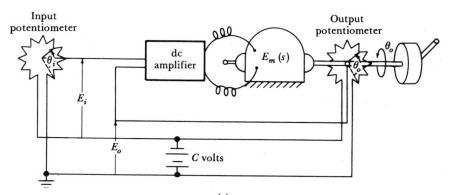

(a)

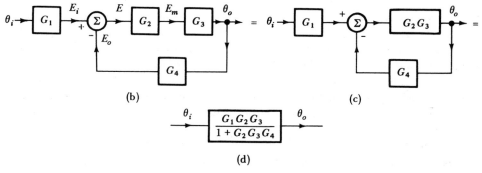

(b)   (c)

(d)

Figure 4.13

But

$$Y(s) = G(s)E(s)$$
$$= G(s)[F(s) - H(s)Y(s)]$$

or

$$Y(s)[1 + G(s)H(s)] = G(s)F(s)$$

---

†This result is true under the assumption that a subsystem represented by one of the blocks does not load (or alter) the transfer function of the preceding subsystem.

and

$$\frac{Y(s)}{F(s)} = \frac{G(s)}{1 + G(s)H(s)} \tag{4.52}$$

Therefore the feedback loop can be replaced by a single block with transfer function shown in Eq. 4.52 (see Fig. 4.12c).

Let us consider an automatic position control system shown in Fig. 4.13a. This system is used to control the angular position of any object such as a tracking antenna, antiaircraft gun mount, or ship position. The input is the desired angular position of the object. This is θ_i, and can be set at a desired angle by positioning the wiper of the potentiometer at the input. The second potentiometer, identical to the first, is placed at the output. The wiper of this potentiometer is mounted on the output shaft. This potentiometer setting θ_o indicates the actual angular position of the object. A voltage proportional to $\theta_i - \theta_o$ appears at the input of the amplifier. The amplifier output is fed to the motor which is coupled to the load. If the desired position θ_i and the actual position θ_o are not identical, a voltage (proportional to $\theta_i - \theta_o$) is amplified and fed to the motor which turns the load until $\theta_o = \theta_i$. In this position $\theta_i - \theta_o = 0$, and there is no input to the motor and the motor stops. It is evident that this system can control the angular position of a remote and heavy object by setting the input potentiometer at the desired position.

We shall draw a block diagram for this system. Let us begin with the input θ_i. The input potentiometer converts this input (angular position) into electrical voltage $E_i = (C/2\pi)\theta_i$. This is shown in Fig. 4.13b by a transfer function $G_1 = C/2\pi$. Similarly, the output potentiometer converts the output angular position θ_o into electrical voltage $E_o = (C/2\pi)\theta_o$. This is shown by the transfer function $G_4 = C/2\pi$. The two wipers are connected to the input of the amplifier. Hence the amplifier input is $E_i - E_o$. The subtraction of E_o from E_i is shown at the summing device (with a negative sign on E_o). The resulting signal $E = E_i - E_o$ is fed to the amplifier with transfer function $G_2 = K$ (gain of the amplifier). The output of the amplifier is E_m which is fed to the motor. The transfer function of the motor and load combination is G_3.

To find the transfer function $\theta_o(s)/\theta_i(s)$, relating the output θ_o to input θ_i, we can use the block-diagram reduction rules developed in earlier (Fig. 4.12). The block diagram in Fig. 4.13b is reduced to that in Fig. 4.13c and then to Fig. 4.13d.

This result can also be obtained directly by writing the system equations. From Fig. 4.13b we have

$$E = G_1\theta_i - G_4\theta_o \tag{4.53a}$$

and

$$\theta_o = G_2G_3E \tag{4.53b}$$

Hence

$$\theta_o = G_2G_3(G_1\theta_i - G_4\theta_o) \tag{4.53c}$$

or

$$\theta_o(1 + G_2G_3G_4) = G_1G_2G_3\theta_i \tag{4.53d}$$

Therefore the transfer function on $T(s)$ which relates the response $\theta_o(s)$ to the input $\theta_i(s)$ is given by

$$T(s) = \frac{\theta_o(s)}{\theta_i(s)} = \frac{G_1 G_2 G_3}{1 + G_2 G_3 G_4} \qquad (4.54)$$

This is exactly the same result as obtained earlier by using reduction techniques. The efficacy of the reduction method is obvious. Several other rules of reduction are available which enable us to handle more complex block diagrams. Mason's rule (developed in the next section) is the most general rule applicable to all complex block diagrams. This rule enables one to determine transfer function(s) (ratio of one variable to the other) of a complex block diagram by inspection. This rule can be applied with great ease to signal flow graphs (a simplified form of block diagrams) developed in next section. For this reason, we shall defer Mason's rule till the next section.

The salient feature of the block-diagram representation is that we focus our attention only on the relevant information about the system that is needed in our analysis. Secondly the information is in a quantitative form (such as a transfer function). The motor has several poles, armature winding, friction at the brushes, and the like. For our purpose, however, the only relevant data is the relationship between the motor input voltage E_m and the output shaft position θ_o. This is given in the form of transfer function $G_3(s)$. Similar observation holds true for other subsystems.

4.12   SIGNAL-FLOW GRAPH REPRESENTATION OF SYSTEMS

The block-diagram representation discussed earlier can be further simplified by what is known as the *signal-flow graph* representation. This form of representation is basically identical to the block-diagram representation. It is

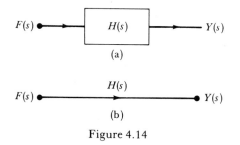

Figure 4.14

merely a simplified representation where each block is represented by a directed line and summing elements are eliminated.

Figure 4.14a shows the block-diagram representation of a system with transfer function $H(s)$. This representation may be further simplified by replacing the block by a path with an arrow indicating the sense in which the signal is transmitted (from the input to the output). Each such path is labeled by the transfer function associated with the path (Fig. 4.14b). This simplified representation is

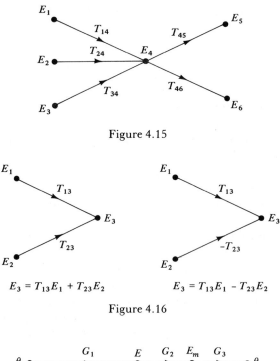

Figure 4.15

$E_3 = T_{13}E_1 + T_{23}E_2$ $E_3 = T_{13}E_1 - T_{23}E_2$

Figure 4.16

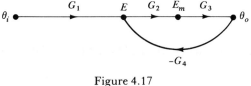

Figure 4.17

known as the *signal-flow-graph representation* of a system. Note that a path starts and ends on some node which represents a signal variable in the system. In Fig. 4.14b the path representing $H(s)$ starts at a node representing signal $F(s)$ and ends on the node representing signal $Y(s)$. The signal that is being transmitted along a path is equal to the signal at node where the path originated, multiplied by the transfer function of the path. In Fig. 4.14 the signal traveling on the path is $F(s) \times H(s)$, that is, $H(s)F(s)$. At each node there may be a number of incoming branches and a number of outgoing branches. The value of the signal variable at any node is equal to the sum of signals at that node due to all incoming branches only. In Fig. 4.15, for example, the node variable E_4 is given by

$$E_4 = T_{14}E_1 + T_{24}E_2 + T_{34}E_3$$

The node variable E_5 and E_6 are given by

$$E_5 = T_{45}E_4$$
$$E_6 = T_{46}E_4$$

Note that a node acts as a summing point for signals on all the incoming branches. A subtraction (or comparison) can be effected merely by placing a negative sign on

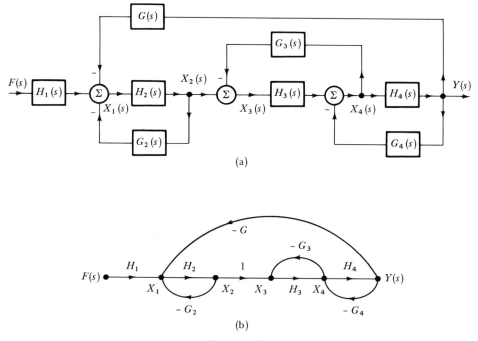

Figure 4.18

the transmittance of the branch whose signal is to be subtracted. This is illustrated in Fig. 4.16. The automatic position-control system shown in Fig. 4.13b, can be represented by a signal-flow graph as shown in Fig. 4.17. Figure 4.18 shows the block diagram and the corresponding signal-flow graph of a more complex system. The economy due to shorthand representation effected by the signal flow graph is now evident. The important data in any block diagram are transfer functions and the interconnections of various subsystems. The signal-flow graph retains only this information and throws away all other unnecessary decor such as boxes and summing circles. It therefore allows us to focus only on the very essence of the system.

4.13   ANALYSIS OF SIGNAL-FLOW GRAPHS

For a given system we are interested in determining transfer functions relating certain response variable(s) to some input(s). For example in the automatic position control system in Fig. 4.13b, there are several response variables such as $\theta_o(s)$, $E_m(s)$, $E(s)$ and so on, and we may be interested in determining transfer functions $\theta_o(s)/\theta_i(s)$, $E_m(s)/\theta_i(s)$ or $E(s)/\theta_i(s)$. Any desired transfer function can be obtained by writing equations at various summing junctions in the block diagram (or nodes in signal-flow graphs). This will give n simultaneous equations in n variables. Application of Cramer's rule then yields the desired transfer function. Consider for example the system in Fig. 4.18a (or Fig. 4.18b). There are five response variables X_1, X_2, X_3, X_4, and Y represented at five junctions or nodes. We now write equations for each of the five junctions in Fig. 4.18a (or Fig. 4.18B).

These are

$$X_1 = H_1F - G_2X_2 - GY$$
$$X_2 = H_2X_1$$
$$X_3 = X_2 - G_3X_4$$
$$X_4 = H_3X_3 - G_4Y$$
$$Y = H_4X_4$$

In this graph, for a given input $F(s)$, we have five unknown variables X_1, X_2, X_3, X_4, and Y and five equations. Therefore by applying Cramer's rule to these equations we can determine transfer function relating any one of these five variables to the input $F(s)$.

There is, however, a better method of determining these transfer functions by inspection of the signal-flow graph. A simple rule due to S. J. Mason allows us to determine the transfer function between two nodes by mere inspection of the signal-flow graph. The rule will be given here without proof.† We shall define a few terms before explaining the rule.

An *independent node* or a *source node* is a node with at least one branch radiating from it but none directed toward it (examples are node θ_i in Fig. 4.17 and node $F(s)$ in Fig. 4.18b). All other nodes are dependent nodes.

A *path* is a branch or a continuous sequence of branches which can be traversed without meeting any branch directed in the opposite direction (examples are path X_1-X_2-X_3-X_4, and so on in Fig. 4.18b).

An *open path* is a path along which no node is met twice (an example is F-X_1-X_2-X_3-X_4-Y in Fig. 4.18b).

A loop is a closed path or a closed loop (example are loops X_1-X_2-X_1, X_3-X_4-X_3, X_4-Y-X_4, and X_1-X_2-X_3-X_4-Y-X_1 in Fig. 4.18b.

A *path-transfer function* is the total transfer function along the path—that is, the product of all branch transfer functions.

A *loop-transfer function* is the total transfer function along the loop.

The formula for the transfer function T_{xy} between an independent or source node X and a dependent node Y is

$$\frac{Y}{X} = T_{xy} = \frac{\sum_k T_k \Delta_k}{\Delta} \tag{4.55}$$

where T_k is the path-transfer function of the kth open path between nodes X and Y and

$$\Delta = 1 - \sum_i L_i + \sum_{i,j} L_i'L_j' - \sum_{i,j,k} L_i''L_j''L_k'' + \cdots \tag{4.56}$$

In this equation L_i represents the loop-transfer function of the ith loop. Thus ΣL_i represents the sum of all individual loop transfer functions. $L_i'L_j'$ represents

---

†For proof see S. J. Mason, "Feedback Theory-Further Properties of Signal Flow Graphs," Proc. IRE 44 (1956) 920–926.

the product of loop-transfer functions of any two nontouching loops. Thus $\Sigma L_i' L_j'$ represents the sum of the products of the loop transmittances of all possible pairs of nontouching loops. Similarly $\Sigma L_i'' L_j'' L_k''$ represents the sum of the products of loop-transfer functions of all possible triplets of nontouching loops, and so forth. The quantity Δ_k is the value of Δ for that part of the graph which does not touch the kth open path.

We illustrate the application of this rule to the flow graph in Fig. 4.17. For this graph there is only one loop with transfer function $-G_2 G_3 G_4$. Hence

$$\Delta = 1 - (-G_2 G_3 G_4) = 1 + G_2 G_3 G_4$$

There is only one open path between input θ_i and the response θ_0. The transfer function T_1 of this path is $G_1 G_2 G_3$. Also when the forward path is removed from the graph, the loop is broken. Hence Δ_1 for this graph is

$$\Delta_1 = 1 - (0) = 1$$

Hence

$$\frac{\theta_0(s)}{\theta_i(s)} = \frac{T_1 \Delta_1}{\Delta} = \frac{G_1 G_3 G_3}{1 + G_2 G_3 G_4}$$

This is identical to the result obtained earlier (Eq. 4.54).

Let us apply this rule to a slightly more complex graph in Fig. 4.18b. For this graph there is only one open path between the output Y and the input F. The transfer function of this path is $H_1 H_2 H_3 H_4$. Therefore

$$T_1 = H_1 H_2 H_3 H_4$$

There are a total of four loops: X_1-X_2-X_1, X_3-X_4-X_3, X_4-Y-X_4, and X_1-X_2-X_3-X_4-Y-X_1. The respective loop transfer functions are $-H_2 G_2$, $-H_3 G_3$, $-H_4 G_4$ and $-H_2 H_3 H_4 G$. Hence

$$\Sigma L_i = -(H_2 G_2 + H_3 G_3 + H_4 G_4 + H_2 H_3 H_4 G)$$

There are two pairs of nontouching loops:

$$X_1\text{-}X_2\text{-}X_1 \quad \text{and} \quad X_3\text{-}X_4\text{-}X_3$$
$$X_1\text{-}X_2\text{-}X_1 \quad \text{and} \quad X_4\text{-}Y\text{-}X_4$$

Hence

$$\Sigma L_i' L_j' = G_2 H_2 G_3 H_3 + G_2 H_2 G_4 H_4$$

There are no combinations of three or more nontouching loops. Hence

$$\Delta = 1 + (G_2 H_2 + G_3 H_3 + G_4 H_4 + H_2 H_3 H_4 G)$$
$$+ (G_2 H_2 G_3 H_3 + G_2 H_2 G_4 H_4) \quad (4.57)$$

Also Δ_1 is the value of Δ for that part of graph which does not touch path F-X_1-X_2-X_3-X_4-Y. Since all loops of the graph touch this path, all the loops are broken if this path is removed. Hence

$$\Delta_1 = 1 - 0 + 0 - 0 + \cdots = 1$$

and

$$T = \frac{H_1 H_2 H_3 H_4}{1 + G_2 H_2 + G_3 H_3 + G_4 H_4 + H_2 H_3 H_4 G + G_2 H_2 G_3 H_3 + G_2 H_2 G_4 H_4}$$

We can find the transfer function relating any node variable to the input in Fig. 4.18. Consider for example, the transfer function relating X_3 to the input F.

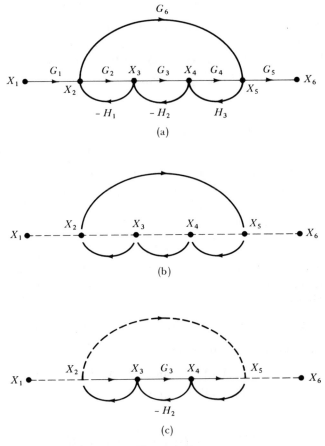

Figure 4.19

The Δ for the graph is already found in Eq. 4.57. There is one open path between X_3 and F with path transfer function $H_1 H_2$. When this path (F-X_1-X_2-X_3) is removed all the loops except one (X_4-Y-X_4) are opened. This loop has a transfer function $-G_4 H_4$. Therefore

$$\Delta_1 = 1 - (-G_4 H_4) = 1 + G_4 H_4$$

and

$$\frac{X_3}{F} = \frac{T_1 \Delta_1}{\Delta}$$

$$= \frac{H_1 H_2 (1 + G_4 H_4)}{1 + G_2 H_2 + G_3 H_3 + G_4 H_4 + H_2 H_3 H_4 G + G_2 H_2 G_3 H_3 + G_2 H_2 G_4 H_4}$$

We shall consider one more flow graph shown in Fig. 4.19a. The transfer function relating the response X_6 to the input X_1 will be found by using Mason's rule. For this graph, there are three loops and two nontouching loops. Hence

$$\Delta = 1 - (-G_2 H_1 - G_3 H_2 + G_4 H_3) + (-G_2 H_1 G_4 H_3)$$
$$= 1 + G_2 H_1 + G_3 H_2 - G_4 H_3 - G_2 G_4 H_1 H_3$$

There are two paths between X_1 and X_6 (shown dotted in Figs. 4.19b and 4.19c):

$$T_1 = G_1 G_2 G_3 G_4 G_5, \qquad \Delta_1 = 1$$
$$T_2 = G_1 G_6 G_5, \qquad \Delta_2 = 1 + G_3 H_2$$

The graphs to compute Δ_1 and Δ_2 are shown in Fig. 4.19b and c respectively. In Fig. 4.19b path 1 (shown dotted) is removed. This opens all the loops and hence $\Delta_1 = 1$. In Fig. 4.19c, the second path (shown dotted) is removed. This opens two outer loops. The center loop with transfer function $-G_3 H_2$ is unbroken. Hence $\Delta_2 = 1 + G_3 H_2$. Therefore

$$\frac{X_6}{X_1} = \frac{G_1 G_2 G_3 G_4 G_5 + G_1 G_6 G_5 (1 + G_3 H_2)}{1 + G_2 H_1 + G_3 H_2 - G_4 H_3 - G_2 G_4 H_1 H_3}$$

4.14 SYSTEM SIMULATION

It is very desirable to simulate a given system in the laboratory in order to study its behavior. This allows us to observe the changes in system characteristics caused by varying different system parameters. The simulation here implies simulation in a mathematical sense. The simulated system is not identical to the given system in specific details. But the two systems will have identical system equations (or transfer functions).

Consider a system whose output y is related to the input f by a differential equation of the form

$$(p^n + a_{n-1} p^{n-1} + \cdots + a_1 p + a_0) y$$
$$= (b_m p^m + b_{m-1} p^{m-1} + \cdots + b_1 p + b_0) f(t) \qquad (4.58)$$

We can easily simulate this equation by using differentiators and/or integrators. The differentiators, however, are undesirable because they tend to accentuate the stray noise.† In addition, integrator is more convenient than differentiator for the

---

†This follows from the fact that ideal differentiator transfer function is $H(s) = s$ and $H(j\omega) = j\omega$. Hence the gain of a differentiator increases linearty with frequency. This causes transmission (and accentuation) of noise spectrum. On the other hand, the ideal integrator transfer function $H(s) = 1/s$ or $H(j\omega) = 1/j\omega$. The gain of integrator decreases with frequency and hence it tends to suppress the noise signal spectrum at higher frequencies.

purpose of simulating differential equations. Hence it is the practice to use only integrators in the laboratory for simulation. Let us first find the transfer function of an ideal integrator. Let y be the output of an integrator when the input is x. Thus

$$y(t) = \int_{-\infty}^{t} x(\tau)\,d\tau$$

$$= \int_{-\infty}^{0} x(\tau)\,d\tau + \int_{0}^{t} x(\tau)\,d\tau$$

$$= y(0) + \int_{0}^{t} x(\tau)\,d\tau \qquad\qquad (4.59a)$$

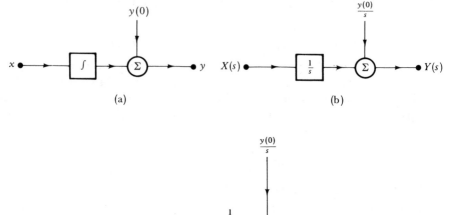

(a)   (b)

(c)

Figure 4.20   Integration with initial conditions.

This equation can be simulated by an arrangement shown in Fig. 4.20a. If the initial condition is zero, then $y(0) = 0$ and the arrangement simplifies to that shown in Fig. 4.21a. Note that in both these figures, the integrators begin integrating at some initial instant $t = 0$.

We can derive the same results in the frequency domain. Taking the Laplace

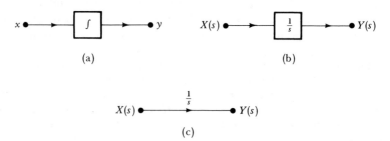

(a)   (b)

(c)

Figure 4.21   Integration with zero initial conditions.

transform of Eq. 4.59a we obtain

$$Y(s) = \frac{y(0)}{s} + \frac{1}{s}X(s) \tag{4.59b}$$

The block diagram for this equation is shown in Fig. 4.20b. Note that this arrangement (Fig. 4.20b) is the frequency domain equivalent of the arrangement in Fig. 4.20a. The transfer function of the integrator is $1/s$. The Laplace Transform of the signal $y(0)$ is $y(0)/s$. Hence the signal $y(0)/s$ in Fig. 4.20b actually represents a dc signal $y(0)$. The corresponding signal flow graph is shown in Fig. 4.20c. The frequency domain representation of the integrator in Fig. 4.21a is shown in Fig. 4.21b and the corresponding signal flow graph is shown in Fig. 4.21c. Note that the transfer function is the ratio of the Laplace transforms of the output and the input when all the initial conditions are zero (see Eq. 4.48). Hence the Transfer function $H(s)$ of an ideal integrator is $1/s$:

$$H(s) = \frac{1}{s} \tag{4.60}$$

Before giving a simulation diagram for a general differential equation 4.58, let us consider a simpler equation

$$(p^3 + a_2 p^2 + a_1 p + a_0)y = f(t) \tag{4.61a}$$

or

$$\dddot{y} + a_2 \ddot{y} + a_1 \dot{y} + a_0 y = f(t) \tag{4.61b}$$

and

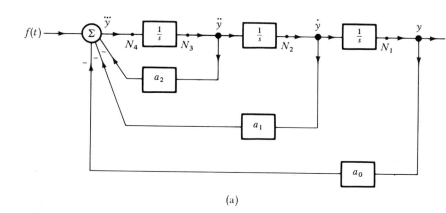

(a)

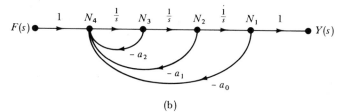

(b)

Figure 4.22

$$\dddot{y} = -a_2\ddot{y} - a_1\dot{y} - a_0y + f(t) \qquad (4.61c)$$

This equation can be simulated by the system shown in Fig. 4.22a. The node equation at the summing junction of this system is seen to be Eq. 4.61c. The variables y, $\dot{y}$, and y are obtained by successive integration of $\dddot{y}$. Note that only integrators, summers and scalar multipliers are used in this scheme. These elements are readily available in the laboratory. The flow graph of the simulated system is shown in Fig. 4.22b. The nodes N_1, N_2, N_3, and N_4 represent signals $Y(s)$, sY, s^2Y, and s^3Y respectively. These are frequency-domain representation of y, $\dot{y}$, $\ddot{y}$, and $\dddot{y}$ respectively. Hence if the initial conditions $y(0)$, $\dot{y}(0)$, and $\ddot{y}(0)$ are nonzero, they must be applied at N_1, N_2, and N_3 respectively (see Fig. 4.20).

Alternately, we could have viewed the simulation problem in the frequency-domain right from the beginning. The transfer function $H(s)$ which relates y to the input $f(t)$ is seen from Eq. 4.61a to be

$$H(s) = \frac{1}{s^3 + a_2 s^2 + a_1 s + a_0} \qquad (4.62a)$$

The simulation problem therefore reduces to simulation of a system whose transfer function is given by Eq. 4.62a. We must remember, however, that we are restricted to use only integrators (components with transfer function $1/s$). Let us divide Eq. 4.62a by s^3 throughout:

$$H(s) = \frac{\dfrac{1}{s^3}}{1 + \dfrac{a_2}{s} + \dfrac{a_1}{s^2} + \dfrac{a_0}{s^3}} \qquad (4.62b)$$

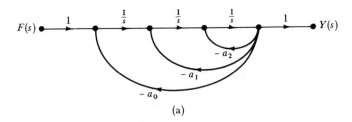

(a)

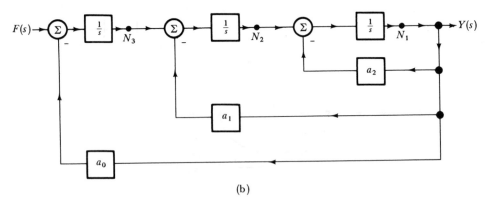

(b)

Figure 4.23

With an eye on Mason's rule, we can now synthesize this transfer function.   The resultant graph must have only one open path between input and the output with transfer function $1/s^3$.   There should be three touching loops with loop-transfer functions a_2/s, a_1/s^2, and a_0/s^3.   This is precisely the situation in Fig. 4.22b.   The same result is also accomplished by the graph in Fig. 4.23a.   The corresponding block diagram is shown in Fig. 4.23b.   Thus the transfer function $H(s)$ in Eq. 4.62a (or the differential Eq. 4.61) can be simulated by two arrangements shown in Figs. 4.22 and 4.23.   It should, however, be noted in Fig. 4.23b, the node $\mathcal{N}_1$ represents y. But the nodes $\mathcal{N}_2$ and $\mathcal{N}_3$ do not represent $\dot{y}$ and $\ddot{y}$.   This is obvious from Fig. 4.23. Unlike Fig. 4.22, the signal appearing at a node in Fig. 4.23b is not the integral of signal at the previous node but has an additional feedback term.   Hence if the given initial conditions are $y(0)$, $y(0)$, $\ddot{y}(0)$, we will have to derive a new set of initial conditions for the variables appearing at the outputs of all the integrators.†

Let us next consider the system whose output y and input f are related by the

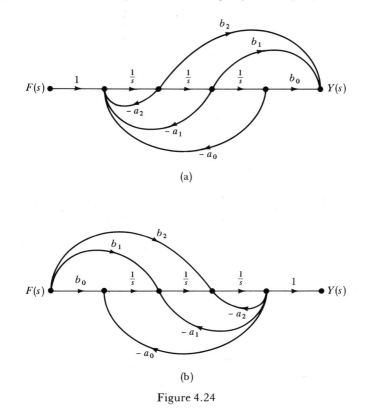

(a)

(b)

Figure 4.24

---

†It can be done by deriving new set of initial conditions for signals appearing at $\mathcal{N}_1$, $\mathcal{N}_2$, and $\mathcal{N}_3$. Let these signals be x_1, x_2, and x_3 respectively. Then from Fig. 4.23b it can be seen that $X_1(s) = Y(s)$, and application of Mason's rule yields $X_2(s) = (s + a_2) Y(s)$ and $X_3(s) = (s^2 + a_2s + a_1) Y(s)$. Taking the inverse Laplace of these equations, we have $x_1 = y_1$, $x_2 = \dot{y} + a_2 y$ and $x_3 = \ddot{y} + a_2\dot{y} + a_1 y$. Hence the appropriate initial conditions at $\mathcal{N}_1$, $\mathcal{N}_2$, and $\mathcal{N}_3$ are $y(0)$, $\dot{y}(0) + a_2 y(0)$, and $\ddot{y}(0) + a_2\dot{y}(0) + a_1 y(0)$, respectively.

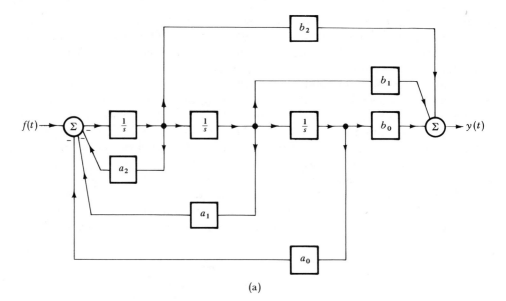

(a)

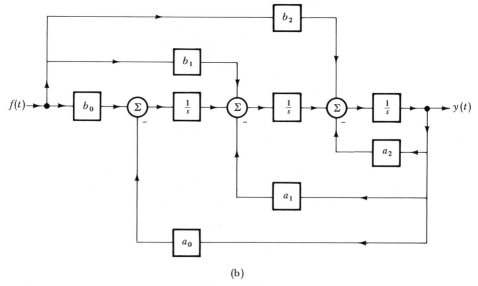

(b)

Figure 4.25

differential equation

$$(p^3 + a_2 p^2 + a_1 p + a_0) y(t) = (b_2 p^2 + b_1 p + b_0) f(t) \qquad (4.63a)$$

We now consider this problem in the frequency domain. The transfer function $H(s)$ of the system is given by

$$H(s) = \frac{Y(s)}{F(s)} = \frac{b_2 s^2 + b_1 s + b_0}{s^3 + a_2 s^2 + a_1 s + a_0} \qquad (4.63b)$$

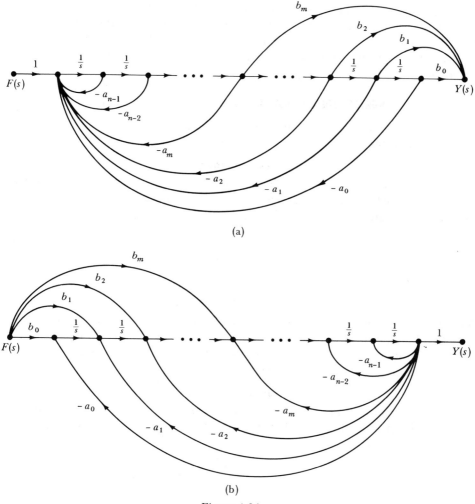

(a)

(b)

Figure 4.26

In order to obtain an appropriate flow graph to synthesize this transfer function, one may employ the same technique as that used to obtain Fig. 4.22b or Fig. 4.23a.

We can rewrite Eq. 4.63b as

$$H(s) = \frac{\dfrac{b_2}{s} + \dfrac{b_1}{s^2} + \dfrac{b_0}{s^3}}{1 + \dfrac{a_2}{s} + \dfrac{a_1}{s^2} + \dfrac{a_0}{s^3}} \tag{4.63c}$$

Comparison of Eq. 4.63c with Eq. 4.62b immediately suggest the appropriate modifications in the flow graph of Fig. 4.22b or Fig. 4.23a in order to synthesize the transfer function in Eq. 4.63c. All we need to do is to create three open paths between the input $F(s)$ and the output $Y(s)$ with transfer functions b_2/s, b_1/s^2, and b_0/s^3, respectively. Moreover these paths must touch all the three loops so that

$\Delta_1 = \Delta_2 = \Delta_3 = 1$.   The signal-flow graphs in Fig. 4.22b and 4.23a can be easily modified as shown in Fig. 4.24a and 4.24b respectively to incorporate these additions. The corresponding block diagrams are shown in Fig. 4.25. We can extend these results easily to the general differential equation,† Eq. 4.58, as shown in Fig. 4.26.

SERIES AND PARALLEL FORM OF REALIZATION

A transfer function (or a differential equation) may also be realized in alternate form known as a series or a parallel form.

Series Realization

In series form, a transfer function is expressed as a product of several transfer functions and each of these component transfer functions is realized separately and their realization are then connected in series. For example, consider

$$H(s) = \frac{5(s + 1)}{s(s + 2)(s + 5)} \tag{4.64a}$$

$$= \left(\frac{5}{s + 2}\right)\left(\frac{s + 1}{s + 5}\right)\left(\frac{1}{s}\right) \tag{4.64b}$$

We can now realize each of these three transfer functions as shown in Fig. 4.27a, b, and c. These individual realizations are now connected in series (Fig. 4.27d) to yield the simulation of the desired transfer function. Figure 4.27e shows the corresponding block diagram representation.

Parallel Realization

A transfer function can be expanded into its partial fractions and each factor can be realized individually and then connected in parallel. Consider the same transfer function (see Eq. 4.64)

$$H(s) = \frac{5(s + 1)}{s(s + 2)(s + 5)}$$

We can expand this into partial fractions as

$$H(s) = \frac{1/2}{s} + \frac{5/6}{s + 2} - \frac{4/3}{s + 5}$$

Each of the three component transfer functions is realized as shown in Figs. 4.28a, b, and c. The three realizations are now connected in parallel as shown in Fig. 4.28d. The corresponding block-diagram representation is shown in Fig. 4.28e.

We cannot realize an individual transfer function for a complex pole because of complex numbers. In such case the two component transfer functions due to the conjugate pair of poles are combined together and are realized as one transfer function.

---

†It is obvious from Fig. 4.26 that we must have $m \leq n$. This restriction is necessary for simulation using only integrators.

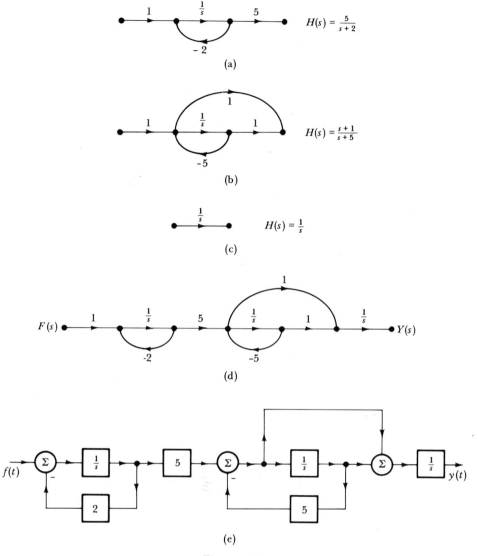

$H(s) = \frac{5}{s+2}$

(a)

$H(s) = \frac{s+1}{s+5}$

(b)

$H(s) = \frac{1}{s}$

(c)

(d)

(e)

Figure 4.27

Example 4.4. Simulate the system shown in Fig. 4.29a. This system can be simulated in several ways. We may realize each of the transfer functions $1/(s + 2)$, $5/(s + 10)$, and $1/(s + 1)$ and connect them as shown in Fig. 4.29b. Another possibility would be to find the transfer function $Y(s)/F(s)$ and realize this transfer function directly. Application of Mason's rule to Fig. 4.29a yields

$$\frac{Y(s)}{F(s)} = \frac{5(s + 1)}{5 + (s + 1)(s + 2)(s + 10)} = \frac{5(s + 1)}{s^3 + 13s^2 + 32s + 25} \qquad (4.65)$$

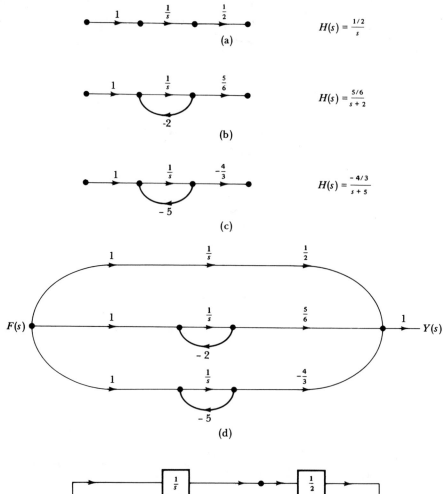

$$H(s) = \frac{1/2}{s}$$

(a)

$$H(s) = \frac{5/6}{s+2}$$

(b)

$$H(s) = \frac{-4/3}{s+5}$$

(c)

(d)

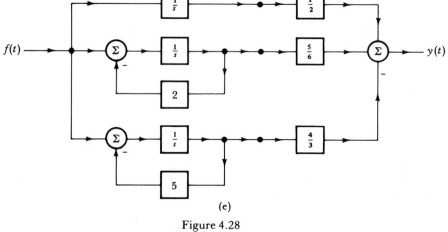

(e)

Figure 4.28

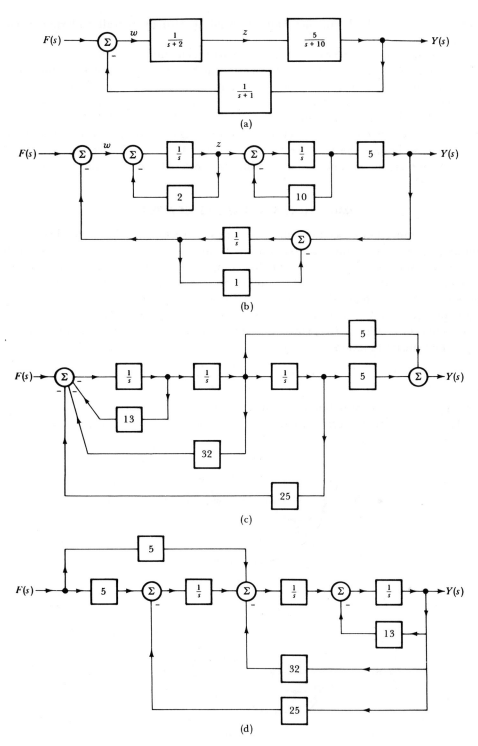

Figure 4.29

We can realize this transfer function by two arrangements found in Fig. 4.25. This is shown in Figs. 4.29c and d. The realization in Fig. 4.29b has the advantage that it retains the structure of the original system. Moreover the variables z and w appearing in the original system are also available (for measurements or observation) in Fig. 4.29b. Such is not the case in Figs. 4.29c and 4.29d. These two realizations are good if we are interested in observing only the output of the system. We may also realize the transfer function in Eq. 4.65 by using series or parallel form. However, this will necessitate additional work of factorizing the denominator of Eq. 4.65.

SIMULATION ON ANALOG COMPUTER

The simulation procedures discussed in this section can be used directly to simulate a linear time-invariant system on an analog computer. The only difference is that the integrators used in analog computers, integrate and reverse the sign of a signal. Hence the transfer function of an integrating element in analog computers is $-(1/s)$ instead of $1/s$. This fact can be easily accounted for by changing signs of signals appearing at alternate integrators. Corresponding sign changes also should be made for feedback and feedforward signals. As an example, consider the transfer function

$$H(s) = \frac{2s + 3}{s^3 + 2s^2 + s + 5}$$

This system can be simulated by using the first cascade form as shown in Fig. 4.30a. If, however, integrators have transfer function $-(1/s)$, the simulation dia-

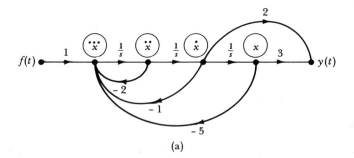

(a)

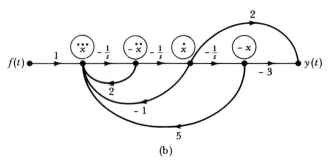

(b)

Figure 4.30

gram will be as shown in Fig. 4.30b. Both the diagrams are identical except for sign changes in the variables appearing at the outputs of alternate integrators. This also causes sign changes in the corresponding feedback and feedforward signals starting from the alternate outputs.

4.15 NATURAL FREQUENCIES OF A SYSTEM

Consider the system in Fig. 4.31. The zero-input response of this system is given by

$$y_x(t) = c_1 e^{\lambda_1 t} + c_2 e^{\lambda_2 t} + \cdots + c_n e^{\lambda_n t}$$

where $\lambda_1, \lambda_2, \ldots, \lambda_n$ are the poles of $H(s)$. This response is sustained by the system on its own without the external driving function. Consequently it is also known as the *natural response*. The complex frequencies $\lambda_1, \lambda_2, \ldots, \lambda_n$ appearing in the natural response are known as the natural frequencies of the response variable y.

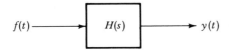

Figure 4.31

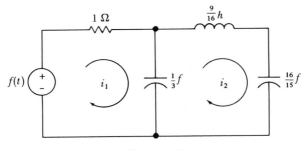

Figure 4.32

A system has several variables or outputs. If λ_k is a natural frequency of any system variable, it is a natural frequency of the system. It might appear that in order to determine natural frequencies of a system we would have to examine all possible transfer functions associated with the system. Fortunately this is not so. We note that a transfer function of a system is obtained by applying Cramer's rule to system equations. The denominator of every transfer function is given by the determinant of the system equation matrix. Hence the natural frequencies of a system can be determined from the determinant of the system equation matrix.

As an example, consider the network in Fig. 4.32. The loop equations of the network are†

---

†These equations are written directly in frequency domain. They are obtained by taking Laplace transform of the time-domain equations.

$$\begin{bmatrix} \left(1 + \dfrac{3}{s}\right) & \dfrac{-3}{s} \\[2ex] \dfrac{-3}{s} & \dfrac{3}{s} + \dfrac{15}{16s} + \dfrac{9s}{16} \end{bmatrix} \begin{bmatrix} I_1(s) \\[2ex] I_2(s) \end{bmatrix} = \begin{bmatrix} F(s) \\[2ex] 0 \end{bmatrix} \qquad (4.66)$$

From Cramer's rule it follows that the denominator of every transfer function of this system will be given by $d(s)$, the determinant of the 2×2 matrix on the left-hand side of Eq. 4.66. This is

$$d(s) = \left(1 + \frac{3}{s}\right)\left(\frac{3}{s} + \frac{15}{16s} + \frac{9s}{16}\right) - \frac{9}{s^2} \qquad (4.67)$$

$$= \frac{9(s^3 + 3s^2 + 7s + 5)}{16s^2} \qquad (4.68)$$

$$= \frac{9(s + 1)(s + 1 - j2)(s + 1 + j2)}{16s^2} \qquad (4.69)$$

The zeros of $d(s)$ are the natural frequencies of the system.†   Hence, the natural frequencies of this network are -1, $-1 + j2$, and $-1 - j2$. We shall obtain the same result by using node equations or any other set of system equations. The equation

$$d(s) = 0$$

is also known as the *characteristic equation* of the system. The roots of the characteristic equation are the natural frequencies of the system.

4.16   SYSTEM STABILITY

The reader is no doubt acquainted with the intuitive concept of stability, which can be illustrated by an example of a right circular cone. If a cone standing on its base is perturbed slightly it will eventually return to its equilibrium position (original position) after the source of disturbance is removed. The cone is said to be in *stable equilibrium*. On the other hand, if it stands on its apex, then the slight disturbance will cause the cone to move farther and farther away from its equilibrium position even after the source of disturbance is removed. In this case the cone is said to be in unstable equilibrium. We use these concepts in defining the stability of a system.

---

†It can be shown that zeros of $d(s)$ yield all natural frequencies with nonzero values. It may or may not give the correct number of natural frequencies with zero value (at the origin). This ambiguity is inherent when a system is described by integrodifferential equations. Such is not the case when a system is described by differential equation. In Chapter 6, where a system is described by differential equations (state equations), it is shown that the determinant $d(s)$ of the state equation matrix gives complete set of natural frequencies.

A linear system is *unstable* if a disturbance, no matter how small, produces a response which grows without limit even after the disturbance is removed. If the response remains within a certain bound, the system is *marginally stable*. If the response eventually vanishes, the system is *asymptotically stable*. When we talk about the response, we mean every possible output or variable in the system. Thus for asymptotically stable system every possible variable or the output eventually vanishes after the disturbance is removed.

When the disturbance is removed the system is in zero-input condition. Hence the stability criterion can be conveniently translated in terms of its zero-input or natural response, as shown below.†

Natural Response	Stability
Grows without limit	Unstable
Remains within a finite bound	Marginally stable
Eventually vanishes	Asymptotically stable

If all the natural frequencies of the system are in the LHP, the system is asymptotically stable. If at least one natural frequency lies in the RHP, the system is unstable. If at least one natural frequency lies along the $j\omega$-axis and all the remaining natural frequencies lie in the LHP, the system is marginally stable. Note that multiple natural frequencies along the $j\omega$-axis give rise to unbounded natural response (see Eq. 2.56a). Hence this will represent unstable system. The network in Fig. 4.32 is an example of asymptotically stable system because all its natural frequencies lie in the LHP.

TESTING FOR STABILITY

The polynomial $d(s)$ can be obtained from the system equations as discussed in Sec. 4.15. In some cases we may be given only a transfer function $H(s)$. The $d(s)$ is simply the denominator‡ of $H(s)$. In most of the cases $d(s)$ is in unfactored form (see Eq. 4.68). In order to determine the natural frequencies (and the stability), $d(s)$ must be factored. This is not very easy for polynomials of order ≥ 3. To determine the system stability, we need not know the exact location of the natural frequencies. It is sufficient to know if there are any natural frequencies in the RHP.

---

†The stability defined here is the zero-input stability and is concerned with the internal dynamics of the system. A stability may also be defined to reflect the external behavior of the system. If for every bounded input the output is bounded for all possible initial conditions, then the system is stable in the BIBO (bounded-input, bounded-output) sense. It can be shown that for linear time-invariant systems, a BIBO stability implies asymptotic stability, and vice versa.

‡In this case we do not have the information about inner structure of the system. It is therefore necessary to make the assumption that $H(p)$ has no common factors. Only when this assumption is made can we determine the system stability.

What we really need is some method of testing a polynomial for location of its zeros. Such a test is provided by the Routh-Hurwitz procedure.

Before discussing this procedure, we shall give one simpler test. A necessary but not sufficient condition for all zeros of a polynomial

$$d(s) = a_0 s^n + a_1 s^{n-1} + \cdots + a_{n-1} s + a_n \tag{4.70}$$

to lie in the LHP is that all the coefficients $a_0, a_1, a_2, \ldots, a_n$ must be nonzero and have the same sign. If $a_0 > 0$, then

$$a_1 > 0, a_2 > 0, \qquad \ldots, a_n > 0 \tag{4.71}$$

To prove this we observe that the polynomial $d(s)$ in Eq. 4.70 has n zeros λ_1, $\lambda_2, \ldots, \lambda_n$ (some of which may be repeated). The zeros may be real or complex with positive or negative real parts. Equation 4.70 can be expressed as

$$d(s) = a_0 (s - \lambda_1)(s - \lambda_2) \cdots (s - \lambda_n) \tag{4.72}$$

Expanding Eq. 4.72 and equating it with Eq. 4.70 yield

$$\frac{a_1}{a_0} = -(\lambda_1 + \lambda_2 + \cdots + \lambda_n)$$

$$\frac{a_2}{a_0} = (\lambda_1 \lambda_2 + \lambda_1 \lambda_3 + \cdots + \lambda_2 \lambda_3 + \lambda_2 \lambda_4 + \cdots + \lambda_{n-1} \lambda_n) \tag{4.73}$$

$$\cdots \cdots \cdots \cdots \cdots \cdots \cdots \cdots \cdots \cdots \cdots \cdots \cdots \cdots$$

$$\frac{a_n}{a_0} = (-1)^n \lambda_1 \lambda_2 \cdots \lambda_n$$

From the set of Eqs. 4.73 we can draw the following important conclusions.

1. If all of the coefficients $a_0, a_1, \ldots, a_n$ are positive real, then the complex zeros, if any, must occur in conjugate pairs.

2. If all of the zeros $\lambda_1, \lambda_2, \ldots, \lambda_n$ are real and negative and if $a_0 > 0$, then it is evident from Eq. 4.73 that all of the coefficients $a_0, a_1, \ldots, a_n$ must be positive real. Similarly, one may establish that if some of the zeros are in the form of complex conjugate pairs with negative real parts, then all of the coefficients must be positive.† It is thus obvious that for all of the zeros to lie in the LHP, a necessary (but not sufficient) condition is that (a) all of the coefficients must be positive, and (b) all the coefficients should be nonzero. To express this succinctly: A necessary condition for all of the zeros to have negative real parts (to lie in the LHP) is that the coefficients of the polynomial $d(s)$ satisfy‡:

$$a_i > 0, \qquad i = 0, 1, 2, \ldots, n \tag{4.74}$$

---

†W. Kaplan, *Operational Methods for Linear Systems* (Reading, Mass.: Addison-Wesley, 1962).

‡The conclusion obviously applies even if all of the coefficients are negative—that is,

$$a_i < 0 \qquad i = 0, 1, 2, \ldots, n$$

The significant point here is that they should all be nonzero and have the same sign.

The above conclusion provides only the necessary condition for stability.   A system satisfying these conditions need not be stable. Consider

$$d(s) = s^3 + 4s^2 - 3s + 2 \tag{4.75}$$

In this case all of the coefficients of $d(s)$ do not have the same signs, and therefore it represents an unstable system.

Similarly, a polynomial

$$d(s) = 2s^3 + 7s + 9 \tag{4.76}$$

represents an unstable system because one of the coefficients a_1 is zero. On the other hand, a polynomial

$$d(s) = 3s^3 + s^2 + 2s + 8 \tag{4.77}$$

also represents an unstable system, since it may be factored as

$$d(s) = (s^2 - s + 2)(3s + 4) \tag{4.78}$$

The term $(s^2 - s + 2)$ represents the zeros in the RHP.

Although in general the condition of Eq. 4.74 is necessary but not sufficient to guarantee the location in the LHP of the zeros, one can easily show that for the first- and second-order equations it serves as a necessary as well as a sufficient condition.

THE ROUTH-HURWITZ CRITERION FOR STABILITY

The condition $a_0 > 0, a_1 > 0, \ldots, a_n > 0$ provides a necessary but not a sufficient condition for the roots of $d(s) = 0$ to lie in the LHP. There are additional conditions, due to Routh, to ensure that the real parts of the roots are negative. This test is known as the *Routh-Hurwitz criterion*. Using this test it is possible to find out exactly the number of roots of a polynomial lying in the LHP and RHP without having to evaluate the actual roots. This test is used in many different forms. The "array form" of the test is particularly useful to us, since it reveals not only whether the system is stable or unstable but also the exact number of roots lying in the LHP and RHP. The proof of this criterion is rather involved. The test procedure will be presented below without proof.

For the nth-order equation with real coefficients,

$$D(s) = a_0 s^n + a_1 s^{n-1} + \cdots + a_{n-1}s + a_n = 0 \tag{4.79}$$

The location of the roots in the complex s-plane is determined by the following procedure. Arrange the coefficients of this equation in two rows as follows:

Then complete the following array (shown here for the sixth-order system):

$$
\begin{array}{cccc}
a_0 & a_2 & a_4 & a_6 \\
a_1 & a_3 & a_5 & 0 \\
b_1 & b_2 & b_3 & 0 \\
c_1 & c_2 & 0 & 0 \\
d_1 & d_2 & 0 & 0 \\
e_1 & 0 & 0 & 0 \\
f_1 & 0 & 0 & 0 \\
0 & 0 & 0 & 0
\end{array}
\tag{4.80}
$$

where the coefficients b, c, e, f, and g are formed as follows:

$$
b_1 = \begin{array}{c} a_0 \quad a_2 \\ \diagdown\!\!\!\!\diagup \\ a_1 \quad a_3 \end{array} = \frac{a_1 a_2 - a_0 a_3}{a_1}
$$

$$
b_2 = \begin{array}{c} a_0 \quad a_4 \\ \diagdown\!\!\!\!\diagup \\ a_1 \quad a_5 \end{array} = \frac{a_1 a_4 - a_0 a_5}{a_1}
$$

$$
c_1 = \begin{array}{c} a_1 \quad a_3 \\ \diagdown\!\!\!\!\diagup \\ b_1 \quad b_2 \end{array} = \frac{b_1 a_3 - a_1 b_2}{b_1}
$$

$$
d_1 = \begin{array}{c} b_1 \quad b_2 \\ \diagdown\!\!\!\!\diagup \\ c_1 \quad c_2 \end{array} = \frac{c_1 b_2 - b_1 c_2}{c_1} \qquad \text{etc.}
$$

In general, any new element is formed from the first column elements of the two rows immediately above the element, and from the two elements above it and in the column to the immediate right of the element. All these coefficients form a determinant like structure, and if we designate α_{jk} as the element in the jth row and kth column, then

$$
\alpha_{jk} = \frac{\alpha_{(j-1),1}\,\alpha_{(j-2),(k+1)} - \alpha_{(j-2),1}\,\alpha_{(j-1),(k+1)}}{\alpha_{(j-1,1)}}
\tag{4.81}
$$

We continue forming such elements until the last row has all zero elements. In the process, the coefficients in any row may be multiplied or divided by a positive number without altering the result. Such operation often simplifies the numerical work of finding the coefficients of succeeding rows.

If all of the terms in the first column (that is, a_0, a_1, b_1, ..., etc.) are of the same sign, all of the roots of the equation lie in the LHP. If there are changes in the sign of the elements in the first column, the total number of changes in sign represents the number of roots in the RHP.

Let us consider the polynomial

$$
d(s) = 2s^4 + s^3 + 12s^2 + 8s + 2
$$

We form the test array according to the rules stated above:

2	12	2
1	8	0
$\dfrac{12 - 16}{1} = -4$	$\dfrac{2 - 0}{1} = 2$	0
$\dfrac{-32 - 2}{-4} = 8.5$	0	0
$\dfrac{17}{8.5} = 2$	0	0
0	0	0

Note that there are two sign changes in the first column, from $+1$ to -4 and from -4 to $+8.5$. Hence, there are two zeros in the RHP and the remaining two zeros must be in the LHP. The system is therefore unstable.

There are certain cases that require special handling. The first case is one in which the first-column term of any row vanishes, but where the remaining terms in this row are not all zero. Under these conditions the first column element in the next row becomes indeterminate. Consider the example

$$d(s) = s^5 + 2s^4 + 2s^3 + 4s^2 + s + 1 = 0 \tag{4.82}$$

The required array is

$$
\begin{array}{ccc}
1 & 2 & 1 \\
2 & 4 & 1 \\
0 & \frac{1}{2} & 0 \\
-\frac{1}{0} & &
\end{array}
\tag{4.83}
$$

Here $b_1 = 0$ and $c_1 = -1/0$, which is indeterminate. We cannot proceed further. There are various ways of handling this situation. One of them is to replace the zero in the first column by an arbitrarily small number δ and proceed as usual. The terms in δ^2 shall be neglected unless there is an uncertainty regarding the relative magnitudes of the derived coefficients. It will be found that the total number of sign changes in the first column is independent of the sign of δ.

Using this procedure we continue the array of Eq. 4.83 as

$$
\begin{array}{ccc}
1 & 2 & 1 \\
2 & 4 & 1 \\
\delta & \frac{1}{2} & 0 \\
(4 - 1/\delta) & 1 & 0 \\
\left(\dfrac{1}{2} - \dfrac{\delta^2}{4\delta - 1}\right) & 0 & 0
\end{array}
\tag{4.84}
$$

We shall neglect the term $\delta^2/(4\delta - 1)$ and also

$$4 - \frac{1}{\delta} \simeq - \frac{1}{\delta}$$

since

$$\delta \ll 1$$

The array thus can be written as

$$
\begin{array}{ccc}
1 & 2 & 1 \\
2 & 4 & 1 \\
\delta & \frac{1}{2} & 0 \\
-1/\delta & 0 & 0 \\
\frac{1}{2} & 0 & 0
\end{array}
\tag{4.85}
$$

Observe that there are in all two sign changes in the first column whether δ is assumed positive or negative. The fifth-order polynomial $d(s)$ therefore has two zeros in the RHP and the remaining three zeros are in the LHP.

Another simple way of handling this situation is to invert the order of the coefficients.† Thus instead of starting the array as shown in Eq. 4.80, we start as follows:

$$
\begin{array}{ccc}
a_5 & a_3 & a_1 \\
a_4 & a_2 & a_0 \\
b_1 & b_2 & b_3
\end{array}
\tag{4.86}
$$

For the polynomial in Eq. 4.82, the inverted array is

$$
\begin{array}{ccc}
1 & 4 & 2 \\
1 & 2 & 1 \\
2 & 1 & 0 \\
\frac{3}{2} & 1 & 0 \\
-\frac{1}{3} & 0 & 0 \\
1 & 0 & 0
\end{array}
\tag{4.87}
$$

We have the same result as before. There is a total of two sign changes. Therefore, $d(s)$ has two roots in the RHP and three roots in the LHP.

A second exception to the standard procedure occurs when all of the coefficients in any of the derived rows vanish. This situation occurs when the corresponding elements in two consecutive rows of an equal number of elements are

---

†See Prob. 4.33 at end of chapter.

proportional. This happens when two diametrically opposite roots, equidistant from the origin, are present—that is, two roots $\sigma, + j\omega,$ and $-\sigma, - j\omega,$ are present. In this case we may complete the array by replacing the row containing zeros by the coefficients of the derivative of the auxiliary polynomial whose coefficients are the numbers in the last nonvanishing row.

We assign the highest power of $d(s)$ to the first row in the RH array and reduce the power by one in successive rows (Eq. 4.89). The auxiliary equation occurs only in the even powers and the power of the auxiliary equation is the power assigned to the last nonvanishing row. Consider for example

$$d(s) = s^4 + 3s^3 + 4s^2 + 6s + 4 = 0 \qquad (4.88)$$

The RH array is

$$
\begin{array}{cccc}
s^4 & 1 & 4 & 4 \\
s^3 & 3 & 6 & 0 \\
s^2 & 2 & 4 & 0 \\
& 0 & 0 & 0
\end{array} \qquad (4.89)
$$

The auxiliary equation is therefore

$$2s^2 + 4$$

The derivative of the auxiliary equation is $4s$. Therefore we should replace the fourth row with element 4. The new modified array is now

$$
\begin{array}{ccc}
1 & 4 & 4 \\
3 & 6 & 0 \\
2 & 4 & 0 \\
4 & 0 & 0 \\
4 & 0 & 0
\end{array} \qquad (4.90)
$$

The roots of the auxiliary equation

$$2s^2 + 4 = 0$$

are also the roots of $d(s) = 0$—that is, $s = \pm j\sqrt{2}$ are roots of $d(s) = 0$. From the array of Eq. 4.90, it is clear that the remaining two roots lie in the LHP and the system is marginally stable.

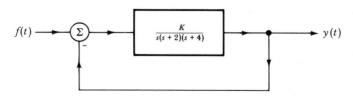

Figure 4.33

The array test discussed above is one form of the Routh-Hurwitz criterion. There are other forms available.

Example 4.5.   Find the range of gain K for which the system in Fig. 4.33 is stable. The transfer function $T(s)$ of the system is given by

$$T(s) = \frac{\dfrac{K}{s(s + 2)(s + 4)}}{1 + \dfrac{K}{s(s + 2)(s + 4)}}$$

$$= \frac{K}{s^3 + 6s^2 + 8s + K}$$

In order to determine the range of K for which this system is stable we must examine the zeros of the denominator polynomial of the transfer function $T(s)$. The Routh-Hurwitz array for this polynomial is given below.

$$
\begin{array}{cc}
1 & 8 \\[4pt]
6 & K \\[4pt]
\dfrac{48 - K}{6} & 0 \\[4pt]
K & 0
\end{array}
$$

It is obvious that there will be no sign changes in the first column when $0 < K < 48$. When K is negative or $K > 48$, the system becomes unstable. The system is marginally stable for $K = 48$.

4.17   INSTABILITY AND OSCILLATIONS

For unstable systems with natural frequencies in the RHP, the natural response grows exponentially. If the natural frequencies are in the RHP along the real axis, the natural response grows monotonically. However, for complex natural frequencies in the RHP, the natural response is exponentially growing sinusoid. Theoretically the natural response of such unstable system grows without limit. If the response becomes too large it may destroy the system. But in many practical cases the growing amplitude of this component causes the changes in the system parameters in such a way as to reduce the instability of the system. The changes in the parameters are such as to shift the natural frequencies in the RHP onto the $j\omega$-axis. The natural response, under such circumstances, grows to a certain value exponentially and then settles down to a constant amplitude as shown in Fig. 4.34.

This is precisely what happens in electronic oscillators and signal generators. These systems are intentionally designed to be unstable to generate the desired frequency. In the beginning the natural response grows exponentially with time, but soon the amplitude is stabilized through the mechanism described earlier. In vacuum tubes the amplification factor μ generally decreases with the signal ampli-

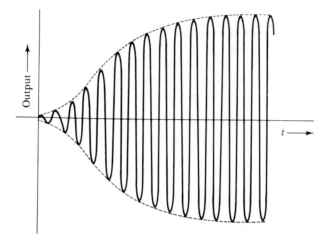

Figure 4.34

tude.   Similarly, in transistors the current gain factor β decreases with signal amplitude.   The decrease in these gain parameters causes the shift of natural frequencies from the RHP toward $j\omega$-axis.   At some signal level the natural frequencies arrive at the $j\omega$-axis and the dynamical equilibrium is reached.   We shall illustrate this point by analyzing the Colpitt oscillator.

COLPITT OSCILLATOR

A transistor Colpitt oscillator is shown in Fig. 4.35a.   A simplified equivalent circuit (model) of the transistor is shown in Fig. 4.35b.   The equivalent circuit of the oscillator is shown in Fig. 4.35c.   This oscillator may be analyzed in several ways.

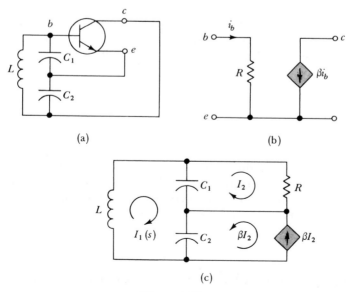

Figure 4.35

In order to determine its natural frequencies we shall write mesh equations and find the mesh equations matrix determinant $d(s)$.

As seen in Fig. 4.35c, there are three meshes. Observe, however, that the currents in the second and the third mesh are I_2 and βI_2. These are not independent currents. Consequently we have only two unknown mesh currents, I_1 and I_2, and we need write only two mesh equations (for meshes 1 and 2). These are

$$
\begin{bmatrix}
Ls + \dfrac{C_1 + C_2}{C_1 C_2 s} & -\dfrac{1}{C_1 s} + \dfrac{\beta}{C_2 s} \\[4mm]
-\dfrac{1}{C_1 s} & R + \dfrac{1}{C_1 s}
\end{bmatrix}
\begin{bmatrix}
I_1(s) \\[4mm]
I_2(s)
\end{bmatrix}
=
\begin{bmatrix}
0 \\[4mm]
0
\end{bmatrix}
$$

The loop-impedance matrix determinant $d(s)$ is given by

$$
\begin{aligned}
d(s) &= \left(Ls + \frac{C_1 + C_2}{C_1 C_2 s}\right)\left(R + \frac{1}{C_1 s}\right) + \frac{1}{C_1 s}\left(-\frac{1}{C_1 s} + \frac{\beta}{C_2 s}\right) \\[3mm]
&= \frac{LC_1 C_2 R s^3 + LC_2 s^2 + R(C_1 + C_2)s + (\beta + 1)}{C_1 C_2 s^2}
\end{aligned}
\tag{4.91}
$$

The natural frequencies are given by the zeros of $d(s)$. The RH array for this polynomial is

$$
\begin{array}{c|cc}
s^3 & LC_1 C_2 R & R(C_1 + C_2) \\[2mm]
s^2 & LC_2 & \beta + 1 \\[2mm]
& R(C_2 - C_1\beta) & 0 \\[2mm]
& \beta + 1 & 0 \\[2mm]
& 0 &
\end{array}
\tag{4.92}
$$

For the zeros to lie in the RHP, we must have sign changes in the first column. Since L, C_1, C_2, R, and β are all positive, the sign change can occur only if the third element in the first column is negative—that is, if

$$
C_1\beta > C_2
$$

or

$$
\beta > \frac{C_2}{C_1}
\tag{4.93}
$$

This represents the condition for growing oscillation. If the transistor current gain β is large enough to satisfy this condition, the circuit will oscillate. If $\beta < C_2/C_1$, the system will be stable. If however, $\beta > C_2/C_1$, the natural frequencies will be in the RHP. For the marginal case,

$$
\beta = \frac{C_2}{C_1}
\tag{4.94}
$$

and the natural frequencies will lie along the $j\omega$-axis. For this case the third row in the RH array (Eq. 4.92) vanishes and hence the auxiliary equation is

$$LC_2s^2 + (\beta + 1) = 0 \qquad (4.95)$$

Substitution of Eq. 4.94 in Eq. 4.95 yields

$$LC_2s^2 + \frac{C_1 + C_2}{C_1} = 0$$

or

$$s^2 + \frac{1}{LC_0} = 0 \qquad \text{where} \quad C_0 = \frac{C_1 C_2}{C_1 + C_2}$$

and

$$s = \pm j \sqrt{\frac{1}{LC_0}}$$

Thus the frequency of oscillation is $1/\sqrt{LC_0}$, provided we choose the current gain β of the transistor large enough to satisfy Eq. 4.93.

In practice, β is chosen large enough so that Eq. 4.93 is satisfied. This causes the natural frequencies to lie in the RHP rather than along $j\omega$-axis. Consequently the response (natural response), which is sinusoidal, grows exponentially in amplitude. As the signal amplitude grows β decreases until such time when the inequality 4.93 is barely satisfied. The natural frequencies have now moved from the RHP to the $j\omega$-axis (due to reduction in β) and the amplitude of oscillation is stabilized.

EXCITATION FOR UNSTABLE SYSTEMS

By definition, an unstable system moves farther away from its equilibrium position if a small disturbance is applied. The disturbance is only momentary and it may be as small as we choose.

The next question is: what happens to an unstable system if it is not disturbed? Theoretically, if an unstable system is not disturbed from its equilibrium state there will be no response. On the other hand, a disturbance, no matter how small, will start a natural response in the system which will eventually grow with time. In practice it is impossible to shield any system from the disturbances of nature.

For example, in electrical circuits, there is always a random-noise voltage present across every element. Similarly, every element is picking up signals by various processes of induction from an electromagnetic field caused by external sources which are always present in space. In a mechanical system, air molecules are constantly exerting a random force. Thus a slight disturbance from these natural phenomena can start a response which eventually grows to a large value. Therefore unstable systems do not need any deliberate excitation to exhibit their growing response. In electronic oscillators the oscillations start without any external deliberate attempt to disturb the system. A small noise voltage, which is always present in a circuit, is enough to start such oscillations.

From the above discussion it is evident that an unstable system is useless as an amplifier or processor of any signal, since the response of such a system to these signals will be swamped or drowned out by the natural response component. Therefore any system used to process signals in any manner must be a stable system. Yet unstable systems are not always liabilities. They can be conveniently used to generate signals of various frequencies. The response of the system are the signals of the natural frequencies of the system. Hence if we set the natural frequencies of a system on the $j\omega$-axis at a frequency $\pm j\omega_0$, then it will generate a response which will be a sinusoidal signal of frequency ω_0. This is the principle of signal generation in electronic oscillators.

PROBLEMS

4.1.   Find the Laplace transforms of the following functions.

(a) $u(t) - u(t - 1)$

(b) $\sin \omega_0 (t - \tau) u(t - \tau)$

(c) $\sin \omega_0 t\, u(t - \tau)$

(d) $\sin \omega_0 (t - \tau) u(t)$

(e) $te^{-t} u(t - \tau)$

(f) $(t - \tau) \sin \omega_0 (t - \tau)$

(g) $\sin \omega_1 t \cos \omega_2 t\, u(t)$

4.2.   Find the inverse Laplace transforms of following functions.

(a) $\dfrac{2s + 5}{s^2 + 7s + 12}$

(b) $\dfrac{2s + 7}{s^2 + 4s + 29}$

(c) $\dfrac{2(s^2 + 3)}{(s + 2)(s^2 + 2s + 5)}$

(d) $\dfrac{4s + 5}{(s + 1)(s + 2)^2}$

(e) $\dfrac{1}{(s + 2)(s + 3)^4}$

(f) $\dfrac{e^{-(s-1)} + 2}{(s - 1)^2 + 4}$

(g) $\dfrac{e^{-s} + e^{-2s} + 1}{(s + 1)(s + 2)}$

(h) $\dfrac{s - a}{(s + a)^2}$

(i) $\dfrac{s^2 + b^2}{(s + a)^2 + b^2}$

4.3.   Find the output y if the transfer function $H(s)$ relating the output y to the input $f(t)$ is given by

$$H(s) = \frac{s + 3}{s^2 + 3s + 2}$$

and if

(a) $f(t) = e^{-3t} u(t)$, $y(0) = 1$, $y'(0) = 2$

(b) $f(t) = e^{-t} u(t)$, $y(0) = 1$, $y'(0) = 2$

(c) $f(t) = e^{t} u(t)$, $y(0) = 1$, $y'(0) = 2$

4.4.   Repeat Prob. 4.3 if

$$H(s) = \frac{s}{s^2 + 4}$$

and if

(a) $f(t) = e^{-t}u(t)$, $y(0) = 1$, $y'(0) = 0$
(b) $f(t) = \cos 2t$, $y(0) = y'(0) = 0$

4.5.   Repeat Prob. 4.3 if

$$H(s) = \frac{3s + 1}{s(s + 1)^2}$$

and if

(a) $f(t) = u(t)$, $y(0) = 0$, $y'(0) = 1$
(b) $f(t) = e^{-t}u(t)$, $y(0) = y'(0) = 0$
(c) $f(t) = te^{-t}u(t)$, $y(0) = y'(0) = 0$

4.6.   Repeat Prob. 4.3 if

$$H(s) = \frac{s + 1}{s^2 + 2s + 1}$$

and if

(a) $f(t) = te^{-t}u(t)$, $y(0) = y'(0) = 0$
(b) $f(t) = u(t)$, $y(0) = 1$, $y'(0) = 0$

4.7.   Repeat Prob. 4.3 if

$$H(s) = \frac{s + 4}{s(s^2 + 3s + 2)}$$

and if

(a) $f(t) = u(t)$, $y(0) = y'(0) = y''(0) = 0$
(b) $f(t) = tu(t)$, $y(0) = y'(0) = y''(0) = 0$

4.8.   Repeat Prob. 4.3 if

$$H(s) = \frac{s^2 + 4s + 5}{s^2 + 3s + 2}$$

and if

(a) $f(t) = e^{-3t}u(t)$, $y(0) = 1$, $y'(0) = 1$
(b) $f(t) = e^{-t}u(t)$, $y(0) = y'(0) = 0$

4.9.   Find the position x of the platform of the vibration absorber in Fig. P-2.8a if $M = 1000$, $D = 1000$, $K = 4000$, and the input $f(t) = 10 \cos 2t$. The initial condition $x(0) = 0$, $x'(0) = 10$.

4.10.   Find the zero-state response $i(t)$ for the parallel resonant circuit in Fig. P-4.10 when the input $f(t)$ is given by

(a) $f(t) = A \cos \omega_0 t$ $\omega_0^2 = \dfrac{1}{LC}$
(b) $f(t) = A \sin \omega_0 t$

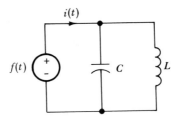

Figure P-4.10

4.11.   Find the transfer function of an ideal delay of T seconds (Fig. P-4.11).

$$f(t) \quad \boxed{\begin{array}{c} \text{Delay} \\ T \end{array}} \quad f(t - T)$$

Figure P-4.11

4.12.   Using initial- and final-value theorem, find the final value of the zero-state response of a system whose transfer function is

$$H(s) \; = \; \frac{s + 4}{s(s^2 + 3s + 2)}$$

and the input is

(a) $u(t)$ (b) $e^{-t}u(t)$

4.13.   The switch K in the network of Fig. P-4.13 is initially in position a, for a long time so as to ensure a steady-state condition. At $t = 0$, the switch is connected to position b. Find the output voltage $v_o(t)$.

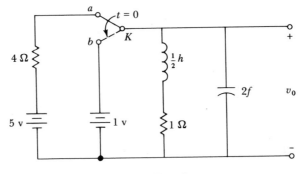

Figure P-4.13

4.14.   Find the voltage v_o in Fig. 2.4 if $f(t) = u(t)$, $v_o(0) = 1$, and $i(0) = 1$.
4.15.   Find the voltage $v_o(t)$ for the network in Fig. P-4.15 if $f(t) = te^{-t}$, $v_C(0) = 2$, and $i_L(0) = 1$.

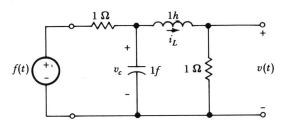

Figure P-4.15

4.16. Find the positions $x_1(t)$, $x_2(t)$ in the mechanical system shown in Fig. P-4.16 if $x_1(0) = x_2(0) = 0$, $x_1'(0) = 10$, and $f(t) = u(t)$.

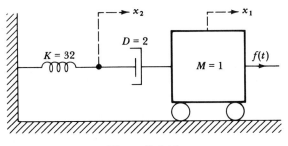

Figure P-4.16

4.17. Linear differential equations can be solved by taking the Laplace transforms of the equations as discussed in Sec. 4.9. Using this technique, solve the following differential equations

(a) $(p + 2)y + 3 = 0$, $y(0) = 2$
(b) $(p^2 + 3p + 2)y + 2e^{-t} = 0$, $y(0) = 1$, $y'(0) = 2$

4.18. Linear differential equations can be solved by taking the Laplace transforms of the equations as discussed in Sec. 4.9. Using this technique, solve simultaneous differential equations in Prob. 2.2a (p. 86), for x_1 and x_2, if

(a) $f(t) = u(t)$, $x_1(0) = 2$ and $x_1'(0) = 1$
(b) $f(t) = u(t)$, $x_1(0) = 1$ and $x_2(0) = 2$

4.19. Repeat Prob. 4.18 for simultaneous differential equations in Prob. 2.2b.
4.20. Repeat Prob. 4.18 for simultaneous differential equations in Prob. 2.2c.
4.21. Repeat Prob. 4.18 for simultaneous differential equations in Prob. 2.2d.
4.22. A block diagram of a certain system is shown in Fig. P-4.22. Find the following transfer function:

(a) $\dfrac{Y(s)}{F(s)}$ (b) $\dfrac{Z(s)}{F(s)}$ (c) $\dfrac{X(s)}{F(s)}$

4.23. Find the transfer function $Y(s)/F(s)$ and $X(s)/F(s)$ for the systems whose signal-flow graphs are shown in Fig. P-4.23.
4.24. For the system in Fig. P-4.24 (see p. 237) find the transfer function $Y(s)/F(s)$

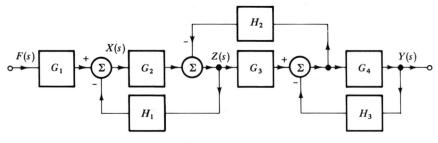

Figure P-4.22

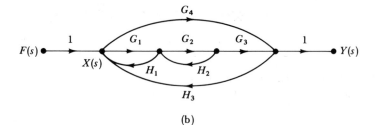

(a)

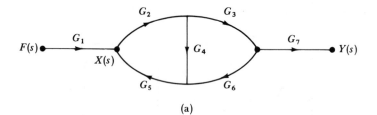

(b)

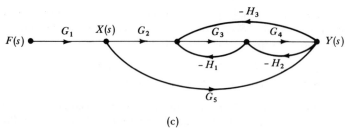

(c)

Figure P-4.23

(a) by using Mason's rule, (b) without using Mason's rule (write equations at each of the junctions and apply Cramer's rule to these simultaneous equations. Comment.

4.25.   A system transfer function is given by

$$H(s) = \frac{s(s + 2)}{(s + 1)(s + 3)(s + 4)}$$

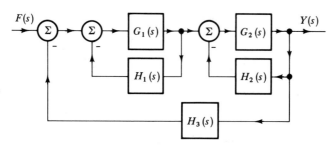

Figure P-4.24

Simulate this system using cascade form (Figs. 4.26a and 4.26b), series form (Fig. 4.27) and parallel form (Fig. 4.28). Draw corresponding simulation diagrams for an analog computer.

4.26. Repeat Prob. 4.25 if

$$H(s) = \frac{2s + 3}{s(s + 2)^2(s + 3)}$$

4.27. Repeat Prob. 4.25 if

$$H(s) = \frac{s^3}{(s + 1)(s + 2)(s + 3)}$$

4.28. Determine the natural frequencies of the systems shown in Fig. P-4.28 (see p. 238).

4.29. A linear model of a single-axis gyro (rate gyro) is shown in Fig. P-4.29 (see p. 239). Determine the natural frequencies of this system.

4.30. A linearized model of an angular position control servo is shown in Fig. P-4.30 (see p. 239). Determine the natural frequencies of this system.

4.31. Determine the range of K for which each of the systems in Fig. P-4.31 (see p. 239) is stable. In each case determine the value K for which the system becomes marginally stable. When the system is marginally stable, determine the values of its poles along the $j\omega$-axis.

4.32. Apply the Routh-Hurwitz criterion to determine whether any of the roots of the following polynomials lie in the RHP.

(a) $s^5 + s^4 + 2s^3 + s + 2$
(b) $s^4 + s^3 + s^2 + 10s + 10$
(c) $8s^4 + 2s^3 + 3s^2 + s + 5$
(d) $10s^4 + 2s^3 + s^2 + 5s + 3$
(e) $s^4 + 5s^3 + 10s^2 + 20s + 24$
(f) $s^6 + 2s^4 + 8s^2 + 2s + 3$
(g) $s^4 + 10s^3 - 8s^2 + 2s + 3$
(h) $s^5 + 2s^4 + 2s^3 + 4s^2 + 11s + 10$

4.33. Show that if a polynomial

$$a_n s^n + a_{n-1} s^{n-1} + \cdots + a_1 s + a_0 = 0$$

has k roots in the RHP, then the inverted polynomial

$$a_0 s^n + a_1 s^{n-1} + \cdots + a_{n-1} s + a_n = 0$$

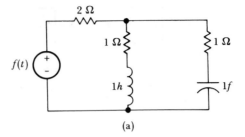

(a)

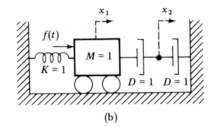

(b)

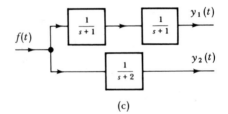

(c)

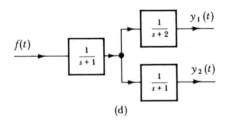

(d)

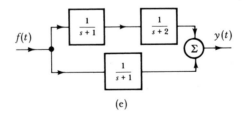

(e)

Figure P-4.28

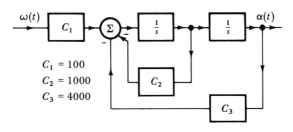

$C_1 = 100$
$C_2 = 1000$
$C_3 = 4000$

Figure P-4.29

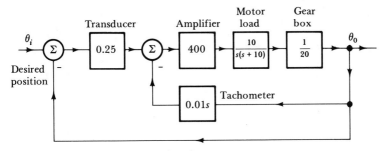

Figure P-4.30

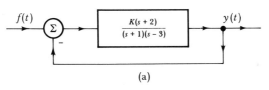

(a)

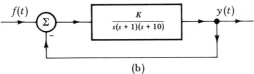

(b)

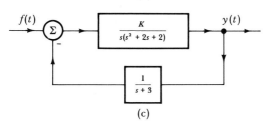

(c)

Figure P-4.31

also has k roots in the RHP. [*Hint:* If a polynomial has a root $s_0 = \sigma_0 + j\omega_0$ in the RHP, then the reciprocal of this root $(1/s_0) = 1/\sigma_0 + j\omega_0$) also lies in the RHP.]

4.34.   Find the condition for oscillation and the natural frequency of oscillation of a transistor Hartley oscillator shown in Fig. P-4.34a.   The equivalent circuit of the transistor is shown in Fig. P-4.34b.   The two inductor coils are coupled with coupling coefficient M.

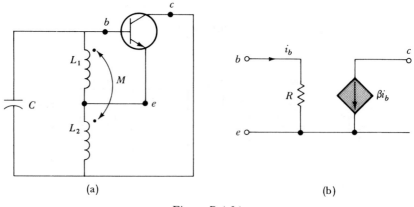

(a)   (b)

Figure P-4.34

5

Feedback and Control

5.1 INTRODUCTION

Design of a control system poses the following problem. We are given a system (called *the plant*) and the input r (Fig. 5.1a). The output y of the system must meet certain given performance criteria. The plant in general is unalterable or fixed. The only way the output can meet the given performance criteria is to select the appropriate input f to the plant. Once the input f is determined we can design a subsystem (*the controller*) which will generate f from the given input r (Fig. 5.1a) The input f generated by the controller, in contrast to the external input r is called the *control input*.

Assume that we have a system that yields the desired performance. In practice, however, the plant characteristics change with time, as a result of aging or replacement of some components, or because of changes in the environment in which the system is operating. Hence for a given control input f the output of the system will also change with time. This condition is very undesirable.

A possible solution to this problem is to apply an input to the plant which is not a predetermined function of time but which will change to counteract the effects of changing plant characteristics and the environment. In short, we must provide a correction at the plant input to account for the undesired changes mentioned above. Unfortunately these changes are generally unpredictable, and raise the question of finding appropriate corrections to the input. There is a simple solution to this problem. The unpredictable changes affect the output, hence the actual output is no longer the desired output. We can observe the actual output and compare it with the desired output. The difference between the actual and the desired output will give an indication of the unpredictable changes in the system, and may therefore be used to correct the input. This requires the output to be fed back to the input for comparison. Thus the input to the controller consists of two components (i) some predetermined external input r (called the *reference input*), and (ii) the output of the system. The output is continuously monitored and fed back to the controller. The input to the plant is therefore continuously adjusted in order to obtain the desired response. Such systems (Fig. 5.1b) are called *feedback systems* because the plant output is fed back to the system. These systems are also called *closed-loop systems* for obvious reasons. The systems which do not use feedback (Fig. 5.1a) are called *open-loop systems*. Superiority of closed-loop systems (when properly designed) over open-loop systems is evident. Feedback is used to achieve,

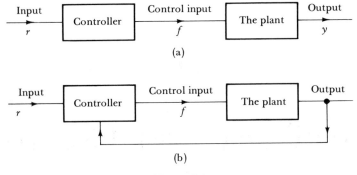

Figure 5.1

with a given system, the desired objective within a given tolerance, despite partial ignorance of the system and the environment.

We can observe thousands of examples of feedback systems around us in everyday life. Most social, economical, educational, and political processes are in fact feedback processes. The human system itself is a fine example of a feedback system; almost all of our actions are the product of feedback mechanism. The human sensors such as eyes, ears, nose, tongue, and touch are continuously monitoring the state of our system. This information is fed back to the brain, which acts as a controller to apply corrected inputs through our motor mechanisms and thus accomplish the desired objective.

It will be very instructive for the reader to consider the process of driving an automobile. Our senses are continuously feeding the information to the brain. Eyes report seeing a child in the road. The brain will immediately apply input to the arms to steer the car away from the child. There is a red light ahead. Again the brain will apply the input to the legs to brake the car. Suddenly ears report an ambulance siren, the brain will apply corrective input to steer the car off the street temporarily. A foul smell is reported by the nose. The brain will again take measures to speed away the car from the spot. In this example, the child, the red light, the ambulance siren, and the foul smell are unpredictable changes in the environment (fed back to the input by our senses). Despite all this, the process of reaching one destination from another is completed because of feedback. The foregoing is an example of *multiple variable feedback*.

The so-called automatic control systems are actually feedback mechanisms. In Chapter 4 we discussed an automatic position control system (Fig. 4.13). In this system the input θ_i is proportional to the desired angle of the object. The output position (angle θ_o) is fed back at the input and is compared with θ_i, the desired angle. If the two angles are not equal, the difference $\theta_i - \theta_o$ is applied to the motor which will turn until $\theta_o = \theta_i$. Under this condition the motor input is zero and the object stays at the desired position. In this example, the controller is simply a comparator. The control input is proportional to $(\theta_i - \theta_o)$, the difference between the desired position and the actual position. A household heating system with a thermostatic control is another familar example of a feedback system.

Almost all technological processes demanding precision utilize feedback mechanisms.

As stated earlier, feedback systems are mainly used to counteract the effects
of unpredictable variations in the system parameters, the load and the environment. All these unpredictable variations may be classified as partial ignorance of
the system. It may be appropriately said that feedback is used in order to achieve
with a given system the desired objectives within a given tolerance, despite partial
ignorance of the system and of the environment.

A feedback system thus has a certain ability for supervision and self-correction
in the face of changes in the plant parameters, and external disturbances (change
in the environment). As an example, consider the feedback amplifier shown in
Fig. 5.2. The forward amplifier gain G is 10,000. One hundredth of the output

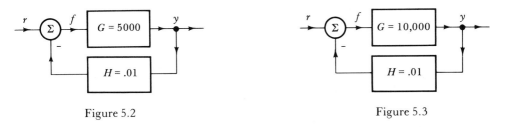

Figure 5.2   Figure 5.3

is fed back to the input ($H = 0.01$). The gain T of the feedback amplifier is
given by

$$T = \frac{G}{1 + GH} = \frac{10,000}{1 + 100} \simeq 99$$

Let us now assume that due to aging or replacement of some transistors the gain
of the forward amplifier dropped to 5000 as shown in Fig. 5.3. The new gain T of
the feedback amplifier is given by

$$T = \frac{G}{1 + GH} = \frac{5000}{1 + 50} \simeq 98$$

Observe that 100% variation in the forward gain G causes only 1% variation† in T.

A practical system is also affected by unwanted disturbances. Examples are
random-noise signals in electronic systems, a gust of wind affecting a tracking
antenna, a meteorite hitting a spacecraft, and the rolling motion of antiaircraft gun
platforms mounted on ships or moving tanks. The feedback properly applied can
reduce or eliminate the effects of such disturbances. Feedback may also be used
to reduce nonlinearities in a system, or control its rise time (or bandwidth).

---

†Note, however, if the signal that is fed back is added to the input, instead of being sub-
tracted, and if $|GH| < 1$, the situation will be exactly opposite. In this case variations in
T are larger than those in G. Such a feedback is called *positive feedback* and is rarely used.

5.2   ANALYSIS OF A SIMPLE CONTROL SYSTEM

The designing of a linear system finally amounts to choosing proper location in the s-plane for the poles and zeros of various transfer functions. In order to understand the role of the pole-zero locations in determining system performance, let us analyze a second-order system.

Consider the position control system in Fig. 4.13. In this system the input is a desired angular position θ_i, the output is the actual angular position θ_o. The

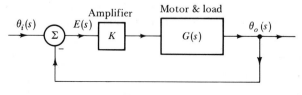

Figure 5.4

output θ_o is fed back to the input where it is compared with the input θ_i (desired position). The difference proportional to $\theta_i - \theta_o$ is amplified and fed to the motor. This arrangement† is shown in Fig. 5.4. Let us consider the system where

$$G(s) = \frac{1}{s(s + 8)} \tag{5.1}$$

and

$$K = 80$$

The transfer function $T(s)$ is given by

$$T(s) = \frac{\theta_o(s)}{\theta_i(s)} = \frac{KG(s)}{1 + KG(s)} \tag{5.2}$$

$$= \frac{80/s(s + 8)}{1 + 80/s(s + 8)}$$

$$= \frac{80}{s^2 + 8s + 80} \tag{5.3}$$

Let us see how this system performs under various inputs.

1. STEP INPUT

When the output position of the control system in Fig. 5.4 is required to change from one position to the other (an antiaircraft gun for example), a step input of an amount equal to the desired change, must be applied. Let us observe

---

†Note that this arrangement is equivalent to that in Fig. 4.13b. In Fig. 4.13b, $G_1 = G_4$ and the transfer function of the arrangement in Fig. 4.13b is identical in form to that in Fig. 5.4.

the behavior of the system for unit step input. If $\theta_i(s) = 1/s$ the output $\theta_o(s)$ is given by (see Eq. 5.3)

$$\theta_o(s) = \frac{80}{s(s^2 + 8s + 80)} = \frac{80}{s(s + 4 - j8)(s + 4 + j8)}$$

$$= \frac{1}{s} + \frac{(\sqrt{5}/4)e^{j153°}}{s + 4 - j8} + \frac{(\sqrt{5}/4)e^{-j153°}}{s + 4 + j8}$$

and

$$\theta_o(t) = \mathcal{L}^{-1}[\theta_o(s)] = 1 + \frac{\sqrt{5}}{2}e^{-4t}\cos(8t + 153°) \qquad (5.4)$$

The output θ_o in Eq. 5.4 is plotted in Fig. 5.5. Note several important features of

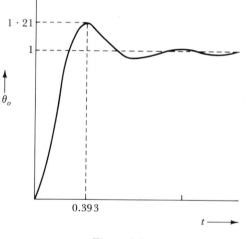

Figure 5.5

this response. It is oscillatory, with an overshoot of 21% where the percent over-shoot (PO) is defined as

$$PO = \frac{\theta_{max} - \theta_{steady\text{-}state}}{\theta_{steady\text{-}state}} \times 100\%$$

The response reaches its peak value at *peak time* $t_p = 0.393$. We are also concerned about the speed of the response. The rise time t_r and the delay time t_d give an in-dication of speed of the response. The rise time t_r is defined as the time required for the response to rise from 10 to 90% of its steady-state value. The delay time t_d is defined as the time required by the response to reach 50% of its steady-state value. From Fig. 5.5 we read $t_r = 0.175$ sec, and $t_d = 0.141$ sec.

The steady-state value of the response is 1. This is exactly what is desired. Hence there is no difference between the desired response and the actual response. Thus the error in the steady state is zero. Theoretically it takes infinite time for the response to reach the desired value of unity. In practice, however, we may con-sider the response to have settled to the final value if it is close to the final value. A

widely accepted measure of closeness is within 2% of the final value. The time required for the response to reach and stay within 2% of the final value is called the
settling time t_s of the system.† In Fig. 5.5 the settling time $t_s \simeq 1$. A good system
has a small overshoot, a small value of t_r and t_s, and a small steady-state error.

2. RAMP INPUT

If the antiaircraft gun in Fig. 5.4 is tracking an enemy plane moving with a
uniform velocity, the gun-position angle must increase linearly with t. In other
words, the input should be a ramp input. Let us find the response of this system
when θ_i is a unit ramp $tu(t)$. The response in this case is

$$\theta_o(s) = \frac{80}{s^2(s^2 + 8s + 80)}$$

$$= \frac{-0.1}{s} + \frac{1}{s^2} + \frac{(1/16)e^{j37°}}{s + 4 - j8} + \frac{(1/16)e^{-j37°}}{s + 4 + j8}$$

and

$$\theta_o(t) = -0.1 + t + \tfrac{1}{8}e^{-4t}\cos(8t + 37°) \tag{5.5}$$

This function is sketched in Fig. 5.6. It can be seen from this figure that the actual
position angle never reaches the desired position angle. There is a steady-state
error of 0.1 radians. In many cases such a small steady-state error may be toler-
able. If, however, a zero steady-state error to a ramp input is required, this system
in its present form is unsatisfactory. We must add some form of compensator to
the system.

Figure 5.6

5.3   DESIGN SPECIFICATIONS

The above discussion gives some ideas regarding possible specifications for
a control system. In general we may be required to design a control system
to meet some or all of the following specifications.

---

†Some authors use the definition of t_s as the time required to reach and stay within
5% of the final value.

1. Sensitivity
   (a) Specified sensitivity to some plant parameter variations.
   (b) Specified sensitivity to certain disturbance inputs.
2. Transient Response
   (a) Specified overshoot to step input.
   (b) Specified rise time t_r and delay time t_d.
   (c) Specified settling time t_s.
3. Steady-state error
   Specified steady-state error to certain expected inputs such as step, ramp, or parabolic inputs.†

In control systems the transient response is specified for the step input. This is because the step input represents a sudden jump discontinuity. Hence if a system has an acceptable transient response for step input, it is likely to have acceptable transient response for most of the practical inputs. Steady-state errors, however, must be specified for typical inputs of the system. For the given system one must determine what kind of input (step, ramp, etc.) are likely to occur, and then specify acceptable steady-state requirements for these inputs.

5.4   A SECOND-ORDER SYSTEM

It is easy to see that the transient response depends upon the location of poles and zeros of the transfer function $T(s)$. For a general case, however, there is no quick way of predicting transient response parameters (PO, t_r, t_s) from the knowledge of poles and zeros of $T(s)$. This means we may have to use cut-and-dry methods to determine the appropriate pole-zero locations of $T(s)$ to meet the given transient specifications. Workers in the field have prepared several charts and plots which show the relationships between various combinations of pole-zero position and the respective transient response. Using these graphs, a designer is able to choose an appropriate range of pole-zero locations for $T(s)$ to meet transient specifications. He can then check whether the system so designed also meets the sensitivity and steady-state error specifications. If not, he may have to consider using some form of compensation.

For a second-order system with no zeros, there is a direct relationship between the pole locations and the transient response. In such a case the pole locations can be immediately determined from the knowledge of transient parameters. If a higher-order system has a pole-zero configuration in which there are two complex poles near the $j\omega$-axis and all remaining poles and zeros are far away from $j\omega$-axis, then it can be shown that the system response is dominated by the two poles close to $j\omega$-axis (dominant poles). The system behaves essentially like a second-order system. Since many higher-order systems in practice fall under this

---

†In many problems the requirement on the error may be in the form of some performance index. If we denote the error by $e(t)$, then we may be required to minimize a certain performance index which represents some average effect of $e(t)$. The performance index may be an integral of $e^2(t)$, $|e(t)|$, or $te^2(t)$, etc., over the entire interval.

category, study of second-order systems is very useful. For this reason we shall now study the behavior of a second-order system in detail.

TRANSIENT RESPONSE OF A SECOND-ORDER SYSTEM

Let us consider a second-order transfer function $T(s)$ given by

$$T(s) = \frac{\omega_n^2}{s^2 + 2\zeta\omega_n s + \omega_n^2} \tag{5.6}$$

It can be seen from Eq. 5.6 that the poles of $T(s)$ are real for $\zeta \geq 1$ and complex for $\zeta < 1$. Oscillatory motion will be observed only for complex poles ($\zeta < 1$). For this reason we shall consider the case where $\zeta < 1$. The poles of $T(s)$ are $\zeta\omega_n \pm j\omega_n \sqrt{1 - \zeta^2}$. These are shown in Fig. 5.8.

The unit step response $y(t)$ is given by

$$y(t) = \mathcal{L}^{-1} \frac{\omega_n^2}{s(s^2 + 2\zeta\omega_n s + \omega_n^2)}$$

$$= \mathcal{L}^{-1} \left[\frac{1}{s} + \frac{\dfrac{1}{2\sqrt{1 - \zeta^2}} e^{j\theta}}{s + \zeta\omega_n - j\omega_n\sqrt{1 - \zeta^2}} + \frac{\dfrac{1}{2\sqrt{1 - \zeta^2}} e^{-j\theta}}{s + \zeta\omega_n + j\omega_n\sqrt{1 - \zeta^2}} \right]$$

where

$$\theta = -(\cos^{-1}\sqrt{1 - \zeta^2} + 180°) \tag{5.7}$$

Hence

$$y(t) = \left[1 + \frac{1}{\sqrt{1 - \zeta^2}} e^{-\zeta\omega_n t} \cos(\omega_n\sqrt{1 - \zeta^2}\, t + \theta) \right] u(t) \tag{5.8}$$

The response y as a function of $\omega_n t$ is shown in Fig. 5.7 for various values of ζ. Note that there is no overshoot† for $\zeta \geq 1$. The overshoot increases as ζ decreases. We shall now relate the response characteristics such as overshoot, rise time t_r, and settling time t_s to the locations of the poles of $T(s)$. The poles of $T(s)$ are $-\zeta\omega_n \pm j\omega_n \sqrt{1 - \zeta^2}$. The pole locations are shown in Fig. 5.8.

The quantity ζ is known as the *damping ratio*. From Fig. 5.7 we observe that smaller values of ζ indicate oscillations with lesser damping. It takes longer time for the oscillations to damp out. On the other hand, larger values of ζ cause quick damping of oscillations.

For $\zeta < 1$, the poles of the transfer function are complex and the response is oscillatory. For $\zeta \geq 1$, the poles of the transfer function are real and the response is monotonic (no overshoot). When $\zeta = 1$, both the poles are real but coincident at $s = -\omega_n$ (see Eq. 5.6). In this case, the system is said to be *critically damped*.

From Fig. 5.7 it can be seen that smaller damping causes higher overshoot

---

†The analysis here is carried out for $\zeta < 1$. For $\zeta \geq 1$ the poles of $T(s)$ are real and the response is a sum of two monotonically decaying exponentials and a constant (unity). Hence there will be no overshoot for $\zeta \geq 1$.

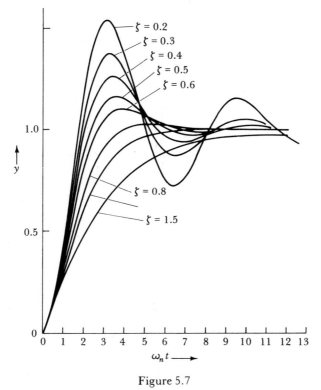

Figure 5.7

and shorter rise time. Critical damping provides the smallest rise time without any overshoot.

From Fig. 5.8 we observe that ζ is constant for all poles lying along a radial line emanating from the origin in the s-plane. Thus each radial line in the s-plane represents a certain value of ζ (Fig. 5.11a). Observe that the distance of either of the poles from the origin is ω_n. The angular frequency of oscillation of the response is $\omega_n \sqrt{1 - \zeta^2}$. This is the imaginary part of the poles of $T(s)$. Let us define

$$\omega_d = \omega_n \sqrt{1 - \zeta^2} \qquad (5.9)$$

where ω_d is known as the *damped natural frequency*. As seen from Eq. 5.8, this is the frequency of oscillation of the response. Also from Eq. 5.8, the oscillatory component decays exponentially as $e^{-\zeta \omega_n t}$. The time constant of the envelope is therefore $1/\zeta \omega_n$. We may consider $1/\zeta \omega_n$ as the time constant τ of the second-order system in Eq. 5.6.

$$\tau = \frac{1}{\zeta \omega_n} \qquad (5.10)$$

The exponential signal decays to about 2% of its initial value in four time constants. Therefore the amplitude of the oscillatory component will be within 2% of the final

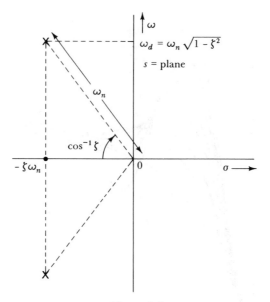

Figure 5.8

value for all $t \geq 4/\zeta\omega_n$. In short, the response settling time t_s is given by†

$$t_s = \frac{4}{\zeta\omega_n} \qquad (5.11a)$$

Note that the real part of the poles of $T(s)$ is $-\zeta\omega_n$. Hence the settling time is inversely related to the real part of the poles of $T(s)$.

Let us now find t_p, the peak time (time when the response reaches the first peak). It can be easily seen that t_p can be obtained from equation

$$\frac{dy}{dt} = 0 \qquad \text{at} \quad t = t_p$$

Straightforward differentiation of Eq. 5.8 yields

$$\frac{dy}{dt} = 0$$

for

$$t = \frac{k\pi}{\omega_d} \qquad k = 0, 1, 2, \ldots$$

The first peak is obtained for $k = 1$. Hence

$$t_p = \frac{\pi}{\omega_d} \qquad (5.11b)$$

---

†This settling time applies to the envelope. The actual settling time t_s for the response to settle down to within 2% of the final value may be somewhat smaller than t_s.

Thus the peak time is inversely related to the imaginary part of the poles of $T(s)$.
     To find the percent of overshoot (PO), we find y at $t = t_p$:

$$\text{PO} = \frac{\text{peak value} - \text{final value}}{\text{final value}} \times 100\%$$

$$= [y(t_p) - 1] \times 100\%$$

$$= \left[1 - \frac{1}{\sqrt{1 - \zeta^2}}\, e^{-\zeta\pi/\sqrt{1-\zeta^2}} \cos(\pi + \theta) - 1 \right] \times 100\%$$

$$= \left(e^{-\zeta\pi/\sqrt{1-\zeta^2}} \right) \times 100\% \tag{5.11c}$$

The percent of overshoot PO as a function of ζ is shown in Fig. 5.9. Note that for a
second-order system (Eq. 5.6), the PO is directly related to the damping ratio ζ.

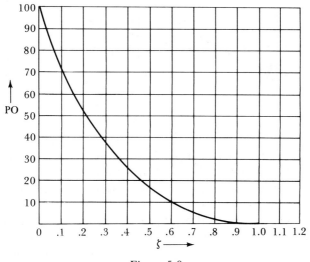

Figure 5.9

We may now proceed to determine similar expressions for t_r, and t_d. Unfortunately
these turn out to be transcendental equations. These equations can be solved on a
digital computer. Results of these computation are shown in Fig. 5.10, where
$\omega_n t_r$, and $\omega_n t_d$ are plotted as a function of ζ.
     For the second-order system (Eq. 5.6), it is evident that the pole locations
determine the transient behavior of the system. It can easily be seen that closer the
poles to the $j\omega$-axis, smaller the value of ζ, and larger the PO. Generally speaking
poles should not be too close to the $j\omega$-axis. Systems whose poles are too close to
the $j\omega$-axis have little safety margin as far as the stability is concerned. A little
change in some system parameter can cause large increase in PO, or worse, may
push the poles in the RHP, thus causing instability. Hence it is generally desirable
to have a large value of ζ (small PO). For a fast response it is desirable to have

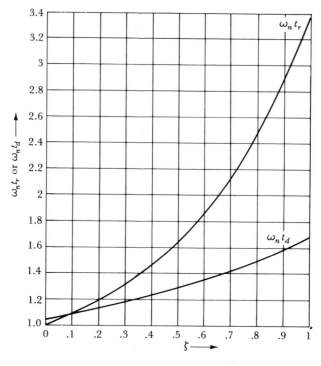

Figure 5.10

small values for t_p, t_r, t_s, t_d, etc. Hence for a fast response and good stability it is desirable to have small values for PO, t_r, t_s, t_p, and t_d.

It is evident from this discussion that all the transient parameters (PO, t_p, t_s, t_r, t_d) are related to the pole location of $T(s)$ of the second-order system. From the point of view of the system design it will be more convenient to relate the important transient parameters PO, t_s, and t_r directly to the s-plane. This can be done as follows. We know that each radial line drawn from the origin in the s-plane represents a constant ζ line. Since the PO is directly related to ζ (Fig. 5.9), each radial line also represents a line of constant PO, as shown in Fig. 5.11a. Thus we can be sure that PO $\leq 16\%$ if we choose $T(s)$ such that both of its pole lie in the region demarcated by two radial lines PO = 16%.

From Eq. 5.11a, we have

$$t_s = \frac{4}{\zeta \omega_n}$$

It can be seen from Fig. 5.8 that $\zeta \omega_n$ represents the real part of the poles of $T(s)$. Hence a constant t_s implies constant value of $\zeta \omega_n$. The contours of constant t_s are, therefore, verticle lines as shown in Fig. 5.11b.

Similarly the rise time is a function of ζ and ω_n as shown in Fig. 5.10. From this figure we can draw a family of contours representing constant t_r, as shown in Fig. 5.11c. All the three sets of contours are superimposed in Fig. 5.11d. This plot

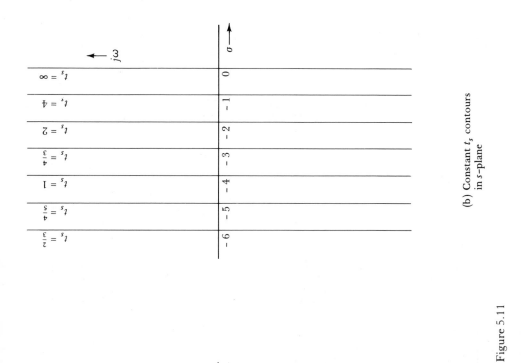

(b) Constant t_s contours in s-plane

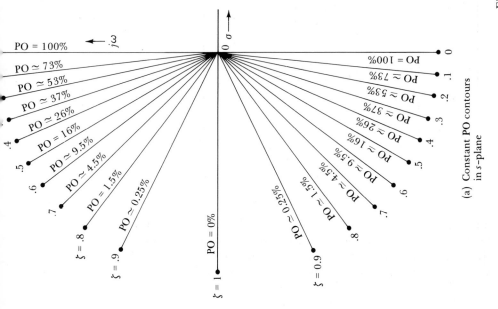

(a) Constant PO contours in s-plane

Figure 5.11

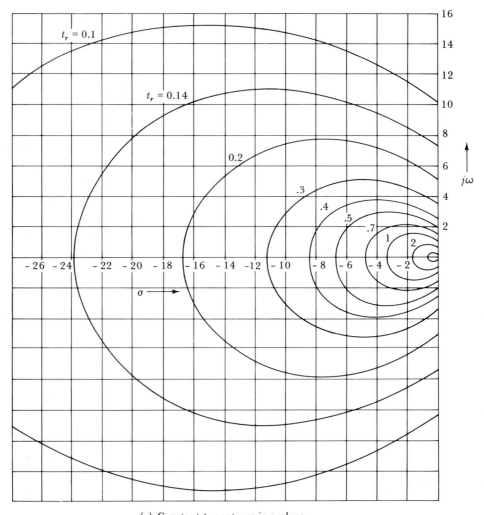

(c) Constant t_r contours in s-plane

Figure 5.11 (*continued*)

allows us to determine by inspection, the important transient characteristics (PO, t_r, t_s) of a second-order system from the knowledge of its pole locations. On the other hand if we are required to synthesize a second-order system to meet a given transient specification, we can find the desired $T(s)$ with the help of this figure (Fig. 5.11d).

As an example consider the position control system in Fig. 5.4. Let the transient specifications for this system be given as

$$PO \leq 16\%$$
$$t_r \leq 0.5 \text{ sec}$$
$$t_s \leq 2 \text{ sec}$$

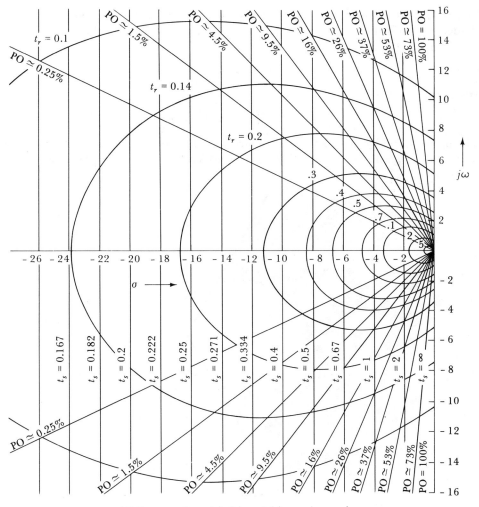

(d) Contours from (a), (b), and (c) superimposed.

Figure 5.11 (*continued*)

We can delineate appropriate contours in Fig. 5.11d to meet the above specifications. The shaded region defined by these contours meets all the three requirements (Fig. 5.12). Hence $T(s)$ must be chosen so that both of its poles lie in the shaded region. The transfer function $T(s)$ for the system in Fig. 5.4 is given by

$$T(s) = \frac{K/s(s + 8)}{1 + K/s(s + 8)} \tag{5.12}$$

$$= \frac{K}{s^2 + 8s + K} \tag{5.13a}$$

It is obvious from this equation that locations of the poles of $T(s)$ can be adjusted by changing the gain K. We must choose the gain K so that the poles lie in the shaded region in Fig. 5.12. The poles of $T(s)$ are the roots of the characteristic equation (see Eq. 5.13a)

$$s^2 + 8s + K = 0 \qquad\qquad (5.13\text{b})$$

Hence the poles are

$$s_1, s_2 = -4 \pm \sqrt{16 - K}$$

The poles s_1 and s_2 (the roots of the characteristic equation) move along a certain path in the s-plane as we vary K from 0 to ∞. When $K = 0$, the poles are -8, 0.

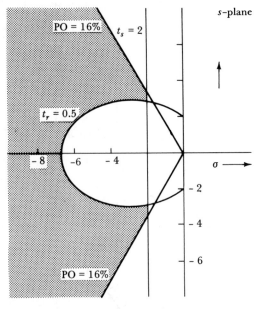

Figure 5.12

For $K < 16$, the poles are real and both poles move towards a value -4 as K varies from 0 to 16. For $K = 16$, both poles coincide at -4. For $K > 16$, the poles become complex:

$$s_1, s_2 = -4 \pm j\sqrt{K - 16}$$

Note that the real part of the poles is -4 for all $K > 16$. Hence the path of the poles is vertical as shown in Fig. 5.13a. One pole moves up and the other (its conjugate) moves down along the vertical line passing through -4. We can label the value of K for several points along these paths as shown in Fig. 5.13. Each of these paths represents a locus of the poles of $T(s)$ or the locus of the roots of characteristic equation of $T(s)$ as K is varied from 0 to ∞. For this reason this set of paths is called the *root locus*. The root locus gives us the information as to how

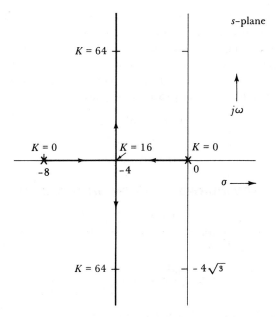

Figure 5.13

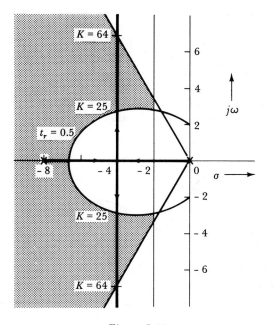

Figure 5.14

the poles of the closed loop transfer function $T(s)$ move as the gain K is varied from 0 to ∞. In our design problem, we must choose a value of K such that the poles of $T(s)$ lie in the shaded region shown in Fig. 5.12. This is most conveniently accomplished by superimposing Fig. 5.13 on Fig. 5.12 as shown in Fig. 5.14. We observe from this figure that the system will meet the given specifications (Eq. 5.12) for $25 \leq K \leq 64$. For $K = 64$, for example, we have

$$\text{PO} = 16\%$$
$$t_r = 0.2 \text{ sec} \tag{5.14}$$
$$t_s = \tfrac{4}{4} = 1 \text{ sec}$$

SECOND-ORDER SYSTEM WITH A ZERO

The second-order transfer function in Eq. 5.6 has no zero. A second-order transfer function with a finite zero is observed frequently in practice. For this reason we shall consider this case.

$$T(s) = \frac{\omega_n^2}{a} \left[\frac{s + a}{s^2 + 2\zeta\omega_n s + \omega_n^2} \right] \tag{5.15}$$

The multiplying constant ω_n^2/a is chosen in order to have a unit steady-state value for unit step input. It will have no influence on the PO, t_r, and t_s. The response $y(t)$ of this system to unit step input is given by

$$y(t) = \mathcal{L}^{-1} \frac{\omega_n^2}{a} \left[\frac{s + a}{s(s^2 + 2\zeta\omega_n s + \omega_n^2)} \right] \tag{5.16a}$$

$$= \mathcal{L}^{-1} \left[\frac{\omega_n^2}{s(s^2 + 2\zeta\omega_n s + \omega_n^2)} \right] + \frac{1}{a} \mathcal{L}^{-1} \left[\frac{s\omega_n^2}{s(s^2 + 2\zeta\omega_n s + \omega_n^2)} \right] \tag{5.16b}$$

Note that the first term of the right-hand side of this equation is $y_0(t)$, the unit step response of the second-order system with no zero (Eq. 5.6). The term inside the second bracket is s times the term in the first bracket. Hence from the time

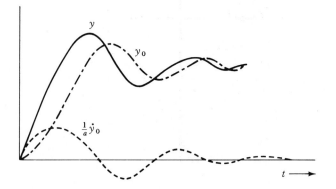

Figure 5.15

differentiation property of the Laplace transform (Eq. 4.35a), the second term is dy_0/dt. Therefore

$$y(t) = y_0(t) + \frac{1}{a}\frac{dy_0}{dt} \tag{5.17}$$

where $y(t)$ is the response of the second-order system with a zero and $y_0(t)$ is the response of the second-order system without a zero. This is shown in Fig. 5.15. It is obvious from this figure that $y(t)$ is faster than and has a larger overshoot than $y_0(t)$. However, since the time constant is still $1/\zeta\omega_n$, the settling time t_s is practically unchanged. The complete analysis of this system (derivation of $y(t)$, t_p, PO) is given in Appendix E. The result of this analysis is shown in Figs. 5.16a and 5.16b. The PO and $\omega_n t_r$ are plotted as a function of $a/\zeta\omega_n$ for various values of ζ.

5.5   HIGHER-ORDER SYSTEMS

We have extensively discussed the relationship between the pole location of $T(s)$ and the transient response when $T(s)$ is of second order. If $T(s)$ has additional poles which are far away to the left of $j\omega$-axis, they have negligible effect on the transient behavior of the system. This is because the time constants of such poles are considerably smaller when compared to the time constant of the complex conjugate poles near the $j\omega$-axis. Consequently, the exponentials arising due to poles far away from the $j\omega$-axis die quickly compared to those arising due to poles located near the $j\omega$-axis. In addition, the coefficients of the former terms are much smaller than unity. Hence they are also very small to begin with and decay rapidly. Consider for example

$$T(s) = \frac{20}{(s + 10)(s^2 + 2s + 2)}$$

The unit-step response of this system is given by

$$y(t) = \mathcal{L}^{-1}\frac{20}{s(s + 10)(s^2 + 2s + 2)}$$

$$= \mathcal{L}^{-1}\left[\frac{1}{s} - \frac{1/41}{s + 10} + \frac{(5/\sqrt{41})\,e^{j129^\circ}}{(s + 1 - j1)} + \frac{(5/\sqrt{41})\,e^{-j129^\circ}}{s + 1 + j1}\right]$$

$$= 1 - 0.0248\,e^{-10t} + 1.56e^{-t}\cos(t + 129^\circ)$$

The pole at $s = -10$ is 10 times as far from the $j\omega$-axis as the complex conjugate poles at $-1 \pm j1$. The pole at -10 gives rise to the term $-0.0248\,e^{-10t}$. It can be seen that this term not only decays rapidly compared to the sinusoidal term $1.56\,e^{-t}\cos(t + 129^\circ)$ but is very small to begin with. Consequently the transient behavior (overshoot, peak time, settling time, etc.) is essentially determined by the complex poles near the origin, $-1 \pm j1$. The poles at $-1 \pm j1$ are called the *dominant poles*.

A criterion commonly used is that any pole which is six times as far from the $j\omega$-axis as the dominant poles contributes negligibly to the step-input response.

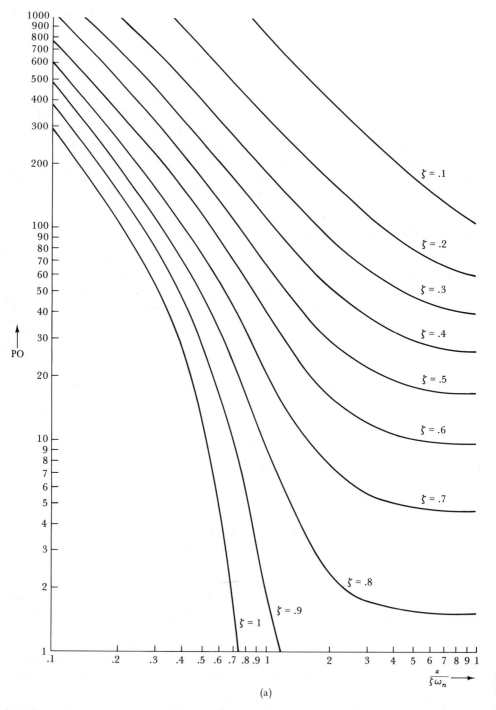

(a)

Figure 5.16

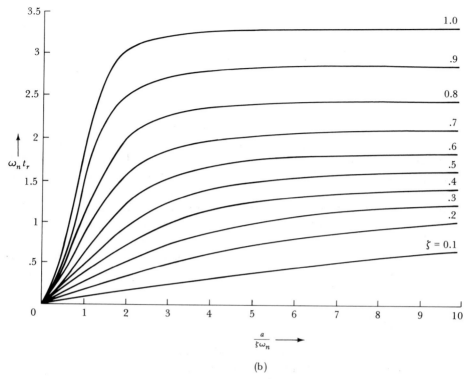

(b)

Figure 5.16 (*continued*)

In other words, we can ignore any pole (real or complex) whose real part is greater than six times the real part of the dominant poles. If a system has a pair of dominant poles and all other poles are at least six times as far away from the $j\omega$-axis as the dominant poles, the system behaves essentially like a second-order system with the dominant poles so far as the transient behavior is concerned.

A pole and a zero which are closely spaced are called a *dipole*. We can easily show that presence of a dipole in $T(s)$ also contributes negligibly to the transient response and consequently may be ignored. Consider

$$T(s) = \left(\frac{2a}{a + \delta}\right)\frac{(s + a + \delta)}{(s + a)(s^2 + 2s + 2)}$$

where $\delta \rightarrow 0$. In this case we have a pair of complex conjugate poles at $-1 \pm j1$ and a dipole with pole at $-a$ and a zero at $-(a + \delta)$. The unit step response of this system is given by

$$y(t) = \mathcal{L}^{-1}\left[\frac{\dfrac{2a}{a + \delta}(s + a + \delta)}{s(s + a)(s + 1 - j1)(s + 1 + j1)}\right]$$

$$= \mathcal{L}^{-1}\left[\frac{1}{s} + \frac{c_1}{s + a} + \frac{c_2 e^{j\theta}}{s + 1 - j1} + \frac{c_2 e^{-j\theta}}{s + 1 + j1}\right]$$

where

$$c_1 \simeq -\frac{\delta}{2a(a^2 - 2a + 2)}$$

$$c_2 \simeq \frac{1}{4\sqrt{2}}$$

It can be seen that since $\delta \to 0$, $c_1 \ll c_2$ and the effect of the dipole may be ignored. The step response is predominantly determined by the poles $-1 \pm j1$.

Because of the feasibility of ignoring poles far away from the $j\omega$-axis, and neglecting the effect of dipoles in $T(s)$, the majority of the pole-zero configuration of practical significance which are encountered in design of feedback control systems reduces to two or three poles with one or two zeros. Workers in the field of control system have investigated such pole-zero configurations, and have prepared plots of the step response of these systems for several values of poles and zeros.[†] From these charts one can evaluate the transient behavior of higher order systems of practical significance.

5.6   ROOT LOCUS

The example in Sec. 5.4 gives a good idea about the utility of the root locus. The root locus tells us about the location of poles of $T(s)$ for any value of the gain K in the forward path.   The root locus is the locus of the roots of the characteristic equation.   For the system in Fig. 5.17, the transfer function $T(s)$ is given by

$$T(s) = \frac{G(s)}{1 + G(s)H(s)}$$

Hence the characteristic equation of the system is

$$1 + G(s)H(s) = 0 \tag{5.18}$$

where the forward-path transfer function, $G(s)$ includes a variable-gain factor K. The quantity $G(s)H(s)$ is known as the open-loop transfer function.[‡]   It can be expressed as

$$G(s)H(s) = K\frac{(s - z_1)(s - z_2)\cdots(s - z_m)}{(s - p_1)(s - p_2)\cdots(s - p_n)} \tag{5.19}$$

where $z_1, z_2, \ldots, z_m$ are the zeros and $p_1, p_2, \ldots, p_n$ are the poles of the open-loop transfer function $G(s)H(s)$.

[†]O. I. Elgerd and W. C. Stephens, "Effects of Closed-Loop Transfer Function Pole and Zero Location on the Transient Response of Linear Control Systems," *AIEE Trans.* 78 Pt. II (Applications and Industry), (May 1959), 121–127.

[‡]If the feedback loop is opened at any point, the transfer function around the loop (the loop-transfer function) is $-G(s)H(s)$.

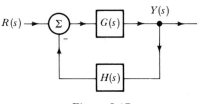

Figure 5.17

The characteristic equation 5.18 can be expressed as

$$G(s)H(s) = -1 = 1e^{j(r\pi)} \qquad r = \pm 1, \pm 3, \ldots \tag{5.20}$$

Hence

$$\angle G(s)H(s) = r\pi, \qquad r = \pm 1, \pm 3, \ldots \tag{5.21a}$$

and

$$|G(s)H(s)| = 1 \tag{5.21b}$$

Therefore all points in the s-plane that satisfy Eq. 5.20 (or Eqs. 5.21a and 5.21b) are the points on the root locus.

Substitution of Eq. 5.19 in Eqs. 5.21a and 5.21b yields following equations

$$\sum_{i=1}^{m} \angle (s - z_i) - \sum_{j=1}^{n} \angle (s - p_j) = r\pi, \qquad r = \pm 1, \pm 3, \ldots \tag{5.22a}$$

and

$$\frac{\prod_{i=1}^{m} |s - z_i|}{\prod_{j=1}^{n} |s - p_j|} = \frac{1}{K} \tag{5.22b}$$

or

$$\frac{\prod_{j=1}^{n} |s - p_j|}{\prod_{i=1}^{m} |s - z_i|} = K \tag{5.22c}$$

A point on the root locus must satisfy both these equations. On the other hand if we wish to test whether a certain point (test point) s_0 in the s-plane lies on the root locus, we must let $s = s_0$ in Eqs. 5.22a and 5.22b and see whether it satisfies these equations.

Let us first consider Eq. 5.22a. This is the angle criterion for the test point. Consider a test point $s = s_0$ in the s-plane (Fig. 5.18a). Let z be some other point in the s-plane (Fig. 5.18a). The directed line segment from the origin to s_0 represents s_0 in polar form.   The length of this line is $|s_0|$, the magnitude of s_0,

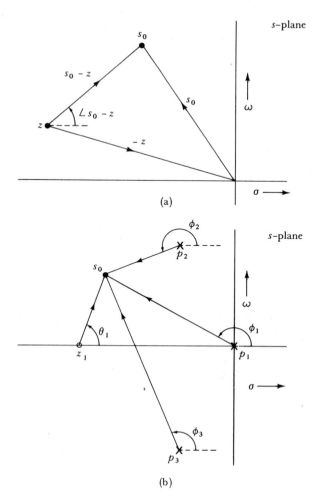

Figure 5.18

and the angle of this line (as measured from the positive real axis) is $\angle s_0$. Similarly, the line from the origin to the point z represents the complex number z in polar form. Obviously the line from z to the origin represents point $-z$ in polar form. Hence the directed line segment drawn from z to s_0 is $s_0 - z$. The angle of this line segment is $\angle s_0 - z$ and the length of this line segment is $|s_0 - z|$. This result points the way to test whether a certain point lies along the root locus. We draw directed line segments from all the poles and zeros of the open-loop transfer function $G(s)H(s)$ to the test point (Fig. 5.18b). If the angle of the directed line segment associated with a zero z_i is θ_i and that associated with a pole p_j is ϕ_j, then the point on root locus must satisfy (see Eq. 5.22a)

$$(\theta_1 + \theta_2 + \cdots + \theta_m) - (\phi_1 + \phi_2 + \cdots + \phi_n) = r\pi \qquad r = \pm 1, \pm 3, \ldots \qquad (5.23a)$$

The value of K corresponding to this point is given by Eq. 5.22c. If α_i is the length of the line segment associated with a zero at z_i and β_j is the length of the line segment associated with a pole at p_j, then (see Eq. 5.22c)

$$K = \frac{\beta_1 \beta_2 \cdots \beta_n}{\alpha_1 \alpha_2 \cdots \alpha_m} \tag{5.23b}$$

We have given here a method of determining whether or not a certain point lies on a root locus. If the point lies on the root locus, the corresponding value of K is determined from Eq. 5.23b. Thus in Fig. 5.13, the value of K at the point $(-4 + j4\sqrt{3})$ is found to be 64. The distance of this point from either of the poles is 8. There are no open-loop zeros. Hence according to Eq. 5.23b

$$K = (8)(8) = 64$$

The method of sketching the root locus by testing several points in the s-plane appears very time-consuming. Fortunately there are several rules due to Evans† which make it possible to sketch root loci with ease. These rules are remarkable for their simplicity and will now be listed.

Rule 1.   Root loci begin on open-loop poles and terminate on open-loop zeros.

Proof.   The characteristic equation is

$$1 + G(s)H(s) = 0$$

Substituting Eq. 5.19 in this equation, we have

$$1 + \frac{K(s - z_1)(s - z_2)\cdots(s - z_m)}{(s - p_1)(s - p_2)\cdots(s - p_n)} = 0 \tag{5.24}$$

or

$$(s - p_1)(s - p_2)\cdots(s - p_n) + K(s - z_1)(s - z_2)\cdots(s - z_m) = 0 \tag{5.25a}$$

The parameter K is varied from 0 to ∞. When $K = 0$, the characteristic equation 5.25a becomes

$$(s - p_1)(s - p_2)\cdots(s - p_n) = 0$$

and

$$s = p_1, p_2, \ldots, p_n$$

Thus for $K = 0$ the roots of the characteristic equation of the closed-loop system are the poles of open-loop transfer function $G(s)H(s)$. Hence the root loci begin at the open-loop poles.

The characteristic equation 5.25a can also be written as

$$\frac{(s - p_1)(s - p_2)\cdots(s - p_n)}{K} + (s - z_1)(s - z_2)\cdots(s - z_m) = 0 \tag{5.25b}$$

---

†W. R. Evans, "Graphical Analysis of Control Systems," *Trans. IRE* 67, (1948), 547–551.

When $K = \infty$, the characteristic equation 5.25b becomes

$$(s - z_1)(s - z_2) \cdots (s - z_m) = 0$$

or

$$s = z_1, z_2, \ldots, z_m$$

Thus for $K = \infty$, the roots of characteristic equation of the closed-loop system are the zeros of the open-loop transfer function $G(s)H(s)$. Hence the root loci terminate on open-loop zeros.†

We shall assume that for the open-loop transfer function $G(s)H(s)$, $m \leq n$. This is true of all practical systems. This means that there are n branches of the root locus beginning on n poles of $G(s)H(s)$.‡ Each of these n branches must terminate on one of the zeros of $G(s)H(s)$. Since $G(s)H(s)$ has only m zeros, where would $n - m$ branches terminate? Actually if we consider all zeros (finite and infinite) of $G(s)H(s)$, they must be n in number. Consider for example

$$G(s)H(s) = \frac{K}{s(s + 8)}$$

Here $n = 2$ and $m = 0$. However we notice that $G(s)H(s) = 0$ for $s = \infty$. Hence $s = \infty$ is a zero of this transfer function. Actually there are two zeros at $s = \infty$. This can be seen from the fact that as $s \to \infty$,

$$G(s)H(s) \to \frac{K}{s^2}$$

This function has a double zero at $s = \infty$. In general, $G(s)H(s)$ has n poles, m finite zeros, and $(n - m)$ zeros at infinity. Hence out of the n root locus branches that start on n poles of $G(s)H(s)$, m branches terminate on the m finite zeros of $G(s)H(s)$ and $(n - m)$ branches terminate at infinity. This can be seen from the root locus for $G(s)H(s) = K/s(s + 8)$ shown in Fig. 5.13. There are two branches of the root locus which begin on the poles of $G(s)H(s)$ and both branches go to infinity (terminate at infinity), since there are no finite zeros of $G(s)H(s)$. Figure 5.19 shows root locus for three different open-loop transfer functions. In Figs. 5.19b and c $n = 2$, and $m = 1$. Hence there are two branches of root locus which begin on the poles of GH. One of the branches terminates on the finite zero and the other goes to ∞ (terminates at ∞).

Rule 2. Root locus along the real axis.

A real-axis segment is a part of root locus if the sum of real-axis poles and zeros of $G(s)H(s)$ that lie to the right of the segment is odd.

Proof. Consider $G(s)H(s)$ with pole-zero configuration shown in Fig. 5.20.

---

†It must also be remembered that root loci are symmetrical with respect to the real axis in the s-plane. This follows from the results in Sec. 4.16.

‡In case $m > n$, there will be m root locus branches of which n begin on the n poles of $G(s)H(s)$ and $m - n$ will begin at ∞, since in this case there are $m - n$ poles at ∞ for $G(s)H(s)$.

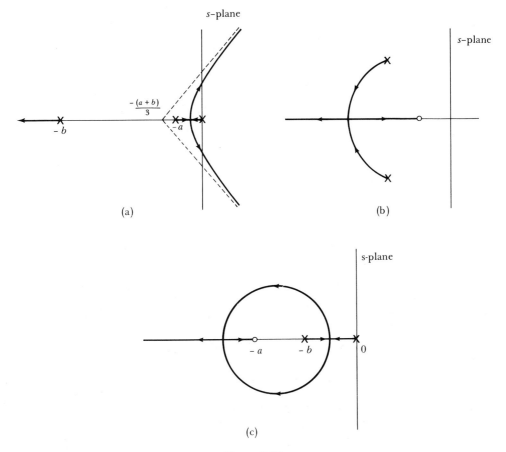

Figure 5.19

Take a test a point s_0 along the real axis. The angles associated with various poles and zeros are shown in Fig. 5.20. We observe that the total angle contribution due to complex conjugate poles is 2π. This is true about every pair of complex conjugate poles or zeros. Hence we may ignore the contribution due to complex poles and zeros. The angles due to real poles and zeros to the left of s_0 are zero, and the angle due to each real pole and real zero to the right of s_0 is π. Therefore if s_0 is a point on the root locus, then

$$\Sigma\, \theta_i \,-\, \Sigma\, \phi_j \,=\, r\pi \qquad (r \text{ odd})$$

where $\Sigma\, \theta_i$ represents the sum of all angles due to zeros to the right of s_0, and $\Sigma\, \phi_j$ is the sum of all angles due to poles to the right of s_0. Note that each of these angles is π. Since the angle π and $-\pi$ represent the same angle, subtraction of angle π is equivalent to addition of π. Hence for s_0 to lie on the root locus, the equivalent criterion is

$$\Sigma\, \theta_i \,+\, \Sigma\, \phi_j \,=\, r\pi \qquad (r \text{ odd})$$

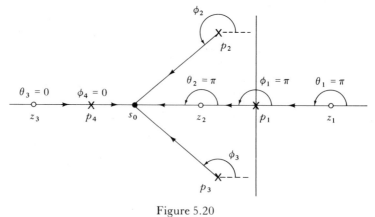

Figure 5.20

This shows that s_0 is on the root locus only if there are a total of odd number of real-axis poles and zeros to the right of s_0. Therefore the entire segment of the real axis between p_4 and z_2 is a part of the root locus. In Fig. 5.13 there is one pole to the right of the entire real-axis segment $(-8, 0)$. Hence this is a part of the root locus. The appropriate real-axis segments that are parts of the root locus can be clearly seen in Fig. 5.19.

   Rule 3.   The $(n - m)$ loci terminate at ∞ along asymptotes at angles $k\pi/(n - m)$ for $k = 1, 3, 5, \ldots,$.

   Proof.   For very large value of s (see Eq. 5.19)

$$G(s)H(s) \simeq K \frac{s^m}{s^n} = \frac{K}{s^{n-m}}$$

Hence the characteristic equation as $s \rightarrow \infty$ is

$$\frac{K}{s^{n-m}} = -1$$

or

$$s^{n-m} = -K = Ke^{jk\pi} \qquad k = 1, 3, 5, \ldots$$

Let

$$s = re^{j\theta}$$

Then

$$r^{n-m} e^{j(n-m)\theta} = Ke^{jk\pi}$$

$$r = (K)^{(1/n-m)}, \qquad \text{and} \quad \theta = \frac{k\pi}{n - m} \qquad k = 1, 3, 5, \ldots \qquad (5.26)$$

Here θ represents angles of the roots of the characteristic equation for large values of s. In other words θ represents the angle of asymptotes of the root locus. Thus when $n - m = 2$ (as in Fig. 5.13), the asymptotic angles are $k\pi/2$, for $k = 1, 3, 5,$

etc. This gives $\theta = \pi/2, 3\pi/2, 5\pi/2, 7\pi/2$, etc. Note that this sequence merely repeats angles, $\pi/2$ and $-\pi/2$ endlessly.

It can be seen from Fig. 5.13 that the two branches terminating at ∞ indeed approach ∞ at angle $\pi/2$ and $-\pi/2$. Similarly in Fig. 5.19a, $n - m = 3$ and the asymptotic angles are $\pi/3, 3\pi/3$, and $5\pi/3$ or $\pm\pi/3$ and π. This is easily verified in Fig. 5.19a.

Rule 4.   The centroid of the asymptotes is σ, given by

$$\sigma = \frac{(p_1 + p_2 + \cdots + p_n) - (z_1 + z_2 + \cdots z_m)}{n - m} \tag{5.27}$$

where the centroid of asymptotes is the point where the asymptotes meet, as shown in Fig. 5.21.

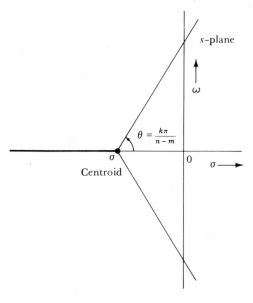

Figure 5.21

Proof.   Let us rewrite $G(s)H(s)$ in Eq. 5.19 in the form

$$G(s)H(s) = K\frac{s^m + \beta_m s^{m-1} + \cdots + \beta_1}{s^n + \alpha_n s^{n-1} + \cdots + \alpha_1} \tag{5.28}$$

where (see Eq. 4.73)

$$\beta_m = -(z_1 + z_2 + \cdots + z_m)$$
$$\alpha_n = -(p_1 + p_2 + \cdots + p_n) \tag{5.29}$$

We can divide the denominator by the numerator in Eq. 5.28 to obtain

$$G(s)H(s) = \frac{K}{s^{n-m} + (\alpha_n - \beta_m)s^{n-m-1} + \cdots}$$

We also have from the binomial expansion theorem that

$$(s - \sigma)^{n-m} = s^{n-m} - \sigma(n - m)s^{n-m-1} + \cdots$$

We are interested in observing the asymptotic behavior of the roots as s becomes very large. In such a case we can ignore all the powers of s lower than $(n - m - 1)$. Hence for asymptotic behavior,

$$G(s)H(s) \simeq \frac{K}{(s - \sigma)^{n-m}}$$

where

$$\sigma = -\frac{\alpha_n - \beta_m}{n - m}$$

$$= \frac{(p_1 + p_2 + \cdots + p_n) - (z_1 + z_2 + \cdots + z_m)}{n - m}$$

The characteristic equation for large values of s is therefore given by

$$\frac{K}{(s - \sigma)^{n-m}} = -1$$

Therefore

$$(s - \sigma)^{n-m} = -K = Ke^{jk\pi} \qquad K = 1, 3, \ldots$$

or

$$s - \sigma = (K)^{1/n-m} e^{jk\pi/n-m}$$

Hence

$$s = \sigma + re^{jk\pi/n-m} \qquad k = 1, 3, \ldots; r = (K)^{1/n-m}$$

The equation represents straight lines originating at σ and having an angle $k\pi/n - m$. This proves the rule stated in Eq. 5.27.

In Fig. 5.13 the value of σ according to Eq. 5.27 is

$$\sigma = \frac{(0 - 8) - (0)}{2} = -4$$

In Fig. 5.19a

$$\sigma = \frac{(0 - a - b) - 0}{3} = -\frac{(a + b)}{3}$$

Rule 5.   Root-locus intersections occur at points which are roots of

$$\frac{d}{ds}[G(s)H(s)] = 0 \tag{5.30}$$

In Fig. 5.13 we observe that the two branches of the root locus meet at -4 and then break away along a vertical line. Similarly in Fig. 5.19a the two branches meet at some point and then break away.   These are the points where root loci meet or cross.   It is evident that these points where root loci meet are characterized by the fact that the characteristic equation has multiple roots.   Hence if s_1 is a meeting point (or breakaway point), the characteristic equation at $s = s_1$ can be expressed as

$$1 + G(s)H(s) = (s - s_1)^k F(s) = 0$$

where

$$k \geq 2$$

Hence

$$\frac{d}{ds}[1 + G(s)H(s)] = (s - s_1)^k F'(s) + k(s - s_1)^{k-1} F(s)$$

The right-hand side is zero at $s = s_1$.   Hence

$$\frac{d}{ds}[1 + G(s)H(s)] = 0 \qquad \text{at} \quad s = s_1$$

But

$$\frac{d}{ds}[1 + G(s)H(s)] = \frac{d}{ds}[G(s)H(s)]$$

and we have $d/ds\,[G(s)H(s)] = 0$ at the points where the root loci cross. (breakaway points).   For the root locus in Fig. 5.13, the root loci across at the root(s) of

$$\frac{d}{ds}\left(\frac{K}{s^2 + 8s + K}\right) = \frac{-K(2s + 8)}{(s^2 + 8s + K)^2} = 0$$

or $2s + 8 = 0$ and $s = -4$.

Thus the breakaway point for the root loci is -4.   This is clearly seen in Fig. 5.13.

Consider the case

$$G(s)H(s) = \frac{K(s + 4)}{s(s + 2)}$$

The root locus (Fig. 5.22) has two branches, starting from the open-loop poles 0, and -2 (rule 1).   One branch terminates on the open-loop zero, -4, the other branch terminates at ∞ (only one asymptote, since $n - m = 1$).   The real axis over 0 to -2 and -4 to $-\infty$, is part of the root locus (rule 2).   The two branches break away from the real axis at a point between 0 and -2, and merge again at a point between -4 and $-\infty$.   To find these points we solve the equation

$$\frac{d}{ds}[G(s)H(s)] = 0$$

or

$$\frac{d}{ds}\left[\frac{K(s + 4)}{s(s + 2)}\right] = K\left[\frac{-(s^2 + 8s + 8)}{s^2(s + 2)^2}\right] = 0$$

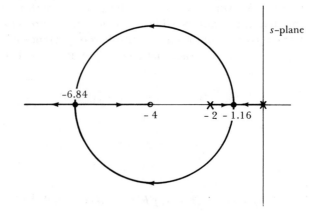

Figure 5.22

This yields

$$s^2 + 8s + 8 = 0$$

The roots are

$$s = -1.16, \qquad -6.84$$

The loci intersect at these points as seen from Fig. 5.22. Note that

$$\frac{d}{ds}\left[\frac{1}{G(s)H(s)}\right] = \frac{\frac{d}{ds}[G(s)H(s)]}{[G(s)H(s)]^2}$$

Hence if

$$\frac{d}{ds}[G(s)H(s)] = 0$$

Then

$$\frac{d}{ds}\left[\frac{1}{G(s)H(s)}\right] = 0$$

We may therefore use either of these conditions to determine the points where the root loci intersect.

   Rule 6. The crossing of the $j\omega$-axis. In some cases (e.g., Fig. 5.19a), the root loci cross the $j\omega$-axis. The points where the loci cross the $j\omega$-axis can be found by applying the Routh-Hurwitz criterion to the characteristic equation. This will now be demonstrated by an example.

   Let us find the root locus for the system shown in Fig. 5.23a.

$$G(s)H(s) = \frac{K}{s(s+2)(s+4)}$$

We first draw the pole-zero map of $G(s)H(s)$ in Fig. 5.23b. Since there are three

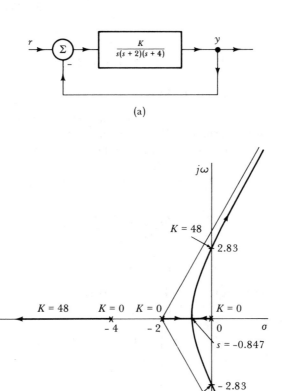

(a)

(b)

Figure 5.23

poles of $G(s)H(s)$, there are three branches of the root locus starting at 0, -2, and -4 (the poles of GH). The real axis in the range $(0, -2)$ and $(-4, -\infty)$ is a part of the root locus (rule 2).

The open-loop transfer function has three poles and no zeros ($n = 3$, $m = 0$). Hence all the three branches of the root locus go to infinity asymptotically at angles $\pi/3$, π, $5\pi/3$ or at angles $60°$, $180°$, and $-60°$. The centroid of the asymptotes is given by

$$\sigma = \frac{(0 - 2 - 4) - 0}{3} = -2$$

The asymptotes are shown in Fig. 5.23b. The breakaway point (root loci crossing) occurs at

$$\frac{d}{ds}\left[\frac{K}{s(s+2)(s+4)}\right] = \frac{-K(3s^2 + 12s + 8)}{s^2(s+2)^2(s+4)^2} = 0$$

This yields

$$3s^2 + 12s + 8 = 0$$

or

$$3(s + 0 \cdot 847)(s + 3 \cdot 15) = 0$$

Out of two values $s = -0.847$ and -3.15, only -0.847 applies in this case (the point -3.15 does not lie on the root locus). Hence the breakaway point is -0.847.

To find the $j\omega$-axis crossing we consider the characteristic equation

$$1 + \frac{K}{s(s + 2)(s + 4)} = 0$$

or

$$\frac{s^3 + 6s^2 + 8s + K}{s(s + 2)(s + 4)} = 0$$

Hence

$$s^3 + 6s^2 + 8s + K = 0$$

This is the characteristic equation. The R-H array is

$$
\begin{array}{c|cc}
s^3 & 1 & 8 \\
s^2 & 6 & K \\
& \dfrac{48 - K}{6} & \\
& K &
\end{array}
$$

Hence some of the roots will lie in the RHP for $K > 48$. For $K = 48$, some roots will be along the $j\omega$-axis ($j\omega$-axis crossing). For $K = 48$, the R-H array becomes

$$
\begin{array}{cc}
1 & 8 \\
6 & 48 \\
0 & 0
\end{array}
$$

The auxiliary equation is

$$6s^2 + 48 = 0$$

or

$$s^2 + 8 = 0$$

and

$$s = j2\sqrt{2} = \pm j2.83$$

Hence the root loci will cross the $j\omega$-axis at $\pm j2.83$ as shown in Fig. 5.23b. Note that the system becomes unstable for $K > 48$.

Rule 7. Angles of departures and arrival. When the poles (zeros) of GH are complex, the root loci depart (arrive) from the poles (zeros) at certain angles. These angles can be obtained by using the angle criterion (Eq. 5.22a). Consider for

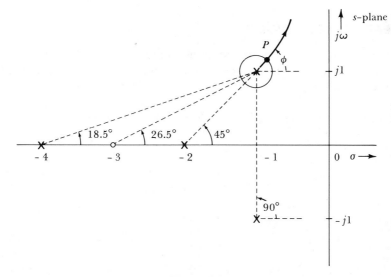

Figure 5.24

example the case

$$G(s)H(s) = \frac{K(s + 3)}{(s + 2)(s + 4)(s^2 + 2s + 2)}$$

The pole-zero configuration for this open-loop transfer function is shown in Fig. 5.24. In order to find the angle of departure ϕ of the branch starting at the pole $-1 + j1$, draw a circle of radius ϵ around $-1 + j1$. Let $\epsilon \to 0$. The root locus will intersect this circle at point P. Hence the point P satisfies the angle criterion (Eq. 5.22a or 5.23a). In order to test this point we draw line segments from all poles and zeros to the point P. Since $\epsilon \to 0$, P is arbitrarily close to $-1 + j1$. Hence we may draw the lines from all the poles and zeros (except $-1 + j1$) to $-1 + j1$. Application of angle criterion to Fig. 5.24 yields

$$26.5 - (90 + 45 + \phi + 18.5) = 180$$

Hence

$$\phi = 53°$$

The angle of arrival can also be computed in the same manner.†

---

†The rules derived here apply to the case when K varies from 0 to ∞. In a similar way we can develop rules for sketching root locus, when K varies from 0 to $-\infty$. In this case the points on root locus must satisfy

$$\angle G(s)H(s) = 2r\pi \qquad r = 0, 1, 2, 3, \ldots$$

Rules 1, 4, 5, 6 are unchanged. Rules 2, 3, and 7 are slightly modified. The reader is encouraged to derive these rules.

5.7   STEADY-STATE ERRORS

As mentioned earlier, control systems generally also have to meet certain steady-state requirements. These requirements impose additional constraints on the closed-loop transfer function $T(s)$. We shall now consider these constraints.

The error is the difference between the desired output (reference r) and the actual output y. Thus

$$e = r - y$$

and

$$
\begin{aligned}
E(s) &= R(s) - Y(s) \\
&= R(s) \left[1 - \frac{Y(s)}{R(s)} \right] \\
&= R(s) [1 - T(s)]
\end{aligned}
$$

(5.31)

The steady-state error e(steady state) is the value of e as $t \to \infty$. This can be easily obtained from the final-value theorem (Eq. 4.42):

$$e(\text{steady state}) = \lim_{s \to 0} sE(s) = \lim_{s \to 0} sR(s) [1 - T(s)]$$

(5.32)

1. For a unit-step input, $R(s) = 1/s$, and e_s, the steady-state error to unit-step input is given by

$$e_s = \lim_{s \to 0} [1 - T(s)] = 1 - T(0)$$

(5.33)

Note that if $T(0) = 1$, the steady-state error to unit-step input is zero.

2. For a unit ramp input $tu(t)$, $R(s) = 1/s^2$, and e_r, the steady-state error to unit ramp, is given by

$$e_r = \lim_{s \to 0} \frac{1 - T(s)}{s}$$

(5.34)

Note that if $T(0) \neq 1$, $e_r = \infty$. Hence for a finite steady-state error to ramp input, $T(0) = 1$, implying zero steady-state error to step input. Assuming

$$T(0) = 1$$

(5.35)

and applying L'Hôpital's rule to Eq. 5.34, we have

$$e_r = \lim_{s \to 0} [-T'(s)] = -T'(0)$$

(5.36)

3. For a unit parabolic input $(t^2/2)u(t)$, $R(s) = 1/s^3$, and e_p, the steady-state error to unit parabolic input is given by

$$e_p = \lim_{s \to 0} \frac{1 - T(s)}{s^2}$$

(5.37)

Observe that $e_p = \infty$ if $T(0) \neq 1$. Assuming $T(0) = 1$, applying L'Hôpital's rule, we obtain

$$e_p = \lim_{s \to 0} \frac{-T'(s)}{2s} \tag{5.38}$$

Again we observe that $e_p = \infty$ if $T'(0) \neq 0$. Thus for e_p to be finite,

$$T(0) = 1$$
$$T'(0) = 0 \tag{5.39}$$

This means that steady-state errors to both the step and the ramp input are zero. Assuming this, application of L'Hôpital's rule to Eq. 5.38 yields

$$e_p = \lim_{s \to 0} \frac{-T''(s)}{2} = \frac{-T''(0)}{2} \tag{5.40}$$

If

$$T(s) = \frac{b_m s^m + b_{m-1} s^{m-1} + \cdots + b_1 s + b_0}{s^n + a_{n-1} s^{n-1} + \cdots + a_1 s + a_0} \tag{5.41}$$

Then from Eq. 5.33 we have

$$e_s = \left(1 - \frac{b_0}{a_0}\right) = \frac{a_0 - b_0}{a_0} \tag{5.42}$$

From Eq. 5.36 we have

$$e_r = -\left(\frac{a_0 b_1 - b_0 a_1}{a_0^2}\right)$$

But for a finite e_r, we must have $e_s = 0$ (Eq. 5.35). Hence $a_0 = b_0$ and

$$e_r = \frac{a_1 - b_1}{a_0} \tag{5.43}$$

Similarly from Eq. 5.40 we have

$$e_p = \frac{a_0 a_1 b_1 + a_0 a_2 b_0 - a_0^2 b_2 - a_1^2 b_0}{a_0^3}$$

for finite e_p it was shown that (see Eq. 5.39), $e_s = e_r = 0$. Therefore

$$a_0 = b_0 \quad \text{and} \quad a_1 = b_1$$

and

$$e_p = \frac{a_2 - b_2}{a_0} \tag{5.44}$$

Thus the steady-state-error performance imposes certain constraints on the closed-loop transfer function $T(s)$.

For a unity feedback system (Fig. 5.25), steady-state-error analysis is greatly simplified. The error signal $E(s)$ in this case is found from Mason's rule to be

$$E(s) = \frac{1}{1 + G(s)} R(s)$$

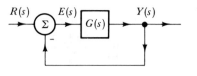

Figure 5.25

For step input, the steady-state-error e_s is given by

$$e_s = \lim_{s \to 0} s \frac{1/s}{1 + G(s)} = \frac{1}{1 + \lim_{s \to 0} G(s)} \tag{5.45}$$

and

$$e_r = \lim_{s \to 0} s \frac{(1/s^2)}{1 + G(s)} = \frac{1}{\lim_{s \to 0} s G(s)} \tag{5.46}$$

$$e_p = \lim_{s \to 0} s \frac{(1/s^3)}{1 + G(s)} = \frac{1}{\lim_{s \to 0} s^2 G(s)} \tag{5.47}$$

We now define

$$K_p = \lim_{s \to 0} G(s) \tag{5.48}$$

$$K_v = \lim_{s \to 0} s G(s) \tag{5.49}$$

$$K_a = \lim_{s \to 0} s^2 G(s) \tag{5.50}$$

where K_p, K_v, and K_a are known respectively as positional error constant, velocity error constant, and the acceleration error constant. In terms of these constants, the steady-state errors can be expressed as

$$e_s = \frac{1}{1 + K_p} \tag{5.51}$$

$$e_r = \frac{1}{K_v} \tag{5.52}$$

$$e_p = \frac{1}{K_a} \tag{5.53}$$

Thus if

$$G(s) = \frac{K}{s(s + 8)}$$

$$K_p = \infty \qquad K_v = \frac{K}{8} \qquad \text{and} \qquad K_a = 0$$

Hence,

$$e_s = \frac{1}{1 + K_p} = 0 \qquad e_r = \frac{1}{K_v} = \frac{8}{K} \qquad e_p = \infty$$

If $G(s)$ has no poles at the origin, then K_p is finite and $K_v = K_a = 0$. Thus for

$$G(s) = \frac{K(s+2)}{(s+1)(s+10)}$$

$$K_p = \frac{K}{5} \qquad \text{and} \qquad K_v = K_a = 0$$

Such a system (where $G(s)$ has no poles at the origin) is designated as type 0 system. For such a system e_s is finite and $e_r = e_p = \infty$. Type 0 system, therefore, may be acceptable when the inputs are step functions. It is clearly not usable when the inputs are ramp or higher-order polynomials in t.

If $G(s)$ has one pole at the origin, then $K_p = \infty$, K_v is finite and $K_a = 0$. Thus for

$$G(s) = \frac{K(s+2)}{s(s+1)(s+10)}$$

$$K_p = \infty \qquad K_v = \frac{K}{5} \qquad \text{and} \qquad K_a = 0$$

Such a system (with one pole at the origin) is designated as Type 1 system. It is obvious that such a system yields zero steady-state error for a step input, finite steady-state error for a ramp input, and infinite steady-state error for parabolic or high-order input.

If $G(s)$ has two poles at the origin, the system is designated as Type 2 system, and $K_p = K_v = \infty$ and $K_a =$ finite. Hence $e_s = e_r = 0$, and e_p is finite.

In general, if $G(s)$ has j poles at the origin, it is a type j system. From the above discussion, it is obvious that for a unity feedback system increasing the number of poles at the origin in $G(s)$ improves the steady-state performance. However, this increases n and thus reduces the magnitude of σ, the centroid of root-locus asymptotes. This causes the root locus to shift toward the $j\omega$-axis with consequent deterioration in the transient performance and in the system stability.

It should be remembered that the results in Eq. 5.48 through Eq. 5.53 apply only to unity feedback systems (Fig. 5.25). Steady-state error specifications in this case are translated in terms of constraints on the open-loop transfer function $G(s)$. On the other hand, the results in Eqs. 5.42 through 5.44 apply to unity as well as nonunity feedback systems, and are obviously more general. Steady-state-error specifications in this case are translated in terms of constraints on the closed-loop transfer function $T(s)$.

Let us consider the system in Fig. 5.4. We have designed this system to meet the following transient specifications:

$$\text{PO} = 16\% \qquad t_r \leq 0.5 \text{ sec} \qquad \text{and} \quad t_s \leq 2 \text{ sec} \qquad (5.54)$$

Let us further specify that the system must meet the following steady-state specifications:

$$e_s = 0 \qquad \text{and} \quad e_r \leq 0.15 \qquad (5.55)$$

The transfer function $T(s)$ is given by (see Eq. 5.13a)

$$T(s) = \frac{K}{s^2 + 8s + K}$$

We immediately observe that

$$a_0 = b_0 = K$$

Hence from Eq. 5.42

$$e_s = 0$$

and

$$e_r = \frac{a_1 - b_1}{a_0} = \frac{8 - 0}{K} = \frac{8}{K}$$

Since the system under consideration has unity feedback, we may determine steady-state errors by simpler relationships in Eq. 5.48 and 5.49. We have

$$K_p = \lim_{s \to 0} G(s) = \lim_{s \to 0} \frac{K}{s(s + 8)} = \infty$$

$$K_v = \lim_{s \to 0} sG(s) = \lim_{s \to 0} \frac{Ks}{s(s + 8)} = \frac{K}{8}$$

Hence

$$e_s = \frac{1}{1 + K_p} = 0 \qquad\qquad (5.56a)$$

$$e_r = \frac{1}{K_v} = \frac{8}{K} \qquad\qquad (5.56b)$$

which agree with the results obtained earlier.

Since

$$e_r \leq 0.15$$

we must have

$$\frac{8}{K} \leq 0.15$$

or

$$K \geq 53.34 \qquad\qquad (5.57)$$

We now turn to Fig. 5.14, and note that poles of $T(s)$ lie in the acceptable region (to meet transient specifications) for $25 < K < 64$. From Eq. 5.57, we must have $K \geq 53.34$ to meet steady-state performance. Therefore to meet both the transient and the steady-state specifications we must set the gain K in the range $53.34 < K < 64$. The lowest steady-state error for a ramp input is obtained for $K = 64$. For this case,

$$e_r = \frac{8}{K} = \frac{8}{64} = 0.125$$

Thus if the system is to meet the transient performance in Eq. 5.54, the minimum steady-state error to a ramp input that can be attained is 0.125. We can do no better. In case we are required to have $e_r < 0.125$ while maintaining the same transient performance (Eq. 5.54), we will have to use some kind of compensation. The performance can also be improved by using multiple variable feedback as discussed in Sec. 5.10.

5.8   COMPENSATION

The synthesis problem (Fig. 5.4) discussed here is a very simple example where the transient and steady-state specifications could be met by simple adjustment of gain K. In many cases it may be impossible to meet both sets of specification (transient and steady state) by simple adjustment of the gain K. We may be able to satisfy one set of specifications or the other but not both. Consider again the system in Fig. 5.4, with the following specifications

$$\text{PO} \leq 16\% \qquad t_r \leq 0.5 \text{ sec} \qquad t_s \leq 2 \text{ sec}$$

and

$$e_s = 0 \qquad e_r \leq 0.05 \tag{5.58}$$

To meet the steady-state specification we must have (see Eq. 5.56)

$$\frac{8}{K} \leq 0.05$$

or

$$K \geq 160$$

But from Fig. 5.14 we observe that for $K > 64$, the poles of $T(s)$ move out of the region acceptable for transient performance. It is therefore obvious that we can meet either the transient or the steady-state specification but not both. In such case we must add some kind of compensation. One possibility is shown in Fig. 5.26. The poles and zeros of the compensator are chosen to modify the root locus to meet all the specifications. A little familiarity with root-locus techniques gives the insight and judgment needed to choose a proper compensator transfer function. To improve transient performance we must increase ζ and ω_n. This implies that the dominant poles of $T(s)$ should be as far to the left of the $j\omega$-axis as possible. This can be done by reshaping the root locus so that it is pushed more to the left of the $j\omega$-axis. The root locus will be shifted to the left, if a compensator transfer function $G_c(s)$ in Fig. 5.26 is so chosen that the centroid of the compensated root locus lies to the left of the centroid of the uncompensated root locus. This can be accomplished if we choose

$$G_c(s) = s + \alpha$$

The centroid σ_u of the uncompensated system is given by (see Eq. 5.27)

$$\sigma_u = \frac{\Sigma p_i - \Sigma z_j}{n - m}$$

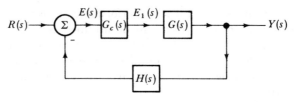

Figure 5.26

Because of addition of a zero at $-\alpha$, the centroid σ_c of the compensated system is given by

$$
\begin{aligned}
\sigma_c &= \frac{\Sigma p_i - (\Sigma z_j - \alpha)}{n - m - 1} \\
&= \frac{(\Sigma p_i - \Sigma z_j) + \alpha}{n - m - 1} \\
&= \frac{\sigma_u(n - m) + \alpha}{n - m - 1} \\
&= \sigma_u + \frac{\sigma_u + \alpha}{n - m - 1}
\end{aligned}
$$

Hence $\sigma_c < \sigma_u$ if $\alpha < -\sigma_u$.

Thus the new centroid σ_c will lie to the left of the old centroid if the compensator zero at $-\alpha$ is located to the right of the old centroid σ_u.

As seen from Fig. 5.26, the plant input $E_1(s)$ is given by

$$
\begin{aligned}
E_1(s) &= (s + \alpha) E(s) \\
&= sE(s) + \alpha E(s)
\end{aligned}
$$

Hence

$$
e_1(t) = \frac{de}{dt} + \alpha e(t)
$$

Thus the control input (input to the plant) is the sum of the derivative of the error signal and α times the error signal. For this reason this control is also known as *proportional plus derivative control*. It is obvious that this type of compensation, by pushing the root locus leftward in the s-plane, improves the transient behavior of the system.

There are two objections to this type of compensation. First, an ideal differentiator is difficult to construct and requires many components which must be critically adjusted. Second, an ideal differentiator tends to accentuate noise signals. This may sometimes saturate electronic amplifiers making the entire system inoperative. For this reason the above type of compensator is avoided. Instead a lead compensator (described below) which approximates proportional plus derivative action is more commonly used.

The root locus can be shifted to the left if a compensator transfer function

$G_c(s)$ in Fig. 5.26 is chosen as

$$G_c(s) = \frac{s + \alpha}{s + \beta} \qquad \beta > \alpha \tag{5.59}$$

From the centroid rule, it can be seen that addition of such a compensator will shift the centroid of the root locus to the left by an amount $(\beta - \alpha)/(n - m)$. Compensator of this form can be realized by the simple RC network shown in Fig. 5.27. This network is known as a *lead network*. The actual pole-zero locations

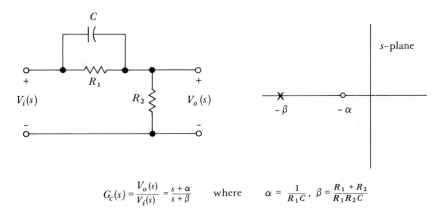

$$G_c(s) = \frac{V_o(s)}{V_i(s)} = \frac{s + \alpha}{s + \beta} \qquad \text{where} \qquad \alpha = \frac{1}{R_1 C}, \ \beta = \frac{R_1 + R_2}{R_1 R_2 C}$$

Figure 5.27   Lead compensator.

(values of α and β) are determined by the cut-and-try method. There is no unique solution. Several sets of values of α and β will steer the root locus into the desired region. The maximum value of β/α that can be chosen is limited by noise considerations. As β/α becomes large, the lead compensator resembles more closely an ideal differentiator, which tends to accentuate noise.

As an example, consider the position control system (Fig. 5.4) with open-loop transfer function

$$G(s) H(s) = \frac{K}{s(s + 8)}$$

The root locus of this system is shown in Fig. 5.13. From this figure it is obvious that

$$(\zeta \omega_n)_{max} = 4$$

and the minimum value of t_s that can be obtained is $4/4 = 1$ sec. Let us use a lead compensator with transfer function

$$G_c(s) = \frac{s + 15}{s + 72}$$

The new open-loop transfer function is

$$G(s) G_c(s) = \frac{K(s + 15)}{s(s + 8)(s + 72)} \tag{5.60}$$

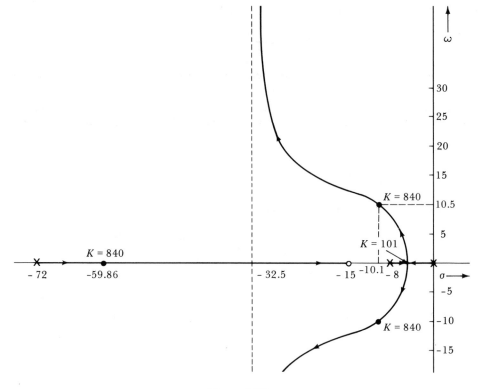

Figure 5.28

The root locus for the compensated system is shown in Fig. 5.28a. Comparison of
Fig. 5.28a with Fig. 5.13 clearly shows that the centroid of the asymptotes is moved
from -4 to -32.5. This pushes the entire root locus far to the left.

For $K = 840$, the roots are $-10.1 \pm j10.5$ and -59.86 (Fig. 5.28a). The
closed-loop transfer function $T(s)$ is

$$T(s) = \frac{G(s)\,G_c(s)}{1 + G(s)\,G_c(s)\,H(s)}$$

$$= \frac{840\,(s + 15)}{s^3 + 80s^2 + 1416s + 12600}$$

$$= \frac{840\,(s + 15)}{(s + 10.1 - j10.5)\,(s + 10.1 + j10.5)\,(s + 59.86)}$$

The poles are $-10.1 \pm j10.5$ and -59.86 as expected (Fig. 5.28). The third pole
-59.86 is about six times as far from the $j\omega$-axis as the real part of the dominant
poles at $-10.1 \pm j10.5$. Hence this pole may be safely ignored in evaluating transient behavior. The damping ratio ζ for the dominant poles is $\cos\,[\tan^{-1}(10.5/$
$10.1)] \cong 0.7$, and $\omega_n = \sqrt{(10.1)^2 + (10.5)^2} = 14.5$. In addition, $T(s)$ has a zero
at -15. Hence we must use Figs. 5.16a and 5.16b to find PO and t_r. In this case

$a/\zeta\omega_n = 15/10.1 \simeq 1.5$. Hence from Figs. 5.16a and b, for $\zeta = 0.7$, we obtain PO $\simeq 11\%$ and $\omega_n t_r = 1.25$. Therefore

$$t_r = \frac{1.25}{\omega_n} = \frac{1.25}{\sqrt{(10.1)^2 + (10.5)^2}} = 0.086$$

The settling time as determined by dominant poles is

$$t_s = \frac{4}{\zeta\omega_n} = \frac{4}{10.1} = 0.397 \text{ sec}$$

To summarize, the system has

$$\text{PO} = 11\% \qquad t_r = 0.086 \text{ sec} \qquad \text{and} \quad t_s = 0.397 \text{ sec}$$

Note that the transient response is improved in every respect.

To obtain the steady-state errors, we have

$$K_p = \infty \qquad K_v = \frac{15K}{(8)(72)} = \frac{(15)(840)}{(8)(72)} = 21.9$$

and†

$$e_s = 0$$

$$e_r = \frac{1}{K_v} = 0.0457$$

Thus the steady-state error e_r is also reduced to 0.0458.

Often the zero of the lead compensator is used to cancel one of the poles of uncompensated open-loop transfer function $G(s)$. This considerably simplifies the design procedure. In our problem we could have chosen the compensator zero at -8, and a pole at appropriate location to meet the remaining specifications. If we choose

$$G_c(s) = \frac{s + 8}{s + 30}$$

the compensated open-loop transfer function is given by

$$G_c(s)G(s) = \frac{K}{s(s + 30)}$$

The root locus for this case is shown in Fig. 5.29. For $K = 600$ we have $\zeta = \cos(52.3°) = 0.61$, $\omega_n = 24.6$. This yields PO $= 9\%$, $t_r = 1.84/24.6 = 0.076$, and $t_s = 4/15 = 0.266$. We also have

$$K_p = \infty \qquad \text{and} \quad K_v = \frac{K}{30} = \frac{600}{30} = 20$$

Hence

$$e_s = 0, e_r = \frac{1}{20} = 0.05$$

_____

†The steady-state errors may also be obtained by using Eqs. 5.42 and 5.43.

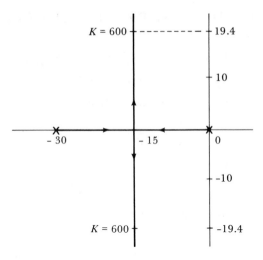

Figure 5.29

The advantages of canceling one of the poles of $G(s)$ with zero of $G_c(s)$ are obvious.† It not only simplifies the sketching of root locus but also simplifies computation of PO, t_r, and t_s.

Cancellation can also be very helpful in cases where the closed-loop system may have a pole so close to the $j\omega$-axis as to make the assumption of dominant poles invalid. In such case a zero of $G_c(s)$ is used to cancel the pole of $G(s)$ which is near the origin (but not at the origin).

We shall now discuss a compensation which is primarily used to improve the steady-state performance.

For unity feedback systems, it was observed in Sec. 5.7 that the steady-state performance of a system is improved by placing an integrator in the forward path of $G(s)$. This increases the system type, thus increasing K_p, K_v, K_a, etc. The compensator in this case is

$$G_c(s) = \frac{1}{s}$$

In this scheme the compensator, which is an ideal integrator, acts as a controller. For this reason this scheme is also known as *integral control*. Design of an ideal integrator necessitates elaborate and expensive equipment. Hence such a compensator is used where cost considerations are not very important. For example, integrating gyroscopes are used for this purpose in aircraft. Here the expense is

---

†Cancellation compensation is also very effective when $G(s)H(s)$ has complex poles near the $j\omega$-axis. In such a case the lead compensator has marginal influence in shifting the root locus. Here cancellation compensation proves very attractive. We can cancel the complex poles of $G(s)H(s)$ near the $j\omega$-axis, leaving only the real poles. The lead compensator may now be used effectively.

justified by the improved performance. In most cases a lag compensator (described below) which closely approximates the behavior of an integrator is used.

A lag compensator transfer function is given by

$$G_c(s) = \frac{s + \alpha}{s + \beta} \qquad \alpha > \beta \tag{5.61a}$$

and

$$G_c(0) = \frac{\alpha}{\beta} \tag{5.61b}$$

For a unity feedback system, addition of a compensator $G_c(s)$ causes all the error constants K_p, K_v, K_a, etc., to be multiplied by $G_c(0)$. Thus the lag compensator increases K_p, K_v, K_a, etc., by a factor ($\alpha/\beta > 1$), thereby reducing steady-state errors.

A lag compensator improves the steady-state performance, but in general degrades the transient performance. Since $\alpha > \beta$, the magnitude of σ, the root-locus centroid, is reduced. This causes the root locus to be shifted toward the $j\omega$-axis, with the consequent deterioration of the transient performance. This side effect of a lag compensator can be made negligible by choosing α and β such that $\alpha - \beta$ is very small, but the ratio α/β is high. This can be achieved by placing both the pole and the zero of $G_c(s)$ close to the origin (α and $\beta \to 0$). Thus if we let $\alpha = 0.1$ and $\beta = 0.01$, the centroid will be shifted only by a negligible amount $(\alpha - \beta)/(n - m) = 0.09/(n - m)$. However, since α/β is 10, all the error constants are increased by a factor of 10. The lag compensator (Eq. 5.61a) can be easily realized by a simple RC network shown in Fig. 5.30.†

The lag compensator adds a dipole to the open-loop transfer function. Since the pole and the zero of a dipole are very close together, they tend to cancel the

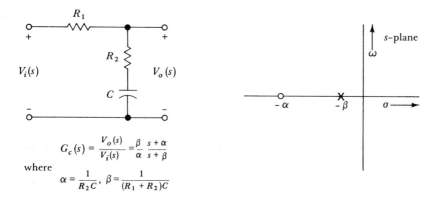

where

$$G_c(s) = \frac{V_o(s)}{V_i(s)} = \frac{\beta}{\alpha} \frac{s + \alpha}{s + \beta}$$

$$\alpha = \frac{1}{R_2 C}, \quad \beta = \frac{1}{(R_1 + R_2)C}$$

Figure 5.30

---

†The transfer function of the RC network in Fig. 5.30 contains an attenuation factor β/α. We can lump this constant with the variable gain K, and treat the compensator as having a transfer function as in Eq. 5.61a.

effect of each other. Hence it is reasonable to expect that the root locus of the compensated system will be almost identical to that of the uncompensated system.†
This is indeed true. We can deduce this result by considering the angle criterion of a test point (Eq. 5.23a). For a test point at a reasonable distance from the dipole, angles due to the pole and the zero of a dipole will be approximately equal and hence will tend to cancel out the mutual effect. Hence a point which satisfies the angle criterion (Eq. 5.23a) for the uncompensated system will also satisfy closely the angle criterion for the compensated system. Thus the root locus for compensated and noncompensated systems will be almost same. Figure 5.31b shows the root locus of the uncompensated system in Fig. 5.31a. Figure 5.31d shows the root

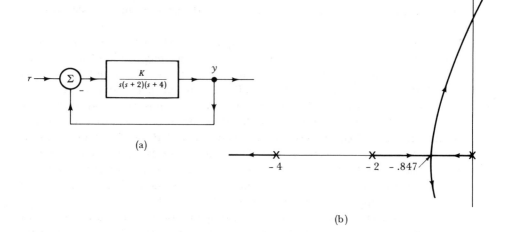

(a)

(b)

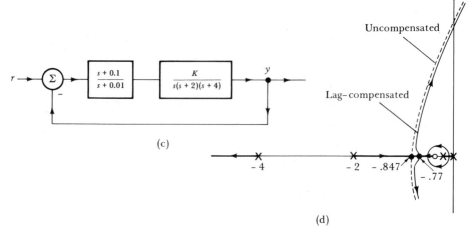

(c)

(d)

Figure 5.31

locus of the same system when a compensator $(s + 0.1)/(s + 0.01)$ is used to improve the steady-state performance. The reader can easily verify that this compensator reduces the steady-state error by a factor approximately $\alpha/\beta = 10$. Since the compensator causes little change in the root locus, the transient performance of the uncompensated system is almost identical to that of the compensated system.

We can improve the transient and the steady-state performance simultaneously by using a combination of lead and lag networks.

5.9 DIRECT SYNTHESIS

This method has both similarities and differences when compared to the root-locus method. In this method we assume a system with configuration in Fig. 5.26 with a compensator $G_c(s)$. From the given specifications (transient and steady state), we determine the desired closed-loop transfer function $T(s)$. We also have

$$T(s) = \frac{G(s)G_c(s)}{1 + G(s)G_c(s)H(s)} \tag{5.62}$$

Since we know $T(s)$, $G(s)$, and $H(s)$, the required compensator transfer function $G_c(s)$ can be determined from Eq. 5.62.

As an example, consider the second-order position control system (Fig. 5.4). This system with compensator added is shown in Fig. 5.32a. The design specifica-

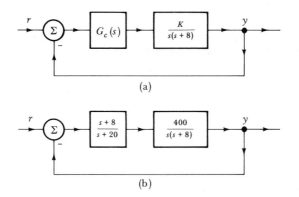

(a)

(b)

Figure 5.32

tions are given as

$$\text{PO} \leq 16\%, \qquad t_r \leq 0.5 \text{ sec}, \qquad \text{and } t_s \leq 2 \text{ sec} \tag{5.63}$$

and

$$e_r \leq 0.05$$

Let us determine the desired $T(s)$. Since e_r is finite, $e_s = 0$. Let

$$T(s) = \frac{b_0}{s^2 + a_1 s + a_0} \tag{5.64}$$

Since $e_r = 0.05$ and $e_s = 0$, we have (see Eq. 5.42)

$$a_0 = b_0$$

Also from Eq. 5.43,

$$0.05 = \frac{a_1}{a_0}$$

and

$$a_1 = 0.05a_0 = \frac{a_0}{20}$$

Hence

$$T(s) = \frac{a_0}{s^2 + \frac{a_0}{20}\,s + a_0} \tag{5.65}$$

The transfer function $T(s)$ in Eq. 5.65 meets the steady-state specifications.   The poles of $T(s)$ are

$$-\frac{a_0}{40} \pm j \sqrt{a_0 - \frac{a_0^2}{1600}} \tag{5.66}$$

The transient specifications in Eq. 5.63 are met if the poles of $T(s)$ lie in the shaded region shown in Fig. 5.12. For $a_0 = 400$, the roots are (see Eq. 5.66)

$$-10 \pm j10\sqrt{3}$$

These roots lie in the desired region (shaded region in Fig. 5.12). Hence we choose $a_0 = 400$ and

$$T(s) = \frac{400}{s^2 + 20s + 400} \tag{5.67}$$

From Fig. 5.32a we have

$$G(s) = \frac{K}{s(s + 8)} \qquad H(s) = 1$$

Substitution of $T(s)$, $G(s)$, and $H(s)$ in Eq. 5.62 gives

$$\frac{400}{s^2 + 20s + 400} = \frac{\dfrac{KG_c(s)}{s(s + 8)}}{1 + \dfrac{KG_c(s)}{s(s + 8)}} = \frac{KG_c(s)}{s^2 + 8s + KG_c(s)}$$

or

$$400[s^2 + 8s + KG_c(s)] = KG_c(s)[s^2 + 20s + 400]$$

and

$$KG_c(s)(s^2 + 20s) = 400(s^2 + 8s)$$

Therefore

$$KG_c(s) = \frac{400(s^2 + 8s)}{s^2 + 20s} = 400\frac{(s + 8)}{s + 20}$$

We can choose

$$K = 400$$

$$G_c(s) = \frac{s + 8}{s + 20}$$

Let us reevaluate the performance of this system:

$$T(s) = \frac{400}{s^2 + 20s + 400}$$

Hence

$$\zeta = 0.5 \qquad \omega_n = 20$$

Hence

$$PO = 16\%$$

$$t_r = \frac{1.62}{20} = 0.081$$

$$t_s = \frac{4}{\zeta\omega_n} = 0.4 \qquad\qquad (5.68)$$

$$e_r = \frac{a_1}{a_0} = 0.05$$

Observe that the compensator is a simple lag network. Note the important feature of this design: The compensator has a zero at $s = -8$ which will cancel the plant pole at $s = -8$. Pole-zero cancellation is the distinguishing feature of this method. One may object to such cancellation on the basis that the plant poles-zeros usually are not known precisely and/or they may change because of aging. We must however remember that these problems exist no matter what method of design is used. Hence these objections will also be valid for any other method of design.

In order to simplify computations in synthesis of $G_c(s)$, one should choose $G_c(s)$ such that the open-loop transfer function has poles only along the real axis.[†] If $G(s)H(s)$ has any complex poles they should be cancelled by corresponding zeros in $G_c(s)$.

5.10 MULTIPLE-VARIABLE FEEDBACK

So far our discussion has been restricted to systems in which only the output is fed back to the input. It is reasonable to expect that the system performance will be improved, if we feed back not just the output but some additional

---

†See for example J. G. Truxal, *Control System Synthesis* (New York: McGraw-Hill, 1955).

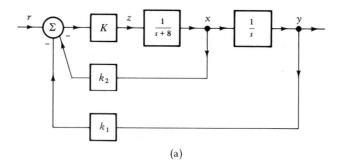

(a)

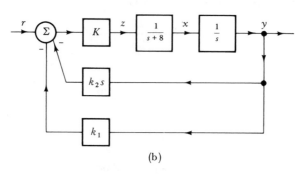

(b)

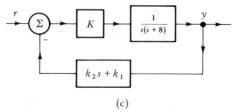

(c)

Figure 5.33

system variables. Consider again the position control system in Fig. 5.4. The open-loop transfer function $G(s)$ is given by

$$G(s) = \frac{K}{s(s + 8)}$$

This system can be represented by an amplifier K, followed by a cascade of two first-order systems with transfer functions $1/(s + 8)$ and $1/s$. The outputs of both these subsystems can now be fed back, as shown in Fig. 5.33a. This type of feedback is obviously more general than the type shown in Fig. 5.4.

The transfer function $T(s)$ of this feedback system can be found from Mason's rule:

$$T(s) = \frac{K/s(s + 8)}{1 + \dfrac{Kk_2}{s + 8} + \dfrac{Kk_1}{s(s + 8)}}$$

$$= \frac{K}{s(s + 8) + Kk_2s + Kk_1}$$

$$= \frac{K}{s^2 + (8 + Kk_2)s + Kk_1} \tag{5.68a}$$

Note that we have three parameters, K, k_1, and k_2, which are adjustable. This gives us three degrees of freedom. We may use this freedom to locate the poles of $T(s)$ at any desired location in the s-plane. Since $T(s)$ has two poles, we use up two degrees of freedom to place the poles of $T(s)$ at a desired location. This leaves one more degree of freedom which can be used to fix one of the error constants. Let us choose $e_s = 0$. To meet this condition we have (see Eq. 5.42)

$$Kk_1 = K$$

This yields $k_1 = 1$. With this value of k_1, $T(s)$ now becomes

$$T(s) = \frac{K}{s^2 + (8 + Kk_2)s + K} \tag{5.68b}$$

Note that we can locate the poles of $T(s)$ anywhere by adjusting K and k_2. Compare this situation with the simple feedback system in Fig. 5.4, where only the parameter K was adjustable. By adjusting K we could locate the poles of $T(s)$ along a certain confined path (the root locus in Fig. 5.13). In the present case of multiple-variable feedback we have the freedom to locate the poles of $T(s)$ not along a confined path but anywhere in the s-plane.

Let the transient specifications be

$$PO \leq 16\% \qquad t_r \leq 0.5 \quad \text{and} \quad t_s \leq 2$$

In order to meet these specifications, the poles of $T(s)$ must lie in the shaded region delineated in Fig. 5.12. Let us choose the poles at $-12 \pm j20$. From Eq. 5.68b it now follows that

$$s^2 + (8 + Kk_2)s + K = (s + 12 - j20)(s + 12 + j20)$$
$$= s^2 + 24s + 544$$

This yields

$$K = 544 \qquad \text{and} \qquad k_2 = 0.0294$$

Using these values in Eq. 5.68b we obtain

$$T(s) = \frac{544}{s^2 + 24s + 544} \tag{5.68c}$$

From Eq. 5.43 we find

$$e_r = \frac{24}{544} = 0.0442$$

From Eq. 5.68c we note that, for this system, $\zeta = 0.514$, and $\omega_n = 23.4$. The transient parameters are found as usual to be

$$PO \cong 15\%, \quad t_s = 0.33, \quad \text{and} \quad t_r = 0.071$$

The steady-state errors are already found to be

$$e_s = 0 \quad \text{and} \quad e_r = 0.0442$$

Compare this performance with that of the system in Fig. 5.4. For the system in Fig. 5.4, the transient performance was inferior. It was also not possible to achieve $e_r < 0.125$ without some compensation.

   This procedure of designing a system with multiple-variable feedback can be applied in the same way to higher-order systems. For an nth-order plant transfer function, there are n feedback signals, and $n + 1$ parameters $(K, k_1, k_2, \ldots, k_n)$ that can be adjusted. This provides us with the freedom of placing all the n poles of $T(s)$ at any desired locations in the s-plane. In addition, we can also fix at least one of the steady-state errors. This type of feedback is more generally known as *state-variable feedback*. The theoretical basis for and general development of this feedback are discussed in the next chapter.

   Often the plant $G(s)$ may be such that the internal outputs (such as output x in Fig. 5.33a) may not be available or accessible for the purpose of feedback. In such case we need to generate the inaccessible variable x, from the accessible variables, such as y Fig. 5.33a. We observe that

$$X(s) = sY(s)$$

Hence x can be generated by passing y through a differentiator (transfer function s). Thus instead of feeding back signal x we may as well feed the signal $sY(s)$. The combined feedback signal $(k_2 x + k_1 y)$ can be generated by passing the signal y through a system with a transfer function $(k_2 s + k_1)$ as shown in Fig. 5.33c.† Note that the differentiator (transfer function s) differentiates the output. In the case of the angular position control system, the differentiator yields angular velocity. In this case the transfer function $k_2 s$ can be realized by a tachometer.

   The signal x can also be generated from the output z (Fig. 5.33b) by passing z through a transfer function $1/(s + 8)$.

ROOT LOCUS FOR MULTIPLE-VARIABLE FEEDBACK SYSTEMS

   The discussion on generating inaccessible state variables, points the way to draw the root locus for multiple-variable feedback systems. We observe that the system in Fig. 5.33c is equivalent to the system in Fig. 5.33a. Hence both systems have the same root locus. We know how to sketch the root locus for the system in

---

   †It follows from the discussion that multiple-variable feedback system is equivalent to a simple output feedback system of the form in Fig. 5.17, provided we choose the feedback transfer function $H(s)$ properly. Generally $H(s)$ turns out to be a polynomial transfer function with numerator of order higher than the denominator. This is generally difficult to synthesize.

Fig. 5.33c, because it is a single variable-feedback system with open-loop transfer function:

$$G(s)H(s) = K \frac{k_2 s + k_1}{s(s + 8)}$$

For given values of k_1 and k_2 we can easily draw the root locus for this case as K varies from 0 to ∞.

We can extend this procedure to a multiple-variable system with a plant transfer function of higher order. Every feedback variable can be generated from the output using an appropriate transfer function. Hence a multiple-variable feedback system can always be converted to a system with a single loop (feedback from the output only).

We shall give here one more example of multiple-variable feedback. The attitude of an aircraft is controlled by three sets of surfaces; elevators, rudder, and ailerons, shown shaded in Fig. 5.34. By adjusting these surfaces one can set the

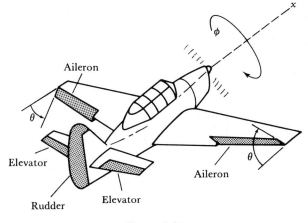

Figure 5.34

aircraft on a desired flight path. We shall discuss here an automatic control system (autopilot) which controls the roll angle ϕ by adjusting aileron surfaces. Deflection of aileron surfaces by an angle θ generates a torque due to air pressure on aileron surfaces. This causes a rolling motion. We shall assume a simplified model where a rolling motion can be considered independent of other motions. The transfer function relating the roll angle ϕ to aileron deflection angle θ is given by[†] (see Prob. 2.6)

$$\frac{\Phi(s)}{\theta(s)} = \frac{a}{s(s + \alpha)} \tag{5.69}$$

[†]J. H. Blakelock, *Automatic Control of Aircrafts and Missiles* (New York: Wiley, 1965), p. 127.

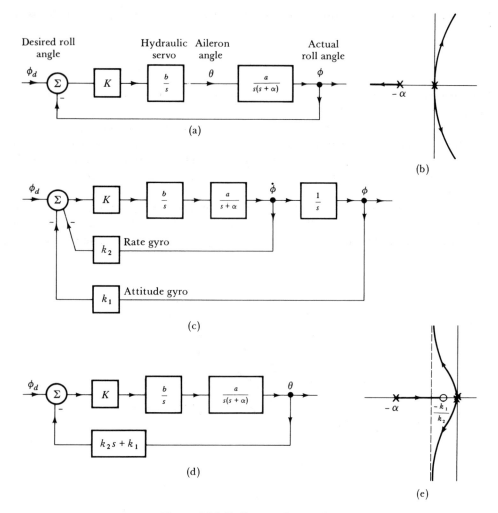

Figure 5.35 Roll control autopilot.

The aileron surfaces are controlled by hydraulic servo with a transfer function† b/s (see Prob. 2.7).

The actual roll angle ϕ is measured by an attitude gyro and fed back at the input as shown in Fig. 5.35a.   The error (difference between the desired roll angle ϕ_d and the actual roll angle) drives the hydraulic servo which in turn changes the aileron deflection. The open-loop transfer function of this system is given by

$$G(s) = \frac{Kab}{s^2(s + \alpha)}$$

---

†Strictly speaking, the hydraulic servo transfer function is of the form $k/(s + \beta)$, where β is rather small. When the hydraulic servo operates with oil under high pressure we are justified in assuming $\beta \cong 0$ (see Prob. 2.7).

The root locus of this system is shown in Fig. 5.35b. It is obvious that the system is unstable. It can be stabilized by using multiple-variable feedback. In this case there are three possible signals that can be fed back. For the purpose of stabilization, however, it suffices to feed back only two signals as shown in Fig. 5.35c. The signals fed back are (i) the roll angle ϕ, and (ii) the roll velocity $\dot{\phi}$. These signals can be obtained by using attitude and rate gyros as shown in Fig. 5.35c. This system is equivalent to the system shown in Fig. 5.35d. The open-loop transfer function of the equivalent system is given by

$$G(s)H(s) = \frac{Kab(k_2 s + k_1)}{s^2(s + \alpha)}$$

This open-loop transfer function has a zero at $-k_1/k_2$ and poles at $0,0$, and $-\alpha$. The root locus of this system is shown† in Fig. 5.35e. Note that the system is now stable for all values of K.

This stabilization could also have been achieved by a series compensator $G_c(s)$ by choosing

$$G_c(s) = k_2 s + k_1$$

From the practical standpoint, this compensation is not as convenient as the one in Fig. 5.35c.

5.11   SENSITIVITY CONSIDERATIONS

The primary reason for using feedback is to reduce the sensitivity of the system to changes in environment (external disturbance) and to changes in the system parameters. We shall now consider both these aspects.

EFFECT OF UNWANTED DISTURBANCE

Let us consider an automatic tracking antenna system. This is a position-control system for which we shall use a field-controlled dc motor (Fig. 5.36a) for tracking. A gust of wind exerts a disturbance torque T_d on the antenna. The system equations are as follows. For the field circuit of the motor (Fig. 5.36a), the loop equation is

$$(R_f + sL_f)I_f(s) = E_f(s) \tag{5.70a}$$

Since the armature current is constant, the generated torque T_g is proportional to the field current (see Eq. 1.24). Hence

$$T_g = K_t I_f(s) \tag{5.70b}$$

The net torque available to rotate the antenna is $T_g + T_d$, the generated torque plus the disturbance torque. The load equation is given by

$$T_g + T_d = (Js^2 + Ds)\theta_0(s) \tag{5.70c}$$

---

†For the root locus to lie in the LHP, we must choose k_1 and k_2 such that the zero at $-k_1/k_2$ is to the right of the pole at $-\alpha$.

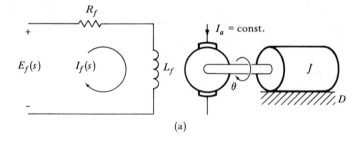

(a)

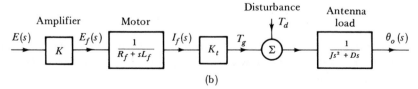

(b)

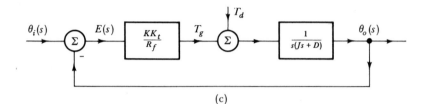

(c)

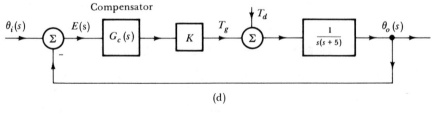

(d)

Figure 5.36

From these equations we can draw a block diagram of the motor and the load as shown in Fig. 5.36b. From the practical point of view the time constant due to mechanical load is much larger compared to the time constant of the field circuit of the motor. Consequently L_f may be ignored. When the feedback is applied, the resultant system is as shown in Fig. 5.36c.

The system has two inputs, θ_i and T_d. Since the system is linear, we can use the principle of superposition to find $\theta_o(s)$. Use of Mason's rule to Fig. 5.36c yields

$$\theta_o(s) = \frac{\dfrac{KK_T/R_f}{s(Js+D)}}{1 + \dfrac{KK_T/R_f}{s(Js+D)}}\,\theta_i(s) + \frac{\dfrac{1}{s(Js+D)}}{1 + \dfrac{KK_T/R_f}{s(Js+D)}}\,T_d(s)$$

$$= \underbrace{\frac{KK_T/R_f}{s(\mathcal{J}s + D) + KK_T/R_f}}_{\text{desired output } \theta_{oi}} \theta_i(s) + \underbrace{\frac{1}{s(\mathcal{J}s + D) + KK_T/R_f}}_{\text{disturbance output } \theta_{od}} T_d(s)$$

The desired response, θ_{oi} and the disturbance output θ_{od} are clearly shown in the above equation. The ratio of the two response components is KK_T/R_f. Hence by making KK_T/R_f large the effect of the undesired component can be reduced. Note, however, increasing KK_T/R_f in general tends to increase the PO. For convenience, let us assume some values. Let $\mathcal{J} = 1$, $D = 5$, $K_T/R_f = 1$. This gives

$$\theta_o(s) = \underbrace{\frac{K}{s(s + 5) + K}}_{\theta_{oi}} \theta_i(s) + \underbrace{\frac{1}{s(s + 5) + K}}_{\theta_{od}} T_d(s)$$

Let us assume the input θ_i and the disturbance T_d both to be unit-step functions. From the final-value theorem it can be easily seen that the steady-state value of θ_{oi} is 1, whereas that of θ_{od} is $1/K$. Thus the effect of the disturbance can be reduced by increasing K. However, the steady-state disturbance at the output can never be made zero. If the design calls for complete suppression of the disturbance output θ_{od} in the steady state, we must use some compensation as shown in Fig. 5.36d. We may try an integral compensator $(G_c(s) = 1/s)$. Unfortunately, such a compensator will cause the system to be unstable. We shall therefore try an integral plus proportional type of compensator

$$G_c(s) = 1 + \frac{\alpha}{s} = \frac{s + \alpha}{s}$$

This yields

$$\theta_o(s) = \underbrace{\frac{K(s + \alpha)}{s^2(s + 5) + K(s + \alpha)}}_{\theta_{oi}} \theta_i(s) + \underbrace{\frac{s}{s^2(s + 5) + K(s + \alpha)}}_{\theta_{od}} T_d(s)$$

From the final-value theorem we immediately find that the steady-state value of θ_{od} is zero. The steady-state value of θ_{oi} is still unity as desired. This compensation would serve the purpose. As mentioned earlier, such a compensator is rather expensive. The lag compensator closely approximates the integral compensator. If by choosing

$$G_c(s) = \frac{s + \alpha}{s + \beta} \qquad \alpha > \beta$$

the reader can easily show that for unit-step disturbance

$$\theta_{od}(\text{steady-state}) = \frac{\beta}{\alpha}\left(\frac{1}{K}\right)$$

By choosing $\alpha/\beta \gg 1$, the steady-state disturbance output can be made as small as desired. The steady-state value of the desired output θ_{oi} is still unity, as required.

Note that this compensator not only reduces the effect of the external disturbance but also reduces the steady-state errors by a factor α/β.

EFFECT OF CHANGES IN SYSTEM PARAMETERS

As noted earlier, the performance of a properly designed feedback system is less sensitive to changes in the system parameters. We shall now develop a method which will allow us to visualize the effect of parameter changes on the system performance. The root locus accomplishes this objective for changes in the gain parameter K. We are now interested in finding a similar locus when K is fixed, but some other system parameter varies. Such a root locus (for a system parameter other than gain K) is known as a root contour.

Let us again consider the tracking antenna system in Fig. 5.36c, with $K_T/R_f = 1$, and $\mathcal{J} = 1$ as shown in Fig. 5.37a. The open-loop transfer function of

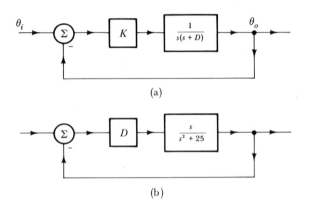

(a)

(b)

Figure 5.37

the system is given by

$$G(s) = \frac{K}{s(s + D)}$$

Let the value of K be set at 25. We shall now find the root contour as the parameter D varies from 0 to ∞. For $K = 25$,

$$G(s) = \frac{25}{s(s + D)}$$

The characteristic equation is given by

$$1 + G(s) = 0$$

or

$$1 + \frac{25}{s(s + D)} = 0$$

This yields

$$s^2 + Ds + 25 = 0 \tag{5.71a}$$

This equation may be rearranged as

$$1 + \frac{Ds}{s^2 + 25} = 0 \tag{5.71b}$$

We observe an interesting fact about this characteristic equation (Eq. 5.71b). It is also a characteristic equation of the system shown in Fig. 5.37b. Hence the roots of Eq. 5.71b lie on the root locus of the system in Fig. 5.37b, as D varies from 0 to ∞. The open-loop transfer function $\hat{G}(s)$ of the system in Fig. 5.37b is given by

$$\hat{G}(s) = \frac{Ds}{s^2 + 25}$$

$$= \frac{Ds}{(s - j5)(s + j5)}$$

This open-loop transfer function has a zero at the origin, and poles at $\pm j5$. The root locus can now easily be sketched (Fig. 5.38a), by using the techniques developed earlier. We are thus able to apply the root-locus techniques to draw a root contour for a system parameter other than the gain K. The root contour in Fig. 5.38a is for a fixed value of $K = 25$. We may draw several such root contours for different values of K. This family of root contours enables us to visualize the effects on system behavior of simultaneous variations of two parameters, K and D.

Let us now observe the effect on the step response of the system when $K = 25$ and when D varies from a value of 4 to 6. From Fig. 5.38a, when $K = 25$ and $D = 4$, we read $\zeta = 0.4$, $\omega_n = 5$. Hence

$$PO = 26\% \quad \text{and} \quad t_s = 2$$

For $\zeta = 0.4$, we read $\omega_n t_r = 1.47$ (Fig. 5.10). Hence

$$t_r = \frac{1.47}{5} = 0.294$$

When $K = 25$ and $D = 6$, we read (Fig. 5.38a) $\zeta = 0.6$ and $\omega_n = 5$. Hence

$$PO = 10\%, \quad t_s = 1.34, \quad \text{and} \quad t_r = 0.372$$

The two responses are shown in Fig. 5.38b. As a result of 50% variation in the value of the parameter D, the transient response has altered somewhat (Fig. 5.38b). But the steady-state value of the output has remained unchanged.

ROOT LOCUS OF A MULTIPLE-LOOP FEEDBACK SYSTEM

The root-contour method discussed here can be used to draw a root locus of a system with more than one feedback loop. Consider for example the multiple-loop system in Fig. 5.33a. The closed-loop transfer function of this system is found

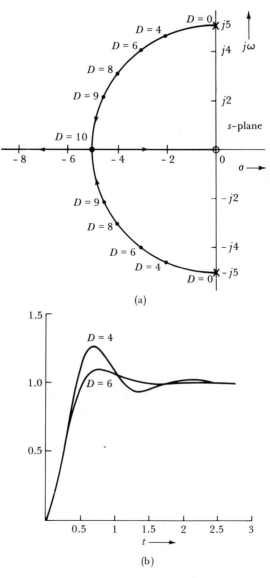

(a)

(b)

Figure 5.38

in Eq. 5.68a. Hence the characteristic equation is given by

$$s^2 + (8 + Kk_2)s + Kk_1 = 0$$

This equation can be rearranged as

$$s^2 + 8s + K(k_2s + k_1) = 0$$

or

$$1 + \frac{K(k_2s + k_1)}{s(s + 8)} = 0$$

It is obvious that the root locus of this system is the same as the root locus of a system whose open-loop transfer function is given by

$$G(s)H(s) = \frac{K(k_2 s + k_1)}{s(s + 8)}$$

For given values of k_1 and k_2 we can easily draw the root locus as K varies from 0 to ∞.

5.12   FREQUENCY-RESPONSE METHOD OF SYNTHESIS

In Appendix C it is shown that the transfer function of a system can be determined from the knowledge of the steady-state response of the system to sinusoidal signals over the frequency range $(0, \infty)$.   The record of such responses is therefore an alternate way of describing a linear time-invariant system. We can thus use the sinusoidal steady-state response to design or synthesize linear systems.

In this section we shall develop techniques of designing feedback systems on the basis of sinusoidal steady-state response description of the plant. This method has certain advantages over the previous methods. In previous methods we had assumed the knowledge of the transfer function of the plant. In practice this information may not be available. In such case we can experimentally determine the sinusoidal steady-state response of the plant and use this data directly to design the appropriate system. Frequency-response methods can also be applied to plants with nonrational transfer functions (plants whose transfer functions are not necessarily ratios of two polynomials in s). For example, a system involving an ideal time delay (dead time) can be conveniently handled by this method.†

It must however be mentioned that the frequency-response method is not as convenient as the previous method so far as transient and steady-state error specifications are concerned. Consequently, both these methods should be considered as complementary rather than as alternatives or as rivals.

The sinusoidal steady-state response is characterized by $|H(j\omega)|$, the magnitude, and $\angle H(j\omega)$, the phase of the response. This data can be expressed in various ways. Bode plots discussed in Appendix C is one possible method. In a Bode plot we represent the log magnitude $[20 \log |H(j\omega)|]$ and the phase $\angle H(j\omega)$ as functions of frequency ω. Alternately, we may plot the complex number $H(j\omega)$ on a complex plane for all values of ω. The point $H(j\omega)$ is represented by a point at a distance $|H(j\omega)|$ from the origin and at an angle $\angle H(j\omega)$ from the real axis. We plot these points for several values of ω in the range 0 to ∞ and join these points in sequence. This is the polar representation of $H(j\omega)$. A third alternative of presenting this data is by using log-magnitude-versus-angle plot $[20 \log |H(j\omega)|$ as a function of $\angle H(j\omega)]$, using ω as a parameter. We shall give an example of all three representations.

Consider the position-control system in Fig. 5.4 with open-loop transfer func-

---

†An ideal delay of T seconds has a nonrational transfer function $H(s) = e^{-sT}$.

tion $G(s)H(s)$ given by

$$G(s)H(s) = \frac{K}{s(s+8)} \tag{5.72a}$$

For the purpose of illustration, let $K = 64$. Thus

$$G(s)H(s) = \frac{64}{s(s+8)} \tag{5.72b}$$

and

$$G(j\omega)H(j\omega) = \frac{64}{j\omega(j\omega+8)} \tag{5.73}$$

We shall now demonstrate the three methods of representing $G(j\omega)H(j\omega)$.

1. BODE PLOTS

We can represent $G(j\omega)H(j\omega)$ by using Bode plots (log magnitude and angle) discussed in Appendix C. These plots are shown in Fig. 5.39.

2. POLAR PLOT

Here we plot points $G(j\omega)H(j\omega)$ in the complex plane [$G(j\omega)H(j\omega)$ plane]. For any given value of ω, the point $G(j\omega)$ $H(j\omega)$ is plotted at a distance $|G(j\omega)H(j\omega)|$ from the origin and at an angle $\angle G(j\omega)H(j\omega)$ from the real axis.

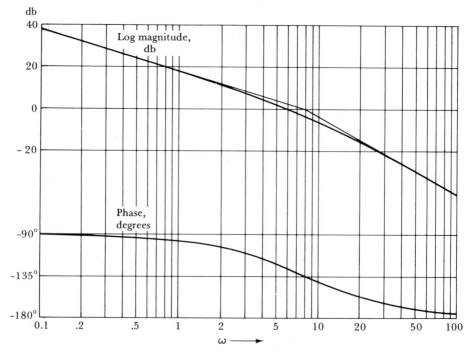

Figure 5.39

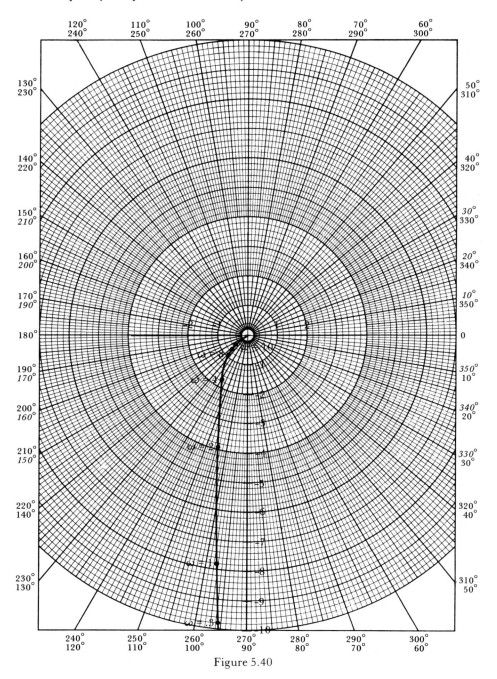

Figure 5.40

Thus for $\omega = 1$, we observe from Bode plots (Fig. 5.39) that $|G(j\omega)H(j\omega)| = 7.94$ and $\angle G(j\omega)H(j\omega) = -97.12°$. This point is plotted in $G(j\omega)H(j\omega)$ plane and labeled $\omega = 1$ (see Fig. 5.40). Similarly for $\omega = 8$, $|G(j\omega)H(j\omega)| = 0.707$, and $\angle H(j\omega)H(j\omega) = -135°$. This point is plotted in Fig. 5.40 and labeled

$\omega = 8$. In this way we plot points for several values of ω in the range 0 to ∞ and join the points in sequence to obtain the polar plot of $G(j\omega)H(j\omega)$. The polar plot is labeled for various values of ω along its path.

3. LOG-MAGNITUDE VERSUS ANGLE PLOT

In this representation the log magnitude is plotted (in rectangular coordinates) as a function of the phase angle. We can use Bode plots directly for this representation. From Fig. 5.39 we observe that at $\omega = 1$ the log magnitude is 17.99 db and the phase is $-97.12°$. We therefore plot a point $(-97.12, 17.99)$ and label it $\omega = 1$ (Fig. 5.41). Similarly at $\omega = 8$ the log magnitude is -3.01 and the phase is $-135°$. Thus we plot another point $(-135, -3.01)$ and label it $\omega = 8$. In this manner we plot points for several values of ω in the range 0 to ∞ and join the points in sequence to obtain the required plot (Fig. 5.41). Note that this plot has all the information of Bode plot (or polar plot). Indeed all these three plots carry identical information. We can use any of the three plots for the purpose of synthesis.

STABILITY: NYQUIST CRITERION

When the open-loop transfer function $G(s)H(s)$ is known the stability of the closed-loop system is easily determined from the roots of the characteristic equation

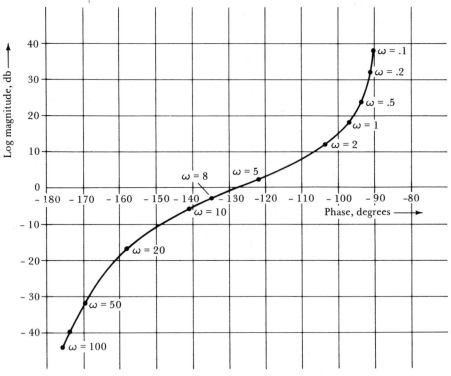

Figure 5.41

$1 + G(s)H(s) = 0$.   If, however, we do not know $G(s)H(s)$, but are given $G(j\omega)H(j\omega)$ in one of the graphical forms in Figs. 5.39, 5.40, or 5.41 instead, how shall we determine stability (or instability) of the closed-loop· system? There is a very simple criterion, known as the *Nyquist criterion,* which allows us to determine the stability of a closed-loop system from the knowledge of $G(j\omega)H(j\omega)$, the frequency response of the open-loop system.

As an example, consider the polar plot corresponding to an open-loop transfer function

$$G(s)H(s) = \frac{K}{s(s + 2)(s + 4)} \tag{5.74a}$$

Let us sketch a polar plot corresponding to this function for $K = 24$:

$$G(j\omega)H(j\omega) = \frac{24}{j\omega(j\omega + 2)(j\omega + 4)} \tag{5.74b}$$

We may easily construct Bode plots for this function (see Fig. 5.44) and then draw the polar plot as shown in Fig. 5.42.

Note that at $\omega = 2.83$, the polar plot crosses the real axis at -0.5. If we increase K in Eq. 5.74a, the magnitudes also increase proportionately. Consequently the polar plot expands directly in proportion to the value of K. If we let $K = 48$ instead of 24, every magnitude in the polar plot will double as shown in Fig. 5.42. We now observe that at $\omega = 2.83$, $G(j\omega)H(j\omega) = -1$. That is,

$$G(j2.83)H(j2.83) = -1$$

Hence

$$1 + G(j2.83)H(j2.83) = 0 \tag{5.75}$$

However, $1 + G(s)H(s) = 0$ is the characteristic equation of the closed-loop system. It is obvious from Eq. 5.75 that the characteristic equation is satisfied for $s = j2.83$. Since (see Eq. 3.38)

$$G(-j\omega)H(-j\omega) = G^*(j\omega)H^*(j\omega)$$
$$G(-j2.83)H(-j2.83) = (-1)^* = -1$$

Hence

$$1 + G(-j2.83)H(-j2.83) = 0$$

and

Note that $s = -j2.83$ is also a root of the characteristic equation of the closed-loop system. Hence $\pm j2.83$ are the natural frequencies of the system. In other words, natural frequencies of the closed-loop system lie on the $j\omega$-axis for $K = 48$. For $K > 48$, the natural frequencies move in the RHP and the system becomes unstable (see root locus in Fig. 5.23b). The polar plot (called *Nyquist plot*) crosses real axis at -1 (at $-1 + j0$) when the natural frequencies are along

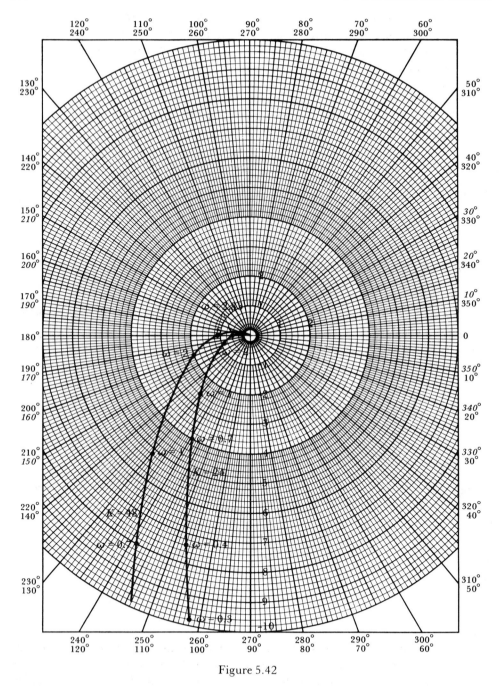

Figure 5.42

the $j\omega$-axis. If $K > 48$, the closed-loop system becomes unstable. The Nyquist plot of the open-loop transfer function for $K > 48$ expands beyond the critical point -1 and will enclose the (critical) point -1 inside it. This is the criterion for instability. When the polar plot of the open-loop transfer function encloses the critical point

$-1 + j0$ inside it, the corresponding closed-loop system becomes unstable.† For the rigorous proof of this statement and some qualifications attached to it, the reader is referred to Appendix F.

Observe that the system whose Nyquist plot appears in Fig. 5.40 will never become unstable for any value of K. As K increases, the Nyquist plot expands without enclosing the point $-1 + j0$. This is because the phase lag of $G(j\omega)\,H(j\omega)$ never exceeds $180°$ and consequently never crosses the negative real axis. We may also verify this fact from the root locus of $K/s(s+8)$ shown in Fig. 5.13. It is evident from this figure that the roots of the characteristic equation remain in LHP for all values of K.

GAIN MARGIN AND PHASE MARGIN

For a stable closed-loop system the Nyquist plot must not enclose the point $-1 + j0$. Even if the plot does not enclose the point -1, but is very close to the point -1, the system is in danger of becoming unstable in case some of the system parameters change. For this reason it is desirable to maintain a certain margin of stability. The closed-loop system becomes unstable if the open-loop transfer function has a phase $-180°$ at a frequency where its gain is greater than unity. Let the open-loop gain magnitude be α_M when its phase is $-180°$ (Fig. 5.43). If $\alpha_M < 1$, the corresponding closed loop system is stable. Smaller the α_M, more the margin of stability. We define gain margin as $1/\alpha_M$, or

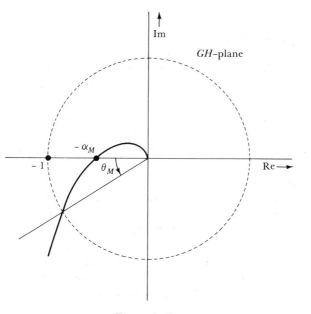

Figure 5.43

---

†Here we are assuming open-loop stability—that is, the open-loop transfer function $G(s)H(s)$ has no poles in the RHP. For a general case refer to Appendix F.

$$\text{Gain margin} = \frac{1}{\alpha_M} \qquad (5.76a)$$

It is more convenient to express gain margin in decibels as

$$\text{Gain margin} = -20 \log_{10}(\alpha_M) \text{ db}$$

We may also consider the point where the open-loop gain magnitude is unity and the phase is $-180 + \theta_M$ (Fig. 5.43). The system is short of being unstable by a phase margin of θ_M. Consequently, we call θ_M the phase margin, or

$$\text{Phase margin} = \theta_M \qquad (5.76b)$$

In order to protect the system from becoming unstable because of variations in system parameters (or in the environment), the system should be designed with reasonable gain and phase margins. Zero-gain or phase margins indicate some of the poles of $T(s)$ lie on the $j\omega$-axis. Positive gain and phase margins indicate poles of $T(s)$ in the LHP. If the margins are too small (positive), it implies that the poles of $T(s)$ are in the LHP but close to the $j\omega$-axis. The transient response of such a system will have large overshoot. On the other hand, very large (positive) gain and phase margins may indicate a sluggish system. There exist an approximate relationship between gain margin, phase margin, and the value of ζ and ω_n of the dominant poles of $T(s)$ of the closed-loop system.† Generally a gain margin of 3 db ($\alpha_M = 0.707$) and a phase margin of about 30° to 60° are considered desirable. Design specifications for transient performance are often given in terms of certain gain and phase margins.

We can express the Nyquist stability criterion in terms of Bode plot or log-magnitude-phase plot. Figure 5.44 shows Bode plots for the open-loop transfer function in Eq. 5.74b. It is easy from these plots to see that at $\omega = 2.83$ the phase is $-180°$ and the magnitude is $0.5 (-6 \text{ db})$. According to the Nyquist criterion the closed-loop system is unstable if the gain is greater than 1 when the phase is $-180°$. In terms of Bode plots we can say that the closed-loop system is unstable if the open-loop transfer function magnitude is > 0 db when the phase is $-180°$. The frequency where the gain is 0 db is called the *gain crossover frequency* ω_g, and the frequency where the phase is $-180°$ is called the *phase crossover frequency* ω_p.

The gain margin gives an indication of the amount by which the gain constant K in the open-loop transfer function can be increased before the system becomes unstable. Thus in Fig. 5.44 the gain margin is 6 db. If we increase K by 6 db (a factor of 2), the gain margin goes to zero and the system becomes unstable. If we can accept a gain margin of 3 db, then we can safely increase K by 3 db (a factor of $\sqrt{2}$). Thus in Fig. 5.44 for a gain margin of 3 db we may increase K from 24 to $24\sqrt{2} = 33.8$ in Eq. 5.74b. Note that increasing K by α db in the open-loop transfer function raises the entire magnitude plot by α db while the phase plot remains unchanged. In Fig. 5.44 we observe that $\omega_g = 1.9$ and $\omega_p =$

---

†J. Holt-Smith in Discussion on "A Simple Correlation Between Closed-Loop Transient Response and Open-Loop Frequency Response," *Proc. Inst. Elec. Engr.* (London), Part II (1953), p. 210.

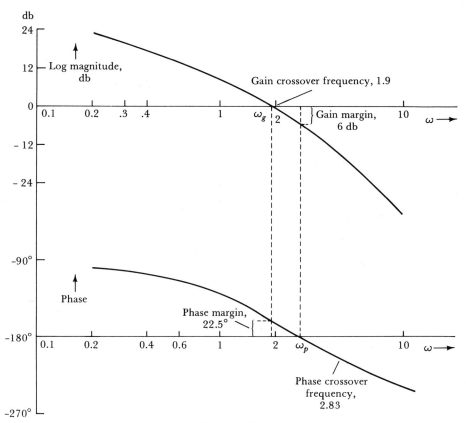

Figure 5.44

2.83. The gain below 0 db at ω_p is the gain margin and the phase above $-180°$ at ω_g is the phase margin. From Fig. 5.44 we read gain margin to be 6 db and the phase margin to be 22.5°. For a stable system, the gain and phase margins both must be positive.

Figure 5.45 shows the log-magnitude versus angle plot for the same open-loop transfer function (Eq. 5.74b). The gain crossover frequency ω_g, the phase cross-over frequency ω_p, the gain margin, and the phase margin are readily seen from this plot. The critical point is (0 db, $-180°$). If this point lies to the left of the plot, the closed loop system is stable. Increasing K by α db shifts the entire curve upward by α.

TRANSIENT PERFORMANCE IN TERMS OF FREQUENCY RESPONSE

In order to design a control system for a specific transient performance, we must know the relationship between the frequency response and the transient response. As before, let us assume a second-order system with closed-loop transfer function

$$T(s) = \frac{\omega_n^2}{s^2 + 2\zeta\omega_n s + \omega_n^2}$$

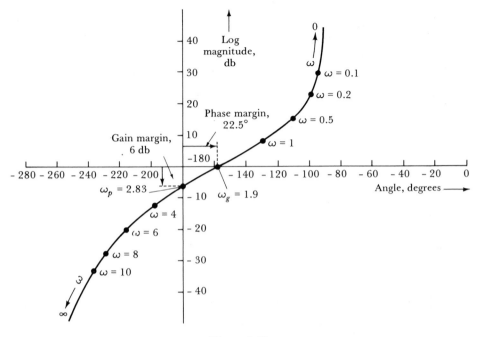

Figure 5.45

Hence

$$T(j\omega) = \frac{\omega_n^2}{(j\omega)^2 + 2j\zeta\omega_n\omega + \omega_n^2}$$

$$= \frac{1}{1 + 2j\zeta\left(\dfrac{\omega}{\omega_n}\right) - \left(\dfrac{\omega}{\omega_n}\right)^2}$$

This is a case of two complex poles. The frequency response (Bode plots) for such a case is shown in Fig. C.4, Appendix C. We notice that the frequency response shows a resonance peak for $\zeta < 0.707$. The peak occurs at a frequency ω_m which is slightly below ω_n (see Fig. 5.46). We also observe that smaller the value of ζ larger is the height of peak (M_m). Hence M_m and ζ are related. The peak occurs (at $\omega = \omega_m$) where the quantity $|1 + 2j\zeta(\omega/\omega_n) - (\omega/\omega_n)^2|$ is minimum. Straightforward differentiation yields

$$\omega_m = \omega_n\sqrt{1 - 2\zeta^2} \qquad \zeta < 0.707 \tag{5.77a}$$

and

$$M_m = -20\log_{10}\left|1 + 2j\zeta\frac{\omega_m}{\omega_n} - \left(\frac{\omega_m}{\omega_n}\right)^2\right|$$

$$= -20\log_{10}2\zeta\sqrt{1 - \zeta^2} \qquad \zeta < 0.707 \tag{5.77b}$$

Thus the peak height M_m and the peak frequency ω_m are related to ζ and ω_n. We can plot M_m as a function of ζ. However since percent overshoot (PO) is also a

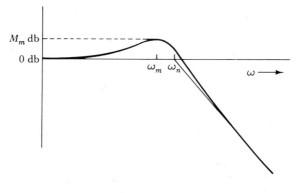

Figure 5.46

function of ζ (Eq. 5.11c), we can plot M_m directly as a function of overshoot using ζ as parameter. This is shown in Fig. 5.47a. Figure 5.47b shows ω_m/ω_n as a function of ζ (Eq. 5.77a). From the frequency response, we can determine M_m and ω_m. From Fig. 5.47a we can determine ζ and percent overshoot. From Fig. 5.47b we can determine ω_n. In addition, the knowledge of ω_n allows us to determine the rise time† t_r from Fig. 5.10. Also,

$$t_s = \frac{4}{\zeta \omega_n}$$

Thus the transient parameters (PO, t_r, t_s) are determined from the knowledge of M_m and ω_m.

The relationship between the frequency response and the transient response given here is true only for a second-order system with complex poles or a higher-order system with two dominant complex poles. In general if the frequency response of a plant is determined experimentally, one cannot be certain about dominant poles, and the above relations between frequency response and transient response should be used only as a guide.

Since we are designing in terms of the open-loop frequency response, our next problem is to relate the open-loop frequency response to the closed-loop system transfer function. This can be done as follows. We first relate the open-loop frequency response to the closed-loop system frequency response. This immediately opens the way to relate the open-loop frequency response to the closed-loop system transient response.

Let us consider a unity feedback system [$H(s) = 1$]. The closed-loop transfer function $T(s)$ is given by

$$T(s) = \frac{G(s)}{1 + G(s)} \tag{5.78a}$$

---

†Rise time can be more accurately estimated from the bandwidth of the closed-loop transfer function $T(j\omega)$. See Sec. 3.16.

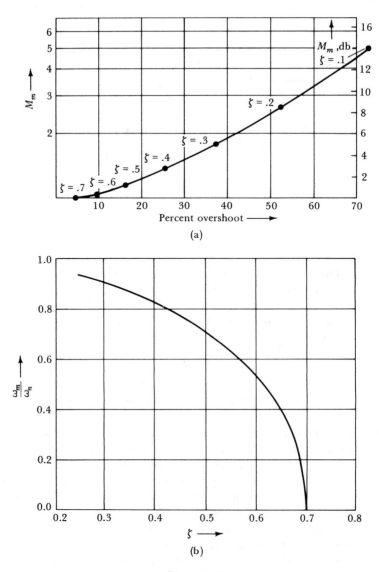

Figure 5.47

and

$$T(j\omega) = \frac{G(j\omega)}{1 + G(j\omega)} \tag{5.78b}$$

where $T(j\omega)$ is the frequency response of the closed-loop system and $G(j\omega)$ is the frequency response of the open-loop system. Let

$$G(j\omega) = x(\omega) + jy(\omega)$$
$$T(j\omega) = M(\omega)e^{j\alpha(\omega)}$$

Then

$$Me^{j\alpha} = \frac{x + jy}{1 + x + jy}$$

and

$$M = \frac{|x + jy|}{|1 + x + jy|} = \left[\frac{x^2 + y^2}{(1 + x)^2 + y^2}\right]^{1/2}$$

Straightforward manipulation of this equation yields

$$\left(x + \frac{M^2}{M^2 - 1}\right)^2 + y^2 = \frac{M^2}{(M^2 - 1)^2} \tag{5.79a}$$

Similarly, it can be shown that

$$\left(x + \frac{1}{2}\right)^2 + \left(y - \frac{1}{2\tan\alpha}\right)^2 = \frac{1}{4}\frac{\tan^2\alpha + 1}{\tan^2\alpha} \tag{5.79b}$$

Let us consider Eq. 5.79a. This is an equation of a circle with its center at $[-M^2/(M^2 - 1), 0]$ in the $G(j\omega)$-plane. Figure 5.48 shows these circles for various values of M. This figure relates the magnitude M of the closed-loop transfer function to the open-loop transfer function $G(j\omega)$. Thus if $G(j\omega) = -2 - j1.85$ (point A) at some frequency $\omega = \omega_1$, $|T(j\omega_1)| = 1.3$, since the point A lies along the circle $M = 1.3$. From Eq. 5.79b we can also draw circles to represent constant values of $\alpha[\angle T(j\omega)]$.

We can superimpose the Nyquist plot on M circles (and α circles) to obtain the closed-loop frequency response $T(j\omega)$. For each point of $G(j\omega)$ (Nyquist plot) we can determine $M(|T(j\omega)|)$, using M circles. We are primarily interested in finding M_m (the maximum value of M) and ω_m, the frequency, where M is maximum. This can be easily done as shown in Fig. 5.49. The circle to which the Nyquist plot is tangent corresponds to the value M_m. The frequency at which the Nyquist plot is tangent to this M circle is ω_m. For the system whose Nyquist plot appears in Fig. 5.49, it can be seen that $M_m = 1.6$ and $\omega_m = 2$. The transient performance for this system can now be estimated from Figs. 5.47a and 5.47b.

From the given transient specifications we determine the required value of M_m and ω_m (see Fig. 5.47). The Nyquist plot in conjunction with M circles suggests how these values of M_m and ω_m may be obtained. In many cases mere change in gain K in the open-loop transfer function may suffice. This expands (or contracts) the Nyquist plot, thus altering the value of M_m and ω_m. If this is not enough we should consider some form of compensation such as lag and/or lead networks.

If the steady-state error is larger than specified, we need to increase the gain K in the open-loop transfer function. We can show that for unity feedback system, steady-state errors are inversely proportional to K. This follows from the fact that (see Eqs. 5.45, 5.46, and 5.47)

$$e_s = \frac{1}{1 + \lim_{s \to 0} G(s)} \cong \frac{1}{\lim_{s \to 0} G(s)} \tag{5.80a}$$

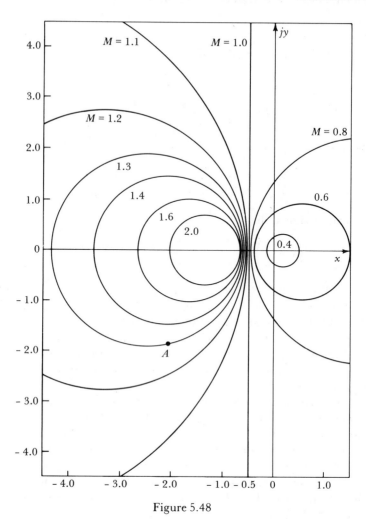

Figure 5.48

and

$$e_r = \lim_{s \to 0} \frac{1}{s[1 + G(s)]} = \frac{1}{\lim_{s \to 0} sG(s)} \qquad (5.80b)$$

Similarly,

$$e_p = \frac{1}{\lim_{s \to 0} s^2 G(s)} \qquad (5.80c)$$

Since the gain K appears in the numerator of $G(s)$, it is obvious from the above equations that steady-state errors are inversely proportional to the gain K. Therefore if a system meets transient specifications but not the steady-state specifications, we must reshape the Nyquist plot such that K is increased yet M_m and ω_m are unchanged.

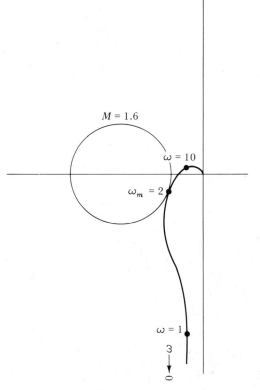

Figure 5.49

COMPENSATION

The compensation game can be played by using Nyquist plots, Bode plots, or log-magnitude-angle plots. Because it proves more convenient than the rest, we shall use the log-magnitude-angle plane. The M circles and α circles on Nyquist plane can be translated on the log-magnitude-angle plane. Figure 5.50 shows constant M and constant-α contours in log magnitude-angle plane. This chart is known as the *Nichols chart* and is available commercially. The log-magnitude-angle plot of $G(j\omega)$ is plotted. This plot is tangential to the constant-M contour corresponding to M_m (Fig. 5.50). The frequency of the point of tangency is ω_m. Increasing the gain K of $G(j\omega)$ amounts to shifting the entire plot vertically up by an amount equal to increase in gain (in decibels). Note that increasing K (shifting the plot up) generally increases both M_m and ω_m. This will cause increase in overshoot and a reduction in t_r. Reducing K has the opposite effect.

Let us consider a unity feedback system with open-loop transfer function

$$G(s) = \frac{128}{s(s + 8)}$$

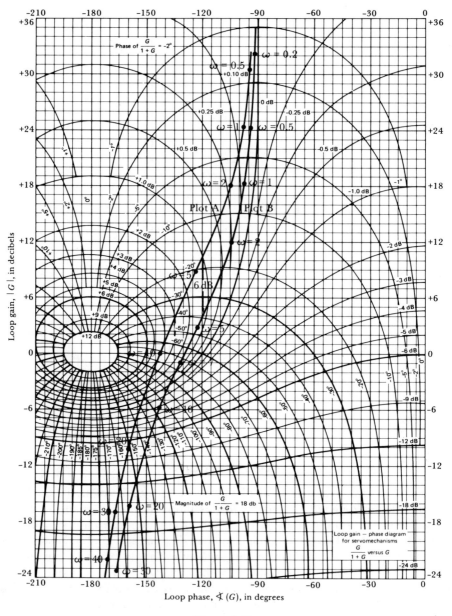

Figure 5.50

To obtain the frequency response $G(j\omega)$, we may sketch Bode plots of this function† and then draw log-magnitude-versus-angle plot as shown in Fig. 5.50 (plot A). We notice that this plot is tangential to $M \cong 4$ db contour at $\omega \cong 10$—that

---

†Digital computers can be conveniently used for this purpose.

is, $\omega_m = 10$. This information can be used to compute the transient response. From Fig. 5.47a we find PO = 30% and $\zeta = 0.35$ for $M_m = 4$ db. Also from Fig. 5.47b, for $\zeta = 0.35$

$$\frac{\omega_m}{\omega_n} \simeq 0.88$$

and

$$\omega_n = \frac{\omega_m}{0.88} = \frac{10}{.88} = 11.3$$

Hence

$$t_s = \frac{4}{\zeta \omega_n} = \frac{4}{(0.35)(11.3)} \cong 1$$

From Fig. 5.10, for $\zeta = 0.35$, $\omega_n t_r = 1.4$. Hence

$$t_r = \frac{1.4}{11.3} = 0.124$$

Thus

$$\text{PO} = 30\% \qquad t_r = 0.124 \qquad \text{and} \qquad t_s = 1$$

Suppose our specifications call for

$$\text{PO} \le 16\% \qquad t_r \le 0.25 \qquad \text{and} \qquad t_s \le 1$$

The above system meets all but the PO specification. Let us see if the situation can be corrected by simple manipulation of gain K. Increase in gain K by $+\alpha$ db shifts the log-magnitude-angle plot up by $+\alpha$ db. For PO $\le 16\%$, we note that $M_m \le 1.3$ db. Examination of Fig. 5.50 indicates that if we shift the log-magnitude-angle plot about 6 db below (plot B), then the plot is tangential to the $M \simeq 1.3$ contour at $\omega_m = 6$. Let us compute t_r and t_s for this situation. From Fig. 5.47a, for $M_m = 1.3$ db, $\zeta = 0.5$ and

$$\frac{\omega_m}{\omega_n} = 0.73$$

Therefore

$$\omega_n = \frac{\omega_m}{0.73} = \frac{6}{.73} = 8.25$$

Hence

$$t_s = \frac{4}{\zeta \omega_n} = \frac{4}{(0.5)(8.25)} = 0.975$$

From Fig. 5.10, $\omega_n t_r = 1.64$ for $\zeta = 0.5$. Therefore

$$t_r = \frac{1.64}{8.25} = 0.199$$

Hence the system meets all the specifications.

Shifting the plot 6 db below implies a reduction in the gain K by 6 db (by a factor of 2). Hence the desired transient performance is achieved by using gain $K = 64$. Therefore

$$G(s) = \frac{64}{s(s + 8)}$$

The advantage of the Nichols chart over the polar plot (Nyquist plot) is obvious from this example. Reducing the gain by 6 db merely shifts the log-magnitude-angle plot vertically by 6 db. But in polar plot we must compress the Nyquist plot by the factor of 2. The Nichols chart allows one to perceive the effect of variation in gain K much more readily than does the polar plot.

In practice, rather than shift the log-magnitude-angle plot up or down, one can use a Nichols chart drawn on a transparent sheet (available commercially). We can place the Nichols chart on top of the log-magnitude-angle plot and shift the transparency up or down until the log-magnitude plot is tangential to the desired M contour.

Transient performance of a system is determined from the values of M_m and ω_m. The steady-state errors are determined by the value of the open-loop transfer function at $\omega = 0$ (see Eqs. 5.80a, 5.80b, and 5.80c). Thus if the gain at dc ($\omega = 0$) is increased by a factor α, all the steady-state error are reduced by the factor α. It is obvious that increasing the value of gain K by a factor α will reduce all steady-state errors by a factor α. We shall now discuss methods of compensation to improve transient and steady-state performance.

LEAD COMPENSATION: IMPROVING TRANSIENT RESPONSE

Let us consider the system whose log-magnitude plot is shown in Fig. 5.51, where $M_m = M_1$ and $\omega_m = \omega_1$. In order to reduce PO, t_r, and t_s we must re-

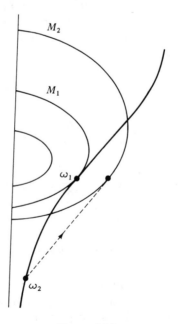

Figure 5.51

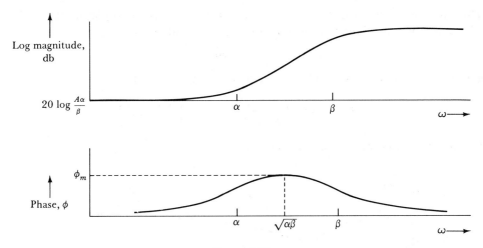

Figure 5.52

duce M_m and increase ω_m. Suppose the desired $M_m = M_2$ and the desired $\omega_m = \omega_2$ (Fig. 5.51). It is obvious that we must reshape the angle magnitude plot so that the frequency ω_2 is tangential to the M_2 contour (Fig. 5.51).

This calls for adding some magnitude and phase (phase lead) at the frequency ω_2 and its vicinity. The point ω_2 and its vicinity must be shifted in northeast direction implying an increase in both phase and magnitude. This can be accomplished by using lead compensation which causes increase in phase angle.

The lead network (Fig. 5.27) has a transfer function $G_c(s)$ of the form

$$G_c(s) = A \frac{s + \alpha}{s + \beta}$$

The constant A appearing in the transfer function is not important at the moment. Since these compensators appear in series with $G(s)$ which has an amplifier with variable gain K, the gain constant A appearing in $G_c(s)$ can always be lumped with K. Hence we are free to choose any value of A we wish and later adjust K by an appropriate factor.

The Bode plots for $G_c(s)$ are shown in Fig. 5.52. We observe that the gain of the compensator increases with frequency and levels out at higher frequencies while the phase increases from zero to some maximum value (less than 90°) at $\omega = \sqrt{\alpha\beta}$ and then goes to zero again. The compensator therefore yields gain as well as phase lead. When a compensator is placed in series with $G(s)$, the open-loop transfer function is $G_c(s)G(s)$ and the log magnitude (or phase) of $G_c(s)G(s)$ is the sum of log magnitudes (and phases) of $G_c(s)$ and $G(s)$. Hence the compensator will increase the gain and cause an increase in phase angle, and will shift the log-magnitude-phase plot in northeasterly direction for frequencies in the range (α, β) and vicinity. This is exactly what is desired. The phase shift of the lead network is given by

$$\phi = \tan^{-1}\left(\frac{\omega}{\alpha}\right) - \tan^{-1}\left(\frac{\omega}{\beta}\right)$$

The maximum value of ϕ that can be obtained is a function of ratio β/α. By setting $d\phi/d\omega = 0$, we can show that the maximum phase shift occurs at

$$\omega = \sqrt{\alpha\beta}$$

and the maximum phase shift at this frequency is ϕ_m is given by

$$\phi_m = \tan^{-1}\left(\frac{\beta - \alpha}{2\sqrt{\alpha\beta}}\right) = \tan^{-1}\left(\frac{\beta/\alpha - 1}{2\sqrt{\beta/\alpha}}\right)$$

Figure 5.53 shows the plot of ϕ_m as a function of β/α.

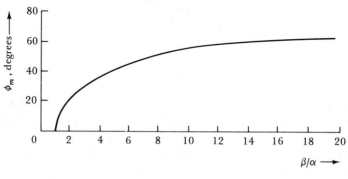

Figure 5.53

Knowing the desired amount of shift of gain and phase (Fig. 5.51), one can choose appropriate values of A, α, and β for the lead compensator (Fig. 5.27).

Earlier we had analyzed the effect of lead compensator using the root-locus (Fig. 5.28). Let us analyze this situation on the Nichols chart. The uncompensated transfer function is

$$G(s) = \frac{K}{s(s + 8)}$$

The compensator transfer function is

$$G_c(s) = \frac{s + 15}{s + 72}$$

Hence for $K = 840$,

$$G_c(s)G(s) = \frac{840(s + 15)}{s(s + 8)(s + 72)}$$

The log-magnitude-angle plots for $G(s)$ and $G_c(s)G(s)$ are shown in Fig. 5.54. Note that the points on uncompensated plot appear to have moved in a northeasterly direction in compensated plot. For the compensated plot $M_m \cong 0.74$ db and $\omega_m \cong 9.5$. From Fig. 5.47a, we read $\zeta = 0.55$, and

$$\omega_m/\omega_n \cong 0.634$$

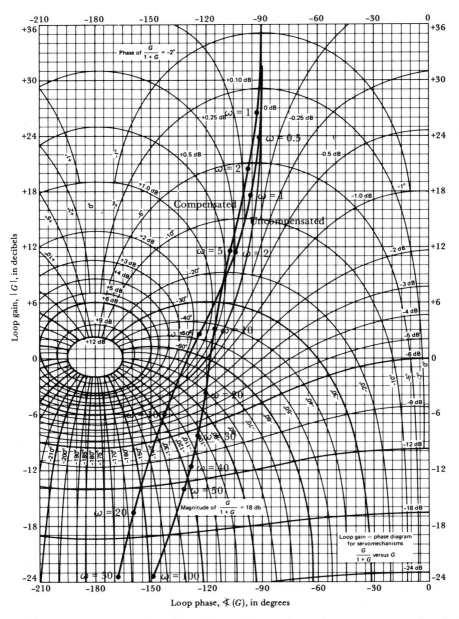

Figure 5.54

Hence

$$\omega_n = \frac{\omega_m}{0.634} = \frac{9.5}{0.634} = 15$$

and from Fig. 5.47a,

$$PO \cong 12\%$$

$$t_r = \frac{1.74}{15} = 0.116 \text{ sec}$$

$$t_s = \frac{4}{\zeta \omega_n} = 0.483 \text{ sec}$$

The compensator has reduced the values of t_r and t_s significantly. The values of t_r and t_s obtained here are somewhat different from those obtained earlier by the root-locus method. There are two reasons for this discrepancy: (1) plotting and reading errors in the Nichols chart and, (2) the compensated system is a third-order system with two dominant poles, and one pole far in the LHP and one zero at -15.

In the frequency-response method the relationship between M_m, ω_m, and ω_n in Fig. 5.47 holds strictly only for a second-order system without a zero. For other systems these relationships should be used only as guides. This example clearly shows that the relationship between frequency response and transient response is more or less qualitative rather than quantitative.

LAG COMPENSATION: IMPROVING STEADY-STATE PERFORMANCE

We have shown earlier that steady-state error depends directly upon the behavior of the open-loop transfer function at very low frequencies (see Eqs. 5.80). If the open-loop transfer function gain is increased by a factor α at very low frequencies ($s \to 0$ or $\omega \to 0$), all the steady-state errors are reduced by a factor of α. This can be conveniently done by using a lag compensator (Fig. 5.30) whose transfer function is

$$G_c(s) = A \frac{s + \alpha}{s + \beta} \qquad \beta < \alpha$$

As argued previously, the value of A may be chosen arbitrarily. Let $A = 1$. This gives

$$G_c(s) = \frac{s + \alpha}{s + \beta}$$

The Bode plots of $G_c(s)$ are shown in Fig. 5.55. As seen from this figure, the gain at high frequencies is unity but increases at lower frequencies and levels off at $20 \log (\alpha/\beta)$ db (gain of α/β) at very low frequencies. This fact is of interest to us. If we want to improve the steady-state performance of the system without affecting its transient performance, we should choose the lag compensator such that α and $\beta \ll \omega_m$. Hence at ω_m the compensator gain is unity, and it will have no effect on the frequency response near about ω_m. This will keep the transient performance unchanged. But the gain at very low frequencies has been increased by a factor α/β. This will reduce all the steady-state errors by a factor of α/β. Note, however, we must choose α and $\beta \ll \omega_m$. As an example consider again the uncompensated system

$$G(s) = \frac{64}{s(s + 8)}$$

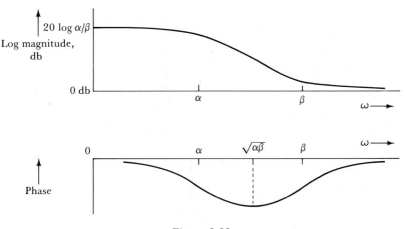

Figure 5.55

From Eq. 5.80 it can be seen that

$$e_s = 0 \qquad \text{and} \qquad e_r = 0.125$$

Suppose we want to reduce e_r by a factor of 10 without altering the transient response. We must choose $\alpha/\beta = 10$ and both α, $\beta \ll \omega_m$. From Fig. 5.54 we found $\omega_m \cong 6$. We shall arbitrarily choose $\beta = 0.01$ and $\alpha = 0.1$. This gives $\alpha/\beta = 10$ and both α, $\beta \ll \omega_m$. Thus

$$G_c(s) = \frac{s + 0.1}{s + 0.01}$$

and

$$G_c(s)\,G(s) = \frac{64\,(s + 0.1)}{s\,(s + 0.01)\,(s + 8)}$$

The log-magnitude-angle plots for $G(s)$ and $G_c(s)\,G(s)$ are shown in Fig. 5.56. Note that the two plots are almost identical for $\omega > 5$. It is also clear from this figure that ω_m, and M_m are unchanged, and hence the transient response is practically unaffected. The gain at very low frequencies ($\omega \to 0$) however, has increased by a factor of 10. Hence the steady-state error is reduced by a factor of 10. We can easily verify this fact analytically from Eq. 5.80; we have

$$e_s = \lim_{s \to 0} \left[\frac{1}{1 + G_c(s)\,G(s)} \right] = 0$$

$$e_r = \frac{1}{\lim_{s \to 0} s G_c(s)\,G(s)} = \frac{1}{80} = 0.0125$$

This is an improvement by a factor of 10 over the uncompensated system.

One can improve the transient and the steady-state performance by using lead and lag compensations simultaneously.

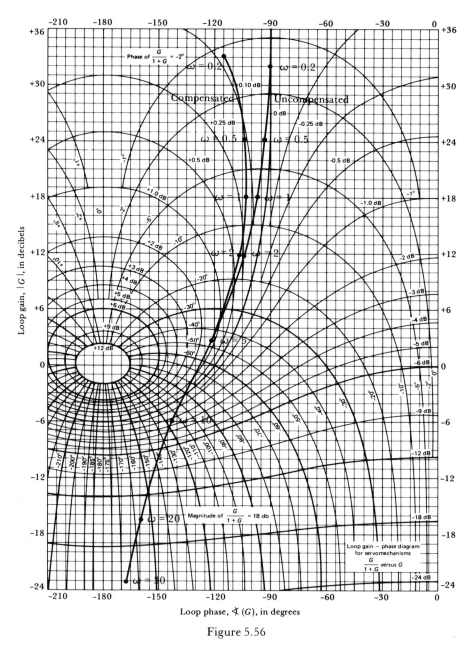

Figure 5.56

SYSTEMS WITH NONUNITY FEEDBACK

The frequency response method discussed here applies to unity feedback systems only. For a system with nonunity feedback,

$$T(s) = \frac{G(s)}{1 + G(s)H(s)}$$

$$= \frac{1}{H(s)} \frac{G(s)H(s)}{1 + G(s)H(s)}$$

$$= \frac{1}{H(s)} \frac{\hat{G}(s)}{1 + \hat{G}(s)}$$

where

$$\hat{G}(s) = G(s)H(s)$$

Thus

$$T(s) = \frac{1}{H(s)} \hat{T}(s)$$

where

$$\hat{T}(s) = \frac{\hat{G}(s)}{1 + \hat{G}(s)}$$

Thus a nonunity feedback system is equivalent to a unity feedback system $[\hat{T}(s)]$ in series with another system with transfer function $1/H(s)$. This fact is used in designing systems with nonunity feedback. The synthesis or the design problem finally boils down to reshaping of the log magnitude-phase plot of $T(s)$ to meet certain design criteria. But the log magnitude (or phase) of $T(s)$ equal to the log magnitude (or phase) of $\hat{T}(s)$ minus the log magnitude (or phase) of $H(s)$. Hence we should adjust or manipulate $\hat{T}(s)$, so that the resultant plot due to $\hat{T}(s)$ and $H(s)$ yields the desired characteristics.

PROBLEMS

5.1. Determine the peak time t_p, the rise time t_r, the settling time t_s, and PO for the unit step input for each of the second-order systems whose transfer functions are given below. In each case determine the steady-state value of the output and sketch the unit-step response.

(a) $H(s) = \dfrac{9}{s^2 + 3s + 9}$

(b) $H(s) = \dfrac{4}{s^2 + 3s + 4}$

(c) $H(s) = \dfrac{10}{s^2 + 10s + 100}$

5.2. A unit-step response of the system shown in Fig. P-5.2a is shown in Fig. P-5.2b. Determine the values of K_1, K_2, and a.

5.3. Figure P-5.3 shows a simple position-control system. The following specifications are imposed on this system: rise time $t_r \le 0.3$, settling time $t_s \le 1$, and PO $\le$ 30%. Which of these specifications can be met by the system?

5.4. Determine the rise time t_r, the settling time t_s, and PO for the unit-step input for each of the second-order systems whose transfer functions are given below. In each

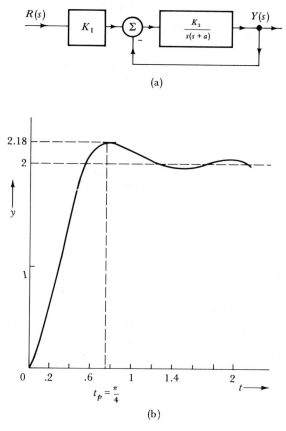

(a)

(b)

Figure P-5.2

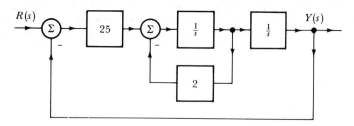

Figure P-5.3

case determine the steady-state value of the output.

(a) $H(s) = \dfrac{10(s + 10)}{s^2 + 10s + 100}$

(d) $H(s) = \dfrac{5}{(s + 20)(s^2 + 6s + 13)}$

(b) $H(s) = \dfrac{20(s + 5)}{s^2 + 18s + 100}$

(e) $H(s) = \dfrac{20(s + 1)}{(s + 1.01)(s + 10)(s^2 + 2s + 2)}$

(c) $H(s) = \dfrac{20}{(s + 10)(s^2 + 2s + 2)}$

5.5.   Sketch the root locus of a closed-loop system for the following open-loop transfer functions.

(a) $\dfrac{K(s+5)}{s(s+3)}$

(g) $\dfrac{K(s+1)}{s(s+4)(s^2+2s+2)}$

(b) $\dfrac{K(s+1)}{s(s+3)(s+5)}$

(h) $\dfrac{K}{s(s+3)(s^2+4s+13)}$

(c) $\dfrac{K(s+1)}{s(s+3)(s+5)(s+7)}$

(i) $\dfrac{K(s+2)}{s(s+3)(s+5)}$

(d) $\dfrac{K}{(s+1)(s+2)(s+6)}$

(j) $\dfrac{K(s+2)(s+4)}{s(s+1)(s+3)}$

(e) $\dfrac{K(s+4)}{(s+1)(s^2+6s+13)}$

(k) $\dfrac{K}{(s+3)(s^2+4s+5)}$

(f) $\dfrac{K}{s(s+3)(s^2+4s+16)}$

5.6.   Given the root locus of a unity feedback system (Fig. P-5.6),

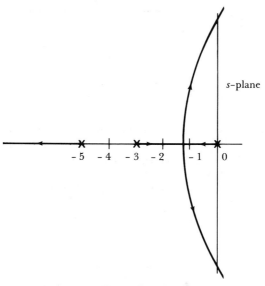

Figure P-5.6

(a) What is the system's open-loop transfer function?
(b) At what value of K does the system have pure sinusoidal oscillations? What is this frequency of oscillation?
(c) What value of K will result in critical damping?
(d) State the limits of K so that PO $\leq$ 10%, $t_r \leq$ 1 sec, and $t_s \leq$ 4 sec.

5.7.   Determine the steady-state errors e_r, e_s, and e_p for unit-step input to each of the systems whose close loop transfer functions are given below.

(a) $\dfrac{10}{s^2+2s+15}$

(b) $\dfrac{100(s+1)}{s^2+2s+100}$

(c) $\dfrac{10(s+2)}{s^3 + s^2 + 10s + 20}$ (d) $\dfrac{100(s+1)}{s^3 + 3s^2 + 100s + 300}$

5.8.   Find the steady-state errors for the unit-step input to each of the systems shown in Fig. P-5.8.

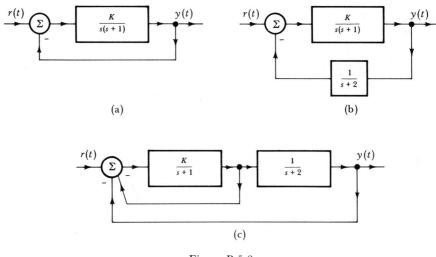

(a)   (b)

(c)

Figure P-5.8

5.9.   A certain system must meet the following specifications: PO $\leq$ 26%, $t_r \leq$ 0.14 sec, and $t_s \leq$ 0.8 sec. Determine the region in the s-plane where the dominant poles of the system may lie in order to satisfy these specifications.

5.10.   (a) For a unity feedback system shown in Fig. P-5.10a, is it possible to meet the following specifications? (1) PO $\leq$ 16%, (2) $t_r \leq$ 0.2 sec, and (3) $t_s \leq$ 0.5 sec. Give reasons for your answer.

(b) If the specifications in part (a) cannot be met, try compensation as shown in

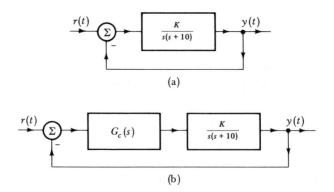

(a)

(b)

Figure P-5.10

Fig. P-5.10b, where

$$G_c(s) = \frac{s + 10}{s + 20}$$

(1) What is the range of K for which the specifications in (a) are met? (2) What is the value of K which will yield the smallest steady-state error e_r while still satisfying all the requirements in part (a)? What is the smallest value of e_r?

5.11. Repeat Prob. 10 for

$$G_c(s) = \frac{s + 15}{s + 20}$$

5.12. (a) For the system in Fig. P-5.10a, is it possible to meet the following specifications? (1) PO $\le 16\%$, (2) $t_r \le 0.2$ sec, (3) $t_s \le 1$ sec, and (4) $e_r \le 0.02$. Give reasons for your answer.

(b) If these specifications cannot be met, what type of compensation would you use? Design the appropriate compensator to meet all the specifications in part (a).

5.13. The roll-angle control autopilot (Fig. 5.35a) was found to be unstable. Devise a suitable series compensator $G_c(s)$ to stabilize this system and to meet the following specifications: PO $\le 26\%$, $t_r \le 2$ sec, $t_s \le 8$ sec, $e_s \le 0$, and $e_r \le 0.2$ sec. The transfer function relating the roll angle ϕ to the aileron angle θ is given as

$$\frac{\Phi(s)}{\theta(s)} = \frac{20}{s(s + 2)}$$

and the transfer function of the hydraulic servo is $0.1/s$.

5.14. A block diagram of a servomotor drive system of an antiaircraft gun is shown in Fig. P-5.14. The input θ_i represents the target angle supplied by a tracking antenna. If the target is moving with a constant velocity, the target angle will vary linearly with t (ramp). Let $\theta_i = 4t$. If the gun is to track within 0.2 degree of the described angle θ_i, a hit is assured. What must be the minimum value of K to accomplish this?

5.15. (a) For a unity feedback system shown in Fig. 5.23, determine (1) the minimum possible rise time, (2) the minimum possible settling time, (3) minimum possible steady-state error e_r.

(b) Can all these minimum values be achieved simultaneously? Explain.

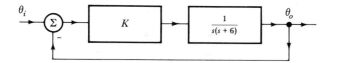

Figure P-5.14

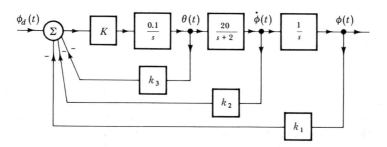

Figure P-5.16

5.16.   Consider the autopilot system in Fig. 5.35a. This system with multiple-variable feedback is shown in Fig. P-5.16. Determine a set of feedback coefficients k_1, k_2, k_3 and the gain K in order to meet the following specifications: PO $\le$ 16%, $t_r \le$ 2 sec, $t_s \le$ 8 sec, and $e_r \le$ 0.2.

5.17.   The lunar excursion module (LEM) shown in Fig. P-5.17a presents the control engineer with a number of interesting problems. These problems arise from the dynamic equations associated with the LEM when operating in near the lunar surface. The absence of atmosphere on the moon eliminates drag forces which are associated with such

(a) "LEM" controlling attitude using jet thrusters

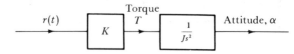

(b) Simulation of lunar excursion module's attitude response.

Figure P-5.17

mediums, and therefore the LEM dynamic equation is given by

$$T = J\ddot{\alpha}$$

where T is jet thruster's torque (see figure), J is the polar moment of inertia about axis perpendicular to plane of motion, and α is the attitude. The jets can be fired in either direction. Assume that the generated thrust (and the torque) are controlled by an electrical signal $r(t)$. The block diagram for the attitude control of LEM therefore appears as shown in Fig. P-5.17b. The constant K can be varied from 0 to ∞. This system is unsatisfactory because it has no damping. Devise a feedback scheme for this system in order to meet the following specifications: PO $\le$ 16%, $t_r \le$ 0.3, $t_s \le$ 1 and $e_s = 0$. Assume $J = 1$.

5.18. An automatic tracking antenna (Fig. P-5.18a) tracks a moving target. Conical scan of the antenna beam is used to detect the presence of a target. The target angle θ_T is detected by the beam reflected off the target. The error between the target angle θ_T and the angle of the antenna axis θ_A is amplified and is fed to a motor-generator set which rotates the antenna until the error is zero ($\theta_A = \theta_T$).

The motor-generator set is shown in Fig. P-5.18b. The block diagram for the azimuth angle servo is shown in Fig. P-5.18c. There will be similar servo loop for the elevation angle. The motor-generator equations are

$$e_g = 100i_f, \qquad e_b = K_b\omega_m, \qquad \text{motor torque } T = K_T i_a$$

where K_b = 3 volts/rad/sec and K_T = 2 newton-meter/amp. Also R_f = 5 ohms, $R_a = R_g =$ 0.5 ohm, and L_f = 0.5 henry. Equivalent moment of inertial of the antenna load at the motor = 100 Newton-meter-sec^2. The gear ratio 400:1. Draw the block diagram of this

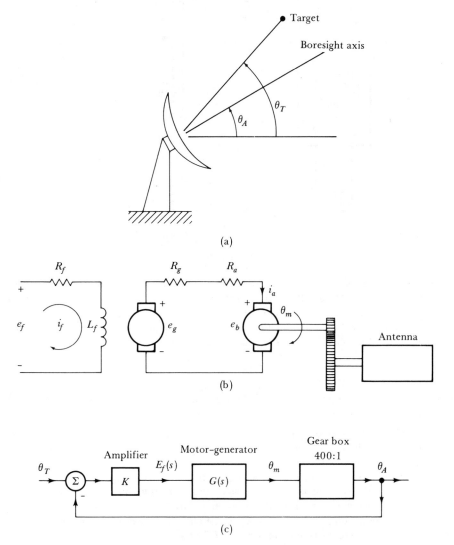

(a)

(b)

(c)

Figure P-5.18

system. Determine PO, t_r, and t_s. Does this system meet the following specifications? $PO \leq 5\%$, $t_r \leq 2$ sec, $t_s \leq 5$ sec. If not, design a compensator to accomplish this. For the compensated system, what is the value of e_r, the steady-state error to unit ramp input? Does this design meet the specification $e_r \leq 0.2$. If not, design an additional compensator.

5.19.   An automatic temperature control system, shown in Fig. P-5.19, uses a thermocouple to detect the actual room temperature T. The desired room temperature T_r is set by adjusting the wiper of the input potentiometer so that the voltage $v_r = K_r T_r$. The thermocouple voltage $v_c = K_r T$. The difference between these two voltages is $K_r(T_r - T)$.

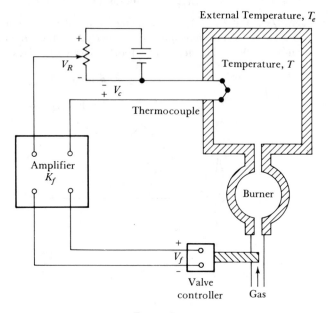

Figure P-5.19

This voltage, which is proportional to the error, is amplified and applied to a valve controller which regulates the gas flow. The heat produced by the gas burner is given by $K_f v_f$, where v_f is the valve-controller voltage. The heat loss due to conduction through the walls is given by $K_c(T - T_e)$, where T_e is the external temperature (outside the walls). The rate of change of the room temperature T is proportional to the net heat input. Hence

$$\frac{dT}{dt} = K_h[K_f v_f - K_c(T - T_e)]$$

Draw the block diagram of this system. The external temperature T_e should be considered as external disturbance input.

    (a) What is the system type?
    (b) What is the steady-state error to a unit step input?
    (c) How can you reduce this error?
    (d) What compensation would you suggest to achieve zero steady-state error?
    (e) Sketch the root locus for this system (without any compensation). Discuss the effect of compensation in part (d) on the transient behavior of the system.
    (f) For the uncompensated system, do the changes in external temperature T_e affect the steady-state value of the room temperature T?

(g) How can you eliminate the influence of variation of T_e on the steady-state value of T?

(h) For the uncompensated system, determine t_r and t_s (*Note:* This is a first-order system. Hence PO = 0. If τ is the time constant then it can be shown that $t_r = 2.2\tau$. Also $t_s = 4\tau$.)

5.20.   A system in Fig. P-5.20 has an input $R(s)$ and a disturbance $D(s)$. Determine the steady-state value of the desired output and disturbance output when both inputs

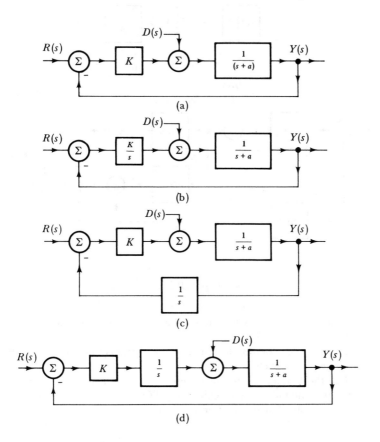

Figure P-5.20

$R(s)$ and $D(s)$ are unit steps $(1/s)$. In order to reduce the effect of the disturbance, an integral compensator $(1/s)$ is placed in three different positions shown in Figs. P-5.20b, c, and d. Analyze and compare these three cases with respect to their transient behavior and ability to suppress disturbance.

5.21.   The system in Fig. P-5.21a is known as a conditional feedback system, and the one in Fig. P-5.21b is known as a model feedback system. The input $D(s)$ is a disturbance. In both these systems show that the effective feedback is zero as long as the plant parameters are unchanged and/or the disturbance is zero. The feedback is effective only when the plant parameters change, and/or the disturbance is nonzero.

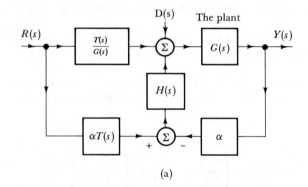

(a)

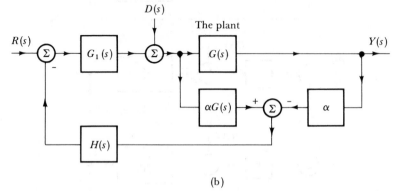

(b)

Figure P-5.21

5.22.   Draw the root locus for the feedback systems shown in Fig. P-5.22 as the gain K is varied from 0 to ∞.

5.23.   Draw a root contour for each of the systems shown in Fig. P-5.23, (see p. 338) as the parameter b varies from 0 to ∞.

5.24.   Use the frequency-response method to design compensator $G_c(s)$ in Fig. P-5.10 in order to meet the following specifications: PO $\leq$ 16%, $t_r \leq 0.2$, and $t_s \leq 0.5$.

5.25.   Solve Prob. 12 using the frequency-response method.

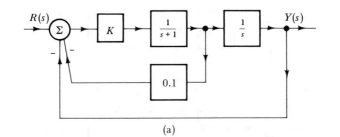

(a)

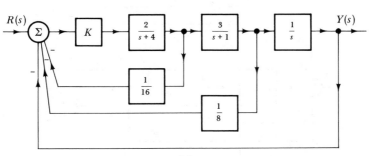

(b)

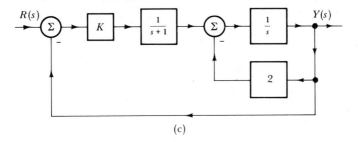

(c)

Figure P-5.22

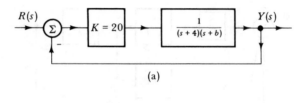

(a)

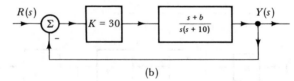

(b)

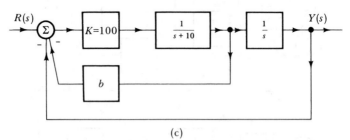

(c)

Figure P-5.23

6

State-Space Analysis

6.1 DESCRIPTION OF A SYSTEM

A system is described by a mathematical model. Broadly speaking, there are two approaches in describing a system: (i) the terminal approach (input-output description), and (ii) internal approach (the state variable description).

THE INPUT-OUTPUT DESCRIPTION

We are given a linear time-invariant system with m inputs f_1, f_2, ..., f_m and k outputs $y_1, y_2, \ldots, y_k$. In the terminal approach, we determine the relationships between the outputs and the inputs. These are in the form of nth-order differential equations or various transfer functions. This method of system description is well suited when we are interested in a few output variables. It proves quite adequate in handling simpler problems such as designing position-control servos, and the like. It is, however, inconvenient in handling more ambitious problems arising in the modern control systems where one is interested in several system variables simultaneously. In these problems we want to look at several variables simultaneously and manipulate them to meet certain design criteria. For these cases we need to describe a system which bares its internal mathematical structure. The terminal approach relates merely the variables at the output terminals to the variables at the input terminals. For more complex problems described above, we need to view the system from inside. The state-variable method is admirably suited for this.

STATE VARIABLE DESCRIPTION

We have observed in Chapter 1 that certain variables, called the *state variables* in a dynamic system, represent a vital key in the analysis problem. These variables have one very important property. If we know all the state variables at some instant t_0 (along with the input(s) at t_0), then every possible output of the system at $t = t_0$ is known. In other words, every possible system output at $t = t_0$ can be expressed in terms of the state variables and the input(s) at $t = t_0$. If we can determine the time behavior of the state variables, then the time behavior of every possible output is known for given inputs. There are two steps in this method of analysis: (i) determine the time behavior of the state variables, and (ii) determine the desired output from the knowledge of the state variables. Accordingly, we have two sets of equations: (i) state equations which relate state variables and the inputs, and (ii) output equations which relate outputs to state variables and inputs. These two sets of equations constitute the internal description of the system. We shall presently see that state equations are n-simultaneous, first-order differential equations in n state variables $x_1, x_2, \ldots, x_n$.

$$\dot{x}_k = g_k(x_1, x_2, \ldots, x_n, f_1, f_2, \ldots, f_m) \qquad k = 1, 2, \ldots, n \qquad (6.1)$$

where $f_1(t), f_2(t), \ldots, f_m(t)$ are the inputs. These n equations represent the general form of state equations. For the case of linear systems, these equations reduce to a simpler linear form:

$$\dot{x}_k = a_{k1}x_1 + a_{k2}x_2 + \cdots + a_{kn}x_n + b_{k1}f_1 + b_{k2}f_2 + \cdots + b_{km}f_m$$
$$k = 1, 2, \ldots, n \quad (6.2)$$

The output equations are of the form

$$y_j = c_{j1}x_1 + c_{j2}x_2 + \cdots + c_{jn}x_n + d_{j1}f_1 + d_{j2}f_2 + \cdots + d_{jm}f_m \qquad j = 1, 2, \ldots, k$$

There are several major advantages in describing a system by first order differential equations:

1. Such equations have been extensively investigated and several analysis techniques are available for solving these equations.
2. This form of representation can be extended to time-varying and non-linear systems whereas such an extension is not easy in the case of loop and node equations. Indeed the first-order differential equations form may be the only effective method of analyzing most of the time-varying and nonlinear systems.
3. This form of representation is most suitable for analog and digital computers.

6.2   STATE EQUATIONS OF A SYSTEM

In the input-output method a system is described by a set of simultaneous differential equations. Generally speaking, these constitute second-order differential equations. The state equations, on the other hand, are all first-order differential equations. It would be desirable if we could break up second-order differential equations into first-order equations. This would allow us to obtain state equations from the input-output set of equations. We shall now show that this indeed can be done. A second-order differential equation can always be expressed as a set of two first-order differential equations in two variables. Consider the differential equation

$$\ddot{x} + a_1\dot{x} + a_0x = f(t) \qquad (6.3)$$

Let us define two variables x_1 and x_2 as

$$x = x_1$$
$$\dot{x} = x_2 \qquad (6.4)$$

Hence

$$\dot{x}_1 = x_2 \tag{6.5a}$$

Substitution of Eq. 6.4 in Eq. 6.3 yields

$$\dot{x}_2 + a_1 x_2 + a_0 x_1 = f(t) \tag{6.5b}$$

Thus the second-order differential equation 6.3 is transformed into two first-order differential equations 6.5a and 6.5b. These equations can be rearranged as

$$\begin{aligned} \dot{x}_1 &= x_2 \\ \dot{x}_2 &= -a_0 x_1 - a_1 x_2 + f(t) \end{aligned} \tag{6.6}$$

or

$$\begin{bmatrix} \dot{x}_1 \\ \dot{x}_2 \end{bmatrix} = \begin{bmatrix} 0 & 1 \\ -a_0 & -a_1 \end{bmatrix} \begin{bmatrix} x_1 \\ x_2 \end{bmatrix} + \begin{bmatrix} 0 \\ 1 \end{bmatrix} f(t) \tag{6.7}$$

Note that this set of equations (Eqs. 6.6 or 6.7) is of the same form as Eq. 6.2 (state equations). Thus the second-order differential equation 6.3 has been' expressed in the form of state equations 6.6 (or Eq. 6.7)† which are two first-order differential equations. Remember that $x_1 = x$ and $x_2 = \dot{x}$.

We now have a systematic method of writing state equations. In the first step we write the input-output equations (such as loop or node equations) which are in general second-order differential equations. These equations can then be reduced to first-order differential equation by the procedure discussed above. It must be remembered that the number of state variables is equal to the number of initial conditions.

We shall illustrate the application of this procedure to mechanical and electrical systems by a few examples.

MECHANICAL SYSTEMS

Consider the mechanical system in Fig. 6.1a. The free body diagram of the system is shown in Fig. 6.1b. The node equation is

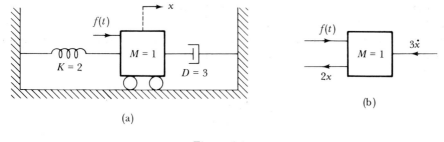

(a)

(b)

Figure 6.1

---

†It will be shown in Sec. 6.3 that extension of the above procedure allows us to express nth-order differential equation by n first-order differential equations (state equations).

$$Mp^2x + Dpx + Kx = f(t)$$

Since $M = 1$, $D = 3$, and $K = 2$, we have

$$\ddot{x} + 3\dot{x} + 2x = f(t) \tag{6.8}$$

Using the results in Eq. 6.6 (or Eq. 6.7) we can express this second-order differential equation by two first-order state equations as

$$\dot{x}_1 = x_2$$
$$\dot{x}_2 = -2x_1 - 3x_2 + f(t) \tag{6.9}$$

or

$$\begin{bmatrix} \dot{x}_1 \\ \dot{x}_2 \end{bmatrix} = \begin{bmatrix} 0 & 1 \\ -2 & -3 \end{bmatrix} \begin{bmatrix} x_1 \\ x_2 \end{bmatrix} + \begin{bmatrix} 0 \\ 1 \end{bmatrix} f(t) \tag{6.10}$$

Note that $x = x_1$ and $\dot{x} = x_2$. Hence the state variables x_1 and x_2 represent the position and the velocity of the mass in Fig. 6.1a. We can easily see that the variables x_1 and x_2 are state variables because the unique solution of Eq. 6.8 requires the knowledge of $x(0)$ and $\dot{x}(0)$ or $x_1(0)$ and $x_2(0)$. Hence x_1 and x_2 represent the state of the system.

Example 6.1. Write the state equations of the mechanical system shown in Fig. 1.9. The equations of motion for this system were found to be (see Eqs. 1.20 and 1.21)

$$(Mp^2 + D_1p + K_1 + K_2)x_1 - K_2x_2 = f(t)$$
$$-K_2x_1 + (D_2p + K_2)x_2 = 0 \tag{6.11}$$

For convenience, we shall assign numerical values to various parameters. Let $M = 1$, $D_1 = 2$, $D_2 = 1/2$, $K_1 = 3$, and $K_2 = 2$. Hence Eqs. 6.11 become

$$\ddot{x}_1 + 2\dot{x}_1 + 5x_1 - 2x_2 = f(t) \tag{6.12a}$$

$$-2x_1 + \tfrac{1}{2}\dot{x}_2 + 2x_2 = 0 \tag{6.12b}$$

Note that Eq. 6.12b is already in the form of a state equation (first-order differential equation) and there is no need to break it further. But Eq. 6.12a is a second-order differential equation and must be broken into two first-order differential equations. We define three variables z_1, z_2, and z_3 as follows:

$$x_1 = z_1$$
$$\dot{x}_1 = z_2$$
$$x_2 = z_3$$

We can immediately write Eq. 6.12a as (see Eq. 6.6)

$$\dot{z}_1 = z_2 \tag{6.13a}$$

$$\dot{z}_2 = -5z_1 - 2z_2 + 2z_3 + f(t) \tag{6.13b}$$

and Eq. 6.12b becomes

$$-2z_1 + \tfrac{1}{2}\dot{z}_3 + 2z_3 = 0$$

or

$$\dot{z}_3 = 4z_1 - 4z_3 \tag{6.13c}$$

We now have three first-order differential equations in three state variables z_1, z_2, and z_3 (Eq. 6.13a, 6.13b, and 6.13c). In matrix form, these equations can be written as

$$\begin{bmatrix} \dot{z}_1 \\ \dot{z}_2 \\ \dot{z}_3 \end{bmatrix} = \begin{bmatrix} 0 & 1 & 0 \\ -5 & -2 & 2 \\ 4 & 0 & -4 \end{bmatrix} \begin{bmatrix} z_1 \\ z_2 \\ z_3 \end{bmatrix} + \begin{bmatrix} 0 \\ 1 \\ 0 \end{bmatrix} f(t) \tag{6.14}$$

Here the state variables z_1, z_2, and z_3 are represented by x_1, $\dot{x}_1$, and x_2 (the position and the velocity of junction 1, and the position of junction 2 in Fig. 1.9.) It should be noted that in this particular system the variable $\dot{x}_2$ (the velocity of junction 2) does not appear as a state variable, because for a unique solution of the response we need to know only $x_1(0)$, $\dot{x}_1(0)$, and $x_2(0)$—that is, the state is completely specified by only three variables. In general the number of state variables in a mechanical system is $\leq 2N$, where N is the number of junctions.

ELECTRICAL SYSTEMS

The procedure for electrical systems is somewhat different than that used for mechanical systems. The position coordinates in electrical systems are charges q for mesh or loop equations (and flux linkages λ for node equations). In mechanical systems the position x is a variable of significance. Such is not the case for position coordinates (q or λ) in electrical systems. The position coordinates may or may not be state variables. Hence certain position coordinates may be unessential. This is because in electrical networks the variables of direct interest are currents and voltages rather than charges (or flux linkages). Of course the currents and voltages can be obtained from the knowledge of charge variables (or flux linkages). But obtaining v and i from q (or λ) may necessitate more information than need be. Consider for example, a series RL circuit shown in Fig. 6.2. The mesh equation

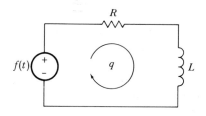

Figure 6.2

for this circuit using the charge variable q is

$$L\ddot{q} + R\dot{q} = f(t) \tag{6.15a}$$

This is a second-order differential equation, and for unique solution we need two initial conditions such as $q(0)$ and $\dot{q}(0)$. The solution will give us the time behavior of q. However, the charge variable q in this circuit is not important. We really need the current i, which is $\dot{q}$. In terms of current $i(i = \dot{q})$, Eq. 6.15a can be expressed as

$$L\frac{di}{dt} + Ri = f(t) \tag{6.15b}$$

This is a first-order differential equation, needing only one initial condition such as $i(0)$ for unique solution. Hence there is only one state variable i. There is nothing wrong with the formulation in Eq. 6.15a. But it has one additional state variable q of no interest, and is therefore unessential.

      However, it must not be concluded from this discussion that the charge variable is always unessential. Consider, for example, the RLC network in Fig. 6.3.

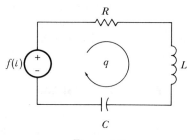

Figure 6.3

The mesh equation is given by

$$L\ddot{q} + R\dot{q} + \frac{1}{C}q = f(t) \tag{6.16}$$

This is a second-order differential equation and cannot be transformed into a first-order differential equation by using the current variable $(i = \dot{q})$. Using current merely changes it to integrodifferential equation (of second order). Hence in this case the two initial conditions are essential. Thus the charge variable here is an essential state variable.

      We therefore observe that the position coordinate q may or may not be essential. In a complex network, some mesh charges will be essential and the remaining mesh charges will be unessential. How do we determine which charge variables are essential and which are unessential? How do we choose state variables in general? There is a simple rule for this. In electrical networks the energy storages are determined by inductor currents and capacitor voltages. It can be shown that at any instant all voltages and currents in a network can be determined from the knowledge of all inductor currents and all capacitor voltages† (and the inputs). From the discussion in Sec. 1.2 it follows that inductor currents and capacitor voltages are indeed a possible set of state variables of a network. This is the clue to our problem. In the circuit in Fig. 6.2, there is only one inductor and no capacitor. Hence the inductor current (which is $\dot{q}$) is the only state variable. If we define $\dot{q} = x$, then Eq. 6.15a becomes

$$L\dot{x} + Rx = f(t)$$

or

$$\dot{x} = -\frac{R}{L}x + \frac{1}{L}f(t)$$

This is the desired state equation (Note that this equation is identical to Eq. 6.15b, since $x = \dot{q} = i$.)

We shall now give a systematic procedure for writing state equations of electrical systems. We can use either mesh (loop) or node (cutset) equations to obtain state equations. We shall give the procedure for mesh (or loop) equations.

Step 1.   Choose all inductor currents and capacitor voltages as state variables.

Step 2.   Write relationship between mesh currents and state variables (these equations actually represent inductor and capacitor currents in terms of state variables and their derivatives).

Step 3.   Write mesh equations.

Step 4.   Eliminate all variables other than state variables (and their derivatives) from the two sets of equations.

We shall illustrate this procedure by an example.

Consider the network in Fig. 6.4. Let us write state equations for this network using the above procedure.

Step 1.   There are two inductors and one capacitor in the network. We shall therefore choose inductor currents x_1, x_2 and the capacitor voltage x_3 as state variables.

Step 2.   Relationship between mesh currents i_1, i_2 and state variables can be written immediately by inspection:

$$x_1 = i_1 \tag{6.17}$$

$$x_2 = i_2 \tag{6.18}$$

$$\tfrac{1}{2}\dot{x}_3 = i_1 - i_2 \tag{6.19}$$

---

†It is assumed that they *are* independent. Under certain conditions these variables may not be independent. For example, if there is a loop formed exclusively by capacitors and voltage sources, then according to KVL, the voltage around the loop is zero and one of the capacitor voltage can be expressed in terms of the remaining capacitor voltages (and sources) in the loop. Hence one of the capacitor voltages is not independent and does not carry new information. Consequently, it cannot be a state variable. Similar situation exists if there is a cut set formed by all inductors and current sources.

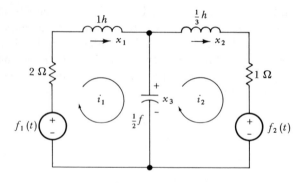

Figure 6.4

Note that these equations represent inductor and capacitor currents in terms of state variables.

Step 3.   The mesh equations are

$$2i_1 + \dot{x}_1 + x_3 = f_1(t) \tag{6.20}$$

$$-x_3 + \tfrac{1}{3}\dot{x}_2 + i_2 = -f_2(t) \tag{6.21}$$

Step 4.   From the two sets of equations above eliminate i_1, i_2, and i_3 to obtain the desired state equations.   Substitution of Eq. 6.17 in Eq. 6.20 yields

$$\dot{x}_1 = -2x_1 - x_3 + f_1(t)$$

Substitution of Eq. 6.18 in Eq. 6.21 yields

$$\dot{x}_2 = -3x_2 + 3x_3 - 3f_2(t)$$

Substitution of Eqs. 6.17 and 6.18 in Eq. 6.19 yields

$$\dot{x}_3 = 2x_1 - 2x_2$$

These are the three state equations. We can express them in a matrix form as

$$\begin{bmatrix} \dot{x}_1 \\ \dot{x}_2 \\ \dot{x}_3 \end{bmatrix} = \begin{bmatrix} -2 & 0 & -1 \\ 0 & -3 & 3 \\ 2 & -2 & 0 \end{bmatrix} \begin{bmatrix} x_1 \\ x_2 \\ x_3 \end{bmatrix} + \begin{bmatrix} 1 & 0 \\ 0 & -3 \\ 0 & 0 \end{bmatrix} \begin{bmatrix} f_1(t) \\ f_2(t) \end{bmatrix} \tag{6.22}$$

We shall consider one more example.

Example 6.2.   Write state equations for the network shown in Fig. 6.5.

Step 1.   There is one inductor and one capacitor in the network. We shall therefore choose inductor current x_1 and the capacitor voltage x_2 as state variables.

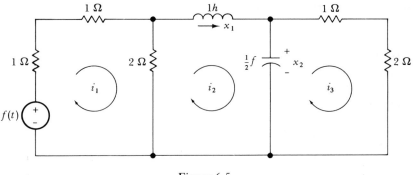

Figure 6.5

Step 2.   The relationship between mesh currents and state variables can be written by inspection.

$$x_1 = i_2 \tag{6.23a}$$

$$\tfrac{1}{2}\dot{x}_2 = i_2 - i_3 \tag{6.23b}$$

Step 3.   Mesh equations are

$$4i_1 - 2i_2 = f(t) \tag{6.24a}$$

$$2(i_2 - i_1) + \dot{x}_1 + x_2 = 0 \tag{6.24b}$$

$$-x_2 + 3i_3 = 0 \tag{6.24c}$$

Step 4.   Eliminate i_1, i_2, and i_3 from Eqs. 6.23 and 6.24 as follows.   From Eq. 6.24b we have

$$\dot{x}_1 = 2(i_1 - i_2) - x_2 \tag{6.25}$$

We can eliminate i_1 and i_2 from this equation by using Eqs. 6.24a and 6.23a to obtain

$$\dot{x}_1 = -x_1 - x_2 + \tfrac{1}{2} f(t) \tag{6.26}$$

Substitution of Eqs. 6.23a and 6.24c in Eq. 6.23b yields

$$\dot{x}_2 = 2x_1 - \tfrac{2}{3}x_2 \tag{6.27}$$

These are the desired state equations.   We can express them in a matrix form as

$$\begin{bmatrix} \dot{x}_1 \\ \dot{x}_2 \end{bmatrix} = \begin{bmatrix} -1 & -1 \\ 2 & -\tfrac{2}{3} \end{bmatrix} \begin{bmatrix} x_1 \\ x_2 \end{bmatrix} + \begin{bmatrix} \tfrac{1}{2} \\ 0 \end{bmatrix} f(t) \tag{6.28}$$

Derivation of state equations from loop equations is considerably facilitated by choosing loops in such a way that only one loop current passes through each of the inductors or capacitors.

We can also derive state equations from node (or cutset) equations. The procedure is identical to that used for mesh (or loop) equations with node voltages replacing mesh currents.

For a linear system with n state variables $x_1, x_2, \ldots, x_n$, m inputs $f_1, f_2, \ldots, f_m$, and k outputs $y_1, y_2, \ldots, y_k$, the state equations can be expressed as

$$
\begin{bmatrix} \dot{x}_1 \\ \dot{x}_2 \\ \cdots \\ \dot{x}_n \end{bmatrix} = \begin{bmatrix} a_{11} & a_{12} & \cdots & a_{1n} \\ a_{21} & a_{22} & \cdots & a_{2n} \\ \cdots & \cdots & & \cdots \\ a_{n1} & a_{n2} & & a_{nn} \end{bmatrix} \begin{bmatrix} x_1 \\ x_2 \\ \cdots \\ x_n \end{bmatrix} + \begin{bmatrix} b_{11} & b_{12} & \cdots & b_{1m} \\ b_{21} & b_{22} & & b_{2m} \\ \cdots & \cdots & & \cdots \\ b_{n1} & b_{n2} & & b_{nm} \end{bmatrix} \begin{bmatrix} f_1 \\ f_2 \\ \cdots \\ f_m \end{bmatrix} \tag{6.29}
$$

We shall define a state vector $\mathbf{x}$ whose components are the state variables $x_1, x_2, \ldots, x_n$. Thus

$$
\mathbf{x} = \begin{bmatrix} x_1 \\ x_2 \\ \vdots \\ x_n \end{bmatrix} \tag{6.30}
$$

The derivative of the state vector is given by

$$
\dot{\mathbf{x}} = \begin{bmatrix} \dot{x}_1 \\ \dot{x}_2 \\ \vdots \\ \dot{x}_n \end{bmatrix} \tag{6.31}
$$

Let us denote the $n \times n$ matrix formed by coefficients a_{ij} in Eq. 6.29 by $\mathbf{A}$ and the $n \times m$ matrix formed by elements b_{jk} be denoted by $\mathbf{B}$:

$$
\mathbf{A} = \begin{bmatrix} a_{11} & a_{12} & \cdots & a_{1n} \\ a_{21} & a_{22} & \cdots & a_{2n} \\ \cdots & \cdots & & \cdots \\ a_{n1} & a_{n2} & \cdots & a_{nn} \end{bmatrix} \qquad \mathbf{B} = \begin{bmatrix} b_{11} & b_{12} & \cdots & b_{1m} \\ b_{21} & b_{22} & \cdots & b_{2m} \\ \cdots & \cdots & & \cdots \\ b_{n1} & b_{n2} & \cdots & b_{nm} \end{bmatrix} \tag{6.32}
$$

We shall define the input vector $\mathbf{f}$ formed by inputs $f_1, f_2, \ldots, f_m$ as

$$
\mathbf{f}(t) = \begin{bmatrix} f_1(t) \\ f_2(t) \\ \vdots \\ f_m(t) \end{bmatrix} \tag{6.33}
$$

The state equation 6.29 can now be expressed as

$$\dot{\mathbf{x}} = \mathbf{A}\mathbf{x} + \mathbf{B}\mathbf{f} \tag{6.34}$$

This is a first-order vector differential equation. The vector $\mathbf{x}$ is n-dimensional. In the input-output methods we deal in nth-order differential equations in scalar variables. The state equation, on the other hand, is a first-order differential equation in an n-dimensional vector variable. Thus instead of dealing with scalars, we shall now deal with vectors. The complexity introduced by vectors is compensated by the fact that the equation is now only of the first order.

THE OUTPUT EQUATION

As mentioned earlier, the state-space analysis is carried in two steps: (i) determine the state of the system, and (ii) determine any output of the system from the knowledge of the state of the system. The state of the system is determined from the solution of state equations. Every possible output of the system can be written in terms of the state of the system (and the inputs). Equations which relate the output to the state of the system are called *output equations*. For linear systems these equations are linear algebraic equations. If $y_1, y_2, \ldots, y_k$ are the outputs, and $f_1, f_2, \ldots, f_m$ are the inputs, then the output equations can be expressed as

$$
\begin{aligned}
y_1 &= c_{11}x_1 + c_{12}x_2 + \cdots + c_{1n}x_n + d_{11}f_1 + d_{12}f_2 + \cdots + d_{1m}f_m \\
y_2 &= c_{21}x_1 + c_{22}x_2 + \cdots + c_{2n}x_n + d_{21}f_1 + d_{22}f_2 + \cdots + d_{2m}f_m \\
&\quad \cdots \cdots \cdots \cdots \cdots \cdots \cdots \cdots \cdots \cdots \cdots \cdots \cdots \cdots \\
y_k &= c_{k1}x_1 + c_{k2}x_2 + \cdots + c_{kn}x_n + d_{k1}f_1 + d_{k2}f_2 + \cdots + d_{km}f_m
\end{aligned} \tag{6.35}
$$

or

$$
\underbrace{\begin{bmatrix} y_1 \\ y_2 \\ \cdots \\ y_k \end{bmatrix}}_{\mathbf{y}} = \underbrace{\begin{bmatrix} c_{11} & c_{12} & \cdots & c_{1n} \\ c_{21} & c_{22} & \cdots & c_{2n} \\ \cdots & \cdots & \cdots & \cdots \\ c_{k1} & c_{k2} & \cdots & c_{kn} \end{bmatrix}}_{\mathbf{C}} \underbrace{\begin{bmatrix} x_1 \\ x_2 \\ \cdots \\ x_n \end{bmatrix}}_{\mathbf{x}} + \underbrace{\begin{bmatrix} d_{11} & d_{12} & \cdots & d_{1m} \\ d_{21} & d_{22} & \cdots & d_{2m} \\ \cdots & \cdots & \cdots & \cdots \\ d_{k1} & d_{k2} & \cdots & d_{km} \end{bmatrix}}_{\mathbf{D}} \underbrace{\begin{bmatrix} f_1 \\ f_2 \\ \cdots \\ f_m \end{bmatrix}}_{\mathbf{f}} \tag{6.36}
$$

Defining the output vector $\mathbf{y}$ as the vector formed by $y_1, y_2, \ldots, y_k$, and matrices $\mathbf{C}$ and $\mathbf{D}$ formed by elements c_{ij} and d_{ij} respectively. Equation 6.36 can be expressed as

$$\mathbf{y} = \mathbf{C}\mathbf{x} + \mathbf{D}\mathbf{f} \tag{6.37}$$

Every possible output of a system can be expressed in terms of state variables (and the input). This fact was demonstrated in Chapter 1 (see Eq. 1.5).

It is evident that once the state of the system is known at some instant, every output at that instant is determined. If we know the time behavior of the state, then the time behavior of every output is determined. The time behavior of the state of the system is obtained by solving state equations by techniques developed in Sec. 6.4.

6.3   STATE EQUATIONS FROM BLOCK DIAGRAM

We may be given a system in block-diagram form where several subsystems are connected in a certain manner. Each subsystem is characterized by its transfer function. We must therefore find some method of characterizing a system by state equations when its transfer function is known.† We shall now develop several such techniques.

It should be noted that the output x_0 and the input x_i of an integrator are related by a first-order differential equation:

$$\dot{x}_0 = x_i \tag{6.38}$$

Consequently the output of each integrator qualifies as a state variable. This is the key to our problem. We can simulate a given transfer function using integrators (as discussed in Sec. 4.14) and label the output of each integrator as a state variable. The input of the kth integrator is $\dot{x}_k$. The kth state equation is obtained by equating the input of the kth integrator to $\dot{x}_k$.

Consider for example a first-order transfer function shown in Fig. 6.6. We simulate this transfer function using an integrator (Fig. 6.7). The output y of this

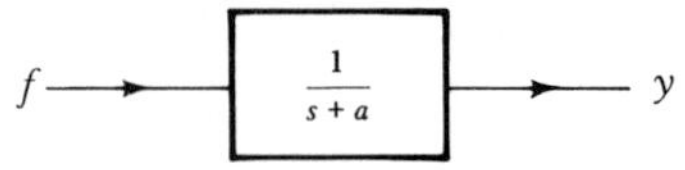

$$f \longrightarrow \boxed{\dfrac{1}{s+a}} \longrightarrow y$$

Figure 6.6

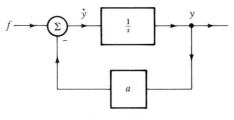

Figure 6.7

integrator is therefore a state variable. The input to the integrator is $\dot{y}$. It can be seen from Fig. 6.7, that

$$\dot{y} = -ay + f \tag{6.39}$$

This is the desired state equation of this system.

---

†Implicit here is the assumption that if there are any common factors in the numerator and the denominator of the transfer operator $H(p)$ they are not canceled in the transfer function $H(s)$. Such a cancellation will cause the loss of information and the resulting state equations will represent only a part of the system.

When the system is of higher order we may use any of the several methods (cascade, series, or parallel) to simulate such a transfer function. We can label the output of each integrator as a state variable. The input equation of each integrator will be a state equation. As an example, consider a transfer function

$$H(s) = \frac{4s + 10}{s^3 + 8s^2 + 19s + 12} \tag{6.40}$$

$$= \frac{4(s + 5/2)}{(s + 1)(s + 3)(s + 4)} \tag{6.41}$$

We may also express $H(s)$ in terms of its partial fractions as

$$H(s) = \frac{1}{s + 1} + \frac{1}{s + 3} - \frac{2}{s + 4} \tag{6.42}$$

We shall now simulate $H(s)$ in each of the three forms suggested by Eqs. 6.40, 6.41, and 6.42.

1. CASCADE REPRESENTATION (PHASE VARIABLES)

The transfer function $H(s)$ in Eq. 6.40 may be simulated by two arrangements discussed in Sec. 4.14 (Fig. 4.26a and 4.26b). We shall consider the form in Fig. 4.26a. Realization of $H(s)$ in this form is shown in Fig. 6.8a. The output of each integrator qualifies as a state variable. Let us label the three integrator outputs as x_1, x_2, and x_3 as shown in Fig. 6.8a. The inputs of these integrators are obviously $\dot{x}_1$, $\dot{x}_2$, and $\dot{x}_3$. From Fig. 6.8a, we observe that

$$\begin{aligned} \dot{x}_1 &= x_2 \\ \dot{x}_2 &= x_3 \\ \dot{x}_3 &= -12x_1 - 19x_2 - 8x_3 + f \end{aligned} \tag{6.43}$$

and the output y is given by

$$y = 10x_1 + 4x_2 \tag{6.44}$$

Thus the state equation and the output equation are

$$\begin{bmatrix} \dot{x}_1 \\ \dot{x}_2 \\ \dot{x}_3 \end{bmatrix} = \begin{bmatrix} 0 & 1 & 0 \\ 0 & 0 & 1 \\ -12 & -19 & -8 \end{bmatrix} \begin{bmatrix} x_1 \\ x_2 \\ x_3 \end{bmatrix} + \begin{bmatrix} 0 \\ 0 \\ 1 \end{bmatrix} f \tag{6.45}$$

and

$$y = \begin{bmatrix} 10 & 4 & 0 \end{bmatrix} \begin{bmatrix} x_1 \\ x_2 \\ x_3 \end{bmatrix} \tag{6.46}$$

State variables obtained from the cascade simulation of this form are known as

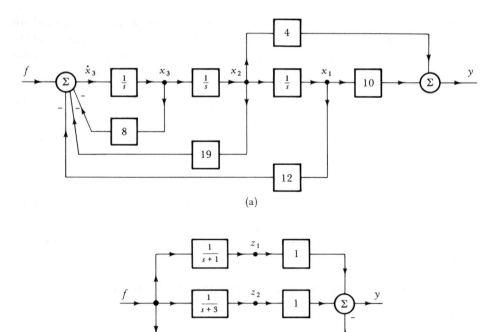

(a)

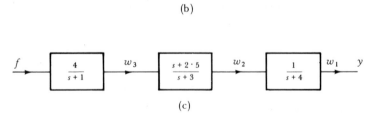

(b)

(c)

Figure 6.8

phase variables. Observe that we could also simulate Eq. 6.40 by an alternate arrangement shown in Fig. 4.26b. This would yield a different set of state variables.

Parallel Representation (Diagonalized Variables)

Let us next consider the parallel form representation (Fig. 6.8b) obtained from simulation of $H(s)$ by its partial fractions (Eq. 6.42). From Fig. 6.8b we observe that there are three first-order transfer functions of the form shown in Fig. 6.6. The output of each first-order subsystem is a state variable. Accordingly we label the outputs of the three subsystems in Fig. 6.8b as z_1, z_2, and z_3. The three state equations are (see Eq. 6.39)

$$\dot{z}_1 = -z_1 + f$$
$$\dot{z}_2 = -3z_2 + f \qquad (6.47)$$
$$\dot{z}_3 = -4z_3 + f$$

and the output y is given by

$$y = z_1 + z_2 - 2z_3 \tag{6.48}$$

or

$$\begin{bmatrix} \dot{z}_1 \\ \dot{z}_2 \\ \dot{z}_3 \end{bmatrix} = \begin{bmatrix} -1 & 0 & 0 \\ 0 & -3 & 0 \\ 0 & 0 & -4 \end{bmatrix} \begin{bmatrix} z_1 \\ z_2 \\ z_3 \end{bmatrix} + \begin{bmatrix} 1 \\ 1 \\ 1 \end{bmatrix} f \tag{6.49a}$$

and

$$y = \begin{bmatrix} 1 & 1 & -2 \end{bmatrix} \begin{bmatrix} z_1 \\ z_2 \\ z_3 \end{bmatrix} \tag{6.49b}$$

Note that in this form the matrix $\mathbf{A}$ is diagonalized and the nonzero elements along the diagonal are the poles of $H(s)$. In addition, all the elements of $\mathbf{B}$ matrix are unity.

Series Representation

The transfer function $H(s)$ can also be simulated in a series form by using Eq. 6.41. One possible series realization is shown in Fig. 6.8c. There are three subsystems in series, and each is of the first order. Consequently their outputs qualify as state variables. Let these variables be w_1, w_2, and w_3. From Fig. 6.8c we can now immediately write

$$\dot{w}_1 = -4w_1 + w_2 \tag{6.50a}$$
$$\dot{w}_2 + 3w_2 = \dot{w}_3 + \tfrac{5}{2}w_3 \tag{6.50b}$$
$$\dot{w}_3 = -w_3 + 4f \tag{6.50c}$$

Elimination of w_3 from Eq. 6.50b using Eq. 6.50c converts these equations into the desired state form:

$$\begin{bmatrix} \dot{w}_1 \\ \dot{w}_2 \\ \dot{w}_3 \end{bmatrix} = \begin{bmatrix} -4 & 1 & 0 \\ 0 & -3 & \tfrac{3}{2} \\ 0 & 0 & -1 \end{bmatrix} \begin{bmatrix} w_1 \\ w_2 \\ w_3 \end{bmatrix} + \begin{bmatrix} 0 \\ 4 \\ 4 \end{bmatrix} f \tag{6.51a}$$

and the output y is w_1:

$$y = \begin{bmatrix} 1 & 0 & 0 \end{bmatrix} \begin{bmatrix} w_1 \\ w_2 \\ w_3 \end{bmatrix} \tag{6.51b}$$

It is obvious from this discussion that a given system can be characterized by several sets of state variables. Notable among these are the phase variables and

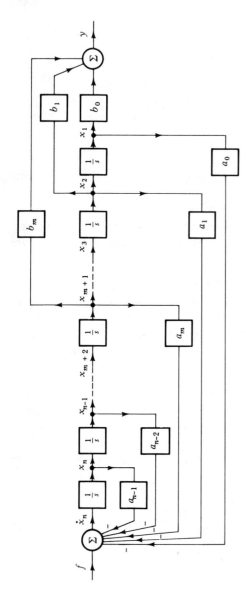

(a)

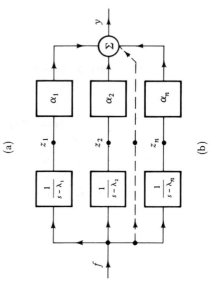

(b)

Figure 6.9

diagonalized variables. State equations in these forms are highly formalized and can be written immediately by inspection of the transfer function. Let us consider the general transfer function

$$H(s) = \frac{b_m s^m + b_{m-1} s^{m-1} + \cdots + b_1 s + b_0}{s^n + a_{n-1} s^{n-1} + \cdots + a_1 s + a_0} \tag{6.52a}$$

We may also express $H(s)$ by its partial fractions as

$$H(s) = \frac{\alpha_1}{s - \lambda_1} + \frac{\alpha_2}{s - \lambda_2} + \cdots + \frac{\alpha_n}{s - \lambda_n} \tag{6.52b}$$

We shall now simulate $H(s)$ in Eq. 6.52a using the form in Fig. 4.26a. This is shown in Fig. 6.9a. The outputs of the n integrators are labeled as n state variables (phase variables) $x_1, x_2, \ldots, x_n$, respectively. The inputs of these integrators are therefore $\dot{x}_1, \dot{x}_2, \ldots, \dot{x}_n$, respectively.

Let us write input equations of all the n integrators. We observe that the input of the last integrator is x_2 and the output is x_1. Therefore

$$\dot{x}_1 = x_2$$

This is the input equation of the last integrator. We can continue this process and write similar equations for all the integrators. This yields†

$$
\begin{aligned}
\dot{x}_1 &= x_2 \\
\dot{x}_2 &= x_3 \\
&\cdots\cdots \\
\dot{x}_{n-1} &= x_n
\end{aligned} \tag{6.53a}
$$

$$\dot{x}_n = -a_{n-1} x_n - a_{n-2} x_{n-1} - \cdots - a_1 x_2 - a_0 x_1 + f$$

and the output y is

$$y = b_0 x_1 + b_1 x_2 + \cdots + b_m x_{m+1} \tag{6.53b}$$

or

$$
\begin{bmatrix} \dot{x}_1 \\ \dot{x}_2 \\ \cdots \\ \dot{x}_{n-1} \\ x_n \end{bmatrix}
=
\begin{bmatrix}
0 & 1 & 0 & \cdots & 0 & 0 \\
0 & 0 & 1 & \cdots & 0 & 0 \\
\cdots & \cdots & \cdots & \cdots & \cdots & \cdots \\
0 & 0 & 0 & \cdots & 0 & 1 \\
-a_0 & -a_1 & -a_2 & \cdots & -a_{n-2} & -a_{n-1}
\end{bmatrix}
\begin{bmatrix} x_1 \\ x_2 \\ \cdots \\ x_{n-1} \\ x_n \end{bmatrix}
+
\begin{bmatrix} 0 \\ 0 \\ \cdots \\ 0 \\ 1 \end{bmatrix} f(t) \tag{6.54a}
$$

and

---

†These n state equations (6.53a) along with the output equation (6.53b) represent the nth-order differential equation 4.58. In other words, the nth-order differential equation 4.58 can be expressed by n simultaneous first-order differential equations (6.53a), and one algebraic equation (output equation 6.53b).

$$y = [b_0 \quad b_1 \quad \cdots \quad b_m \quad 0 \quad \cdots \quad 0] \begin{bmatrix} x_1 \\ x_2 \\ \vdots \\ \vdots \\ x_n \end{bmatrix} \qquad (6.54b)$$

Observe that these equations (state equations and the output equation) can be written immediately by inspection of $H(s)$. This form of representation is known as Kalman's first form. The use of alternate simulation scheme shown in Fig. 4.26b for $H(s)$, yields another set of state and output equations (Kalman's second form). See Prob. 6.21.

Figure 6.9b shows simulation of $H(s)$ in Eq. 6.52b by its partial fractions. The outputs of the n subsystems are the state variables $z_1, z_2, \ldots, z_n$. The state equations and the output for this arrangement can be seen from Fig. 6.9b to be†

$$\begin{bmatrix} \dot{z}_1 \\ \dot{z}_2 \\ \cdots \\ \dot{z}_{n-1} \\ \dot{z}_n \end{bmatrix} = \begin{bmatrix} \lambda_1 & 0 & \cdots & 0 & 0 \\ 0 & \lambda_2 & \cdots & 0 & 0 \\ \cdots & \cdots & \cdots & \cdots & \cdots \\ 0 & 0 & \cdots & \lambda_{n-1} & 0 \\ 0 & 0 & \cdots & 0 & \lambda_n \end{bmatrix} \begin{bmatrix} z_1 \\ z_2 \\ \cdots \\ z_{n-1} \\ z_n \end{bmatrix} + \begin{bmatrix} 1 \\ 1 \\ \cdots \\ 1 \\ 1 \end{bmatrix} f(t) \qquad (6.55a)$$

and

$$y = [\alpha_1 \quad \alpha_2 \quad \cdots \quad \alpha_{n-1} \quad \alpha_n] \begin{bmatrix} z_1 \\ z_2 \\ \cdots \\ z_{n-1} \\ z_n \end{bmatrix} \qquad (6.55b)$$

Note that the **A** matrix is a diagonal matrix with the poles $\lambda_1, \lambda_2, \ldots, \lambda_n$ as its diagonal elements. All the elements of **B** matrix are unity. The **C** matrix elements are the coefficients in the partial fraction expansion.

Example 6.3. Write state equations and the output equation for the system shown in Fig. 6.10a.

We may attack this problem in several ways. Let us represent the second-order transfer function in the forward path as shown in Fig. 6.10b (series arrangement). The system now has three first-order subsystems. The output of each subsystem qualifies as a state variable. Let x_1, x_2, and x_3 be the outputs of these subsystems (Fig. 6.10b). The three desired state equations can now be immediately written from Fig. 6.10b as

$$\dot{x}_1 = x_2$$
$$\dot{x}_2 = -3x_2 + 3(-x_3 + f)$$
$$\dot{x}_3 = -x_3 + x_1$$

and

$$y = x_1$$

---

†Here we are assuming the poles $\lambda_1, \lambda_2, \ldots, \lambda_n$ to be distinct. For multiple poles, the arrangement is slightly modified.

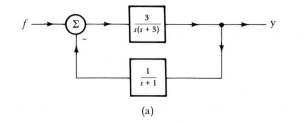

(a)

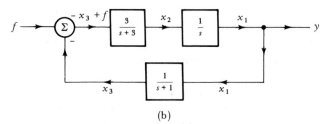

(b)

Figure 6.10

or

$$\begin{bmatrix} \dot{x}_1 \\ \dot{x}_2 \\ \dot{x}_3 \end{bmatrix} = \begin{bmatrix} 0 & 1 & 0 \\ 0 & -3 & -3 \\ 1 & 0 & -1 \end{bmatrix} \begin{bmatrix} x_1 \\ x_2 \\ x_3 \end{bmatrix} + \begin{bmatrix} 0 \\ 3 \\ 0 \end{bmatrix} f(t)$$

and

$$y = \begin{bmatrix} 1 & 0 & 0 \end{bmatrix} \begin{bmatrix} x_1 \\ x_2 \\ x_3 \end{bmatrix}$$

Another approach to this problem would be to find the transfer function $H(s)$ relating y to f. Using Mason's rule, it is seen that

$$H(s) = \frac{3/s(s+3)}{1 + \dfrac{3}{s(s+1)(s+3)}} = \frac{3(s+1)}{s^3 + 4s^2 + 3s + 3}$$

We can now write state equations for this transfer function by using any one of the methods discussed earlier. Any of these simulations, however, yields state variables which are artificially defined and are not available for observation, measurement, or manipulation. On the other hand the state variables defined in Fig. 6.10b, have a definite physical significance, and are available for observation.

SIMULATION OF A TRANSFER FUNCTION ON A DIGITAL COMPUTER

A given transfer function can be conveniently simulated on a digital computer by means of state equations. Any one of the procedures to write state equations from a given transfer function may be used. The phase variables prove particularly convenient. In this manner a response(s) of a given system to any input(s) can be obtained on a digital computer.

6.4   SOLUTION OF STATE EQUATIONS

State equations of a linear system are n-simultaneous linear differential equations of first order. We have studied the techniques of solving simultaneous linear differential equations in Chapters 2–4. The same techniques can be applied to state equations without any modification. However, since we are interested in n state variables we are also interested in finding the solution to all these n variables. Consequently it is more convenient to carry out the solution in the framework of matrix notation. Although the solution of state equations that we are now going to develop may superficially appear different than the techniques studied in Chapters 2–4, a little reflection will show that they are basically the same.

These equations can be solved both in the time domain and in the frequency domain. The frequency-domain solution requires fewer new concepts and consequently is easier to deal with than the time-domain solution. For this reason we shall first consider the frequency-domain solution.

FREQUENCY-DOMAIN SOLUTION OF STATE EQUATIONS

The kth state equation (Eq. 6.29) is of the form

$$\dot{x}_k = a_{k1}x_1 + a_{k2}x_2 + \cdots + a_{kn}x_n + b_{k1}f_1 + b_{k2}f_2 + \cdots + b_{km}f_m \quad (6.56a)$$

We shall take the Laplace transform of this equation. Let

$$x_k(t) \leftrightarrow X_k(s)$$

then

$$\dot{x}_k \leftrightarrow sX_k(s) - x_k(0)$$

and let

$$f_i(t) \leftrightarrow F_i(s)$$

The Laplace transform of Eq. 6.56a yields

$$sX_k(s) - x_k(0) = a_{k1}X_1(s) + a_{k2}X_2(s) + \cdots + a_{kn}X_n(s) + b_{k1}F_1(s)$$
$$+ b_{k2}F_2(s) + \cdots + b_{km}F_m(s) \quad (6.56b)$$

Taking the Laplace transforms of all n state equations we obtain

$$s\underbrace{\begin{bmatrix} X_1(s) \\ X_2(s) \\ \cdots \\ X_n(s) \end{bmatrix}}_{\mathbf{X}(s)} - \underbrace{\begin{bmatrix} x_1(0) \\ x_2(0) \\ \cdots \\ x_n(0) \end{bmatrix}}_{\mathbf{x}(0)} = \underbrace{\begin{bmatrix} a_{11} & a_{12} & \cdots & a_{1n} \\ a_{21} & a_{22} & \cdots & a_{2n} \\ \cdots & \cdots & \cdots & \cdots \\ a_{n1} & a_{n2} & & a_{nn} \end{bmatrix}}_{\mathbf{A}} \underbrace{\begin{bmatrix} X_1(s) \\ X_2(s) \\ \cdots \\ X_n(s) \end{bmatrix}}_{\mathbf{X}(s)}$$

$$+ \underbrace{\begin{bmatrix} b_{11} & b_{12} & \cdots & b_{1m} \\ b_{21} & b_{22} & \cdots & b_{2m} \\ \cdots & \cdots & \cdots & \cdots \\ b_{n1} & b_{n2} & & b_{nm} \end{bmatrix}}_{\mathbf{B}} \underbrace{\begin{bmatrix} F_1(s) \\ F_2(s) \\ \cdots \\ F_m(s) \end{bmatrix}}_{\mathbf{F}(s)} \quad (6.57a)$$

Defining the vectors as shown above,† we have

$$s\mathbf{X}(s) - \mathbf{x}(0) = \mathbf{AX}(s) + \mathbf{BF}(s)$$

or

$$s\mathbf{X}(s) - \mathbf{AX}(s) = \mathbf{x}(0) + \mathbf{BF}(s)$$

and

$$(s\mathbf{I} - \mathbf{A})\mathbf{X}(s) = \mathbf{x}(0) + \mathbf{BF}(s) \tag{6.57b}$$

Therefore

$$\mathbf{X}(s) = (s\mathbf{I} - \mathbf{A})^{-1}[\mathbf{x}(0) + \mathbf{BF}(s)] \tag{6.58a}$$
$$= \mathbf{\Phi}(s)[\mathbf{x}(0) + \mathbf{BF}(s)] \tag{6.58b}$$

Where we define for convenience

$$\mathbf{\Phi}(s) = (s\mathbf{I} - \mathbf{A})^{-1} \tag{6.59}$$

Thus from Eq. 6.58b,

$$\mathbf{X}(s) = \mathbf{\Phi}(s)\mathbf{x}(0) + \mathbf{\Phi}(s)\mathbf{BF}(s) \tag{6.60a}$$

and

$$\mathbf{x}(t) = \underbrace{\mathcal{L}^{-1}[\mathbf{\Phi}(s)]\mathbf{x}(0)}_{\text{zero-input component}} + \underbrace{\mathcal{L}^{-1}[\mathbf{\Phi}(s)\mathbf{BF}(s)]}_{\text{zero-state component}} \tag{6.60b}$$

The matrix $\mathbf{\Phi}(s)$ is known as the *resolvant matrix*.

Equation 6.60b gives the desired solution. Observe the two components of the solution. The first component depends only on the initial state $\mathbf{x}(0)$ and is zero when $\mathbf{x}(0) = 0$. Hence this is zero-input component. In a similar manner, we see that the second component is a function of the input only and is zero when the input is zero. This is therefore a zero-state component.

Example 6.4.   Find the state vector $\mathbf{x}(t)$ for the system whose state equation is given as

$$\dot{\mathbf{x}} = \mathbf{Ax} + \mathbf{B}f(t)$$

where

$$\mathbf{A} = \begin{bmatrix} -12 & \frac{2}{3} \\ -36 & -1 \end{bmatrix} \qquad \mathbf{B} = \begin{bmatrix} \frac{1}{3} \\ 1 \end{bmatrix} \qquad f(t) = u(t)$$

and the initial conditions are: $x_1(0) = 2$, $x_2(0) = 1$. From Eq. 6.58b, we have

$$\mathbf{X}(s) = \mathbf{\Phi}(s)[\mathbf{x}(0) + \mathbf{BF}(s)]$$

Let us first find $\mathbf{\Phi}(s)$. We have

$$(s\mathbf{I} - \mathbf{A}) = s\begin{bmatrix} 1 & 0 \\ 0 & 1 \end{bmatrix} - \begin{bmatrix} -12 & \frac{2}{3} \\ -36 & -1 \end{bmatrix} = \begin{bmatrix} s + 12 & -\frac{2}{3} \\ 36 & s + 1 \end{bmatrix}$$

---

†In all the future development of this chapter, the reader is assumed to have some understanding of matrix algebra. Appendix D provides the necessary background.

and

$$\mathbf{\Phi}(s) = (s\mathbf{I} - \mathbf{A})^{-1} = \begin{bmatrix} \dfrac{s+1}{(s+4)(s+9)} & \dfrac{2/3}{(s+4)(s+9)} \\[3mm] \dfrac{-36}{(s+4)(s+9)} & \dfrac{s+12}{(s+4)(s+9)} \end{bmatrix} \qquad (6.61a)$$

and $\mathbf{x}(0)$ is given to be

$$\mathbf{x}(0) = \begin{bmatrix} 2 \\ 1 \end{bmatrix}$$

Also $F(s) = \dfrac{1}{s}$ and

$$\mathbf{BF}(s) = \begin{bmatrix} \dfrac{1}{3} \\[2mm] 1 \end{bmatrix} \dfrac{1}{s} = \begin{bmatrix} \dfrac{1}{3s} \\[2mm] \dfrac{1}{s} \end{bmatrix}$$

Therefore

$$\mathbf{x}(0) + \mathbf{BF}(s) = \begin{bmatrix} 2 + \dfrac{1}{3s} \\[2mm] 1 + \dfrac{1}{s} \end{bmatrix} = \begin{bmatrix} \dfrac{6s+1}{3s} \\[2mm] \dfrac{s+1}{s} \end{bmatrix}$$

and

$$\mathbf{X}(s) = \mathbf{\Phi}(s)[\mathbf{x}(0) + \mathbf{BF}(s)]$$

$$= \begin{bmatrix} \dfrac{s+1}{(s+4)(s+9)} & \dfrac{2/3}{(s+4)(s+9)} \\[3mm] \dfrac{-36}{(s+4)(s+9)} & \dfrac{s+12}{(s+4)(s+9)} \end{bmatrix} \begin{bmatrix} \dfrac{6s+1}{3s} \\[2mm] \dfrac{s+1}{s} \end{bmatrix}$$

$$= \begin{bmatrix} \dfrac{2s^2 + 3s + 1}{s(s+4)(s+9)} \\[3mm] \dfrac{s - 59}{(s+4)(s+9)} \end{bmatrix}$$

$$= \begin{bmatrix} \dfrac{1/36}{s} - \dfrac{21/20}{s+4} + \dfrac{136/45}{s+9} \\[3mm] \dfrac{-63/5}{s+4} + \dfrac{68/5}{s+9} \end{bmatrix}$$

The inverse Laplace transform of this equation yields

$$
\begin{bmatrix} x_1(t) \\ x_2(t) \end{bmatrix} = \begin{bmatrix} \left(\dfrac{1}{36} - \dfrac{21}{20} e^{-4t} + \dfrac{136}{45} e^{-9t} \right) u(t) \\ \left(-\dfrac{63}{5} e^{-4t} + \dfrac{68}{5} e^{-9t} \right) u(t) \end{bmatrix}
$$

(6.61b)

The Output

The output equation is given by

$$
\mathbf{y} = \mathbf{Cx} + \mathbf{Df}
$$

and

$$
\mathbf{Y}(s) = \mathbf{CX}(s) + \mathbf{DF}(s)
$$

Substitution of Eq. 6.58b in this equation yields

$$
\begin{aligned}
\mathbf{Y}(s) &= \mathbf{C}\{\mathbf{\Phi}(s)[\mathbf{x}(0) + \mathbf{BF}(s)]\} + \mathbf{DF}(s) \\
&= \underbrace{\mathbf{C\Phi}(s)\mathbf{x}(0)}_{\substack{\text{zero-input} \\ \text{response}}} + \underbrace{[\mathbf{C\Phi}(s)\mathbf{B} + \mathbf{D}]\mathbf{F}(s)}_{\substack{\text{zero-state} \\ \text{response}}}
\end{aligned}
$$

(6.62)

The zero-state response $\mathbf{Y}_f(s)$ is given by

$$
\mathbf{Y}_f(s) = [\mathbf{C\Phi}(s)\mathbf{B} + \mathbf{D}]\mathbf{F}(s)
$$

(6.63a)

Note that the transfer function of a system is defined under zero-state condition (see Eq. 4.48). Comparison of Eq. 6.63a with Eq. 4.50c shows that the matrix $\mathbf{C\Phi}(s)\mathbf{B} + \mathbf{D}$ is the transfer-function matrix $\mathbf{H}(s)$ of the system which relates the responses $y_1, y_2, \ldots, y_k$ to the inputs $f_1, f_2, \ldots, f_m$:

$$
\mathbf{H}(s) = \mathbf{C\Phi}(s)\mathbf{B} + \mathbf{D}
$$

(6.63b)

and (for the zero-state case)

$$
\mathbf{Y}_f(s) = \mathbf{H}(s)\mathbf{F}(s)
$$

(6.64)

The matrix $\mathbf{H}(s)$ is a $k \times m$ matrix (k being the number of outputs, m the number of inputs) with elements $H_{ij}(s)$ as seen from Eq. 4.50. The function $H_{ij}(s)$ is the transfer function which relates the output $y_i(t)$ to the input $f_j(t)$.

Let us consider a system whose state equation is

$$
\begin{bmatrix} \dot{x}_1 \\ \dot{x}_2 \end{bmatrix} = \begin{bmatrix} 0 & 1 \\ -2 & -3 \end{bmatrix} \begin{bmatrix} x_1 \\ x_2 \end{bmatrix} + \begin{bmatrix} 1 & 0 \\ 1 & 1 \end{bmatrix} \begin{bmatrix} f_1(t) \\ f_2(t) \end{bmatrix}
$$

(6.65a)

and the output equation is

$$
\begin{bmatrix} y_1 \\ y_2 \\ y_3 \end{bmatrix} = \begin{bmatrix} 1 & 0 \\ 1 & 1 \\ 0 & 2 \end{bmatrix} \begin{bmatrix} x_1 \\ x_2 \end{bmatrix} + \begin{bmatrix} 0 & 0 \\ 1 & 0 \\ 0 & 1 \end{bmatrix} \begin{bmatrix} f_1(t) \\ f_2(t) \end{bmatrix}
$$

(6.65b)

In this case

$$\mathbf{A} = \begin{bmatrix} 0 & 1 \\ -2 & -3 \end{bmatrix} \quad \mathbf{B} = \begin{bmatrix} 1 & 0 \\ 1 & 1 \end{bmatrix} \quad \mathbf{C} = \begin{bmatrix} 1 & 0 \\ 1 & 1 \\ 0 & 2 \end{bmatrix} \quad \mathbf{D} = \begin{bmatrix} 0 & 0 \\ 1 & 0 \\ 0 & 1 \end{bmatrix} \quad (6.65c)$$

and

$$\Phi(s) = (s\mathbf{I} - \mathbf{A})^{-1} = \begin{bmatrix} s & -1 \\ 2 & s+3 \end{bmatrix}^{-1} = \begin{bmatrix} \dfrac{s+3}{(s+1)(s+2)} & \dfrac{1}{(s+1)(s+2)} \\ \dfrac{-2}{(s+1)(s+2)} & \dfrac{s}{(s+1)(s+2)} \end{bmatrix}$$

$$(6.66)$$

Hence the transfer-function matrix $\mathbf{H}(s)$ is given by

$$\mathbf{H}(s) = \mathbf{C}\Phi(s)\mathbf{B} + \mathbf{D}$$

$$= \begin{bmatrix} 1 & 0 \\ 1 & 1 \\ 0 & 2 \end{bmatrix} \begin{bmatrix} \dfrac{s+3}{(s+1)(s+2)} & \dfrac{1}{(s+1)(s+2)} \\ \dfrac{-2}{(s+1)(s+2)} & \dfrac{s}{(s+1)(s+2)} \end{bmatrix} \begin{bmatrix} 1 & 0 \\ 1 & 1 \end{bmatrix} + \begin{bmatrix} 0 & 0 \\ 1 & 0 \\ 0 & 1 \end{bmatrix}$$

$$= \begin{bmatrix} \dfrac{s+4}{(s+1)(s+2)} & \dfrac{1}{(s+1)(s+2)} \\ \dfrac{s+4}{s+2} & \dfrac{1}{s+2} \\ \dfrac{2(s-2)}{(s+1)(s+2)} & \dfrac{s^2+5s+2}{(s+1)(s+2)} \end{bmatrix} \qquad (6.67)$$

and

$$\mathbf{Y}_f(s) = \mathbf{H}(s)\mathbf{F}(s)$$

The ijth element of the transfer-function matrix in Eq. 6.67 represents the transfer function which relates the output $y_i(t)$ to the input $f_j(t)$. Thus the transfer function which relates the output y_3 to the input f_2 is $H_{32}(s)$, and is

$$H_{32}(s) = \frac{s^2 + 5s + 2}{(s+1)(s+2)}$$

CHARACTERISTIC ROOTS (EIGENVALUES) OF A MATRIX

It is interesting to observe that the denominator of every transfer function in Eq. 6.67 is $(s + 1)(s + 2)$, with the exception of $H_{21}(s)$, and $H_{22}(s)$, where cancellation of the factor $(s + 1)$ occurs. This is no coincidence. It can be easily seen that the denominator of every element of $\Phi(s)$ is $|s\mathbf{I} - \mathbf{A}|$ because $\Phi(s) = (s\mathbf{I} - \mathbf{A})^{-1}$, and the inverse of a matrix has its determinant in the denominator (see Eq. D.32b). Since $\mathbf{C}$, $\mathbf{B}$, and $\mathbf{D}$ are matrices with constant elements, it can be seen from Eq. 6.63b that the denominator of $\Phi(s)$ will also be the denominator of $H(s)$. Hence the denominator of every element of $H(s)$ is $|sI - A|$, except for the possible cancellation of some factors as mentioned earlier. In other words, poles of all transfer functions of the system are zeros of the polynomial $|sI - A|$. We therefore conclude that the zeros of the polynomial $|sI - A|$ are the natural frequencies of the system. Hence the natural frequencies of the system are the roots of equation

$$|sI - A| = 0 \tag{6.68a}$$

Since $|sI - A|$ is an nth-order polynomial in s with n roots $\lambda_1, \lambda_2, \ldots, \lambda_n$, we can write Eq. 6.68a as

$$
\begin{aligned}
|sI - A| &= s^n + a_{n-1}s^{n-1} + \cdots + a_1 s + a_0 \\
&= (s - \lambda_1)(s - \lambda_2)\cdots(s - \lambda_n) = 0
\end{aligned}
\tag{6.68b}
$$

For the problem under discussion,

$$
\begin{aligned}
|sI - A| &= \begin{vmatrix} s & 0 \\ 0 & s \end{vmatrix} - \begin{vmatrix} 0 & 1 \\ -2 & -3 \end{vmatrix} \\
&= \begin{vmatrix} s & -1 \\ 2 & s + 3 \end{vmatrix} \\
&= s^2 + 3s + 2 \tag{6.69a} \\
&= (s + 1)(s + 2) \tag{6.69b}
\end{aligned}
$$

Hence

$$\lambda_1 = -1 \quad \text{and} \quad \lambda_2 = -2$$

Equation 6.68 is known as the characteristic equation of the matrix A and $\lambda_1, \lambda_2, \ldots, \lambda_n$ are the characteristic roots of A. The term *eigenvalue*, meaning "characteristic value" in German, is also commonly used in the literature.

In conclusion, we have shown that the natural frequencies of a system are the eigenvalues (characteristic roots) of the matrix A.

At this point the reader will recall that if $\lambda_1, \lambda_2, \ldots, \lambda_n$ are the poles of transfer function, then the zero-input response is of the form

$$y_x(t) = c_1 e^{\lambda_1 t} + c_2 e^{\lambda_2 t} + \cdots + c_n e^{\lambda_n t} \tag{6.70}$$

This fact is also obvious from Eq. 6.62. The denominator of every element of the zero-input response matrix $C\Phi(s)x(0)$ is $|sI - A| = (s - \lambda_1)(s - \lambda_2)\cdots(s - \lambda_n)$. Therefore the partial-fraction expansion and subsequent inverse Laplace transform will yield the zero-input component of the form in Eq. 6.70.

TIME-DOMAIN SOLUTION OF STATE EQUATION

The state equation is

$$\dot{x} = Ax + Bf \tag{6.71}$$

This is a first-order vector differential equation. We are tempted here to apply the techniques of first-order scalar equation. In this we shall not be disappointed. We shall now see that the solution of Eq. 6.71 can be found along the lines used in Chapter 2 to find the solution of a first-order scalar differential equation (Eq. 2.59).

$$\dot{x} = ax + f \tag{6.72}$$

In Chapter 2 we found the solution of this equation to be (see Eq. 2.63b)

$$x = e^{at}x(0) + \int_0^t e^{a(t-\tau)}f(\tau)\,d\tau \tag{6.73}$$

We shall now show that the solution of the vector differential equation 6.71 is identical to Eq. 6.73 in form, when the scalar quantities are replaced by the corresponding vector quantities. It will be shown that the solution of Eq. 6.71 is given by

$$\mathbf{x}(t) = e^{\mathbf{A}t}\mathbf{x}(0) + \int_0^t e^{\mathbf{A}(t-\tau)}\mathbf{B}\mathbf{f}(\tau)\,d\tau \tag{6.74}$$

Before proceeding further we must define the exponential of a matrix appearing in Eq. 6.74. The exponential of a matrix is defined by an infinite series identical to that used in defining an exponential of a scalar. We shall define

$$e^{\mathbf{A}t} = \mathbf{I} + \mathbf{A}t + \frac{\mathbf{A}^2 t^2}{2!} + \frac{\mathbf{A}^3 t^3}{3!} + \cdots + \frac{\mathbf{A}^n t^n}{n!} + \cdots + \tag{6.75a}$$

$$= \sum_{k=0}^{\infty} \frac{\mathbf{A}^k t^k}{k!} \tag{6.75b}$$

Thus if

$$\mathbf{A} = \begin{bmatrix} 0 & 1 \\ 2 & 1 \end{bmatrix}$$

$$\mathbf{I} = \begin{bmatrix} 1 & 0 \\ 0 & 1 \end{bmatrix}$$

$$\mathbf{A}t = \begin{bmatrix} 0 & 1 \\ 2 & 1 \end{bmatrix} t = \begin{bmatrix} 0 & t \\ 2t & t \end{bmatrix}$$

$$\frac{\mathbf{A}^2 t^2}{2!} = \begin{bmatrix} 0 & 1 \\ 2 & 1 \end{bmatrix}\begin{bmatrix} 0 & 1 \\ 2 & 1 \end{bmatrix}\frac{t^2}{2} = \begin{bmatrix} 2 & 1 \\ 2 & 3 \end{bmatrix}\frac{t^2}{2} = \begin{bmatrix} t^2 & \dfrac{t^2}{2} \\ t^2 & \dfrac{3t^2}{2} \end{bmatrix}$$

and so on.

It can be shown that the infinite series in Eq. 6.75 is absolutely and uniformly convergent for all values of t. Consequently it can be differentiated or integrated term by term. Thus to find $(d/dt)e^{\mathbf{A}t}$, we differentiate the series on the right-hand side of Eq. 6.75a, term by term:

$$\frac{d}{dt}e^{\mathbf{A}t} = \mathbf{A} + \mathbf{A}^2 t + \frac{\mathbf{A}^3 t^2}{2!} + \frac{\mathbf{A}^4 t^3}{3!} + \cdots + \cdots \qquad (6.76a)$$

$$= \mathbf{A}\left[\mathbf{I} + \mathbf{A}t + \frac{\mathbf{A}^2 t^2}{2!} + \frac{\mathbf{A}^3 t^3}{3!} + \cdots + \cdots\right]$$

$$= \mathbf{A}e^{\mathbf{A}t} \qquad (6.76b)$$

Note that the infinite series on the right-hand side of Eq. 6.76a may also be expressed as

$$\frac{d}{dt}e^{\mathbf{A}t} = \left[\mathbf{I} + \mathbf{A}t + \frac{\mathbf{A}^2 t^2}{2} + \frac{\mathbf{A}^3 t^3}{3} + \cdots + \right]\mathbf{A}$$

$$= e^{\mathbf{A}t}\mathbf{A}$$

Hence

$$\frac{d}{dt}e^{\mathbf{A}t} = \mathbf{A}e^{\mathbf{A}t} = e^{\mathbf{A}t}\mathbf{A} \qquad (6.77)$$

Also note that from the definition 6.75a it follows that

$$e^0 = \mathbf{I} \qquad (6.78a)$$

If we premultiply or postmultiply the infinite series for $e^{\mathbf{A}t}$ (Eq. 6.75a) by an infinite series for $e^{-\mathbf{A}t}$, we find that

$$(e^{-\mathbf{A}t})(e^{\mathbf{A}t}) = (e^{\mathbf{A}t})(e^{-\mathbf{A}t}) = \mathbf{I} \qquad (6.78b)$$

Coming back to our problem of solving the first-order vector differential Eq. 6.71 we shall find that we can proceed exactly the same way as used in Chapter 2 to solve the first-order scalar differential Eq. 6.72.

It is shown in Appendix D (Eq. D.40) that

$$\frac{d}{dt}(\mathbf{AB}) = \frac{d\mathbf{A}}{dt}\mathbf{B} + \mathbf{A}\frac{d\mathbf{B}}{dt}$$

Using this relationship we observe that

$$\frac{d}{dt}[e^{-\mathbf{A}t}\mathbf{x}] = \left(\frac{d}{dt}e^{-\mathbf{A}t}\right)\mathbf{x} + e^{-\mathbf{A}t}\dot{\mathbf{x}}$$

$$= -e^{-\mathbf{A}t}\mathbf{A}\mathbf{x} + e^{-\mathbf{A}t}\dot{\mathbf{x}} \qquad (6.79)$$

We now premultiply both sides of Eq. 6.71 by $e^{-\mathbf{A}t}$ to yield

$$e^{-\mathbf{A}t}\dot{\mathbf{x}} = e^{-\mathbf{A}t}\mathbf{A}\mathbf{x} + e^{-\mathbf{A}t}\mathbf{B}\mathbf{f} \qquad (6.80a)$$

or

$$e^{-\mathbf{A}t}\dot{\mathbf{x}} - e^{\mathbf{A}t}\mathbf{A}\mathbf{x} = e^{-\mathbf{A}t}\mathbf{B}\mathbf{f} \qquad (6.80b)$$

A glance at Eq. 6.79 shows that the left-hand side of Eq. 6.80b is $\dfrac{d}{dt}\,[e^{-\mathbf{A}t}\mathbf{x}]$. Hence

$$\frac{d}{dt}\,[e^{-\mathbf{A}t}\mathbf{x}] = e^{-\mathbf{A}t}\mathbf{Bf}$$

Integration of both sides of this equation from 0 to t yields

$$e^{-\mathbf{A}t}\mathbf{x}\,\Big|_0^t = \int_0^t e^{-\mathbf{A}\tau}\mathbf{Bf}(\tau)\,d\tau \qquad (6.81a)$$

or

$$e^{-\mathbf{A}t}\mathbf{x}(t) - \mathbf{x}(0) = \int_0^t e^{-\mathbf{A}\tau}\mathbf{Bf}(\tau)\,d\tau \qquad (6.81b)$$

Hence

$$e^{-\mathbf{A}t}\mathbf{x} = \mathbf{x}(0) + \int_0^t e^{-\mathbf{A}\tau}\mathbf{Bf}(\tau)\,d\tau \qquad (6.81c)$$

Premultiplying Eq. 6.81c by $e^{\mathbf{A}t}$ and using Eq. 6.78b, we have

$$\mathbf{x}(t) = \underbrace{e^{\mathbf{A}t}\mathbf{x}(0)}_{\text{zero-input component}} + \underbrace{\int_0^t e^{\mathbf{A}(t-\tau)}\mathbf{Bf}(\tau)\,d\tau}_{\text{zero-state component}} \qquad (6.82a)$$

This is the desired solution. The first term on the right-hand side does not depend upon the input and vanishes when the initial state is zero. Hence it is a zero-input component. The second term by similar argument is seen to be zero-state component.

The results of Eq. 6.82a can be expressed in a more convenient way in terms of matrix convolution. We can define convolution of two matrices in a manner similar to the multiplication of two matrices except that multiplication of two elements is replaced by their convolution. For example,

$$\begin{bmatrix} f_1 & f_2 \\ f_3 & f_4 \end{bmatrix} * \begin{bmatrix} g_1 & g_2 \\ g_3 & g_4 \end{bmatrix} = \begin{bmatrix} (f_1 * g_1 + f_2 * g_3) & (f_1 * g_2 + f_2 * g_4) \\ (f_3 * g_1 + f_4 * g_3) & (f_3 * g_2 + f_4 * g_4) \end{bmatrix}$$

Using this definition of convolution, we can express Eq. 6.82a as

$$\mathbf{x}(t) = e^{\mathbf{A}t}\mathbf{x}(0) + e^{\mathbf{A}t} * \mathbf{Bf}(t) \qquad (6.82b)$$

Note that the limits of convolution integral (Eq. 6.82a) are from 0 to t. Hence all the elements of $e^{\mathbf{A}t}$ in the convolution term of Eq. 6.82b are implicitly assumed to be multiplied by $u(t)$.

The result in Eq. 6.82 can be easily generalized for any initial value of t. It is left as an exercise for the reader to show that the solution of the state equation can be expressed as

$$\mathbf{x}(t) = e^{\mathbf{A}(t-t_0)}\mathbf{x}(t_0) + \int_{t_0}^t e^{\mathbf{A}(t-\tau)}\mathbf{Bf}(\tau)\,d\tau \qquad (6.83)$$

Computation of $e^{\mathbf{A}t}$

As seen from Eq. 6.82, we need to compute $e^{\mathbf{A}t}$ in order to find the solution of the state equation. One can compute $e^{\mathbf{A}t}$ from the definition in Eq. 6.75a. A little thought will immediately convince us that this is not a very pleasant method. Computation of $\mathbf{A}^2$, $\mathbf{A}^3$, $\mathbf{A}^4$ is very laborious. Moreover, this gives the answer as an infinite series† and not as a closed form, which is more desirable. Furthermore, substitution of such an infinite series in Eq. 6.83 will make life even more miserable. Anyone who tries to solve a problem this way is bound to say in exasperation, "There must be a better way!" A better way there is. There are several methods of obtaining $e^{\mathbf{A}t}$ in a closed form which involve little labor. We shall give here just one method. If we compare Eq. 6.82a with Eq. 6.60b, we immediately observe that

$$e^{\mathbf{A}t} = \mathcal{L}^{-1}[\mathbf{\Phi}(s)] \tag{6.84a}$$

$$= \mathcal{L}^{-1}[(s\mathbf{I} - \mathbf{A})^{-1}] \tag{6.84b}$$

and

$$\mathcal{L}[e^{\mathbf{A}t}] = \mathbf{\Phi}(s) = (s\mathbf{I} - \mathbf{A})^{-1} \tag{6.84c}$$

The matrix $e^{\mathbf{A}t}$ is called *state transition matrix* (STM) and is denoted by $\boldsymbol{\phi}(t)$. Thus

$$\boldsymbol{\phi}(t) \leftrightarrow \mathbf{\Phi}(s)$$

This result (Eq. 6.84c) can also be proved independently.

There are several other ways of computing $e^{\mathbf{A}t}$. Two alternate procedures are given in the Appendix D-6.

Let us apply the time-domain technique to the system in Example 6.4. We have solved this problem earlier by using frequency-domain techniques. In this case

$$e^{\mathbf{A}t} = \mathcal{L}^{-1}[(s\mathbf{I} - \mathbf{A})^{-1}] = \mathcal{L}^{-1}\mathbf{\Phi}(s)$$

From Eq. 6.61a we have

$$e^{\mathbf{A}t} = \mathcal{L}^{-1} \begin{bmatrix} \dfrac{s+1}{(s+4)(s+9)} & \dfrac{2/3}{(s+4)(s+9)} \\[4mm] \dfrac{-36}{(s+4)(s+9)} & \dfrac{s+12}{(s+4)(s+9)} \end{bmatrix}$$

$$= \mathcal{L}^{-1} \begin{bmatrix} \left(\dfrac{-3/5}{s+4} + \dfrac{8/5}{s+9}\right) & \left(\dfrac{2/15}{s+4} - \dfrac{2/15}{s+9}\right) \\[4mm] \left(\dfrac{-36/5}{s+4} + \dfrac{36/5}{s+9}\right) & \left(\dfrac{8/5}{s+4} - \dfrac{3/5}{s+9}\right) \end{bmatrix}$$

---

†The infinite series can be expressed in a closed form. Part of the difficulty is in recognizing the series.

$$
=
\begin{bmatrix}
\left(\dfrac{-3}{5}e^{-4t} + \dfrac{8}{5}e^{-9t}\right) & \dfrac{2}{15}(e^{-4t} - e^{-9t}) \\[4mm]
\dfrac{36}{5}(-e^{-4t} + e^{-9t}) & \left(\dfrac{8}{5}e^{-4t} - \dfrac{3}{5}e^{-9t}\right)
\end{bmatrix}
\tag{6.85}
$$

The zero-input component is given by (see Eq. 6.82b)

$$
e^{\mathbf{A}t}\mathbf{x}(0) =
\begin{bmatrix}
\left(-\dfrac{3}{5}e^{-4t} + \dfrac{8}{5}e^{-9t}\right) & \dfrac{2}{15}(e^{-4t} - e^{-9t}) \\[4mm]
\dfrac{36}{5}(-e^{-4t} + e^{-9t}) & \left(\dfrac{8}{5}e^{-4t} - \dfrac{3}{5}e^{-9t}\right)
\end{bmatrix}
\begin{bmatrix} 2 \\ 1 \end{bmatrix}
$$

$$
=
\begin{bmatrix}
\left(\dfrac{-16}{15}e^{-4t} + \dfrac{46}{15}e^{-9t}\right)u(t) \\[4mm]
\left(\dfrac{-64}{5}e^{-4t} + \dfrac{69}{5}e^{-9t}\right)u(t)
\end{bmatrix}
\tag{6.86a}
$$

Note the presence of $u(t)$ in Eq. 6.86a, indicating that the response begins at $t = 0$.
The zero-state component is $e^{\mathbf{A}t} * \mathbf{Bf}$ (see Eq. 6.82b)

$$
\mathbf{Bf} =
\begin{bmatrix} \tfrac{1}{3} \\ 1 \end{bmatrix} u(t) =
\begin{bmatrix} \tfrac{1}{3}u(t) \\ u(t) \end{bmatrix}
$$

and

$$
e^{\mathbf{A}t} * \mathbf{Bf}(t) =
\begin{bmatrix}
(-\tfrac{3}{5}e^{-4t} + \tfrac{8}{5}e^{-9t})u(t) & \tfrac{2}{15}(e^{-4t} - e^{-9t})u(t) \\[3mm]
\tfrac{36}{5}(-e^{-4t} + e^{-9t})u(t) & (\tfrac{8}{5}e^{-4t} - \tfrac{3}{5}e^{-9t})u(t)
\end{bmatrix}
*
\begin{bmatrix} \tfrac{1}{3}u(t) \\ u(t) \end{bmatrix}
$$

Note again the presence of the term $u(t)$ in every element of $e^{\mathbf{A}t}$. This is because
the limits of convolution integral are from 0 to t (Eq. 6.82). Thus

$$
e^{\mathbf{A}t} * \mathbf{Bf}(t) =
\begin{bmatrix}
\left(\dfrac{-3}{5}e^{-4t} + \dfrac{8}{5}e^{-9t}\right)u(t) * \dfrac{1}{3}u(t) + \dfrac{2}{15}(e^{-4t} - e^{-9t})u(t) * u(t) \\[4mm]
\dfrac{36}{5}(-e^{-4t} + e^{-9t})u(t) * \dfrac{1}{3}u(t) + \left(\dfrac{8}{5}e^{-4t} - \dfrac{3}{5}e^{-9t}\right)u(t) * u(t)
\end{bmatrix}
$$

$$
=
\begin{bmatrix}
-\dfrac{1}{15}e^{-4t}u(t) * u(t) + \dfrac{2}{5}e^{-9t}u(t) * u(t) \\[4mm]
-\dfrac{4}{5}e^{-4t}u(t) * u(t) + \dfrac{9}{5}e^{-9t}u(t) * u(t)
\end{bmatrix}
$$

The convolution integrals above are now found from the convolution table
(Table 2.2, pair 2). This yields,

$$e^{\mathbf{A}t} * \mathbf{B}f(t) = \begin{bmatrix} -\frac{1}{60}(1 - e^{-4t})u(t) + \frac{2}{45}(1 - e^{-9t})u(t) \\ -\frac{1}{5}(1 - e^{-4t})u(t) + \frac{1}{5}(1 - e^{-9t})u(t) \end{bmatrix}$$

$$= \begin{bmatrix} \left(\frac{1}{36} + \frac{1}{60}e^{-4t} - \frac{2}{45}e^{-9t}\right)u(t) \\ \frac{1}{5}(e^{-4t} - e^{-9t})u(t) \end{bmatrix} \tag{6.86b}$$

The sum of the two components (Eq. 6.86a and Eq. 6.86b) gives the desired solution for $\mathbf{x}(t)$.

$$\mathbf{x}(t) = \begin{bmatrix} x_1(t) \\ x_2(t) \end{bmatrix} = \begin{bmatrix} \left(\frac{1}{36} - \frac{21}{20}e^{-4t} + \frac{136}{45}e^{-9t}\right)u(t) \\ \left(\frac{-63}{5}e^{-4t} + \frac{68}{5}e^{-9t}\right)u(t) \end{bmatrix} \tag{6.86c}$$

This checks out with the solution obtained by using frequency-domain method (see (Eq. 6.61b). Once the state variables x_1 and x_2 are found for all time $t \geq 0$, all the remaining variables determined from the output equation.

The Output

The output equation is given by

$$\mathbf{y}(t) = \mathbf{C}\mathbf{x}(t) + \mathbf{D}f(t)$$

Substitution of the solution for $\mathbf{x}$ (Eq. 6.82b) in this equation yields

$$\mathbf{y}(t) = \mathbf{C}[e^{\mathbf{A}t}\mathbf{x}(0) + e^{\mathbf{A}t} * \mathbf{B}f(t)] + \mathbf{D}f(t) \tag{6.87a}$$

Since the elements of $\mathbf{B}$ are constants, we can express

$$e^{\mathbf{A}t} * \mathbf{B}f(t) = e^{\mathbf{A}t}\mathbf{B} * f(t)$$

Using this result, Eq. 6.87a becomes

$$\mathbf{y}(t) = \mathbf{C}[e^{\mathbf{A}t}\mathbf{x}(0) + e^{\mathbf{A}t}\mathbf{B} * f(t)] + \mathbf{D}f(t) \tag{6.87b}$$

It will be recalled that convolution of $f(t)$ with unit impulse $\delta(t)$ yields $f(t)$:

$$f(t) * \delta(t) = \delta(t) * f(t) = f(t)$$

Let us define an $m \times m$ diagonal matrix $\boldsymbol{\delta}(t)$ such that all its diagonal elements are unit impulse function. It is then obvious that

$$\boldsymbol{\delta}(t) * f(t) = f(t)$$

and Eq. 6.87b can be expressed as

$$\mathbf{y}(t) = \mathbf{C}[e^{\mathbf{A}t}\mathbf{x}(0) + e^{\mathbf{A}t}\mathbf{B} * f(t)] + \mathbf{D}\boldsymbol{\delta}(t) * f(t) \tag{6.88a}$$

$$= \underbrace{\mathbf{C}e^{\mathbf{A}t}\mathbf{x}(0)}_{\substack{\text{zero-input} \\ \text{response}}} + \underbrace{[\mathbf{C}e^{\mathbf{A}t}\mathbf{B} + \mathbf{D}\boldsymbol{\delta}(t)] * f(t)}_{\substack{\text{zero-state} \\ \text{response}}} \tag{6.88b}$$

Using the notation $\boldsymbol{\phi}(t)$ for $e^{\mathbf{A}t}$, Eq. 6.88b may be expressed as

$$\mathbf{y}(t) = \underbrace{\mathbf{C}\boldsymbol{\phi}(t)\mathbf{x}(0)}_{\substack{\text{zero-input}\\\text{response}}} + \underbrace{[\mathbf{C}\boldsymbol{\phi}(t)\mathbf{B} + \mathbf{D}\boldsymbol{\delta}(t)] * \mathbf{f}(t)}_{\substack{\text{zero-state}\\\text{response}}} \tag{6.88c}$$

Consider now the zero-state response

$$\mathbf{y}_f(t) = [\mathbf{C}\boldsymbol{\phi}(t)\mathbf{B} + \mathbf{D}\boldsymbol{\delta}(t)] * \mathbf{f}(t) \tag{6.89a}$$
$$= \mathbf{h}(t) * \mathbf{f}(t) \tag{6.89b}$$

where

$$\mathbf{h}(t) = \mathbf{C}\boldsymbol{\phi}(t)\mathbf{B} + \mathbf{D}\boldsymbol{\delta}(t) \tag{6.90}$$

The matrix $\mathbf{h}(t)$ is a $k \times m$ matrix and is known as the *impulse response matrix*. The reason for this designation is obvious. The *ij*th element of $\mathbf{h}(t)$ is $h_{ij}(t)$, which represents the zero-state response y_i when the input $f_j(t) = \delta(t)$ and all other inputs (and all the initial conditions) are zero. It can also be seen from Eq. 6.64 and 6.89b that

$$\mathcal{L}[\mathbf{h}(t)] = \mathbf{H}(s)$$

As an example, let us consider the system described by Eqs. 6.65a and 6.65b, for this system:

$$\boldsymbol{\phi}(t) = e^{\mathbf{A}t} = \mathcal{L}^{-1}\boldsymbol{\Phi}(s)$$

This problem has been solved earlier by using frequency-domain techniques. From Eq. 6.66 we have

$$\boldsymbol{\phi}(t) = \mathcal{L}^{-1}\begin{bmatrix} \dfrac{s+3}{(s+1)(s+2)} & \dfrac{1}{(s+1)(s+2)} \\[3mm] \dfrac{-2}{(s+1)(s+2)} & \dfrac{s}{(s+1)(s+2)} \end{bmatrix}$$

$$= \mathcal{L}^{-1}\begin{bmatrix} \dfrac{2}{s+1} - \dfrac{1}{s+2} & \dfrac{1}{s+1} - \dfrac{1}{s+2} \\[3mm] \dfrac{-2}{s+1} + \dfrac{2}{s+2} & \dfrac{-1}{s+1} + \dfrac{2}{s+2} \end{bmatrix}$$

$$= \begin{bmatrix} 2e^{-t} - e^{-2t} & e^{-t} - e^{-2t} \\[2mm] -2e^{-t} + 2e^{-2t} & -e^{-t} + 2e^{-2t} \end{bmatrix}$$

Also $\boldsymbol{\delta}(t)$ is diagonal $m \times m$ or 2×2 matrix:

$$\boldsymbol{\delta}(t) = \begin{bmatrix} \delta(t) & 0 \\ 0 & \delta(t) \end{bmatrix}$$

Substitution of matrices $\boldsymbol{\phi}(t)$, $\boldsymbol{\delta}(t)$, $\mathbf{C}$, $\mathbf{D}$, and $\mathbf{B}$ (Eq. 6.65c) in Eq. 6.90, we have

$$\mathbf{h}(t) = \begin{bmatrix} 1 & 0 \\ 1 & 1 \\ 0 & 2 \end{bmatrix} \begin{bmatrix} 2e^{-t} - e^{-2t} & e^{-t} - e^{-2t} \\ -2e^{-t} + 2e^{-2t} & -e^{-t} + 2e^{-2t} \end{bmatrix} \begin{bmatrix} 1 & 0 \\ 1 & 1 \end{bmatrix} + \begin{bmatrix} 0 & 0 \\ 1 & 0 \\ 0 & 1 \end{bmatrix} \begin{bmatrix} \delta(t) & 0 \\ 0 & \delta(t) \end{bmatrix}$$

$$= \begin{bmatrix} 3e^{-t} - 2e^{-2t} & e^{-t} - e^{-2t} \\ \delta(t) + 2e^{-2t} & e^{-2t} \\ -6e^{-t} + 8e^{-2t} & \delta(t) - 2e^{-t} + 4e^{-2t} \end{bmatrix} \tag{6.91}$$

The reader can easily verify that the transfer-function matrix $\mathbf{H}(s)$ in Eq. 6.67 is the Laplace transform of the unit-impulse response matrix $\mathbf{h}(t)$ in Eq. 6.91.

6.5   SIMULATION

In this section we shall consider simulation of system equations (state equations and output equations) using integrators, summers, and scalar multipliers. To begin with, let us consider a single first-order differential equation:

$$\dot{x} = ax + bf$$

The system with this equation can be simulated as shown in Fig. 6.11. (The initial condition can be applied at N in Fig. 6.11).

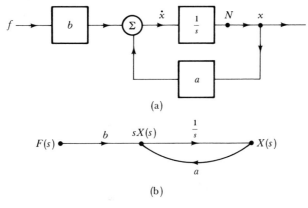

(a)

(b)

Figure 6.11

We shall next consider a system with a single input $f(t)$ and a single output $y(t)$ and 2 state variables x_1 and x_2. The system equations are

$$\dot{x}_1 = a_{11}x_1 + a_{12}x_2 + b_1 f$$
$$\dot{x}_2 = a_{21}x_1 + a_{22}x_2 + b_2 f$$

and

$$y = c_1 x_1 + c_2 x_2 + df$$

Figure 6.12a shows the block diagram of the simulated system and Fig. 6.12b is the

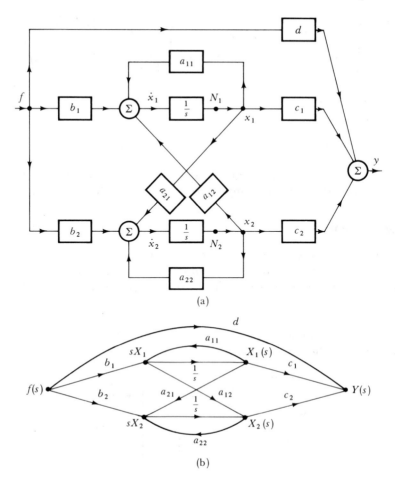

(a)

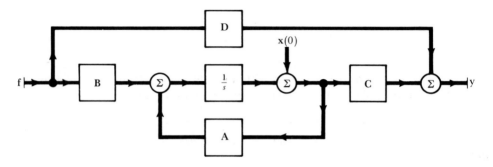

(b)

Figure 6.12

Figure 6.13

corresponding signal-flow graph. The initial conditions $x_1(0)$ and $x_2(0)$ should be applied at $\mathcal{N}_1$ and $\mathcal{N}_2$ in Fig. 6.12a (and at X_1 and X_2 in Fig. 6.12b).

This simulation procedure can be easily extended to a general multiple-input, multiple-output systems with n state variables.

The general philosophy of simulating system equations can be expressed graphically as shown in Fig. 6.13. The heavy connecting lines indicate that we are dealing with vectors in general (multiple-input, multiple-output). The system equations are

$$\dot{\mathbf{x}} = \mathbf{Ax} + \mathbf{Bf}$$
$$\mathbf{y} = \mathbf{Cx} + \mathbf{Df}$$

6.6   LINEAR TRANSFORMATION OF STATE VECTOR

In Chapter 1 we noted that the state of a system can be specified in several possible ways. This fact is compatible with our observation in Sec. 6.3 that a given system may be described by several sets of state variables. The sets of all possible state variables must necessarily be related—that is, if we are given one set of state variables we can find any other set. We are here particularly interested in a linear type of relationship. Let $x_1, x_2, \ldots, x_n$ and $w_1, w_2, \ldots, w_n$ be two different sets of state variables specifying the same system. Let these sets be related by linear equations as

$$w_1 = p_{11}x_1 + p_{12}x_2 + \cdots + p_{1n}x_n$$
$$w_2 = p_{21}x_1 + p_{22}x_2 + \cdots + p_{2n}x_n \qquad (6.92\text{a})$$
$$\cdots\cdots\cdots\cdots\cdots\cdots\cdots\cdots\cdots$$
$$w_n = p_{n1}x_1 + p_{n2}x_2 + \cdots + p_{nn}x_n$$

or

$$\underbrace{\begin{bmatrix} w_1 \\ w_2 \\ \cdots \\ w_n \end{bmatrix}}_{\mathbf{w}} = \underbrace{\begin{bmatrix} p_{11} & p_{12} & \cdots & p_{1n} \\ p_{21} & p_{22} & \cdots & p_{2n} \\ \cdots\cdots\cdots\cdots\cdots \\ p_{n1} & p_{n2} & \cdots & p_{nn} \end{bmatrix}}_{\mathbf{P}} \underbrace{\begin{bmatrix} x_1 \\ x_2 \\ \cdots \\ x_n \end{bmatrix}}_{\mathbf{x}} \qquad (6.92\text{b})$$

Defining the vector $\mathbf{w}$ and matrix $\mathbf{P}$ as shown above, we can write Eq. 6.92b as

$$\mathbf{w} = \mathbf{Px} \qquad (6.92\text{c})$$

and

$$\mathbf{x} = \mathbf{P}^{-1}\mathbf{w} \qquad (6.92\text{d})$$

Thus the state vector $\mathbf{x}$ is transformed into another state vector $\mathbf{w}$ through linear transformation in Eq. 6.92c.

It is obvious that if we know $\mathbf{w}$ we can determine $\mathbf{x}$ from Eq. 6.92d, provided

$\mathbf{P}^{-1}$ exists. This is equivalent to saying that $\mathbf{P}$ is a nonsingular matrix† ($|\mathbf{P}| \neq 0$).
Thus if $\mathbf{P}$ is a nonsingular matrix, then vector $\mathbf{w}$ defined by Eq. 6.92c is also a state
vector.

Consider the state equation of a system

$$\dot{\mathbf{x}} = \mathbf{Ax} + \mathbf{Bf} \tag{6.93a}$$

Let us define a new state vector $\mathbf{w}$:

$$\mathbf{w} = \mathbf{Px} \tag{6.93b}$$

Then

$$\mathbf{x} = \mathbf{P}^{-1}\mathbf{w}$$

and

$$\dot{\mathbf{x}} = \mathbf{P}^{-1}\dot{\mathbf{w}}$$

The state equation 6.93a now becomes

$$\mathbf{P}^{-1}\dot{\mathbf{w}} = \mathbf{AP}^{-1}\mathbf{w} + \mathbf{Bf}$$

or

$$\dot{\mathbf{w}} = \mathbf{PAP}^{-1}\mathbf{w} + \mathbf{PBf} \tag{6.93c}$$
$$= \hat{\mathbf{A}}\mathbf{w} + \hat{\mathbf{B}}\mathbf{f} \tag{6.93d}$$

where

$$\hat{\mathbf{A}} = \mathbf{PAP}^{-1} \tag{6.94a}$$

and

$$\hat{\mathbf{B}} = \mathbf{PB} \tag{6.94b}$$

Equation 6.93c is the state equation for the same system, but of the state vector $\mathbf{w}$.
The output equation is also modified. Let the original output equation be

$$\mathbf{y} = \mathbf{Cx} + \mathbf{Df}$$

In terms of the new state variable $\mathbf{w}$, this becomes

$$\mathbf{y} = \mathbf{C}(\mathbf{P}^{-1}\mathbf{w}) + \mathbf{Df}$$
$$= \hat{\mathbf{C}}\mathbf{w} + \mathbf{Df}$$

where

$$\hat{\mathbf{C}} = \mathbf{CP}^{-1}$$

Example 6.5.   State equations of a certain system are

$$\begin{bmatrix} \dot{x}_1 \\ \dot{x}_2 \end{bmatrix} = \begin{bmatrix} 0 & 1 \\ -2 & -3 \end{bmatrix} \begin{bmatrix} x_1 \\ x_2 \end{bmatrix} + \begin{bmatrix} 1 \\ 2 \end{bmatrix} f(t) \tag{6.95a}$$

Find the state equations for this system when the new state variables w_1 and w_2 are

---

†This condition is equivalent to saying that all the n equations in Eq. 6.92a are linearly
independent—that is, none of the n equations can be expressed as a linear combination of
remaining equations.

defined as

$$w_1 = x_1 + x_2$$
$$w_2 = x_1 - x_2$$

or

$$\begin{bmatrix} w_1 \\ w_2 \end{bmatrix} = \begin{bmatrix} 1 & 1 \\ 1 & -1 \end{bmatrix} \begin{bmatrix} x_1 \\ x_2 \end{bmatrix} \tag{6.95b}$$

In this problem

$$\mathbf{A} = \begin{bmatrix} 0 & 1 \\ -2 & -3 \end{bmatrix} \qquad \mathbf{P} = \begin{bmatrix} 1 & 1 \\ 1 & -1 \end{bmatrix}$$

Hence the state equation for the state variable $\mathbf{w}$ is given by

$$\dot{\mathbf{w}} = \hat{\mathbf{A}}\mathbf{w} + \hat{\mathbf{B}}\mathbf{f}$$

where (see Eq. 6.94)

$$\hat{\mathbf{A}} = \mathbf{PAP}^{-1} = \begin{bmatrix} 1 & 1 \\ 1 & -1 \end{bmatrix} \begin{bmatrix} 0 & 1 \\ -2 & -3 \end{bmatrix} \begin{bmatrix} 1 & 1 \\ 1 & -1 \end{bmatrix}^{-1}$$

$$= \begin{bmatrix} 1 & 1 \\ 1 & -1 \end{bmatrix} \begin{bmatrix} 0 & 1 \\ -2 & -3 \end{bmatrix} \begin{bmatrix} \frac{1}{2} & \frac{1}{2} \\ \frac{1}{2} & -\frac{1}{2} \end{bmatrix}$$

$$= \begin{bmatrix} -2 & 0 \\ 3 & -1 \end{bmatrix}$$

and

$$\hat{\mathbf{B}} = \mathbf{PB} = \begin{bmatrix} 1 & 1 \\ 1 & -1 \end{bmatrix} \begin{bmatrix} 1 \\ 2 \end{bmatrix} = \begin{bmatrix} 3 \\ -1 \end{bmatrix}$$

Hence

$$\begin{bmatrix} \dot{w}_1 \\ \dot{w}_2 \end{bmatrix} = \begin{bmatrix} -2 & 0 \\ 3 & -1 \end{bmatrix} \begin{bmatrix} w_1 \\ w_2 \end{bmatrix} + \begin{bmatrix} 3 \\ -1 \end{bmatrix} f(t)$$

This is the desired state equation for the state vector $\mathbf{w}$. Solution of this equation needs the knowledge of initial state $\mathbf{w}(0)$. This can be obtained from the given initial state $\mathbf{x}(0)$ by using Eq. 6.95b.

INVARIANCE OF EIGENVALUES

We have seen (Sec. 6.4) that the poles of all possible transfer functions of a system are the eigenvalues of matrix $\mathbf{A}$. If we transform state vector from $\mathbf{x}$ to $\mathbf{w}$, the variables $w_1, w_2, \ldots, w_n$ are linear combinations of $x_1, x_2, \ldots, x_n$ and therefore may be considered as outputs. Hence the poles of transfer functions relating

$w_1, w_2, \ldots, w_n$ to various inputs must also be eigenvalues of matrix $\mathbf{A}$. On the other hand, we may view the system as defined by Eq. 6.93c. This means the poles of the transfer functions must be eigenvalues of $\mathbf{A}$. It therefore follows that the eigenvalues of matrix $\mathbf{A}$ remain unchanged under linear transformation of variables represented by Eq. 6.92. Thus the eigenvalues of matrix $\mathbf{A}$ and matrix $\hat{\mathbf{A}}(\hat{\mathbf{A}} = \mathbf{PAP}^{-1})$ are identical. This implies that the characteristic equations of $\mathbf{A}$ and $\hat{\mathbf{A}}$ are also identical. This result can also be proved alternately as follows.

Consider the matrix $\mathbf{P}(s\mathbf{I} - \mathbf{A})\mathbf{P}^{-1}$. We have

$$\mathbf{P}(s\mathbf{I} - \mathbf{A})\mathbf{P}^{-1} = \mathbf{P}s\mathbf{I}\mathbf{P}^{-1} - \mathbf{PAP}^{-1} = s\mathbf{PIP}^{-1} - \hat{\mathbf{A}} = s\mathbf{I} - \hat{\mathbf{A}}$$

Taking the determinants of both sides, we obtain

$$|\mathbf{P}|\,|s\mathbf{I} - \mathbf{A}|\,|\mathbf{P}^{-1}| = |s\mathbf{I} - \hat{\mathbf{A}}|$$

The determinants $|\mathbf{P}|$ and $|\mathbf{P}^{-1}|$ are reciprocals of each other. Hence it follows that

$$|s\mathbf{I} - \mathbf{A}| = |s\mathbf{I} - \hat{\mathbf{A}}| \tag{6.96}$$

This is the desired result. We have shown that the characteristic equations of $\mathbf{A}$ and $\hat{\mathbf{A}}$ are identical. The roots of the characteristic equation are the characteristic roots or eigenvalues. Hence eigenvalues of $\mathbf{A}$ and $\hat{\mathbf{A}}$ are identical.

In Example 6.5 the matrix $\mathbf{A}$ is given as

$$\mathbf{A} = \begin{bmatrix} 0 & 1 \\ -2 & -3 \end{bmatrix}$$

The characteristic equation is

$$|s\mathbf{I} - \mathbf{A}| = \begin{vmatrix} s & -1 \\ 2 & s+3 \end{vmatrix} = s^2 + 3s + 2 = 0$$

Also

$$\hat{\mathbf{A}} = \begin{bmatrix} -2 & 0 \\ 3 & -1 \end{bmatrix}$$

and

$$|s\mathbf{I} - \hat{\mathbf{A}}| = \begin{bmatrix} s+2 & 0 \\ -3 & s+1 \end{bmatrix} = s^2 + 3s + 2 = 0$$

This verifies that the characteristic equations of $\mathbf{A}$ and $\hat{\mathbf{A}}$ are identical. Note that the eigenvalues (characteristic roots λ_1 and λ_2 are given by

$$s^2 + 3s + 2 = (s + 1)(s + 2) = 0$$

Hence

$$\lambda_1 = -1 \quad \text{and} \quad \lambda_2 = -2$$

DIAGONALIZATION OF MATRIX A

For several reasons it is very desirable to have matrix $\mathbf{A}$ to be diagonal. If $\mathbf{A}$ is not diagonal, we can transform the state variables such that the resulting matrix $\hat{\mathbf{A}}$ is diagonal. One can easily show that for any diagonal matrix $\mathbf{A}$, the diagonal elements of this matrix must necessarily be $\lambda_1, \lambda_2, \ldots, \lambda_n$ (the eigenvalues) of the matrix. Consider the diagonal matrix $\mathbf{A}$:

$$\mathbf{A} = \begin{bmatrix} a_1 & 0 & 0 & 0 \\ 0 & a_2 & 0 & 0 \\ \multicolumn{4}{c}{\cdots\cdots\cdots\cdots} \\ 0 & 0 & 0 & a_n \end{bmatrix}$$

The characteristic equation is given by

$$|\, s\mathbf{I} - \mathbf{A}\,| = \begin{bmatrix} (s - a_1) & 0 & 0 & \cdots & 0 \\ 0 & (s - a_2) & 0 & \cdots & 0 \\ \multicolumn{5}{c}{\cdots\cdots\cdots\cdots\cdots\cdots\cdots\cdots} \\ 0 & 0 & 0 & \cdots & (s - a_n) \end{bmatrix} = 0$$

or

$$(s - a_1)(s - a_2)\cdots(s - a_n) = 0$$

Hence the eigenvalues of $\mathbf{A}$ are $a_1, a_2, \ldots, a_n$. The nonzero elements (diagonal elements) of a diagonal matrix are therefore its eigenvalues $\lambda_1, \lambda_2, \ldots, \lambda_n$. We shall denote the diagonal matrix by a special symbol $\mathbf{\Lambda}$.

$$\mathbf{\Lambda} = \begin{bmatrix} \lambda_1 & 0 & 0 & \cdots & 0 \\ 0 & \lambda_2 & 0 & \cdots & 0 \\ 0 & 0 & \lambda_3 & \cdots & 0 \\ \multicolumn{5}{c}{\cdots\cdots\cdots\cdots\cdots\cdots} \\ 0 & 0 & 0 & \cdots & \lambda_n \end{bmatrix} \tag{6.97}$$

Let us now consider the transformation of state vector such that the resulting matrix $\hat{\mathbf{A}}$ is a diagonal matrix $\mathbf{\Lambda}$.

Consider the system

$$\dot{\mathbf{x}} = \mathbf{A}\mathbf{x} + \mathbf{B}\mathbf{f}$$

We shall assume that $\lambda_1, \lambda_2, \ldots, \lambda_n$, the eigenvalues of $\mathbf{A}$, are distinct (no repeated or multiple roots). Let us transform the state vector $\mathbf{x}$ to a new state vector $\mathbf{z}$ using the transformation

$$\mathbf{z} = \mathbf{P}\mathbf{x} \tag{6.98a}$$

Then, following the development of Eq. 6.93c, we have

$$\dot{\mathbf{z}} = \mathbf{P}\mathbf{A}\mathbf{P}^{-1}\mathbf{z} + \mathbf{P}\mathbf{B}\mathbf{f} \tag{6.98b}$$

We desire the transformation to be such that $\mathbf{PAP^{-1}}$ is a diagonal matrix $\mathbf{\Lambda}$ given in Eq. 6.97. Or

$$\dot{\mathbf{z}} = \mathbf{\Lambda z} + \hat{\mathbf{B}}\mathbf{f} \tag{6.98c}$$

Hence

$$\mathbf{\Lambda} = \mathbf{PAP^{-1}} \tag{6.99a}$$

or

$$\mathbf{\Lambda P} = \mathbf{PA} \tag{6.99b}$$

We know $\mathbf{\Lambda}$ and $\mathbf{A}$. Hence Eq. 6.99b can be solved to determine $\mathbf{P}$. Consider again the system in Example 6.5. In this case

$$\mathbf{A} = \begin{bmatrix} 0 & 1 \\ -2 & -3 \end{bmatrix}$$

We found $\lambda_1 = -1$ and $\lambda_2 = -2$. Hence

$$\mathbf{\Lambda} = \begin{vmatrix} -1 & 0 \\ 0 & -2 \end{vmatrix}$$

and Eq. 6.99b becomes

$$\begin{bmatrix} -1 & 0 \\ 0 & -2 \end{bmatrix} \begin{bmatrix} p_{11} & p_{12} \\ p_{21} & p_{22} \end{bmatrix} = \begin{bmatrix} p_{11} & p_{12} \\ p_{21} & p_{22} \end{bmatrix} \begin{bmatrix} 0 & 1 \\ -2 & -3 \end{bmatrix}$$

Equating the four elements on two sides, we obtain

$$-p_{11} = -2p_{12} \tag{6.100a}$$

$$-p_{12} = p_{11} - 3p_{12} \tag{6.100b}$$

$$-2p_{21} = -2p_{22} \tag{6.100c}$$

$$-2p_{22} = p_{21} - 3p_{22} \tag{6.100d}$$

The reader will immediately recognize that Eqs. 6.100a and 6.100b are identical. Similarly, Eqs. 6.100c and 6.100d are identical. Hence two equations may be discarded. This leaves us with only two equations (Eqs. 6.100a and c) in four unknowns. Does this mean there is no solution for this problem? Actually this only means that there is no unique solution. There are infinite number of solutions. We can assign any value to p_{11} and p_{21} to yield one possible solution.† If $p_{11} = k_1$ and $p_{21} = k_2$, then, from Eq. 6.100a and c, we have $p_{12} = k_1/2$ and $p_{22} = k_2$. Thus

_____

†If, however, we want state equations in the diagonalized form as in Eq. 6.55a, where all the elements of $\hat{\mathbf{B}}$ matrix are unity, there is a unique solution. This is because equation $\hat{\mathbf{B}} = \mathbf{PB}$, where all the elements of $\hat{\mathbf{B}}$ are unity, imposes additional constraints. In the present example this will yield $p_{11} = \frac{1}{2}$, $p_{12} = \frac{1}{4}$, $p_{21} = \frac{1}{3}$, and $p_{22} = \frac{1}{3}$. The relationship between $\mathbf{z}$ and $\mathbf{x}$ is

$$z_1 = \tfrac{1}{2}x_1 + \tfrac{1}{4}x_2 \quad \text{and} \quad z_2 = \tfrac{1}{3}x_1 + \tfrac{1}{3}x_2$$

$$\mathbf{P} = \begin{bmatrix} k_1 & \dfrac{k_1}{2} \\ k_2 & k_2 \end{bmatrix} \tag{6.100e}$$

We may assign any values to k_1 and k_2. For convenience let $k_1 = 2$, and $k_2 = 1$. This yields

$$\mathbf{P} = \begin{bmatrix} 2 & 1 \\ 1 & 1 \end{bmatrix} \tag{6.100f}$$

The transformed variables (Eq. 6.98a) are

$$\begin{bmatrix} z_1 \\ z_2 \end{bmatrix} = \begin{bmatrix} 2 & 1 \\ 1 & 1 \end{bmatrix} \begin{bmatrix} x_1 \\ x_2 \end{bmatrix} = \begin{bmatrix} 2x_1 + x_2 \\ x_1 + x_2 \end{bmatrix} \tag{6.101}$$

Thus the new state variables z_1 and z_2 are related to x_1 and x_2 by Eq. 6.101. The system equation with $\mathbf{z}$ as the state vector is given by (see Eq. 6.98c)

$$\dot{\mathbf{z}} = \mathbf{\Lambda z} + \hat{\mathbf{B}}\mathbf{f}$$

where

$$\hat{\mathbf{B}} = \mathbf{PB} = \begin{bmatrix} 2 & 1 \\ 1 & 1 \end{bmatrix} \begin{bmatrix} 1 \\ 2 \end{bmatrix} = \begin{bmatrix} 4 \\ 3 \end{bmatrix}$$

Hence

$$\begin{bmatrix} \dot{z}_1 \\ \dot{z}_2 \end{bmatrix} = \begin{bmatrix} -1 & 0 \\ 0 & -2 \end{bmatrix} \begin{bmatrix} z_1 \\ z_2 \end{bmatrix} + \begin{bmatrix} 4 \\ 3 \end{bmatrix} f(t) \tag{6.102a}$$

or

$$\begin{aligned} \dot{z}_1 &= -z_1 + 4f(t) \\ \dot{z}_2 &= -2z_2 + 3f(t) \end{aligned} \tag{6.102b}$$

Note the distinctive nature of these state equations. Each state equation involves only one variable and therefore can be solved by itself. A general state equation has derivative of one state variable equal to a linear combination of all state variables. Such is not the case with the diagonalized matrix $\mathbf{\Lambda}$. Each state variable z_i is so chosen that it is uncoupled from the rest of the variables. Hence a system with n eigenvalues is split into n decoupled systems each with an equation of the form

$$\begin{aligned} \dot{z}_i &= \lambda_i z_i + \text{(input terms)} \\ &= \lambda_i z_i + \zeta(t) \end{aligned}$$

This fact can also easily be seen from a simulated system. The simulated system block diagram is shown in Fig. 6.14a. The corresponding signal-flow graph is shown in Fig. 6.14b. In contrast, consider the original state equations (see

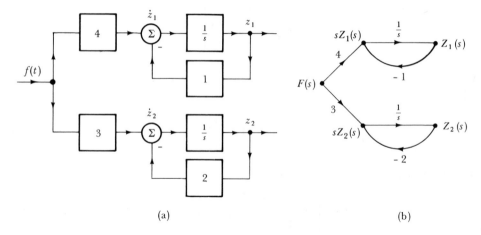

(a)   (b)

Figure 6.14

Eq. 6.95a)

$$\dot{x}_1 = x_2 + f(t)$$
$$\dot{x}_2 = -2x_1 - 3x_2 + 2f(t)$$

The simulated block diagram for these equations is shown in Fig. 6.15a.   The corresponding signal-flow graph is shown in Fig. 6.15b. It can be easily seen from Fig. 6.14, that the states z_1 and z_2 are decoupled where as states x_1 and x_2 (Fig. 6.15) are coupled. It should be remembered that Fig. 6.14 and 6.15 both are simulations of the same system.†

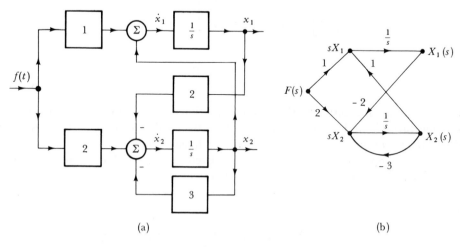

(a)   (b)

Figure 6.15

---

†Here we have only a simulated state equation. The outputs are not shown. The outputs are linear combinations of state variables (and inputs). Hence the output equation can be easily incorporated in these diagrams (see Fig. 6.12 or 6.13).

Let us turn our attention to the solution of these decoupled state equations (Eq. 6.102b). These equations are scalar first-order differential equations of the form of Eq. 2.35. Hence the solution of Eq. 6.102b is given by (see Eq. 2.63)

$$z_1 = z_1(0)e^{-t} + 4e^{-t} * f(t)$$
$$z_2 = z_2(0)e^{-2t} + 3e^{-2t} * f(t)$$

(6.103)

The initial state $z_1(0)$, $z_2(0)$ can be obtained from the given initial state $x_1(0)$, $x_2(0)$ using Eq. 6.101.

Alternately, we may use the general solution

$$\mathbf{z} = e^{\Lambda t}\mathbf{z}(0) + e^{\Lambda t} * \mathbf{f}(t)$$

(6.104)

The exponential of a diagonal matrix is very easily found. It is left as an exercise for the reader to show that $e^{\Lambda t}$ is also a diagonal matrix given by[†]

$$e^{\Lambda t} = \begin{bmatrix} e^{\lambda_1 t} & 0 & 0 & \cdots & 0 \\ 0 & e^{\lambda_2 t} & 0 & \cdots & 0 \\ \multicolumn{5}{c}{\dotfill} \\ 0 & 0 & 0 & \cdots & e^{\lambda_n t} \end{bmatrix}$$

(6.105)

Substitution of Eq. 6.105 in Eq. 6.104 yields the same result as obtained in Eq. 6.103.

6.7 STABILITY

It was shown earlier that the natural frequencies of a system are the roots of the characteristic equation (see Eq. 6.68a)

$$|s\mathbf{I} - \mathbf{A}| = 0$$

One can apply the Routh-Hurwitz criterion to the characteristic polynomial to determine the system stability.

Example 6.6.   Determine the range of the gain K for which the system in Fig. 6.16 is stable.

This system consists of three first-order subsystems. The output of each of the subsystems is therefore a state variable (Fig. 6.16). The state equations for the system can now be written by inspection of Fig. 6.16 as

$$\dot{x}_1 = x_2$$
$$\dot{x}_2 = -x_2 + K(-x_1 - x_3 + f_2)$$
$$\dot{x}_3 = -2x_3 + f_1 - (x_2 + x_3)$$

[†]This is true when all the eigenvalues $\lambda_1, \lambda_2, \ldots, \lambda_n$ are distinct. For repeated or multiple eigenvalues, the result is slightly modified.

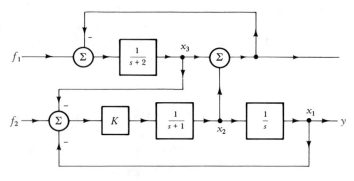

Figure 6.16

or

$$\begin{bmatrix} \dot{x}_1 \\ \dot{x}_2 \\ \dot{x}_3 \end{bmatrix} = \begin{bmatrix} 0 & 1 & 0 \\ -K & -1 & -K \\ 0 & -1 & -3 \end{bmatrix} \begin{bmatrix} x_1 \\ x_2 \\ x_3 \end{bmatrix} + \begin{bmatrix} 0 & 0 \\ 0 & K \\ 1 & 0 \end{bmatrix} \begin{bmatrix} f_1 \\ f_2 \end{bmatrix}$$

The characteristic polynomial $d(s)$ is given by

$$d(s) = |s\mathbf{I} - \mathbf{A}| = \begin{bmatrix} s & -1 & 0 \\ K & s+1 & K \\ 0 & 1 & s+3 \end{bmatrix} = s[(s+1)(s+3) - K] + K(s+3)$$

or

$$d(s) = s^3 + 4s^2 + 3s + 3K$$

The R-H array is

$$
\begin{array}{cc}
1 & 3 \\
4 & 3K \\
\dfrac{12 - 3K}{4} & 0 \\
3K & 0
\end{array}
$$

There will be no sign changes in the first column for $0 < K < 4$. Hence the system is stable only when K is in the range $(0, 4)$.

PHASE-SHIFT OSCILLATOR

A transistor phase-shift oscillator is shown in Fig. 6.17a. Using the transistor model (equivalent circuit) shown in Fig. 6.17b, the equivalent circuit for the oscillator can be drawn as shown in Fig. 6.17c. We shall use state equations to analyze this network. The state equations are obtained directly from the three-node equa-

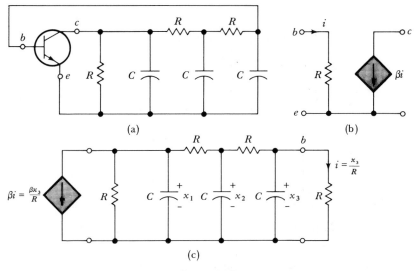

Figure 6.17

tions of this network:

$$C\dot{x}_1 + \frac{x_1}{R} + \frac{x_1 - x_2}{R} + \frac{\beta x_3}{R} = 0$$

$$C\dot{x}_2 + \frac{x_2 - x_1}{R} + \frac{x_2 - x_3}{R} = 0$$

$$C\dot{x}_3 + \frac{x_3 - x_2}{R} + \frac{x_3}{R} = 0$$

These are the state equations and can be expressed in matrix form as

$$\begin{bmatrix} \dot{x}_1 \\ \dot{x}_2 \\ \dot{x}_3 \end{bmatrix} = \begin{bmatrix} \dfrac{-2}{RC} & \dfrac{1}{RC} & \dfrac{-\beta}{RC} \\ \dfrac{1}{RC} & \dfrac{-2}{RC} & \dfrac{1}{RC} \\ 0 & \dfrac{1}{RC} & \dfrac{-2}{RC} \end{bmatrix} \begin{bmatrix} x_1 \\ x_2 \\ x_3 \end{bmatrix}$$

Letting $1/RC = a$, the **A** matrix can be written as

$$\mathbf{A} = \begin{bmatrix} -2a & a & -\beta a \\ a & -2a & a \\ 0 & a & -2a \end{bmatrix}$$

The natural frequencies of this network are the roots of the equation $|s\mathbf{I} - \mathbf{A}| = 0$,

or

$$\begin{vmatrix} s & 0 & 0 \\ 0 & s & 0 \\ 0 & 0 & s \end{vmatrix} - \begin{vmatrix} -2a & a & -\beta a \\ a & -2a & a \\ 0 & a & -2a \end{vmatrix} = 0$$

That is,

$$\begin{vmatrix} s + 2a & -a & \beta a \\ -a & s + 2a & -a \\ 0 & -a & s + 2a \end{vmatrix} = 0$$

or

$$(s + 2a)[(s + 2a)^2 - a^2] + a[-a(s + 2a) + \beta a^2] = 0$$

or

$$s^3 + 6as^2 + 10a^2 s + (\beta + 4)a^3 = 0$$

To investigate the location of roots of this equation, we use the Routh-Hurwitz array test:

$$
\begin{array}{cc}
1 & 10a^2 \\[6pt]
6a & (\beta + 4)a^3 \\[6pt]
\dfrac{(56 - \beta)a^2}{6} & 0
\end{array}
\tag{6.106}
$$

Since parameters a and β are positive, it is evident from the above array, that the system will be stable for $\beta < 56$, and will be unstable (roots in the RHP) for $\beta > 56$. For $\beta = 56$, two roots will be on the $j\omega$-axis and the network will function as an oscillator. The condition for oscillation is therefore

$$\beta \geq 56 \tag{6.107}$$

To find the frequency of oscillation, we let $\beta = 56$ in the array of Eq. 6.106. The last row vanishes and the auxiliary equation is given by the second row (for $\beta = 56$):

$$6as^2 + 60a^3 = 0$$

or

$$s^2 + 10a^2 = 0$$

$$s = \pm j\sqrt{10}a = \pm j\frac{\sqrt{10}}{RC}$$

and the frequency of oscillation ω_0 is given by

$$\omega_0 = \frac{\sqrt{10}}{RC}$$

6.8   STATE-VARIABLE FEEDBACK

We indirectly touched upon state-variable feedback in the preceding chapter. State variables contain the information about every possible output or system variable. It is therefore reasonable to expect that the system performance will be improved, if we feed back at the input not just the output but all the states. This is indeed true. The superiority of this type of feedback was demonstrated in the last chapter. We shall now give the general development of this scheme.

Let us consider an nth-order plant, and feed back all the n states, $x_1, x_2, \ldots,$ x_n to the input as shown in Fig. 6.18. There is an adjustable gain K, in the forward path. The input is r, and the plant input is f.

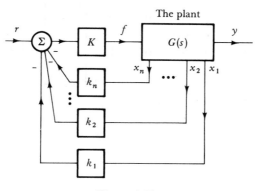

Figure 6.18

Let the plant-transfer function $G(s)$ be given by

$$\frac{Y(s)}{F(s)} = G(s) = \frac{b_m s^m + b_{m-1} s^{m-1} + \cdots + b_1 s + b_0}{s^n + a_{n-1} s^{n-1} + \cdots + a_1 s + a_0} \qquad (6.108)$$

Let us represent the plant by phase variables (see Sec. 6.3). The state equations and the output equation for the plant are

$$\dot{\mathbf{x}} = \mathbf{A}\mathbf{x} + \mathbf{B}f \qquad (6.109a)$$

and

$$\mathbf{y} = \mathbf{C}\mathbf{x} \qquad (6.109b)$$

Since we are using phase variables, these equations are (see Eq. 6.54a and 6.54b)

$$
\begin{bmatrix} \dot{x}_1 \\ \dot{x}_2 \\ \vdots \\ \dot{x}_{n-1} \\ \dot{x}_n \end{bmatrix}
=
\begin{bmatrix}
0 & 1 & 0 & 0 & 0 \\
0 & 0 & 1 & 0 & 0 \\
\vdots & \vdots & \vdots & \vdots & \vdots \\
0 & 0 & 0 & 0 & 1 \\
-a_0 & -a_1 & -a_2 & a_{n-2} & a_{n-1}
\end{bmatrix}
\begin{bmatrix} x_1 \\ x_2 \\ \vdots \\ x_{n-1} \\ x_n \end{bmatrix}
+
\begin{bmatrix} 0 \\ 0 \\ \vdots \\ 0 \\ 1 \end{bmatrix} f
\qquad (6.110a)
$$

and

$$y = [b_0 \quad b_1 \cdots b_m \quad 0 \cdots 0] \begin{bmatrix} x_1 \\ x_2 \\ \cdot \\ \cdot \\ \cdot \\ x_n \end{bmatrix} \qquad (6.110b)$$

From Fig. 6.18 it can be seen that

$$f = K[r - (k_1 x_1 + k_2 x_2 + \cdots + k_n x_n)]$$

Hence Eq. 6.110a becomes

$$\begin{bmatrix} \dot{x}_1 \\ \dot{x}_2 \\ \cdots \\ \dot{x}_{n-1} \\ \dot{x}_n \end{bmatrix} = \begin{bmatrix} 0 & 1 & 0 & \cdots & 0 & 0 \\ 0 & 0 & 1 & \cdots & 0 & 0 \\ \cdots & \cdots & \cdots & \cdots & \cdots & \cdots \\ 0 & 0 & 0 & \cdots & 0 & 1 \\ -a_0 & -a_1 & -a_2 & \cdots & a_{n-2} & a_{n-1} \end{bmatrix} \begin{bmatrix} x_1 \\ x_2 \\ \cdots \\ x_{n-1} \\ x_n \end{bmatrix} + \begin{bmatrix} 0 \\ 0 \\ \cdots \\ 0 \\ 1 \end{bmatrix} K\left(r - \sum_{i=1}^{n} k_i x_i\right)$$

or

$$\begin{bmatrix} \dot{x}_1 \\ \dot{x}_2 \\ \cdots \\ \dot{x}_{n-1} \\ \dot{x}_n \end{bmatrix} =$$

$$\begin{bmatrix} 0 & 1 & 0 & 0 & 0 \\ 0 & 0 & 1 & 0 & 0 \\ \cdots & \cdots & \cdots & \cdots & \cdots \\ 0 & 0 & 0 & 0 & 1 \\ -(a_0 + Kk_n) & -(a_1 + Kk_{n-1}) & -(a_2 + Kk_{n-2}) & -(a_{n-2} + Kk_2) & -(a_{n-1} + Kk_1) \end{bmatrix}$$

$$\begin{bmatrix} x_1 \\ x_2 \\ \cdots \\ x_{n-1} \\ x_n \end{bmatrix} + \begin{bmatrix} 0 \\ 0 \\ \vdots \\ 0 \\ 1 \end{bmatrix} Kr \quad (6.111)$$

The output y is given by Eq. 6.110b.

Equation 6.111 is the state equation and Eq. 6.100b is the output equation of the closed loop system in Fig. 6.18. These equations are in phase-variable form. Hence the transfer function $T(s)$ of the closed-loop system can be immediately written by inspection of Eqs. 6.111 and 6.110 b.

$$\frac{Y(s)}{R(s)} = T(s)$$

$$= \frac{K(b_m s^m + b_{m-1} s^{m-1} + \cdots + b_1 s + b_0)}{s^n + (a_{n-1} + Kk_n)s^{n-1} + (a_{n-2} + Kk_{n-1})s^{n-2} + \cdots + (a_1 + Kk_2)s + (a_0 + Kk_1)}$$

$$(6.112)$$

Comparison of Eq. 6.112 with Eq. 6.108 shows that the zeros of $T(s)$ and $G(s)$ are identical. The poles of $T(s)$ are functions of the feedback coefficients $k_1, k_2, \ldots, k_n$ and the gain K. By proper choice of these coefficients we can place the poles of $T(s)$ anywhere we wish. We can also use the freedom to choose $k_1, k_2, \ldots, k_n$ and K to meet the steady-state error requirements. All in all, we have $(n + 1)$ undertermined coefficients. This gives us $(n + 1)$ degrees of freedom. Since $T(s)$ has n poles, we use n degrees of freedom to position these n poles in order to obtain the desired transient performance. This leaves only one degree of freedom which may be used to fix one of the steady-state error constants.

The general development of state-variable feedback here is carried out by assuming the phase-variable representation of the system. In other words, phase variables have been fed back at the input. The general conclusions drawn here, however, are valid for any possible set of state variables. This is because all sets of state variables are linearly related. Hence if we use state variables other than phase variables, it will only cause the feedback coefficients $k_1, k_2, \ldots, k_n$ to have different values.

We shall consider one example of state-variable feedback. Consider a plant with transfer function

$$G(s) = \frac{K}{s(s + 2)(s + 4)}$$

There are three state variables which are fed back to the input as shown in Fig. 6.19. The transfer function $T(s)$ is given by (see Mason's rule)

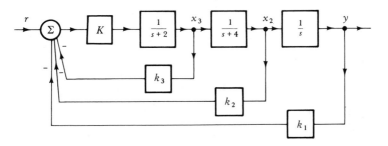

Figure 6.19

$$T(s) = \frac{K/s(s+2)(s+4)}{1 + \dfrac{Kk_3}{s+2} + \dfrac{Kk_2}{(s+2)(s+4)} + \dfrac{Kk_1}{s(s+2)(s+4)}}$$

$$= \frac{K}{s(s+2)(s+4) + Kk_3 s(s+4) + Kk_2 s + Kk_1}$$

$$= \frac{K}{s^3 + (6 + Kk_3)s^2 + (8 + 4Kk_3 + Kk_2)s + Kk_1} \qquad (6.113)$$

Let the design specifications be

$$PO \le 16\%$$
$$t_r \le 0.4$$
$$t_s \le 0.8$$
$$e_s = 0$$

If we choose dominant poles at $-5 \pm j5\sqrt{3}$, $\zeta = 0.5$, $\omega_n = 10$. Hence PO = 16%, and $t_r = \dfrac{1.62}{10} = 0.162$, $t_s = \dfrac{4}{\zeta\omega_n} = 0.8$, and the transient requirements are met. The third pole is chosen along the real axis far to the left of the $j\omega$-axis, so that it will have negligible effect on the transient. This pole should be chosen about six times the real part of the dominant poles. This gives a value of $5 \times 6 = 30$. Thus the poles of $T(s)$ are arbitrarily chosen to be $-5 \pm j5\sqrt{3}$ and -30. Hence

$$T(s) = \frac{K}{(s + 5 - j5\sqrt{3})(s + 5 + j5\sqrt{3})(s + 30)}$$

$$= \frac{K}{s^3 + 40s^2 + 400s + 3000}$$

Comparison of this equation with Eq. 6.113 yields

$$s^3 + (6 + Kk_3)s^2 + (8 + 4Kk_3 + Kk_2)s + Kk_1 = s^3 + 40s^2 + 400s + 3000 \qquad (6.114)$$

In addition, the steady-state requirement $e_s = 0$ immediately yields $k_1 = 1$ (see Eq. 5.42). Hence Eq. 6.114 yields (with $k_1 = 1$)

$$6 + Kk_3 = 40 \qquad (6.115a)$$
$$8 + 4Kk_3 + Kk_2 = 400 \qquad (6.115b)$$
$$K = 3000 \qquad (6.115c)$$

Solution of these equations yields

$$k_2 = 0.086, \qquad k_3 = 0.0113$$

Thus the feedback coefficients are $k_1 = 1$, $k_2 = 0.086$, and $k_3 = 0.0113$. The gain $K = 3000$, and

$$T(s) = \frac{3000}{s^3 + 40s^2 + 400s + 3000}$$

The steady-state error for unit ramp input can now be computed from $T(s)$ (see Eq. 5.43) as

$$e_r = \frac{400}{3000} = 0.134$$

INACCESSIBLE STATE VARIABLES

In the development of state-variable feedback we have implicitly assumed that all the state variables are accessible for the purpose of feeding back. Often certain state variables of the plant are inaccessible. In such cases it may be possible to generate inaccessible state variables from the accessible state variables. Consider the system in Fig. 6.19. Let us assume that the state variable x_2 is not accessible. We can generate x_2 from y by realizing that

$$X_2(s) = sY(s)$$

Hence x_2 can be generated by passing y through a transfer function s. The feedback signal $k_2x_2 + k_1y$ can be generated by passing y through a transfer function $(k_2s + k_1)$ as shown in Fig. 6.20. If x_3 were accessible, we may generate x_2 by passing x_3 through a transfer function $1/(s + 4)$.

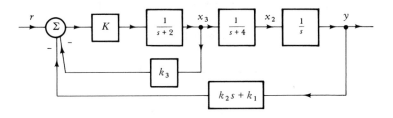

Figure 6.20

PROBLEMS

6.1.   Find the transfer-function matrix for the system shown in Fig. P-6.1. (The subsystem with transfer function $1/s - \lambda_1$ has zero input). Explain if this transfer-function matrix gives complete information about the system. From this transfer-function matrix alone can you determine the form of zero-input response y_1 and y_2? Exactly what is the nature of the information provided by this transfer-function matrix? Under what conditions does the transfer-function matrix also provide the complete information about the system?

6.2.   Convert each of the following second-order differential equations into set of two first-order differential equations (state equations). State which of the sets represent nonlinear equations.

(a) $\ddot{y} + 10\dot{y} + 2y = f(t)$
(b) $\ddot{y} + 4y = f(t)$
(c) $\ddot{y} + 2e^y\dot{y} + \log y = f(t)$
(d) $\ddot{y} + \phi_1(y)\dot{y} + \phi_2(y)y = f(t)$

6.3.   Write state equations for the force-mass system in Fig. 1.1.
6.4.   Write state equations for the network in Fig. 2.20.
6.5.   Write state equations for the auto suspension system shown in Fig. P-6.5.

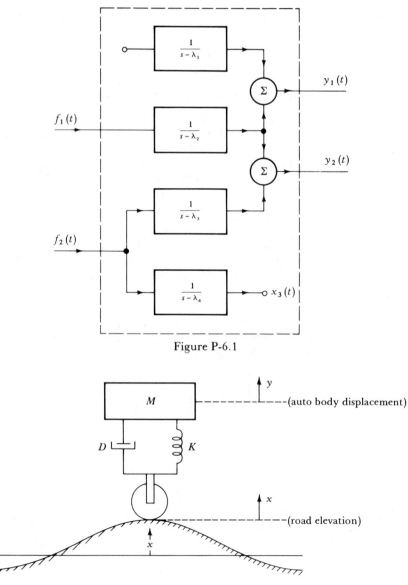

Figure P-6.1

Figure P-6.5

(*Hint:* the input in this case is the road elevation x.   Therefore x and its derivative terms should be considered as input functions which are known.)

  6.6.   (a) Write equation of motion of a pendulum of mass m and length l without any approximations or assumptions. (b) Write state equations of the pendulum in part (a). (c) Now assume small motion about the vertical and write the corresponding state equations.

  6.7.   Write state equations for the network in Fig. 2.5.

  6.8.   Write state equations for the electrical network in Fig. P-6.8.

  6.9.   Write state equations for the network in Fig. 2.6.

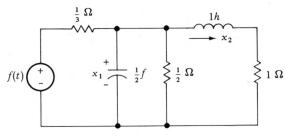

Figure P-6.8

6.10. Write state equations of the armature-controlled motor with load as shown in Fig. 1.10.

6.11. Write state equations of the field-controlled motor with load as shown in Fig. 5.36a.

6.12. Write state equations of the series motor with load as shown in Fig. P-6.12.

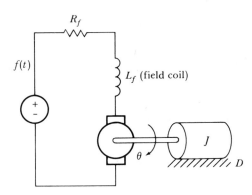

Figure P-6.12

6.13. Write state equations for the system in Fig. P-1.12.

6.14. Write state equations and the output equation of the system shown in Fig. 4.27e.

6.15. Write state equations and the output equation of the system shown in Fig. 4.28e.

6.16. Write state equations and the output equation of the system shown in Fig. 4.29a.

6.17. Write state equations and the output equation of the system shown in Fig. P-6.1.

6.18. Write state equations and the output equation of a single-axis gyro (rate gyro) whose linear model is shown in Fig. P-6.18.

6.19. Write at least three different sets of state equations and the output equation for a system shown in Fig. P-6.19 if

(a) $H(s) = \dfrac{3s + 10}{s^2 + 7s + 12}$

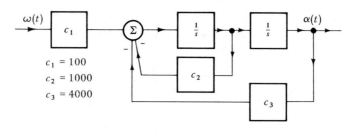

Figure P-6.18

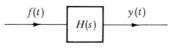

Figure P-6.19

(b) $H(s) = \dfrac{2s + 9}{s^2 + 4s + 29}$

(c) $H(s) = \dfrac{5s^3 + 32s^2 + 122s + 60}{s(s + 3)(s^2 + 4s + 20)}$

6.20.   Repeat Prob. 6.19 if

(a) $H(s) = \dfrac{4s}{(s + 1)(s + 2)^2}$

(b) $H(s) = \dfrac{4s^3 + 16s^2 + 23s + 13}{(s + 1)^3(s + 2)}$

6.21.   Transfer function $H(s)$ of a general form in Eq. 6.52a has been represented

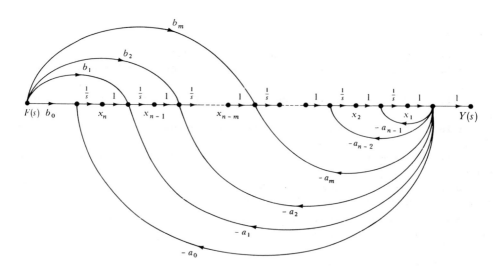

Figure P-6.21

by state equations based on simulation of $H(s)$ in the form shown in Fig. 4.26a. We can use alternate form in Fig. 4.26b to write state equations. Label the output of each integrator as a state variable in a manner shown in Fig. P-6.21. Now write state equations.

6.22. Find $e^{\mathbf{A}t}$ when the matrix $\mathbf{A}$ is given by

(a) $\begin{bmatrix} -1 & 1 \\ 0 & -2 \end{bmatrix}$ (b) $\begin{bmatrix} 0 & 2 \\ -1 & -2 \end{bmatrix}$

(c) $\begin{bmatrix} -4 & -3 \\ 1 & 0 \end{bmatrix}$ (d) $\begin{bmatrix} 4 & 3 \\ -3 & 4 \end{bmatrix}$

(e) $\begin{bmatrix} 2 & 0 & 0 \\ 0 & 1 & 0 \\ 0 & 0 & 3 \end{bmatrix}$ (f) $\begin{bmatrix} 0 & 1 & 0 \\ 0 & 0 & 1 \\ 0 & -2 & -3 \end{bmatrix}$

6.23. Find the state vector $\mathbf{x}(t)$ if

$$\dot{\mathbf{x}} = \mathbf{Ax} + \mathbf{Bf}$$

where

$$\mathbf{A} = \begin{bmatrix} 0 & 2 \\ -1 & -3 \end{bmatrix} \quad \text{and} \quad \mathbf{B} = \begin{bmatrix} 0 \\ 1 \end{bmatrix} \quad f(t) = 0$$

and

$$\mathbf{x}(0) = \begin{bmatrix} 2 \\ 1 \end{bmatrix}$$

6.24. Repeat Prob. 6.23 if

$$\mathbf{A} = \begin{bmatrix} -5 & -6 \\ 1 & 0 \end{bmatrix} \quad \mathbf{B} = \begin{bmatrix} 1 \\ 0 \end{bmatrix} \quad f(t) = \sin 100t$$

and

$$\mathbf{x}(0) = \begin{bmatrix} 5 \\ 4 \end{bmatrix}$$

Label the zero-input and zero-state components.

6.25. Repeat Prob. 6.23 if

$$\mathbf{A} = \begin{bmatrix} -2 & 0 \\ 1 & -1 \end{bmatrix} \quad \mathbf{B} = \begin{bmatrix} 1 \\ 0 \end{bmatrix} \quad f(t) = u(t)$$

$$\mathbf{x}(0) = \begin{bmatrix} 0 \\ -1 \end{bmatrix}$$

6.26.   Find the state vector $\mathbf{x}(t)$ if

$$\dot{\mathbf{x}} = \mathbf{Ax} + \mathbf{Bf}$$

where

$$\mathbf{A} = \begin{bmatrix} -1 & 1 \\ 0 & -2 \end{bmatrix} \qquad \mathbf{B} = \begin{bmatrix} 1 & 1 \\ 0 & 1 \end{bmatrix} \qquad \mathbf{f} = \begin{bmatrix} u(t) \\ \delta(t) \end{bmatrix}$$

$$\mathbf{x}(0) = \begin{bmatrix} 1 \\ 2 \end{bmatrix}$$

Label the zero-input and the zero-state components.

6.27.   Find the response y if

$$\dot{\mathbf{x}} = \mathbf{Ax} + \mathbf{Bf}(t)$$
$$y = \mathbf{Cx} + \mathbf{Df}(t)$$

where

$$\mathbf{A} = \begin{bmatrix} -3 & 1 \\ -2 & 0 \end{bmatrix} \qquad \mathbf{B} = \begin{bmatrix} 1 \\ 0 \end{bmatrix} \qquad \mathbf{C} = [0 \quad 1] \qquad \mathbf{D} = 0$$

and

$$f(t) = u(t) \qquad \mathbf{x}(0) = \begin{bmatrix} 2 \\ 0 \end{bmatrix}$$

Label the zero-input and the zero-state components.

6.28.   Repeat Prob. 6.27 if

$$\mathbf{A} = \begin{bmatrix} -1 & 1 \\ -1 & -1 \end{bmatrix} \qquad \mathbf{B} = \begin{bmatrix} 0 \\ 1 \end{bmatrix} \qquad \mathbf{C} = [1 \quad 1] \qquad \mathbf{D} = 1$$

$$f(t) = u(t) \qquad \mathbf{x}(0) = \begin{bmatrix} 2 \\ 1 \end{bmatrix}$$

6.29.   Consider the transfer function $H(s)$ in Prob. 6.19a. Write a set of state equations and the output equation for this system and verify that $H(s) = \mathbf{C}\boldsymbol{\phi}(s)\mathbf{B} + \mathbf{D}$.

6.30.   Find the transfer-function matrix $\mathbf{H}(s)$ for the system in Prob. 6.27.

6.31.   Find the transfer-function matrix $\mathbf{H}(s)$ for the system in Prob. 6.28.

6.32.   Find the transfer-function matrix $\mathbf{H}(s)$ for the system

$$\dot{\mathbf{x}} = \mathbf{Ax} + \mathbf{Bf}$$
$$y = \mathbf{Cx} + \mathbf{Df}$$

where

$$\mathbf{A} = \begin{bmatrix} 0 & 1 \\ -1 & -2 \end{bmatrix} \qquad \mathbf{B} = \begin{bmatrix} 0 & 1 \\ 1 & 0 \end{bmatrix} \qquad \mathbf{f} = \begin{bmatrix} f_1(t) \\ f_2(t) \end{bmatrix}$$

$$\mathbf{C} = \begin{bmatrix} 1 & 2 \\ 4 & 1 \\ 1 & 1 \end{bmatrix} \qquad \mathbf{D} = \begin{bmatrix} 0 & 0 \\ 0 & 0 \\ 1 & 0 \end{bmatrix}$$

6.33. Repeat Prob. 6.23 through 6.28 using the time-domain method.

6.34. Find the unit impulse response matrix $\mathbf{h}(t)$ for the system in Prob. 6.19a using Eq. 6.90.

6.35. Find the unit impulse response matrix $\mathbf{h}(t)$ for the system in Prob. 6.28.

6.36. Find the unit impulse response matrix $\mathbf{h}(t)$ for the system in Prob. 6.32.

6.37. Find the characteristic roots (eigenvalues) for each of the six matrices in Prob. 6.22.

6.38. State equations of a certain system are given as

$$\dot{x}_1 = x_2 + 2f(t)$$
$$\dot{x}_2 = -x_1 - x_2 + f(t)$$

Define a new state vector w as

$$w_1 = x_2$$
$$w_2 = x_2 - x_1$$

Find the state equations of the system with $\mathbf{w}$ as the state vector. Determine the characteristic roots (eigenvalues) of the $\mathbf{A}$ matrix in the original and the transformed state equations.

6.39. State equations of a certain system are

$$\dot{x}_1 = x_2$$
$$\dot{x}_2 = -2x_1 - 3x_2 + 2f(t)$$

(a) Determine a new state vector $\mathbf{w}$ (in terms of vector $\mathbf{x}$), so that the resulting state equations are in a diagonalized form.

(b) If the output $\mathbf{y}$ is given by

$$\mathbf{y} = \mathbf{Cx} + \mathbf{Df}$$

where

$$\mathbf{C} = \begin{bmatrix} 1 & 1 \\ -1 & 2 \end{bmatrix} \qquad \text{and} \quad \mathbf{D} = 0$$

Determine the output $\mathbf{y}$ in terms of the new state vector $\mathbf{w}$.

6.40. Given a system

$$\dot{\mathbf{x}} = \begin{bmatrix} 0 & 1 & 0 \\ 0 & 0 & 1 \\ 0 & -2 & -3 \end{bmatrix} \mathbf{x} + \begin{bmatrix} 0 \\ 0 \\ 1 \end{bmatrix} f(t)$$

Determine a new state vector $\mathbf{w}$ for this system so that the state equations are diagonalized.

6.41. State equations of a certain system are given in diagonalized form:

$$\dot{\mathbf{x}} = \begin{bmatrix} -1 & 0 & 0 \\ 0 & -3 & 0 \\ 0 & 0 & -2 \end{bmatrix} \mathbf{x} + \begin{bmatrix} 1 \\ 1 \\ 1 \end{bmatrix} f(t)$$

The output is given by

$$y = [1 \quad 3 \quad 1]\mathbf{x}$$

Determine the output y when

$$f(t) = u(t)$$

and

$$\mathbf{x}(0) = \begin{bmatrix} 1 \\ 2 \\ 1 \end{bmatrix}$$

7

Discrete-Time Systems

7.1 INTRODUCTION

So far we have discussed systems where the input and the output signals were functions of a continuous variable t. The signals associated with these systems are represented as $f(t)$, $g(t)$, ..., etc., to bring out the fact that these are functions of continuous variable t. In contrast, we have discrete-time systems where the inputs and the outputs are defined at discrete instants of time. This is the case in digital computers where the inputs and the outputs are represented typically by a sequence of numbers appearing at discrete instants t_1, t_2, ..., etc. (Fig. 7.1a). This is an example of a discrete-time signal. Since discrete-time signals are characterized by discrete-time sequences, we can denote such signals by symbols $f(t_k)$, $g(t_k)$, ..., etc., where $t_k (k = 0, \pm 1, \pm 2, ...)$ are instants at which the signals are defined. The interval between t_k, and t_{k+1} is T_k. Thus

$$T_k = t_{k+1} - t_k$$

This interval T_k may be constant or may vary with k. We shall restrict our discussion for the case where T_k is constant for all k. With this assumption $t_k = kT$, where T is the interval between successive instants. The discrete-time signals can now be described as $f(kT)$, $g(kT)$, ..., etc. Thus a discrete-time signal $f(kT)$ is a function of t and its values are defined only at discrete instants $t = 0, \pm T, \pm 2T$, ..., etc. We may further simplify our notation by using $f(k)$ instead of $f(kT)$ to represent a discrete-time signal. This notation proves more convenient in mathematical manipulation. For this reason we shall use the notation $f(k)$, $g(k)$, ..., etc., to represent discrete-time signals.† It is understood that $f(j)$ represents the value of the signal at $t = jT$ (Fig. 7.1).

The signal in Fig. 7.1b is a continuous-time signal. However, its values change only at discrete instants. It can therefore be characterized by a discrete-time sequence $f(k)$. The same is true of the signal in Fig. 7.1c. Such signals can be conveniently handled by using discrete-time techniques. We shall come across this situation later while studying sampled-data systems.

Systems whose input and output signals are discrete-time functions are dis-

---

†Although we are using the notation $f(k)$ to represent a function of t, it actually suggests a function of variable k. This latter view is more rigorous, and may be used if desired.

397

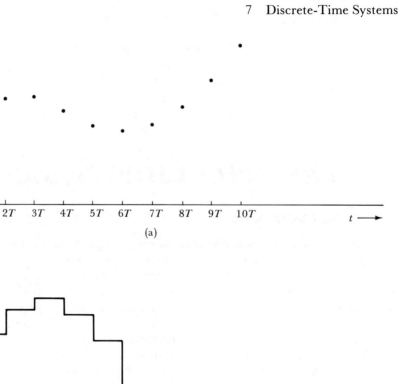

(a)

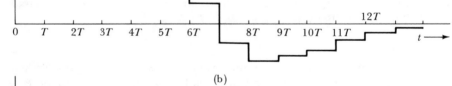

(b)

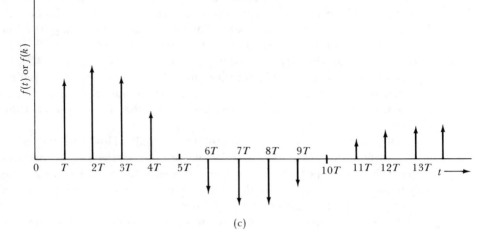

(c)

Figure 7.1

crete-time systems. A digital computer is an example of a discrete-time system. In practice discrete-time systems (such as a digital computer) are often used in conjunction with continuous-time systems. For example, in automatic control systems continuous-time plants are often controlled by digital computers. Such systems are called *hybrid systems*. We shall first develop techniques of analyzing discrete-time systems. These techniques with slight modification can also be applied to hybrid systems. For this reason we shall begin with discrete-time systems and then proceed to hybrid systems.

7.2   MODELING DISCRETE-TIME SYSTEMS

We shall give here one example of discrete-time variables. A person deposits a certain amount of money in a bank regularly. Let the deposit be $f(k)$ dollars at $t = kT$, where T may represent a certain fixed time interval, such as a month. The bank pays an interest of β dollars/dollar for a period T. The interest is compounded every month (period T). Let us determine the principal $y(k)$ at the beginning of the kth month.†

This problem can be solved by considering $y(k)$, the principal at $t = kT$. This is the sum of (i) the principal at $t = (k - 1)T$, (ii) the interest on this principal for the period T, and (iii) the deposit at $t = kT$. Hence

$$y(k) = y(k - 1) + \beta y(k - 1) + f(k)$$
$$= (1 + \beta)y(k - 1) + f(k) \tag{7.1a}$$

or

$$y(k) - (1 + \beta)y(k - 1) = f(k) \tag{7.1b}$$

This equation relates the response (the principal) at discrete instants kT, and $(k - 1)T$ to the input (the deposit) at $t = kT$. Such an equation is known as a *difference equation*. Equation 7.1 is a first-order difference equation. It is interesting to compare a first-order difference equation 7.1a to a first-order differential equation

$$\dot{y} = ay + f \tag{7.2}$$

Since

$$\dot{y} \simeq \frac{y(k) - y(k - 1)}{T}$$

provided T is small, we can express Eq. 7.2 as

$$\frac{y(k) - y(k - 1)}{T} \simeq ay(k) + f(k)$$

or

$$y(k) = \frac{1}{1 - aT} y(k - 1) + \frac{T}{1 - aT} f(k) \tag{7.3}$$

---

†We shall assume that the deposit is made at $t = kT^-$ (that is, at the instant just preceding $t = kT$).

Compare this equation with Eq. 7.1a. It is apparent that a differential equation can be approximated by a difference equation. The approximation improves as T becomes smaller. The well-known Runge-Kutta procedures in digital computer work do indeed use difference equations to solve differential equations.

Before going into a general theory of difference equations, let us consider the possibility of simulating a system described (or modeled) by a difference equation. We used integrators to simulate differential equations. The key element used in simulating difference equation is a *delayor*. This is a device which delays the signal by T seconds. If $f(k)$ and $y(k)$ represent the input and the output of a

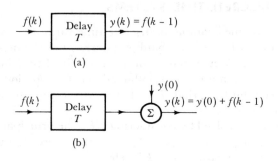

(a)

(b)

Figure 7.2

delayor (Fig. 7.2a), then

$$y(k) = f(k - 1) \qquad (7.4)$$

The initial condition on the output can be incorporated as shown in Fig. 7.2b.

The time delay can be achieved by using a system which stores the input data and releases it T seconds later. This can be accomplished by a system with memory. Digital computers can be conveniently used for this purpose.

Consider a first-order difference equation

$$y(k + 1) - ay(k) = f(k) \qquad (7.5a)$$

This is a linear difference equation because it satisfies the linearity conditions discussed in Chapter 1. Equation 7.5a can be expressed as

$$y(k + 1) = ay(k) + f(k) \qquad (7.5b)$$

This equation can be simulated by the arrangement shown in Fig. 7.3. The figure is self-explanatory.

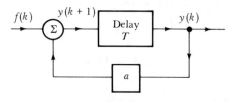

Figure 7.3

So far the development of discrete-time systems appears to parallel that of continuous-time systems. We are therefore encouraged to follow the same procedure in development as we did for continuous-time systems. We wonder

    (i) If a concept of transfer operator (or transfer function) is also applicable to linear† discrete-time systems.

    (ii) If discrete-time systems can be analyzed by time-domain as well as frequency-domain (transform) techniques.

    (iii) If concepts of system state and state-space analysis can be applied to discrete-time systems.

The answer is affirmative to all these questions. In all these respects discrete-time systems follow closely the development of continuous-time systems. We shall now consider each of these problems in detail.

7.3   TRANSFER OPERATOR OF A DISCRETE-TIME SYSTEM

Transfer operator concept applies to linear system only when they are time-invariant. Similarly, the transfer operator concept is meaningful only for linear time-invariant discrete-time systems. If the coefficients of the terms $y(k)$ in difference equation are constants, the difference equation is said to be *time-invariant*.

In continuous-time systems we used an operator p to denote the operator of differentiation. For discrete-time systems we shall use an operator E to denote the operation for advancing the sequence by one time interval. Thus

$$E[f(k)] = f(k + 1)$$
$$E^2[f(k)] = f(k + 2)$$
$$\cdots\cdots\cdots\cdots\cdots \tag{7.6}$$
$$E^n[f(k)] = f(k + n)$$

Using this notation, the first-order difference equation 7.5 can be expressed as

$$E[y(k)] - ay(k) = f(k)$$

or

$$(E - a)y(k) = f(k) \tag{7.7}$$

Following is an example of a second-order difference equation:

$$y(k + 2) + a_1 y(k + 1) + a_0 y(k) = b_1 f(k + 1) + b_0 f(k)$$

Using operational notation, this equation can be expressed as

$$(E^2 + a_1 E + a_0)y(k) = (b_1 E + b_0)f(k) \tag{7.8}$$

A general nth-order difference equation is given by

$$y(k + n) + a_{n-1}y(k + n - 1) + \cdots + a_1 y(k + 1) + a_0 y(k) = b_m f(k + m)$$
$$+ b_{m-1} f(k + m - 1) + \cdots + b_1 f(k + 1) + b_0 f(k) \tag{7.9a}$$

---

†Linear time-invariant systems.

Using operational notation, this general nth-order difference equation can be expressed as

$$(E^n + a_{n-1}E^{n-1} + \cdots + a_1 E + a_0)y(k)$$
$$= (b_m E^m + b_{m-1}E^{m-1} + \cdots + b_1 E + b_0)f(k) \quad (7.9b)$$

For a time-invariant system, coefficients a_i's and b_i's are constant. It must never be forgotten that Eq. 7.9b is not an algebraic equation but is an operational equation.

We shall define the inverse operator as follows. If $D(E)$ is a polynomial in operator E, then the equation

$$y(k) = \frac{1}{D(E)} [f(k)] \qquad (7.10a)$$

by definition means

$$D(E)y(k) = f(k) \qquad (7.10b)$$

and the equation

$$y(k) = \frac{N(E)}{D(E)} [f(k)] \qquad (7.11a)$$

by definition means

$$D(E)y(k) = N(E)f(k) \qquad (7.11b)$$

As a result, Eq. 7.9 can be expressed as

$$y(k) = \frac{b_m E^m + b_{m-1}E^{m-1} + \cdots + b_1 E + b_0}{E^n + a_{n-1}E^{n-1} + \cdots + a_1 E + a_0} f(k) \qquad (7.12)$$

We shall now define the *transfer operator,*† $H(E)$, relating output $y(k)$ to input $f(k)$ as

$$y(k) = H(E)f(k) \qquad (7.13)$$

The reader is again reminded that this is not an algebraic equation but an operational equation. The transfer operator $H(E)$ operates on the input sequence $f(k)$ to yield the output sequence $y(k)$.

We can represent a discrete-time system by a block diagram labeled with the corresponding transfer operator $H(E)$ as shown in Fig. 7.4a.

To find the transfer operator for the compound-interest system (Eq. 7.1), we shift Eq. 7.1b forward by T seconds. This yields

$$y(k + 1) - (1 + \beta)y(k) = f(k + 1)$$

or

$$[E - (1 + \beta)]y(k) = Ef(k)$$

and

$$y(k) = \frac{E}{E - (1 + \beta)} f(k) \qquad (7.14a)$$

---

†This is also known as a *pulse transfer operator.*

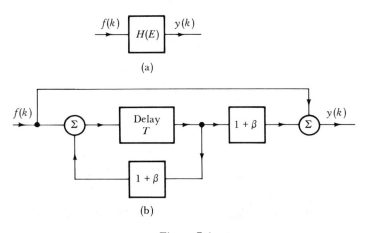

Figure 7.4

Hence the transfer operator $H(E)$ relating the response $y(k)$ to the input $f(k)$ is given by

$$H(E) = \frac{E}{E - (1 + \beta)} \tag{7.14b}$$

The arrangement in Fig. 7.3 simulates Eq. 7.5 given by

$$(E - a)y(k) = f(k)$$

or

$$y(k) = \frac{1}{E - a} f(k)$$

Hence the transfer operator of the system in Fig. 7.3 is given by

$$H(E) = \frac{1}{E - a} \tag{7.15}$$

To simulate the compound interest system (Eq. 7.14b), we note that the transfer operator for this system can be expressed as

$$H(E) = 1 + \frac{(1 + \beta)}{E - (1 + \beta)}$$

This transfer operator is given by 1 plus $(1 + \beta)$ times the transfer operator of the form in Eq. 7.15. It can therefore be simulated as shown in Fig. 7.4b.

Note that the operator $1/E$ represents a delayor. This can be easily seen by considering a system with transfer operator $1/E$ (Fig. 7.5). The output $y(k)$ and

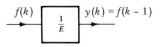

Figure 7.5

the input $f(k)$ are related by

$$y(k) = \frac{1}{E} f(k)$$

or

$$Ey(k) = f(k)$$

Therefore

$$y(k + 1) = f(k)$$

and

$$y(k) = f(k - 1)$$

It is evident that the operator $1/E$ implies a delay of T seconds.

7.4   TIME-DOMAIN ANALYSIS OF DISCRETE-TIME SYSTEMS

Let us consider a system in Fig. 7.6. The output $y(k)$ and the input $f(k)$ of this system are related by difference equation

$$(E^n + a_{n-1}E^{n-1} + \cdots + a_1 E + a_0)y(k) = (b_m E^m + b_{m-1}E^{m-1} + \cdots + b_0)f(k) \tag{7.16}$$

This is a linear difference equation with constant coefficients. As with linear differential equation, the solution of this difference equation also has two components— the zero-input component and the zero-state component.

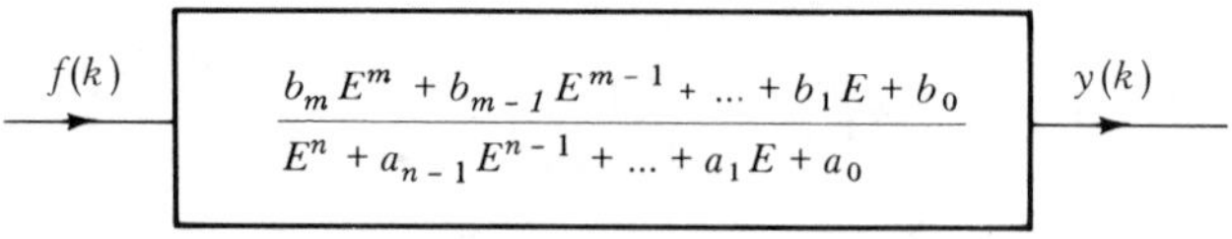

Figure 7.6

ZERO-INPUT COMPONENT

The zero-input component is the solution of Eq. 7.16 with $f(k) = 0$. Thus the zero-input response $y(k)$ is the solution of

$$(E^n + a_{n-1}E^{n-1} + \cdots + a_1 E + a_0)y(k) = 0 \tag{7.17}$$

The procedure for solving this difference equation is similar to that used for the differential equation in Chapter 2 (Eq. 2.33a). The laws that applied to the differential operator p also apply to the shift operator E. Let us first consider the first-order difference equation

$$(E - \gamma)y(k) = 0 \tag{7.18}$$

or

$$Ey(k) = \gamma y(k)$$

or

$$y(k + 1) = \gamma y(k)$$

therefore

$$\frac{y(k + 1)}{y(k)} = \gamma$$

The ratio $y(k + 1)$ to $y(k)$ is γ. This means that the sequence $y(k)$ is a geometric progression with a common ratio γ. Hence $y(k)$ must be of the form

$$y(k) = c\gamma^k \tag{7.19a}$$

where c is an arbitrary constant which may be determined from the initial condition. Letting $k = 0$ in Eq. 7.19a, we have

$$y(0) = c$$

Hence

$$y(k) = y(0)\gamma^k \tag{7.19b}$$

This is the solution of the first-order difference equation 7.18. To find the solution of nth-order difference equation 7.17 we express the operator polynomial in E in a factored form as

$$(E - \gamma_1)(E - \gamma_2)\cdots(E - \gamma_n)y(k) = 0 \tag{7.20}$$

Using argument similar to that used for the differential equation (Eq. 2.45), we conclude that Eq. 7.20 is satisfied by

$$(E - \gamma_r)y(k) = 0 \qquad r = 1, 2, \ldots, n \tag{7.21}$$

and the solution of Eq. 7.20 is given by

$$y(k) = c_1\gamma_1^k + c_2\gamma_2^k + \cdots + c_n\gamma_n^k \tag{7.22}$$

where the constants $c_1, c_2, \ldots, c_n$ may be determined from initial conditions such as $y(0), y(1), \ldots, y(n - 1)$, as shown below.
    From Eq. 7.22,

$$y(0) = c_1 + c_2 + \cdots + c_n$$
$$y(1) = c_1\gamma_1 + c_2\gamma_2 + \cdots + c_n\gamma_n$$
$$y(2) = c_1\gamma_1^2 + c_2\gamma_2^2 + \cdots + c_n\gamma_n^2$$
$$y(n - 1) = c_1\gamma_1^{n-1} + c_2\gamma_2^{n-1} + \cdots + c_n\gamma_n^{n-1}$$

Hence

$$
\begin{bmatrix} y(0) \\ y(1) \\ y(2) \\ \cdots \\ y(n-1) \end{bmatrix}
=
\begin{bmatrix} 1 & 1 & \cdots & 1 \\ \gamma_1 & \gamma_2 & \cdots & \gamma_n \\ \gamma_1^2 & \gamma_2^2 & \cdots & \gamma_n^2 \\ \cdots & & & \cdots \\ \gamma_1^{n-1} & \gamma_2^{n-1} & \cdots & \gamma_n^{n-1} \end{bmatrix}
\begin{bmatrix} c_1 \\ c_2 \\ c_3 \\ \cdots \\ c_n \end{bmatrix}
$$

and

$$
\begin{bmatrix} c_1 \\ c_2 \\ c_3 \\ \cdots \\ c_n \end{bmatrix} = \begin{bmatrix} 1 & 1 & \cdots & 1 \\ \gamma_1 & \gamma_2 & \cdots & \gamma_n \\ \gamma_1^2 & \gamma_2^2 & \cdots & \gamma_n^2 \\ \cdots\cdots\cdots\cdots\cdots\cdots\cdots \\ \gamma_1^{n-1} & \gamma_2^{n-1} & \cdots & \gamma_n^{n-1} \end{bmatrix}^{-1} \begin{bmatrix} y(0) \\ y(1) \\ y(2) \\ \cdots \\ y(n-1) \end{bmatrix} \tag{7.23}
$$

For the case of multiple roots,

$$
(E - \gamma_1)'(E - \gamma_{r+1})\cdots(E - \gamma_n)y(k) = 0
$$

it can be shown that the solution is given by

$$
y(k) = \gamma_1^k(c_1 + c_2 k + \cdots + c_r k^{r-1}) + c_{r+1}\gamma_{r+1}^k + \cdots + c_n\gamma_n^k \tag{7.24}
$$

The arbitrary constants $c_1, c_2, \ldots, c_n$ can be determined from initial conditions.

NATURAL RESPONSE AND THE STABILITY

The zero-input response (Eq. 7.22) is a natural response of the system. It is instructive to study the form of the natural response of the system. A pole at γ of the transfer operator gives rise to the term $c\gamma^k$ in the natural response. If $|\gamma| < 1$ then γ^k decays with time (decays with k). Similarly, if $|\gamma| > 1$ then γ^k grows with time.

The stability concept discussed in Chapter 4 for continuous time systems can also be applied to discrete-time systems. If the zero-input response eventually vanishes, the system is asymptotically stable. If it grows with time, the system is unstable. If it does not vanish but remains within a finite bound, the system is marginally stable. From the above discussion it is obvious that a discrete-time system is asymptotically stable if all the poles of the transfer operator (roots of characteristic equation) have a magnitude less than unity. This is equivalent to saying that all poles lie within a circle centered at the origin and with unit radius. If any of the poles lies outside this unit circle, the system is unstable. If at least one of the poles lies on this unit circle and all the remaining poles lie inside the circle, the system is marginally stable. If a characteristic equation is in unfactored form, we can determine the system stability by a method similar to Routh-Hurwitz test. This test, known as the Jury test, indicates the number of roots that lie outside the unit circle.[†]

The form of natural response γ^k can be studied more conveniently if we let

$$
\gamma = e^{\lambda T} \tag{7.25a}
$$

so that

$$
\lambda = \frac{1}{T}\ln\gamma \tag{7.25b}
$$

[†]E. I. Jury, *Theory and Application of the Z-Transform Method* (New York: Wiley, 1964).

and

$$\gamma^k = e^{\lambda kT}$$

From Eq. 7.25 it follows that if γ is positive and real, then λ is real and if γ is negative real, or complex, λ is complex. In this case let

$$\lambda = \alpha + j\beta \tag{7.26}$$

Then

$$\gamma^k = e^{(\alpha+j\beta)kT} = e^{\alpha kT} (\cos \beta kT + j \sin \beta kT) \tag{7.27}$$

Both the real and the imaginary components of the natural response are oscillatory functions with exponentially growing or decaying amplitudes, depending upon whether $\alpha > 0$ or < 0 (or $|\gamma| >$ or < 1).

It is also evident that

$$\text{Oscillation frequency} = \beta \tag{7.28}$$

Various pole locations and corresponding natural responses are shown in Fig. 7.7a. As γ moves radially, the frequency of oscillation β remains unchanged but α varies, indicating the variation in the rate of change of amplitude with time. As γ moves toward the origin, α decreases and the amplitude decays faster with time. Similarly, if γ moves along a circle of constant radius, α remains constant but β varies, indicating variation in frequency of oscillation. This is clearly shown in Fig. 7.7b and c.

Discussion on resonance (Sec. 2.5) is also applicable to discrete-time systems. As $\alpha \to 0$, the system will exhibit strong resonance characteristics at radian frequency β. The system response will be large to discrete time input of the form $\cos (\beta kT + \theta)$. For $\alpha = 0$, the response goes to infinity. Thus

$$\text{Resonance frequency} \to \beta \quad \text{as} \quad \alpha \to 0 \tag{7.29}$$

Example 7.1. Find the zero-input response of a system with transfer operator

$$H(E) = \frac{E(7E - 2)}{(E - 0.5)(E - 0.2)}$$

when the initial conditions (initial state) are given as $y(0) = 2$ and $y(1) = 4$.

The zero-input response of this system can be immediately written as

$$y(k) = c_1 (0.5)^k + c_2 (0.2)^k$$

where c_1 and c_2 are given by (see Eq. 7.23)

$$\begin{bmatrix} c_1 \\ c_2 \end{bmatrix} = \begin{bmatrix} 1 & 1 \\ 0.5 & 0.2 \end{bmatrix}^{-1} \begin{bmatrix} 2 \\ 4 \end{bmatrix} = \begin{bmatrix} \dfrac{-2}{3} & \dfrac{10}{3} \\ \dfrac{5}{3} & \dfrac{-10}{3} \end{bmatrix} \begin{bmatrix} 2 \\ 4 \end{bmatrix} = \begin{bmatrix} 12 \\ -10 \end{bmatrix}$$

and

$$y(k) = 12(0.5)^k - 10(0.2)^k \tag{7.30}$$

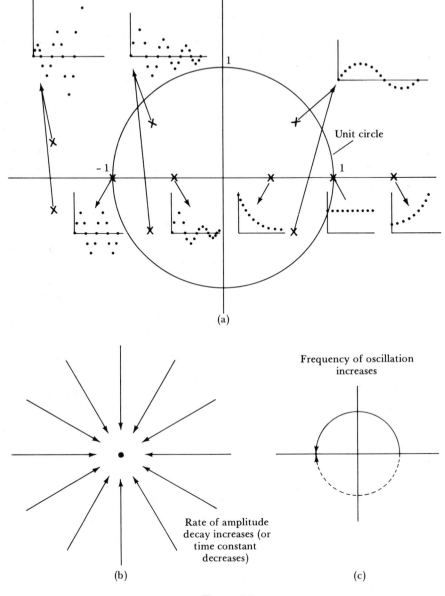

(a)

(b)   Figure 7.7   (c)

It is more convenient to express powers to the natural base e. This is done by observing that

$$0.5 = e^{-0.69} \quad \text{and} \quad 0.2 = e^{-1.6}$$

Hence

$$y(k) = 12e^{-0.69k} - 10e^{-1.6k}$$

ZERO-STATE COMPONENT

The zero-state component in a linear continuous-time system is given by a convolution integral. For discrete-time systems the zero-state component is given by a summation whose structure is similar to convolution integral. This summation is called *convolution summation*.

Let us define a discrete-time function $\partial(k)$ which is counterpart of $\delta(t)$ in continuous-time systems. We shall define the unit function $\partial(k)$:

$$\partial(k) = \begin{cases} 1 & k = 0 \\ 0 & k \neq 0 \end{cases} \tag{7.31}$$

Any discrete-time function $f(k)$ can always be expressed in terms of unit function $\partial(k)$:

$$f(k) = f(0)\partial(k) + f(1)\partial(k - 1) + f(2)\partial(k - 2) + \cdots \tag{7.32}$$

Let the zero-state response of the system to unit input $\partial(k)$ be $h(k)$. For a linear time-invariant system, the response of the system to input $c\partial(k - j)$ will obviously be c times $h(k - j)$. That is,

$$\partial(k) \rightarrow h(k)$$
$$c\partial(k - j) \rightarrow ch(k - j)$$

Using this result, $y(k)$, the zero-state response to $f(k)$ in Eq. 7.32 can immediately be written as

$$y(k) = f(0)h(k) + f(1)h(k - 1) + f(2)h(k - 2) + \cdots \tag{7.33}$$

Note that since $\partial(k)$ begins at $k = 0$, $h(k)$ must be zero at for $k < 0$ for physical systems. Obviously $h(k - j) = 0$ for $j > k$. Consequently, we need to carry the summation 7.33 over $k + 1$ terms only. Thus

$$y(k) = f(0)h(k) + f(1)h(k - 1) + f(2)h(k - 2) + \cdots + f(k)h(0)$$

$$= \sum_{j=0}^{k} h(k - j)f(j) \tag{7.34a}$$

If we change the summation index from j to $k - j$, we have

$$y(k) = \sum_{j=0}^{k} f(k - j)h(j) \tag{7.34b}$$

The two sums in Eqs. 7.34a and 7.34b are equivalent and are called *convolution summation*.† We shall use the notation‡

---

†The reader should compare this equation with Eq. 2.98, which is also a convolution sum representing the approximate response of a continuous-time system.

‡The limits on the convolution summation are $j = -\infty$ to ∞ for a general case where both $f_1(k)$ and $f_2(k)$ exist for k over the entire range. However, when $f_1(k)$ and $f_2(k)$ are zero for negative values of k, the limits can be replaced by $j = 0$ to k. This situation is similar to that observed for the convolution integral.

$$f_1(k) * f_2(k) = \sum_{j=0}^{k} f_1(j)f_2(k-j) \tag{7.35}$$

Using this notation, the complete solution $y(k)$ can be expressed as

$$y(k) = \sum_{j=0}^{k} c_j\gamma_j^k + f(k) * h(k) \tag{7.36}$$

The next step is to obtain $h(k)$, the zero-state response of the system to unit input $\partial(k)$.

If we were to follow a procedure similar to that used in differential equations, we should begin with equation

$$y(k) = \frac{1}{E - \gamma} f(k) \tag{7.37}$$

and find $h(k)$ for this. Higher-order transfer operator $H(E)$ can then be expanded by partial fractions of the form Eq. 7.37. The response $h(k)$ will be the sum of the responses corresponding to each partial fraction. This procedure will work. However, in discrete-time systems it proves more convenient to begin with a system of the form

$$y(k) = \frac{E}{E - \gamma} f(k) \tag{7.38}$$

or

$$(E - \gamma)y(k) = Ef(k)$$

and

$$y(k + 1) = \gamma y(k) + f(k + 1) \tag{7.39}$$

This equation can be solved as follows.

Let $k = -1$ in Eq. 7.39. This yields

$$y(0) = \gamma y(-1) + f(0) \tag{7.40}$$

In our discussion we shall assume causal inputs, $f(k) = 0$ for $k < 0$. Hence $y(k)$ is also causal. Consequently $y(-1) = 0$, and Eq. 7.40 becomes

$$y(0) = f(0) \tag{7.41}$$

Letting $k = 0$ in Eq. 7.39 and using Eq. 7.41, we obtain

$$y(1) = \gamma f(0) + f(1) \tag{7.42}$$

Similarly, by letting $k = 1, 2, \ldots$, etc., we obtain

$$y(2) = \gamma^2 f(0) + \gamma f(1) + f(2)$$
$$y(3) = \gamma^3 f(0) + \gamma^2 f(1) + \gamma f(2) + f(3)$$
$$\cdots \cdots \cdots \cdots \cdots \cdots \cdots \cdots \cdots \cdots$$
$$y(k) = \gamma^k f(0) + \gamma^{k-1} f(1) + \cdots + \gamma f(k-1) + f(k) \tag{7.43}$$

For the input $f(k) = \partial(k)$,

$$f(0) = 1 \qquad f(1) = f(2) = f(3) = \cdots = f(k) = 0$$

and

$$y(k) = h(k) = \gamma^k \tag{7.44}$$

Thus $h(k)$, the unit-function response of the system in Eq. 7.38, is γ^k. Note, that this is a zero-state component. The zero-input component (Eq. 7.19b) is given by $y(0)\gamma^k$. We can use this result to obtain $h(k)$ for the nth-order system:

$$y(k) = \frac{b_m E^m + b_{m-1} E^{m-1} + \cdots + b_1 E + b_0}{E^n + a_{n-1} E^{n-1} + \cdots + a_1 E + a_0} f(k) \tag{7.45a}$$

$$= \frac{b_m E^m + b_{m-1} E^{m-1} + \cdots + b_1 E + b_0}{(E - \gamma_1)(E - \gamma_2) \cdots (E - \gamma_n)} f(k) \tag{7.45b}$$

Partial-fraction expansion of $H(E)$ is permissible for reasons similar to those used for expanding $H(p)$ by partial fractions. We want to expand $H(E)$ by factors of the form $cE/E - \gamma$. This can always be done (see Example 7.3) by expanding $H(k)/E$ rather than $H(k)$. This yields

$$y(k) = \left[k_1 \frac{E}{E - \gamma_1} + k_2 \frac{E}{E - \gamma_2} + k_n \frac{E}{E - \gamma_n} \right] f(k) \tag{7.45c}$$

and

$$h(k) = k_1 \gamma_1^k + k_2 \gamma_2^k + \cdots + k_n \gamma_n^k$$

$$= \sum_{j=1}^{n} k_j \gamma_j^k \tag{7.46}$$

and the complete solution is given by

$$y(k) = \underbrace{\sum_{j=1}^{n} c_j \gamma_j^k}_{\substack{\text{zero-input} \\ \text{response}}} + \underbrace{\sum_{j=0}^{n} h(k-j) f(j)}_{\substack{\text{zero-state} \\ \text{response}}} \tag{7.47}$$

where the c_j's may be determined from initial conditions (Eq. 7.23). Table 7.1 lists pairs of $H(E)$ and corresponding $h(k)$. For a physical system, $h(k) = 0$, for $k < 0$. This is implied in Table 7.1. It is helpful to define a function $u(k)$ as

$$u(k) = \begin{cases} 1 & k \geq 0 \\ 0 & k < 0 \end{cases} \tag{7.48}$$

In Table 7.1 it is understood that all $h(k)$ are multiplied by $u(k)$.

Example 7.2.   Let us solve the compound-interest problem discussed earlier. In this case $y(k)$, and $f(k)$, the principal and the deposit respectively at the beginning of the kth month are related by (see Eq. 7.14)

$$y(k) = \frac{E}{E - (1 + \beta)} f(k)$$

where β is the monthly interest per dollar. The solution of this equation can be

TABLE 7.1

	$H(E)$	$h(k)$
1a	$\dfrac{1}{E - \gamma}$	$\gamma^{k-1}u(k - 1)$
1b	$\dfrac{1}{E - e^{\lambda T}}$	$e^{\lambda(k-1)T}u(k-1)$
2a	$\dfrac{E}{E - \gamma}$	γ^{k}
2b	$\dfrac{E}{E - e^{\lambda T}}$	$e^{\lambda kT}$
3	$c\dfrac{E}{E - \gamma} + c^*\dfrac{E}{E - \gamma^*}$ $c = re^{j\theta},\ \gamma = e^{(\alpha+j\beta)T}$	$2re^{\alpha kT}\cos(\beta kT + \theta)$
4a	$\dfrac{E}{(E - \gamma)^2}$	$k\gamma^{k-1}$
4b	$\dfrac{E}{(E - e^{\lambda T})^2}$	$ke^{\lambda(k-1)T}$
5a	$\dfrac{E}{(E - \gamma)^n}$	$\dfrac{1}{n!}k(k-1)(k-2)\cdots(k-n+1)\gamma^{k-n+1}$
5b	$\dfrac{E}{(E - e^{\lambda T})^n}$	$\dfrac{1}{n!}k(k-1)(k-2)\cdots(k-n+1)e^{\lambda(k-n+1)T}$

immediately written as

$$y(k) = y(0)(1 + \beta)^k + h(k) * f(k)$$

where $y(0)$ is the initial principal and

$$h(k) = (1 + \beta)^k$$

Hence

$$y(k) = y(0)(1 + \beta)^k + (1 + \beta)^k * f(k)$$

$$= y(0)(1 + \beta)^k + \sum_{j=0}^{k}(1 + \beta)^{k-j}f(j)$$

Let us assume a monthly deposit to be a fixed amount D,—that is, $f(k) = Du(k)$. This yields

$$y(k) = y(0)(1 + \beta)^k + D\sum_{j=0}^{k}(1 + \beta)^{k-j}$$

$$= \underbrace{y(0)(1 + \beta)^k}_{\substack{\text{zero-input}\\\text{component}}} + \underbrace{D[(1 + \beta)^k + (1 + \beta)^{k-1} + \cdots + (1 + \beta) + 1]}_{\text{zero-state component}}$$

The terms inside the bracket (zero-state component) represents a geometrical progression with common ratio $(1 + \beta)$ and the last term $(1 + \beta)^k$. Hence

$$y(k) = y(0)(1 + \beta)^k + D\frac{(1 + \beta)^{k+1} - 1}{(1 + \beta) - 1}$$

$$= (1 + \beta)^k y(0) + \frac{(1 + \beta)^{k+1} - 1}{\beta}D$$

For a case where $y(0) = 100$ and $D = 100$, $\beta = 0.005$, $y(k)$ becomes

$$y(k) = 100(1.005)^k + 20{,}000[(1.005)^{k+1} - 1]$$

This function is plotted in Fig. 7.8.

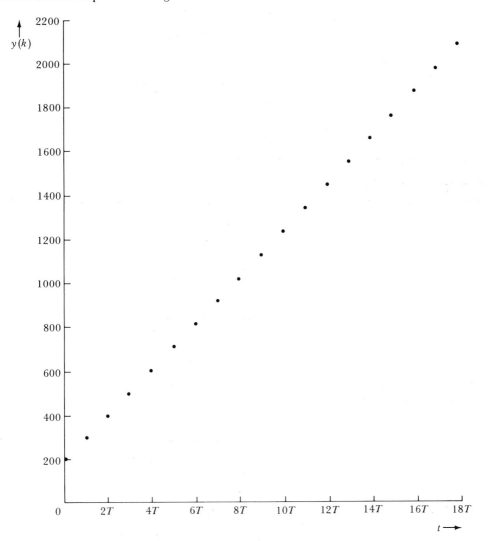

Figure 7.8

CONVOLUTION SUMMATION TABLES

For continuous-time systems, we had prepared a table in which convolution integral was listed for several pairs of functions. Analysis of discrete-data system is greatly facilitated by such a table for discrete summation. It will avoid much of the headache associated with finding a close-form expression for the series such as the one encountered in Example 7.2. Table 7.2 lists convolution summation in the close form for several discrete-time function pairs.

In Example 7.2, $h(k) = (1 + \beta)^k$ and $f(k) = Du(k)$. Hence the zero-state response is given by

$$y_f(k) = h(k) * f(k)$$
$$= (1 + \beta)^k * Du(k)$$
$$= D[(1 + \beta)^k * u(k)$$

From Table 7.2 (pair 2a), we have

$$y_f(k) = D\left\{\frac{1}{1 - (1 + \beta)} [1 - (1 + \beta)^{k+1}]\right\}$$
$$= D\frac{(1 + \beta)^{k+1} - 1}{\beta}$$

which is the same result as obtained earlier.

Example 7.3. Find the response of the system in Example 7.1, when the input is $u(k)$ and the initial conditions are $y(0) = 2$ and $y(1) = 4$. For this system

$$\frac{H(E)}{E} = \frac{7E - 2}{(E - 0.5)(E - 0.2)}$$
$$= \frac{5}{E - 0.5} + \frac{2}{E - 0.2}$$

and

$$H(E) = 5\frac{E}{E - 0.5} + 2\frac{E}{E - 0.2}$$

Hence

$$h(k) = 5(0.5)^k + 2(0.2)^k$$

The zero-input response $y_x(k)$ is already determined in Example 7.1 (Eq. 7.30). The zero-state response is given by

$$y_f(k) = h(k) * u(k)$$
$$= [5(0.5)^k + 2(0.2)^k] * u(k)$$
$$= 5(0.5)^k * u(k) + 2(0.2)^k * u(k)$$

From Table 7.2 (pair 2a), we obtain

$$y_f(k) = \frac{5}{0.5}(1 - 0.5^{k+1}) + \frac{2}{0.8}(1 - 0.2^{k+1})$$
$$= 10[1 - 0.5(0.5)^k] + 2.5[1 - 0.2(0.2)^k]$$
$$= 12.5 - 5(0.5)^k - 0.5(0.2)^k$$

TABLE 7.2
Convolution Summation

No.	$f_1(k)$	$f_2(k)$	$f_1(k) * f_2(k) = f_2(k) * f_1(k)$
1	$\partial(k)$	$f(k)$	$f(k)$
2a	γ^k	$u(k)$	$(1 - \gamma^{k+1})/(1 - \gamma)$
2b	$e^{\lambda kT}$	$u(k)$	$[1 - e^{\lambda(k+1)T}]/(1 - e^{\lambda T})$
3	$u(k)$	$u(k)$	$k + 1$
4a	$(\gamma_1)^k$	$(\gamma_2)^k$	$[(\gamma_1)^{k+1} - (\gamma_2)^{k+1}]/(\gamma_1 - \gamma_2) \qquad \gamma_1 \neq \gamma_2$
4b	$e^{\lambda_1 kT}$	$e^{\lambda_2 kT}$	$[e^{\lambda_1(k+1)T} - e^{\lambda_2(k+1)T}]/(e^{\lambda_1 T} - e^{\lambda_2 T}) \qquad \lambda_1 \neq \lambda_2$
5a	γ^k	γ^k	$(k + 1)\gamma^k$
5b	$e^{\lambda kT}$	$e^{\lambda kT}$	$(k + 1)e^{\lambda kT}$
6a	γ^k	k	$\gamma\left[\gamma^k - 1 + \dfrac{k}{1 - \gamma}\right](\gamma - 1)$
6b	$e^{\lambda kT}$	k	$e^{\lambda T}\left[e^{\lambda kT} - 1 + \dfrac{k}{1 - e^{\lambda T}}\right](e^{\lambda T} - 1)$
7	k	k	$\frac{1}{3}(k - 1)k(k + 1)$
8	$e^{\alpha kT}\cos(\beta kT + \theta)$	$e^{\lambda kT}$	$\dfrac{e^{\alpha(k+1)T}\cos[\beta(k + 1)T + \theta - \phi] - e^{\lambda(k+1)T}\cos(\theta - \phi)}{\sqrt{e^{2\alpha T} + e^{2\lambda T} - 2e^{(\alpha+\lambda)T}\cos\beta T}}$

where

$$\phi = \tan^{-1}[e^{\alpha T}\sin\beta T/(e^{\alpha T}\cos\beta T - e^{\lambda T})]$$

The complete response is given by $y_x(k) + y_f(k)$, where $y_x(k)$ is found in Eq. 7.30. Hence

$$y(k) = 12.5 + 7(0.5)^k - 10.5(0.2)^k \tag{7.49}$$

7.5   THE z-TRANSFORM

Linear discrete-time (and hybrid) systems can also be analyzed by transform method analogous to the transform method used in analysis of linear continuous-time systems. The Laplace (or Fourier) transform converts differential equations into algebraic equations. The particular kind of transform (known as the z-transform) used for discrete-time systems converts difference equations into algebraic equations.

For a discrete-time sequence $f(k)$, we define $F(z)$, the z-transform of $f(k)$, as

$$F(z) = \sum_{k=0}^{\infty} f(k)z^{-k} \tag{7.50}$$

The summation in Eq. 7.50 is carried out over all values of k. If the sequence $f(k)$ exists for k in the range $-\infty$ to ∞ the summation is carried out for $k = -\infty$ to ∞. This is the *two-sided* or the *bilateral* z-transform. For causal sequences (sequences beginning at some finite time say $t = 0$), the sequence exists only for $k \geq 0$. Hence the summation 7.50 is carried out for $k = 0$ to ∞. This is the *unilateral* z-transform. The term z-transform in our future discussion will mean unilateral z-transform unless otherwise stated. Symbolically the z-transform of $f(k)$ will be denoted by $\mathfrak{z}[f(k)]$. Thus

$$\mathfrak{z}[f(k)] = F(z) = \sum_{k=0}^{\infty} f(k)z^{-k}$$

and

$$\mathfrak{z}[u(k)] = \sum_{k=0}^{\infty} (1)\, z^{-k}$$

$$= 1 + \left(\frac{1}{z}\right) + \left(\frac{1}{z}\right)^2 + \left(\frac{1}{z}\right)^3 + \cdots \tag{7.51}$$

We recall that

$$\frac{1}{1-x} = 1 + x + x^2 + x^3 + \cdots \text{for } |x| < 1$$

Hence the series on the right-hand side of Eq. 7.51 can be expressed in close form as

$$\mathfrak{z}[u(k)] = \frac{1}{1-\dfrac{1}{z}} = \frac{z}{z-1}, \quad |z| > 1$$

The condition $|z| > 1$ represents the region of convergence. In a similar way we find

$$\mathfrak{z}[e^{\lambda kT}] = \sum_{k=0}^{\infty} e^{\lambda kT} z^{-k}$$

$$= 1 + \left(\frac{e^{\lambda T}}{z}\right) + \left(\frac{e^{\lambda T}}{z}\right)^2 + \left(\frac{e^{\lambda T}}{z}\right)^3 + \cdots$$

$$= \frac{1}{1-\dfrac{e^{\lambda T}}{z}} = \frac{z}{z-e^{\lambda T}} \qquad |z| > e^{\lambda T}$$

Table 7.3 lists $f(k)$ and the corresponding $F(z)$ for several discrete-time sequences $f(k)$. All the sequences $f(k)$ are causal—that is $f(k) = 0$ for $k < 0$. To be more rigorous, we should multiply every $f(k)$ in Table 7.3 by $u(k)$. Thus for pair 3, for example, $f(k)$ is $ku(k)$, and so on. The inverse z-transform of $z/z - e^{\lambda T}$ in pair 7b, should be read as $e^{\lambda kT}u(k)$, and so on.

    Compare Table 7.3 with Table 7.1. It is evident that $h(k)$ is inverse z-transform of $H(z)$. We shall later prove this result rigorously. Thus

$$\mathfrak{z}[h(k)] = H(z) \tag{7.52a}$$

and

$$h(k) = \mathfrak{z}^{-1}[H(z)] \tag{7.52b}$$

TABLE 7.3

No.	$f(k), f(k) = 0, k < 0$	$F(z)$
1	$\partial(k)$	1
2	$u(k)$	$\dfrac{z}{z-1}$
3	k	$\dfrac{z}{(z-1)^2}$
4	k^2	$\dfrac{z(z+1)}{(z-1)^3}$
5	k^3	$\dfrac{z(z^2+4z+1)}{(z-1)^4}$
6a	$\gamma^{k-1}u(k-1)$	$\dfrac{1}{z-\gamma}$
6b	$e^{\lambda(k-1)T}u(k-1)$	$\dfrac{1}{z-e^{\lambda T}}$
7a	γ^k	$\dfrac{z}{z-\gamma}$
7b	$e^{\lambda kT}$	$\dfrac{z}{z-e^{\lambda T}}$
8a	$\cos\beta kT$	$\dfrac{z(z-\cos\beta T)}{z^2-2z\cos\beta T+1}$
8b	$\sin\beta kT$	$\dfrac{z\cos\beta T}{z^2-2z\cos\beta T+1}$
9a	$k\gamma^{k-1}$	$\dfrac{z}{(z-\gamma)^2}$
9b	$ke^{\lambda(k-1)T}$	$\dfrac{z}{(z-e^{\lambda T})^2}$
10a	$e^{\alpha kT}\cos\beta kT$	$\dfrac{z(z-e^{\alpha T}\cos\beta T)}{z^2-2ze^{\alpha T}\cos\beta T+e^{2\beta T}}$
10b	$e^{\alpha kT}\sin\beta kT$	$\dfrac{ze^{\alpha T}\sin\beta T}{z^2-2ze^{\alpha T}\cos\beta T+e^{2\beta T}}$
11	$2re^{\alpha kT}\cos(\beta kT+\theta)$	$\dfrac{cz}{z-\gamma}+\dfrac{c^*z}{z-\gamma^*}$ $c=re^{j\theta},\ \gamma=e^{(\alpha+j\beta)T}$
12a	$\dfrac{1}{n!}k(k-1)(k-2)\cdots(k-n+1)\gamma^{k-n+1}$	$\dfrac{z}{(z-\gamma)^n}$
12b	$\dfrac{1}{n!}k(k-1)(k-2)\cdots(k-n+1)e^{\lambda(k-n+1)T}$	$\dfrac{z}{(z-e^{\lambda T})^n}$

We shall use the double-arrow notation used for Fourier and Laplace transforms. Thus a z-transform pair will be represented as

$$f(k) \leftrightarrow F(z)$$

7.6   SOME PROPERTIES OF THE z-TRANSFORM

1.   LINEARITY PROPERTY

If

$$f(k) \leftrightarrow F(z)$$
$$g(k) \leftrightarrow G(z)$$

then

$$c_1 f(k) + c_2 g(k) \leftrightarrow c_1 F(z) + c_2 G(z) \qquad (7.53)$$

The proof is trivial. This result can be extended to more than two functions.

2.   TIME-SHIFTING THEOREM

If

$$f(k) \leftrightarrow F(z)$$

then

$$f(k + 1) \leftrightarrow z[F(z) - f(0)]$$

Proof.

$$\mathfrak{z}[f(k + 1)] = \sum_{k=0}^{\infty} f(k + 1)z^{-k}$$

$$= z \sum_{k=0}^{\infty} f(k + 1)z^{-(k+1)}$$

$$= z \sum_{j=1}^{\infty} f(j)z^{-j}$$

$$= z \left[\sum_{j=0}^{\infty} f(j)z^{-j} - f(0) \right]$$

$$= z[F(z) - f(0)] \qquad (7.54a)$$

We can easily extend this result as

$$f(k + 2) \leftrightarrow z^2 F(z) - z^2 f(0) - z f(1) \qquad (7.54b)$$

and

$$f(k + n) \leftrightarrow z^n F(z) - z^n \sum_{k=0}^{n-1} f(k)z^{-k} \qquad (7.54c)$$

Thus

$$E[f(k)] \leftrightarrow z[F(z) - f(0)] \qquad (7.55a)$$
$$E^2[f(k)] \leftrightarrow z^2 F(z) - z^2 f(0) - z f(1) \qquad (7.55b)$$

and

$$E^n [f(k)] \leftrightarrow z^n F(z) - z^n \sum_{k=0}^{n-1} f(k)z^{-k} \qquad (7.55c)$$

Similarly, we can show that

$$f(k - 1) \leftrightarrow \frac{1}{z} F(z) \qquad (7.56a)$$

and

$$f(k - n) \leftrightarrow z^{-n} F(z) \qquad (7.56b)$$

It should be remembered that we are dealing with causal sequences. Hence $f(k) = 0$ for $k < 0$ and $f(k - 1)$ is zero for $k < 1$. The sequence $f(k - 1)$ begins at $k = 1$. In order to avoid inadvertent error caused by this aspect Eqs. 7.56 should be expressed as

$$f(k - 1)u(k - 1) \leftrightarrow \frac{1}{z} F(z) \qquad (7.56c)$$

$$f(k - n)u(k - n) \leftrightarrow z^{-n} F(z) \qquad (7.56d)$$

3. CONVOLUTION PROPERTY

If

$$f_1(k) \leftrightarrow F_1(z)$$
$$f_2(k) \leftrightarrow F_2(z)$$

then

$$f_1(k)* f_2(k) \leftrightarrow F_1(z)F_2(z)$$

Proof. By definition,

$$\mathfrak{z}\left[\sum_{j=0}^{\infty} f_1(j)f_2(k - j)\right] = \sum_{k=0}^{\infty} z^{-k} \sum_{j=0}^{\infty} f_1(j) f_2(k - j)$$

Interchanging the order of summation on the right-hand side, we obtain

$$\mathfrak{z}\left[\sum_{j=0}^{\infty} f_1(j)f_2(k - j)\right] = \sum_{j=0}^{\infty} f_1(j) \sum_{k=0}^{\infty} z^{-k} f_2(k - j)$$

The second summation on the right-hand side is seen from Eq. 7.56b to be $z^{-j} F_2(z)$. Hence

$$\mathfrak{z}\left[\sum_{j=0}^{\infty} f_1(j)f_2(k - j)\right] = \sum_{j=0}^{\infty} f_1(j)z^{-j} F_2(z) \qquad (7.57)$$

$$= F_1(z)F_2(z)$$

Note that $f_1(k)$ and $f_2(k)$ are both assumed to be causal sequences. Hence these sequences have zero values for negative arguments. Therefore $f_2(k - j) = 0$ when $j > k$. Hence the upper limit of the summation on the left-hand side of Eq. 7.57b may be changed to $j = k$. Thus

$$\sum_{j=0}^{k} f_1(j)f_2(k-j) \leftrightarrow F_1(z)F_2(z)$$

or

$$f_1(k) * f_2(k) = f_2(k) * f_1(k) \leftrightarrow F_1(z)F_2(z)$$

7.7   TRANSFORM ANALYSIS OF DISCRETE-TIME SYSTEMS

Consider a discrete-time system with transfer operator

$$H(E) = \frac{b_m E^m + b_{m-1}E^{m-1} + \cdots + b_0}{E^n + a_{n-1}E^{n-1} + \cdots + a_0}$$

The response $y(k)$ and the input $f(k)$ are therefore related by

$$(E^n + a_{n-1}E^{n-1} + \cdots + a_0)\, y(k) = (b_m E^m + b_{m-1}E^{m-1} + \cdots + b_0)f(k)$$

To take the z-transform of this equation, we use the time-shifting property (Eq. 7.55). Assuming all initial conditions to be zero, we obtain

$$(z^n + a_{n-1}z^{n-1} + \cdots + a_0)Y(z) = (b_m z^m + b_{m-1}z^{m-1} + \cdots + b_0)F(z) \qquad (7.58)$$

and

$$Y(z) = H(z)F(z) \qquad (7.59a)$$

also

$$y(k) = \mathfrak{z}^{-1}[H(z)F(z)] \qquad (7.59b)$$

This is the zero-state component of the response. If the initial state (initial conditions) were not zero, we would have additional terms in Eq. 7.58 arising due to nonzero initial conditions. The response due to those terms would be zero-input response.

Let us consider a system with input $\partial(k)$. Because $\mathfrak{z}[\partial(k)] = 1$, the output which is by definition $h(k)$, is given by (Eq. 7.59b).

$$h(k) = \mathfrak{z}^{-1}[H(z)]$$

and

$$h(k) \leftrightarrow H(z)$$

Note that from convolution property, we can write Eq. 7.59b as

$$y(k) = h(k) * f(k)$$

We have already derived this result independently.

TRANSFER FUNCTION OF IDEAL DELAY

An ideal delay of T seconds has the property of delaying input by T seconds. If the input is $f(k)$, the output is $f(k-1)u(k-1)$. Hence

$$Y(z) = \mathfrak{z}[f(k-1)]$$

From Eq. 7.56c we have

$$Y(z) = \frac{1}{z} F(z)$$

Hence

$$H(z) = \frac{Y(z)}{F(z)} = \frac{1}{z}$$

Hence the transfer functions of an ideal delay of one time interval T seconds is $1/z$.

We now demonstrate the application of z-transform in analyzing discrete-time systems.

Example 7.4. The z-transfer function of a system is given by

$$H(z) = \frac{z(7z - 2)}{(z - 0.2)(z - 0.5)} \qquad (7.60a)$$

The input $f(k) = u(k)$ and the initial conditions are $y(0) = 2$, $y(1) = 4$. Find the output $y(k)$.

The zero-input component of $y_x(k)$ may be found as in Example 7.1:

$$y_x(k) = 12e^{-0.69k} - 10e^{-1.6k}$$

and the zero-state component is found by using Eq. 7.59. Since $u(k) \leftrightarrow z/z - 1$,

$$Y_f(z) = \left(\frac{z(7z - 2)}{(z - 0.2)(z - 0.5)}\right)\left(\frac{z}{z - 1}\right) \qquad (7.60b)$$

$$\frac{Y_f(z)}{z} = \frac{z(7z - 2)}{(z - 1)(z - 0.2)(z - 0.5)}$$

$$\equiv \frac{12.5}{z - 1} + \frac{-0.5}{z - 0.2} + \frac{-5}{z - 0.5}$$

and

$$Y_f(z) = 12.5 \frac{z}{z - 1} - 0.5 \frac{z}{z - 0.2} - 5 \frac{z}{z - 0.5}$$

Hence from Table 7.3 (pair 7) we have

$$y_f(k) = 12.5 - 0.5(0.2)^k - 5(0.5)^k \qquad (7.61a)$$

The complete response is obtained by adding $y_x(k)$ in Eq. 7.25a and $y_f(k)$ in Eq. 7.61a). Hence

$$y(k) = 12.5 + 7(0.5)^k - 10.5(0.2)^k \qquad (7.61b)$$

This is the same result as obtained earlier (see Eq. 7.49).

Alternate Procedure

We may solve this problem alternately as follows. Given

$$y(k) = \frac{E(7E - 2)}{(E - 0.2)(E - 0.5)} f(k)$$

$$= \frac{E(7E - 0.2)}{E^2 - 0.7E + 0.1} f(k)$$

and

$$(E^2 - 0.7E + 0.1)y(k) = (7E^2 - 2E)f(k)$$

We can solve this difference equation by taking its z-transform of both sides and using Eq. 7.55 and substituting initial conditions.

7.8   INVERSE z-TRANSFORM

As in the case of other transforms, extensive tables of z-transform pairs are available. The inverse z-transform may be obtained from these tables. In general, the inverse z-transform can be effected by one of the following methods (1) partial-fraction expansion, (2) power series expansion, and (3) inversion integral. Partial-fraction expansion method has already been demonstrated. We shall now discuss the power series expansion method.

POWER-SERIES-EXPANSION METHOD

This method is based upon the fact that $F(z)$ by definition is really a power series in z. Thus

$$F(z) = \sum_{k=0}^{\infty} f(k)z^{-k}$$

$$= f(0) + \frac{f(1)}{z} + \frac{f(2)}{z^2} + \cdots + \cdots$$

It is obvious from this equation that if $F(z)$ is expressed as a power series in z, then the coefficient of the term z^{-k} is $f(k)$. Hence expansion of $F(z)$ in power series in z yields directly the discrete-time function $f(k)$. If

$$F(z) = \frac{z^m + a_{m-1}z^{m-1} + \cdots + a_1z + a_0}{b_nz^n + b_{n-1}z^{n-1} + \cdots + b_1z + b_0}$$

then we can expand $F(z)$ in power series by long division as

$$F(z) = Az^{m-n} + Bz^{m-n-1} + \cdots$$

where

$$A = f(n - m), B = f(n - m + 1), \ldots, \text{etc.}$$

As an example, consider $Y(z)$ in Eq. 7.60b:

$$Y(z) = \frac{z^2(7z - 2)}{(z - 0.2)(z - 0.5)(z - 1)}$$

$$= \frac{7z^3 - 2z^2}{z^3 - 1.7z^2 + .8z - 0.1}$$

Using long division, we can write

$$Y(z) = 7 + 9.9z^{-1} + 11.2z^{-2} + \cdots +$$

Hence

$$y(0) = 7, \qquad y(1) = 9.9, y(2) = 11.2, \ldots, \text{etc.}$$

Compare this solution with the close form solution in Eq. 7.61a.

INVERSION INTEGRAL

Those who are familiar with theory of complex variables will recognize $F(z)$ defined in Eq. 7.50 as a Laurent series. The inverse z-transform can therefore be immediately written as

$$f(k) = \frac{1}{2\pi j} \int_{\Gamma} F(z) z^{n-1} \, dz$$

where the complex integral is performed around the circle which lies in the region of convergence of $F(z)$. For more discussion on this reader may refer to Jury or a text on theory of complex variables.†

7.9   SIMULATION OF DISCRETE-TIME SYSTEMS

The transfer function of a linear time-invariant discrete-time system or a discrete-time model of a sampled-data system is of the form

$$H(z) = \frac{b_m z^m + b_{m-1} z^{m-1} + \cdots + b_1 z + b_0}{z^n + a_{n-1} z^{n-1} + \cdots + a_1 z + a_0} \qquad (7.62)$$

This is exactly of the same form as that of a general transfer function $H(s)$ of a linear continuous-time system with variable z replacing the variable s. We can therefore simulate $H(z)$ in the same manner as $H(s)$ with z replacing s. We can use the signal-flow graphs and block diagrams to represent discrete-time model and use the techniques developed in Sec. 4.14 to realize $H(z)$. In general, we may use series, parallel, and Kalman's forms to simulate $H(z)$.

In a continuous-time system the key element in simulation is an integrator which has a transfer function $1/s$. It is obvious that for a discrete-time transfer function the key element in simulation must have a transfer function $1/z$. As seen earlier (Fig. 7.5), this is an ideal delay of T seconds. Consider a first-order transfer function

$$\frac{Y(z)}{F(z)} = \frac{1}{z + a} \qquad (7.63)$$

This transfer function is realized by the signal-flow graph shown in Fig. 7.9a. The corresponding block diagram is shown in Fig. 7.9b.

We can now simulate a general transfer function in Eq. 7.62 by series, parallel, or cascade (Kalman's forms) realizations. We shall demonstrate these realizations by an example.

---

†E. I. Jury, *Theory and Application of the z-Transform Method* (New York: Wiley, 1964).

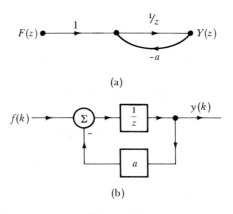

(a)

(b)

Figure 7.9

Example 7.5. Simulate a discrete-time model whose transfer function is given by

$$T(z) = \frac{0.368z + 0.264}{(z - 1)(z - 0.368)} \qquad (7.64a)$$

$$= \left(\frac{1}{z - 1}\right)\left(\frac{0.368z + 0.264}{z - 0.368}\right) \qquad (7.64b)$$

$$= \frac{1}{z - 1} - \frac{0.635}{z - 0.368} \qquad (7.64c)$$

$$= \frac{0.368z + 0.264}{z^2 - 1.368z + 0.368} \qquad (7.64d)$$

These three forms (Eqs. 7.64b, c, and d) are suitable for use in simulation of $T(z)$ by series, parallel, and Kalman's forms respectively. Using the techniques developed in Sec. 4.14, the appropriate forms are realized in Fig. 7.10. Both, signal-flow graphs and block diagrams are shown in Fig. 7.10.

Note that the basic element used in these simulations is a delay element. A time delay can be obtained by using a device which has a memory or which can store and then release the information. A digital computer can be conveniently used for this purpose.

7.10 DIGITAL PROCESSING OF ANALOG SIGNALS

Digital systems can process only discrete-time signals. However, a continuous time signal can be converted into a discrete-time signal by using the principle of sampling. The sample values of a continuous signal $f(t)$, represent a discrete-time signal $f(k)$. This conversion is known as *analog-to-digital* (A/D) *conversion*. We can now process $f(k)$ using discrete-time systems. The output $y(k)$ will be a discrete-time signal and may be considered as samples of an appropriate continuous-time signal $y(t)$. We can construct $y(t)$ from $y(k)$ by the principle discussed in Chapter 3. This conversion is known as *digital-to-analog*

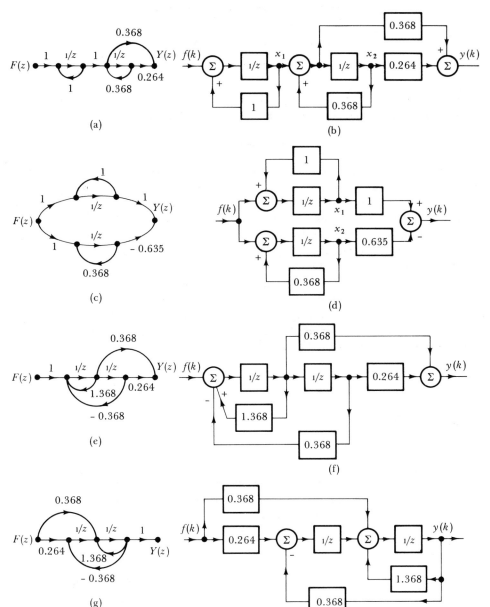

Figure 7.10

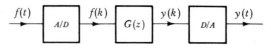

Figure 7.11

(*D/A*) *conversion*. Thus using A/D and D/A converters, we can process continuous-time signals using discrete-time systems as shown in Fig. 7.11.

CONVERTERS

The A/D converter converts analog signals into digital signals. This is done in three steps: (i) sampling, (ii) quantizing, and (iii) coding. The analog signal is first sampled at a rate equal to or greater than the Nyquist rate. The samples are in the form of narrow pulses. Next, the sample values are quantized and then coded into binary numbers as discussed in Sec. 3.18 (Fig. 3.34). Quantizing and coding are required when the discrete-time system $G(z)$ in Fig. 7.11 is a digital computer. Since the computers use discrete-data in binary form, signal samples cannot be fed directly but must be quantized and coded (Fig. 7.12a).

The D/A converter (Fig. 7.12b) converts the discrete data, generally a digital computer output, into an analog signal. This can be done by converting the binary sequence into sequence of impulses. We can construct the appropriate continuous-time signal from the samples (sequence of impulses) by passing the

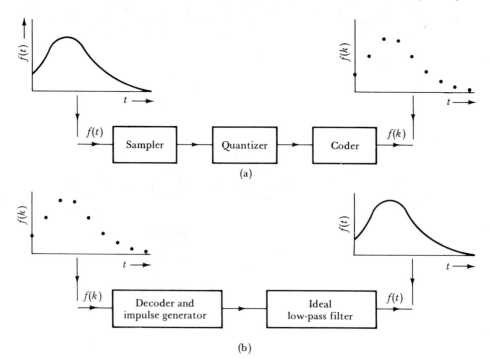

Figure 7.12

samples through an ideal low-pass filter as discussed in Sec. 3.17. Ideal filters, in general, are too expensive. Hence in practice nonideal filters such as a zero-order hold circuit or a first-order hold circuit (discussed in Sec. 7.12) are commonly used.

The system in Fig. 7.11 processes a continuous-time signals. We shall now find the transfer function of this system. For a linear system when the input is e^{st}, the output contains a term $H(s)e^{st}$, where $H(s)$ is the transfer function. We shall use this observation to determine the transfer function of the system in Fig. 7.11. Let the input $f(t) = e^{st}$. The output of the A/D converter will be $f(k) = e^{skT}$, where T is the sampling interval. This discrete-time sequence is now fed to the discrete-time system $G(z)$. The output $Y(z)$ is given by

$$Y(z) = G(z)\,\mathfrak{Z}[e^{skT}]$$
$$= \frac{zG(z)}{z - e^{sT}}$$

and

$$\frac{Y(z)}{z} = \frac{G(z)}{z - e^{sT}}$$

Let

$$G(z) = \frac{N(z)}{(z - \gamma_1)(z - \gamma_2)\cdots(z - \gamma_n)} \tag{7.65}$$

Then

$$\frac{Y(z)}{z} = \frac{N(z)}{(z - \gamma_1)(z - \gamma_2)\cdots(z - \gamma_n)(z - e^{sT})}$$
$$= \frac{c_1}{z - \gamma_1} + \frac{c_2}{z - \gamma_2} + \cdots + \frac{c_n}{z - \gamma_n} + \frac{a}{z - e^{sT}}$$

Note that the coefficient a is given by

$$a = \frac{N(z)}{(z - \gamma_1)(z - \gamma_2)\cdots(z - \gamma_n)}\Big|_{z = e^{sT}}$$
$$= G(z)\big|_{z = e^{sT}}$$
$$= G(e^{sT})$$

Hence

$$Y(z) = c_1\frac{z}{z - \gamma_1} + c_2\frac{z}{z - \gamma_2} + \cdots + c_n\frac{z}{z - \gamma_n} + G(e^{sT})\frac{z}{z - e^{sT}}$$

and

$$y(k) = c_1(\gamma_1)^k + c_2(\gamma_2)^k + \cdots + c_n(\gamma_n)^k + G(e^{sT})e^{skT} \tag{7.66}$$

This signal $y(k)$ is now passed through the D/A converter. The last term in $y(k)$ (Eq. 7.66) will give rise to the term $G(e^{sT})e^{st}$ in the output $y(t)$. This term is $H(s)e^{st}$. Hence

$$H(s) = G(e^{sT}) = G(z)\big|_{z = e^{sT}} \tag{7.67}$$

This is a very important result. The system in Fig. 7.11 behaves as a continuous-time system whose transfer function $H(s)$ is given by Eq. 7.67. Thus if

$$G(z) = \frac{z}{z - 0.5}$$

In Fig. 7.11, then the system will have a continuous-time transfer function $H(s)$ given by

$$H(s) = \frac{e^{sT}}{e^{sT} - 0.5}$$

where T is the sampling interval.

7.11   DIGITAL FILTERING

A system shown in Fig. 7.11 can be used to filter continuous-time signals. These filters are known as *digital filters*. Let us consider a second-order filter

$$G(z) = \frac{z^2}{(z - \gamma)(z - \gamma*)} \tag{7.68}$$

The continuous-time transfer function $H(s)$ of the system in Fig. 7.11 is given by

$$H(s) = G(e^{sT}) = \frac{e^{2sT}}{(e^{sT} - \gamma)(e^{sT} - \gamma*)} \tag{7.69}$$

We shall find the magnitude characteristic $|H(j\omega)|$ of this filter. From Eq. 7.69,

$$H(j\omega) = \frac{e^{2j\omega T}}{(e^{j\omega T} - \gamma)(e^{j\omega T} - \gamma*)} \tag{7.70}$$

Let

$$\gamma = e^{(\alpha + j\beta)T} \tag{7.71}$$

Then

$$H(j\omega) = \frac{e^{2j\omega T}}{[e^{j\omega T} - e^{(\alpha + j\beta)T}][e^{j\omega T} - e^{(\alpha - j\beta)T}]}$$

$$= \frac{1}{1 - 2e^{\alpha T}\cos\beta T e^{-j\omega T} + e^{2\alpha T}e^{-2j\omega T}}$$

and

$$|H(j\omega)| = \frac{1}{[(1 - 2e^{\alpha T}\cos\omega T \cos\beta T + e^{2\alpha T}\cos\omega T)^2}$$
$$+ (2e^{\alpha T}\cos\beta T \sin\omega T - e^{2\alpha T}\sin 2\omega T)^2]^{\frac{1}{2}}$$

plot of $20\log|H(j\omega)|$ as function of ω (Bode plot) is shown in Fig. 7.13, for $T = 10^{-3}$, $\beta = 250\pi$ and several values of α. Note the similarity of the frequency response of this system to that of a second-order continuous-time system. The resonance frequency ω_r approaches $\beta(=250\pi)$ as $\alpha \to 0$ (see Eq. 7.28). Thus the resonance frequency ω_r is given by

$$\omega_r = 250\pi = 785 \text{ rps}$$

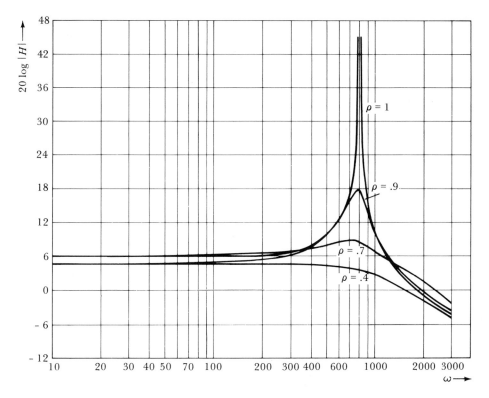

Figure 7.13

Note that for $\alpha = 0$, the gain at the resonance frequency is infinity.   Also $\alpha > 0$ represents an unstable system ($|\gamma| > 1$).   Hence for stable system α must be negative.

In Sec. 3.17 it was shown that the samples of a signal contain the complete information about the signal only if the sampling rate is at least twice the maximum frequency f_m Hz of the signal.   The minimum sampling rate (Nyquist rate) for signal is therefore $2f_m$ samples/sec where f_m is the highest significant frequency of the signal.   The corresponding time interval (Nyquist interval) $T = 1/2f_m$ sec.   In our present case $T = 10^{-3}$.   Therefore, $f_m = 500$ Hz $= 3141.6$ rad/sec. Hence for $T = 10^{-3}$ the input signal $f(t)$ should not contain frequency above 3141.6 rad/sec.   The results derived for this digital filter are meaningful for $\omega \leq 3141.6$ rps.   Beyond this frequency, an aliasing effect appears and the system no longer acts as a linear system.   Consequently, the frequency response plot in Fig. 7.13 is meaningful only for $\omega \leq 3141.6$ rps.   The response of this filter for input signals of frequency $\omega > 3141.6$ will be zero, because the ideal low-pass filter in the D/A converter (Fig. 7.11) has a cutoff frequency of 3141.6 rps.†

---

†The magnitude function $|H(j\omega)|$ is periodic in ω with repetition frequency $f = 2f_m$ Hz $= 6283.2$ rps.   However, an ideal D/A converter (Fig. 7.12b) consists of an ideal low-pass filter with cutoff frequency f_m Hz—that is, 500 Hz $= 3141.6$ rps.   (see Sec 3.17). Hence the response $y(t)$ will be zero for input signals of frequency $\omega > 3141.6$ rps.

The digital filter in Fig. 7.11 can be realized using delay elements as discussed in Sec. 7.9. A digital computer can be conveniently used for this purpose.

There are several techniques of designing digital filters for specified frequency responses. For more information, the reader is referred to Gold and Rader.†

A filter with frequency response similar to that of Butterworth filter (see Fig. 3.28) can be realized by a digital filter in Fig. 7.11, if we choose

$$| G(z) |^2 = \frac{\tan^{2n}(\omega_c T/2)}{\tan^{2n}(\omega_c T/2) + (-1)^n [(z - 1)/(z + 1)]^{2n}}$$

where ω_c is the filter bandwidth.

The frequency response $| H(j\omega) |$ of this filter is given by (see Eq. 7.67)

$$| H(j\omega) |^2 = | G(e^{j\omega T}) |^2 = \frac{\tan^{2n}(\omega_c T/2)}{\tan^{2n}(\omega_c T/2) + (-1)^n [(e^{j\omega T} - 1)/(e^{j\omega T} + 1)]^{2n}}$$

$$= \frac{1}{1 + \dfrac{\tan^{2n}(\omega T/2)}{\tan^{2n}(\omega_c T/2)}}$$

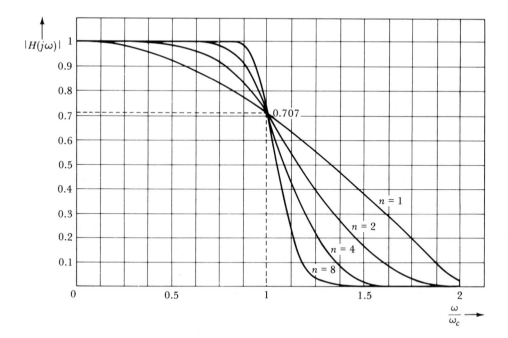

Figure 7.14

†Gold and C. M. Radar, *Digital Processing of Signals* (New York: McGraw-Hill, 1969).

This frequency response is shown in Fig. 7.14 for several values of n, and $\omega_c T = \pi/2$. Compare this response to that in Fig. 3.28. Note that the bandwidth of the filter is ω_c regardless of the value of n. Since $\omega_c T = \pi/2$, $T = \pi/2\omega_c$, and the highest frequency that can be processed by this filter is f_m, given by $1/2T$ Hz or π/T rps—that is, $2\omega_c$ rps.

ADVANTAGES OF DIGITAL FILTERING

A digital system is realized by delay elements which can be provided by memory (storage) elements of a digital computer. Following are some of the significant advantages of digital over analog filters.

1. Digital filters have a greater degree of precision.
2. Very-low-frequency filters, if realized by continuous time systems, need prohibitively bulky components. Such is not the case with digital filters.
3. Time-varying and adaptive filters are readily realized by digital filters.
4. Greater variety of filters can be realized by digital systems. In continuous-time filters, realization of certain filters requires negative inductances. These filters can be easily realized by digital systems.
5. Digital filters can be time-shared and therefore can serve a number of inputs simultaneously.
6. Digital filters are flexible. Their characteristics can be easily altered.
7. Potentially lower cost.

7.12   SAMPLED-DATA (HYBRID) SYSTEMS

Sampled-data systems or hybrid systems contain discrete-time systems (such as digital computers) as well as continuous-time systems as subsystems. In these systems, discrete-time as well as continuous-time systems are interconnected. Such systems are increasingly becoming more important. Consider for example a fire-control system. In this case, the problem is to search and track a moving target and fire a projectile for a direct hit. The data obtained from the search and tracking radar is discrete-time data because of a scanning operation which results in sampling of azimuth, elevation, and the velocity of the target. This discrete-time data is now fed to a digital computer which performs extensive computations. The computer output is then fed to a continuous plant, such as a gun mount, which accordingly positions itself at a certain position and fires.

Another example is attitude-control problem in a spacecraft where the information about the actual spacecraft attitude is fed back to a digital computer which acts as a controller. The computer generates the corrective input (control input) to be applied to the spacecraft which is a continuous-time system. In automatic periodic quality check in production line, the discrete data obtained from the periodic check may be fed to a digital computer which generates corrective signal (control input) to be used to control the continuous-time plant. In complex control systems, use of digital computer as a controller or a compensator for continuous-time plants is growing rapidly.

   In time-sharing systems, where for economic reasons certain facilities are shared by several systems, the signals are by nature discrete-time or sampled. In regulator type control systems, where an output variable must be maintained at a constant value, the external disturbance and plant parameter variation are so slow that continuous monitoring (or feedback) is unnecessary. It is adequate to sample the output periodically and to feed back this discrete data. In such cases feedback transducers, data-processing facilities, and possibly long and expensive feedback communication facilities, etc., can be shared among several control systems.

   Figure 7.15 shows two typical sampled-data systems. The system in Fig.

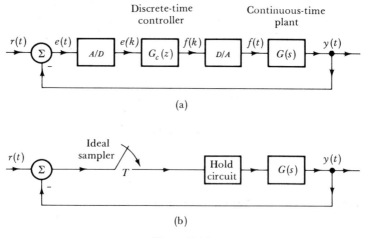

(a)

(b)

Figure 7.15

7.15a contains a digital system with transfer function $G_c(z)$ whereas the system in Fig. 7.15b has no digital subsystem. Since these systems in general contain both digital systems (discrete-time systems) and continuous-time systems, they are also called *hybrid systems*. Note that in Fig. 7.15a each subsystem is interfaced with appropriate (A/D or D/A) converters. For the sake of convenience the converters are often omitted from the diagrams.

   Figure 7.15b contains a sampler and a hold circuit. The latter is a form of low-pass filter. We shall now discuss each of these components.

IDEAL SAMPLER

   Sampled-data systems may contain one or more samplers which sample a continuous-time signals. A sampler is a switch which closes periodically every T seconds for a very short time ϵ (Fig. 7.16a). If we apply a continuous-time

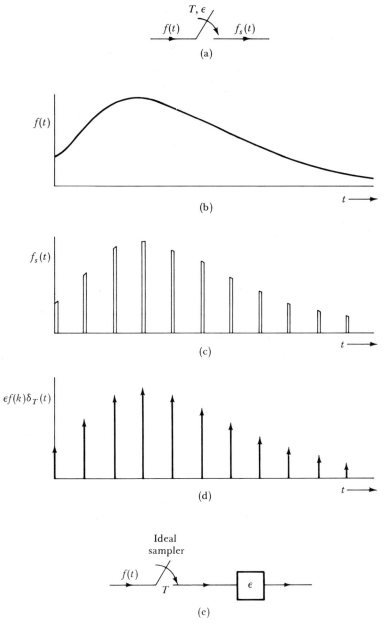

Figure 7.16

signal $f(t)$ at the input of a practical sampler (Fig. 7.16a), the output will be a
sequence of narrow pulses as shown in Fig. 7.16c. If the width ϵ is very small, the
pulses may be considered as impulses. The kth pulse has an area $\epsilon f(k)$, where
$f(k)$ is $f(t)$ at $t = kT$. Hence $f_s(t)$, the output of the sampler, may be expressed as

$$f_s(t) = \epsilon[f(0)\delta(t) + f(1)\delta(t - T) + \cdots + f(k)\delta(t - kT) + \cdots$$
$$= \epsilon f(k)\delta_T(t)$$

where $\delta_T(t)$ is a uniform train of impulses of unit strength (see Fig. 3.17a). The practical sampler can be represented as shown in Fig. 7.16e. The sampler here is an ideal sampler whose output is $f(k)\delta_T(t)$ followed by a multiplier ϵ. We can lump the multiplier ϵ with the system that follows the sampler. Hence for computational purposes we can assume that the sampler is an ideal sampler whose output is a sequence of impulses spaced T seconds apart and the strength of the kth impulse is $f(k)$. A sampler output can thus be described by a discrete-time sequence $f(k)$.

In analyzing sampled data systems we shall often come across signals which are represented by sequences of impulses. This type of signal is a continuous-time signal because it can be applied directly to a continuous-time system. On the other hand, it may be considered a discrete-time signal because it can be described by a discrete-time sequence. Note, however, that it is not a discrete-time signal of the form shown in Fig. 7.1a, nor can it be applied directly to digital systems without conversion into binary form. The interesting fact remains, however, that such a signal can be described by a discrete time sequence $f(k)$. Consequently discrete-time techniques such as z-transformation, etc., can be applied to such signals. Consider for example the system in Fig. 7.19a. The input to $G(s)$ is a sequence of impulses discribed by a sequence $f(k)$. The output $y(t)$ when sampled yields a sequence impulses described by a sequence $y(k)$. These two sequences can be related by a z-transfer function as

$$G(z) = \frac{\mathfrak{z}[y(k)]}{\mathfrak{z}[f(k)]}$$

The z-transfer function $G(z)$ can be easily determined by a method discussed later. Once $G(z)$ is known the output sequence $y(k)$ be determined as

$$y(k) = \mathfrak{z}^{-1}[G(z)F(z)]$$

Hence the system in Fig. 7.19a can be analyzed by discrete-time techniques, although its main component $G(s)$ is a continuous-time system.

THE HOLD CIRCUIT

This is a crude form of low-pass filter. We shall discuss here a zero-order hold (ZOH) circuit shown in Fig. 7.17a. It can be easily seen that the transfer function of this circuit is given by

$$G_{ho}(s) = (1 - e^{-sT})\left(\frac{1}{s}\right)$$

$$= \frac{1 - e^{-sT}}{s} \tag{7.72}$$

Let us observe the unit impulse response of this device. When the input is $\delta(t)$, the input to the integrator in Fig. 7.17a is $[\delta(t) - \delta(t - T)]$. Hence the out-

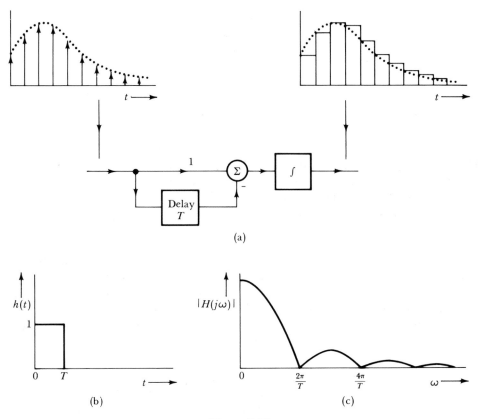

Figure 7.17

put $h(t)$ is given by

$$h(t) = \int_0^t [\delta(\tau) - \delta(\tau - T)] \, d\tau$$

$$= u(t) - u(t - T) \qquad (7.73)$$

The unit impulse response $h(t)$ is a rectangular pulse shown in Fig. 7.17b. It is obvious that if an impulse train $f(k)\delta_T(t)$ is applied to ZOH, the output will be a sequence of rectangular pulses (box car signal) as shown in Fig. 7.17a. Thus a ZOH circuit reconstructs a crude replica of a continuous signal from its samples. It therefore acts as nonideal low-pass filter. Let us find the filter characteristic of a ZOH. From Eq. 7.72 we have

$$H(j\omega) = \frac{1 - e^{-j\omega T}}{j\omega}$$

and

$$|H(j\omega)| = \frac{|1 - \cos\omega T + j\sin\omega T|}{|\omega|}$$

$$= \frac{\sqrt{2\,(1 - \cos\omega T)}}{|\,\omega\,|}$$

$$= T\left|\frac{\sin\left(\dfrac{\omega T}{2}\right)}{\dfrac{\omega T}{2}}\right|$$

$$= T\left|\mathrm{Sa}\left(\frac{\omega T}{2}\right)\right|$$

The magnitude $|H(j\omega)|$ is plotted in Fig. 7.17c. It can be seen that ZOH is indeed a crude low-pass filter. One can use more sophisticated hold circuits such as a first-order hold (FOH) or fractional hold circuits. These have a better filter characteristics but are more complex.

The ZOH circuit in Fig. 7.17a can be separated into two components. The input is a sequence of impulses and the output of the first component is also a sequence of impulses. The second component is the integrator. Note that the input and the output of the first section can be described by discrete time functions. We can therefore assign a z-transfer function to this section. The delay element has a z-transfer function $1/z$. Hence the ZOH circuit is equivalent to the system shown in Fig. 7.18a. The integrator is a continuous-time system with

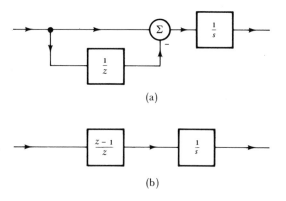

(a)

(b)

Figure 7.18

transfer function $1/s$. The ZOH circuit is therefore equivalent to the system shown in Fig. 7.18b. Since ZOH is followed by a continuous-time system, it is possible to lump $1/s$ part with the continuous-time system.

z-TRANSFER FUNCTION ASSOCIATED WITH A CONTINUOUS-TIME SYSTEM

In sampled-data systems one often comes across a situation shown in Fig. 7.19. The continuous-time input is sampled. The sampled output is a train of impulses which may be applied directly to a continuous-time system $G(s)$ as in Fig. 7.19a or may be applied to $G(s)$ through a hold circuit. The output $y(t)$

of $G(s)$ is sampled. The sampled output is therefore a sequence of impulses. The sampled input and the sampled output can be considered as discrete-time signals and may be described by signals $f(k)$ and $y(k)$ respectively. It is possible to find the z-transfer function relating $y(k)$ to $f(k)$ for a given $G(s)$. This may be easily accomplished as follows. Let $G(z)$ represent† the z-transfer function relating $y(k)$ to $f(k)$ in Fig. 7.19a. By definition, $G(z)$ is the z-transform of the output se-

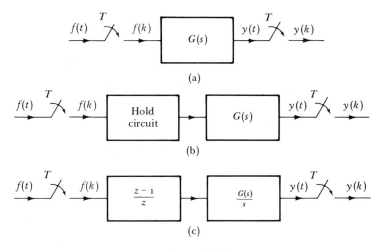

(a)

(b)

(c)

Figure 7.19

quence $y(k)$ when the input sequence $f(k) = \partial(k)$. Note that the input sequence $f(k)$ is a train of impulse where kth impulse has a strength $f(k)$. The sequence $\partial(k)$ in this context will represent a single impulse $\delta(t)$ at the origin. When $\delta(t)$ is applied to $G(s)$, the output $y(t) = g(t)$ where $g(t)$ is the unit impulse response of $G(s)$. Hence the output sequence $y(k)$ is therefore given by

$$y(k) = g(kT)$$

Hence

$$G(z) = \partial[y(k)] = \partial[g(kT)] \tag{7.74}$$

We can determine $G(z)$ either from Table 7.3 or from the definition 7.50 of the z-transform. As an example, let

$$G(s) = \frac{1}{s - \lambda}$$

---

†This notation is somewhat improper. By convention $G(z)$ implies a function $G(s)$ where s is replaced by z. Here, however $G(z)$ represents an entirely different function from $G(s)$. It would be proper to use some such notation as $\mathcal{G}(z)$. However, script letters are usually difficult to read and present some psychological barrier. Hence we have retained the incorrect notation $G(z)$ as preferable of the two evils.

Then

$$g(t) = e^{\lambda t}$$

and

$$g(kT) = e^{\lambda kT}$$

Hence

$$G(z) = \mathfrak{z}[e^{\lambda kT}]$$

From Table 7.3 we have

$$G(z) = \frac{z}{z - e^{\lambda T}} \tag{7.75}$$

Thus the output sequence $y(k)$ and the input sequence $f(k)$ in Fig. 7.19a are related by the z-transfer function in Eq. 7.75 when $G(s) = 1/s - \lambda$. Thus $G(s)$ and $G(z)$ form a pair. Several such pairs are shown in Table 7.4. The system in Fig. 7.19b can be handled as follows. Consider the case of a zero-order hold circuit. It can be represented by a system shown in Fig. 7.18b. We can lump $1/s$ with $G(s)$ as shown in Fig. 7.19c. The z-transfer function corresponding to $G(s)/s$ can now be found from Table 7.4. Let this be $M(z)$. The transfer function relating $y(k)$ to $f(k)$ is now given by $[(z - 1)/z]M(z)$.

TABLE 7.4

No.	$G(s)$	$g(t)$	$g(k) = g(kT)$	$G(z)$
1	K	$K\delta(t)$	$K\partial(k)$	K
2	$\dfrac{1}{s}$	$u(t)$	$u(k)$	$\dfrac{z}{z-1}$
3	$\dfrac{1}{s^2}$	t	kT	$\dfrac{Tz}{(z-1)^2}$
4	$\dfrac{1}{s^3}$	$\dfrac{t^2}{2}$	$\dfrac{k^2T^2}{2}$	$\dfrac{T^2z(z+1)}{2(z-1)^3}$
5	$\dfrac{1}{s-\lambda}$	$e^{\lambda t}$	$e^{\lambda kT}$	$\dfrac{z}{z-e^{\lambda T}}$
6	$\dfrac{1}{(s-\lambda)^2}$	$te^{\lambda t}$	$kTe^{\lambda kT}$	$\dfrac{Tze^{\lambda T}}{(z-e^{\lambda T})^2}$
7	$\dfrac{\beta}{(s-\alpha)^2 + \beta^2}$	$e^{\alpha t}\sin \beta t$	$e^{\alpha kT}\sin \beta kT$	$\dfrac{ze^{\alpha T}\sin \beta T}{z^2 - 2ze^{\alpha T}\cos \beta T + e^{2\alpha T}}$
8	$\dfrac{s-\alpha}{(s-\alpha)^2 + \beta^2}$	$e^{\alpha t}\cos \beta t$	$e^{\alpha kT}\cos \beta kT$	$\dfrac{z(z - e^{\alpha T}\cos \beta T)}{z^2 - 2ze^{\alpha T}\cos \beta T + e^{2\alpha T}}$

Let us recapitulate our results. A train of impulses where the kth impulse has a strength $f(k)$ is applied at the input of a continuous-time system with transfer function $G(s)$. The output $y(t)$ of the system is, in general, a continuous-time signal. We have shown that the discrete-time sequence $y(k)$ formed by values of

$y(t)$ at $t = kT$ is related to the discrete-time sequence $f(k)$ by the transfer function $G(z)$ where $G(z)$ is listed in Table 7.4 for a given $G(s)$. Thus

$$Y(z) = G(z)F(z)$$

The reader is encouraged to ponder upon this result and ensure that he has the proper understanding of its implications.

Table 7.4 may also be interpreted as follows. If a continuous-time signal $g(t)$ has a Laplace transform $G(s)$, then the z-transform of its samples $g(k)$ is $G(z)$. Both these interpretations prove useful in handling sampled data systems.

If a continuous-time signal $r(t)$ is applied to the input of an ideal sampler, the output will be sequence of impulses where kth impulse has a strength $r(k) = r(kT)$. The z-transform of the sampler output is $\mathfrak{z}[r(kT)]$. We note from Eq. 7.74 that the z-transform of $r(k)$ can be obtained directly from Table 7.4 if we know $r(t)$ or $R(s)$. Thus if the input $r(t)$ is given by

$$r(t) = t$$

$R(z)$, the z-transform of the sampler output, is found from Table 7.4 to be

$$R(z) = \frac{Tz}{(z - 1)^2}$$

TRANSFER FUNCTION OF A SAMPLED-DATA SYSTEM

Consider a sampled-data system Fig. 7.20a. The input to $G(s)$ is a sequence of impulses described by a discrete-time function $e(k)$

$$e(k) = r(k) - y(k) \tag{7.76}$$

where $r(k)$ and $y(k)$ are the values of $r(t)$ and $y(t)$ respectively at $t = kT$. The z-transform of Eq. 7.76 yields

$$E(z) = R(z) - Y(z) \tag{7.77a}$$

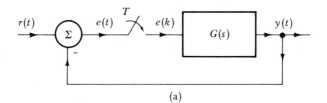

(a)

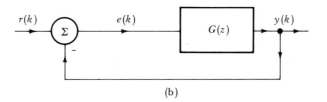

(b)

Figure 7.20

However, output samples $y(k)$ and the input signal $e(k)$ are related by z-transfer function $G(z)$, where $G(z)$ is found from Table 7.4 for a given $G(s)$. Thus

$$Y(z) = G(z) E(z) \tag{7.77b}$$

Equations 7.76, 7.77a and 7.77b immediately suggest a discrete-time model of the system as shown in Fig. 7.20b. This model predicts the system behavior only at sampling instants. The response during the sampling instants can be found by a simple modification of this process. The z-transfer function of the system in Fig. 7.20b is easily found to be

$$T(z) = \frac{Y(z)}{R(z)} = \frac{G(z)}{1 + G(z)} \tag{7.78}$$

This is the strategy we use in analyzing sampled-data system.

Figure 7.21 shows some examples of sampled-data systems. Since we focus our attention on the output samples at $0, T, 2T, \ldots$, it proves convenient to introduce a fictitious sampler at the output as shown in Fig. 7.21. In each of the systems in Fig. 7.21, we shall determine the z-transfer function which relates the output samples $y(k)$ to the input samples $r(k)$. This will be the transfer function of the discrete-time model of the system.

For Fig. 7.21a, the input $r(k)$ and the output $y(k)$ are related by $G(z)$, where $G(z)$ can be found from Table 7.4 for a given $G(s)$. Thus

$$Y(z) = G(z)R(z)$$

Hence

$$T(z) = \frac{Y(z)}{R(z)} = G(z)$$

In Fig. 7.21b, a similar argument shows that

$$Y(z) = GH(z)R(z)$$

and

$$T(z) = \frac{Y(z)}{R(z)} = GH(z)$$

where $GH(z)$ is the function obtained from Table 7.4 corresponding to $G(s)H(s)$. Note that

$$GH(z) \neq G(z)H(z)$$

Thus if

$$G(s) = \frac{1}{s + 2} \quad \text{and} \quad H(s) = \frac{1}{s}$$

from Table 7.4,

$$G(z) = \frac{z}{z - e^{-2T}} \quad \text{and} \quad H(z) = \frac{z}{z - 1}$$

and

$$G(z)H(z) = \frac{z^2}{(z - 1)(z - e^{-2T})}$$

Configuration

Output $Y(z)$

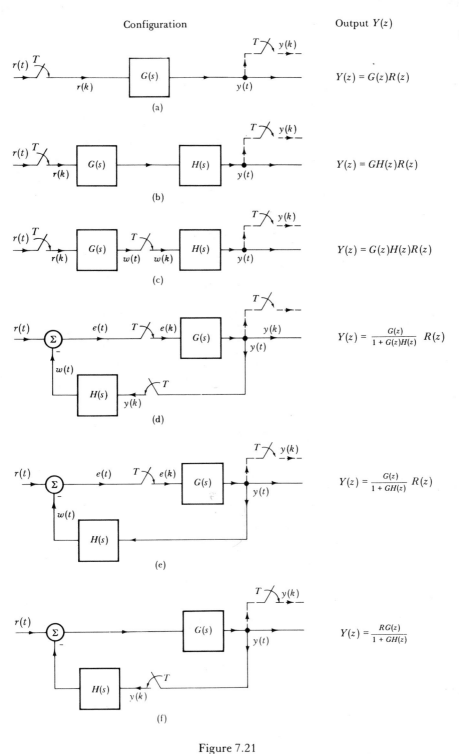

$Y(z) = G(z)R(z)$

(a)

$Y(z) = GH(z)R(z)$

(b)

$Y(z) = G(z)H(z)R(z)$

(c)

$Y(z) = \dfrac{G(z)}{1 + G(z)H(z)} \, R(z)$

(d)

$Y(z) = \dfrac{G(z)}{1 + GH(z)} \, R(z)$

(e)

$Y(z) = \dfrac{RG(z)}{1 + GH(z)}$

(f)

Figure 7.21

441

However,

$$G(s)H(s) \;=\; \frac{1}{s(s+2)} \;=\; \frac{1}{2}\left[\frac{1}{s} - \frac{1}{s+2}\right]$$

and from Table 7.4 we find

$$GH(z) \;=\; \frac{1}{2}\left[\frac{z}{z-1} - \frac{z}{z-e^{-2T}}\right] \;=\; \frac{z(1-e^{-2T})}{2(z-1)(z-e^{-2T})}$$

Thus

$$GH(z) \neq G(z)H(z)$$

In Fig. 7.21c we observe that

$$W(z) \;=\; G(z)R(z)$$

and

$$Y(z) \;=\; H(z)W(z)$$
$$=\; G(z)H(z)R(z)$$

Hence

$$T(z) \;=\; \frac{Y(z)}{R(z)} \;=\; G(z)H(z)$$

For Fig. 7.21d we observe that

$$E(z) \;=\; R(z) - W(z)$$

and

$$W(z) \;=\; H(z)Y(z)$$
$$Y(z) \;=\; G(z)E(z)$$
$$=\; G(z)[R(z) - W(z)]$$
$$=\; G(z)[R(z) - H(z)Y(z)]$$

Hence

$$Y(z) \;=\; \frac{G(z)}{1 + G(z)H(z)} R(z)$$

and

$$T(z) \;=\; \frac{Y(z)}{R(z)} \;=\; \frac{G(z)}{1 + G(z)H(z)}$$

For Fig. 7.21e,

$$E(z) \;=\; R(z) - GH(z)E(z)$$

Hence

$$E(z) \;=\; \frac{1}{1 + GH(z)} R(z)$$

and

$$Y(z) \;=\; G(z)E(z)$$
$$=\; \frac{G(z)}{1 + GH(z)} R(z)$$

and

$$T(z) = \frac{Y(z)}{R(z)} = \frac{G(z)}{1 + GH(z)}$$

For Fig. 7.21f, we observe that $Y(z)$ (the output of the sampler) has 2 components, (i) due to input $r(t)$, and (ii) the component caused by $y(k)$ fed back to $H(s)$. Hence

$$Y(z) = RG(z) - GH(z)Y(z)$$

and

$$Y(z) = \frac{RG(z)}{1 + GH(z)}$$

Note that

$$RG(z) \neq R(z)G(z)$$

Consequently we cannot separate $R(z)$ from $RG(z)$. Therefore it is not possible to write a z-transfer function relating $y(k)$ to the input $r(k)$. Analysis and synthesis of such systems involving only one sampler which is located in a feedback loop is little more difficult.

7.13   RESPONSE DURING SUCCESSIVE SAMPLING INSTANTS: MODIFIED z-TRANSFORM

Our analysis of sampled-data systems has been restricted only to discrete instants (sampling instants). Hence these methods give response values only at sampling instants. We shall now modify our technique in order to enable us to determine the response during sampling instants.

Consider a system shown in Fig. 7.22a. The input signal $f(t)$ is sampled to obtain train of impulses characterized by the discrete-time sequence $f(k) = f(kT)$. The samples of the continuous-time output $y(t)$ at $t = 0, T, 2T, \ldots$, is characterized by the discrete-time sequence $y(k) = y(kT)$. We have shown that the sequence $y(k)$ is related to the sequence $f(k)$ by the transfer function $G(z)$, where $G(z)$ is found from Table 7.4 for a given $G(s)$.

Our aim is to investigate the values of $y(t)$ during sampling instants. This can be done by considering another sequence represented by values of $y(t)$ at $t = \mu T$, $(1 + \mu)T$, $(2 + \mu)T, \ldots$, etc., where $0 < \mu < 1$. It is obvious that this new sequence of points lies somewhere between the previous sequence at $t = 0, T, 2T$, $\ldots$, etc., as shown in Fig. 7.22b. By assigning appropriate value to μ $(0 < \mu < 1)$ we can comb the entire range of t in this manner.

Let $G(z, \mu)$ be the transfer function which relates the new output sequence $y[(k + \mu)T]$ to the input sequence $f(k)$. By definition, $G(z, \mu)$ is the z-transform of the output sequence $y[(k + \mu)T]$ when the input $f(k) = \partial(k)$. Since $f(k)$ is a train of impulse, $\partial(k)$ represents a unit impulse at $t = 0$. Hence $G(z, \mu)$ is the z-transform of the output sequence $y[(k + \mu)T]$ when the input to $G(s)$ is unit impulse. If $g(t)$ is the unit impulse response of $G(s)$, then

$$y[(k + \mu)T] = g[(k + \mu)T] \qquad 0 < \mu < 1$$

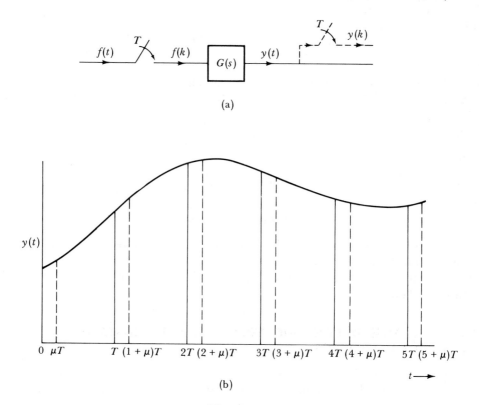

(a)

(b)

Figure 7.22

and

$$G(z, \mu) = \mathfrak{z}\{g[(k + \mu)T]\} = \sum_{k=0}^{\infty} [g(k + \mu)T]z^{-k} \qquad (7.79)$$

Let us consider

$$G(s) = \frac{1}{s - \lambda}$$

$$g(t) = e^{\lambda t}$$

and

$$G(z, \mu) = \mathfrak{z}[e^{\lambda(k + \mu)T}]$$
$$= e^{\lambda \mu T} \mathfrak{z}[e^{\lambda k T}]$$

$$= e^{\lambda \mu T} \frac{z}{z - e^{\lambda T}} = \frac{ze^{\lambda \mu T}}{z - e^{\lambda T}}$$

In this manner we can prepare a table (Table 7.5) which lists $G(z, \mu)$ corresponding to given $G(s)$.†

TABLE 7.5

No.	$G(s)$	$G(z, \mu)$
1	K	0
2	$\dfrac{1}{s}$	$\dfrac{z}{z - 1}$
3	$\dfrac{1}{s^2}$	$\dfrac{Tz[\mu(z - 1) + 1]}{(z - 1)^2}$
4	$\dfrac{1}{s^n}$	$z \lim_{a \to 0} \left[\dfrac{(-1)^{n-1}}{(n - 1)!} \dfrac{\partial^{n-1}}{\partial a^{n-1}} \left(\dfrac{e^{-a\mu T}}{z - e^{-aT}} \right) \right]$
5	$\dfrac{1}{s - \lambda}$	$\dfrac{ze^{\lambda\mu T}}{z - e^{\lambda T}}$
6	$\dfrac{1}{(s - \lambda)^2}$	$\dfrac{Te^{\lambda\mu T}z[e^{\lambda T} + \mu(z - e^{\lambda T})]}{(z - e^{\lambda T})^2}$
7	$\dfrac{\beta}{(s - \alpha)^2 + \beta^2}$	$\dfrac{e^{\alpha\mu T}z[z \sin \beta\mu T + e^{\alpha T} \sin(1 - \mu)\beta T]}{z^2 - 2ze^{\alpha T} \cos \beta T + e^{2\alpha T}}$
8	$\dfrac{s - \alpha}{(s - \alpha)^2 + \beta^2}$	$\dfrac{e^{\alpha\mu T}z[z \cos \beta\mu T - e^{\alpha T} \cos (1 - \mu)\beta T]}{z^2 - 2ze^{\alpha T} \cos \beta T + e^{2\alpha T}}$
9	$\dfrac{1 - e^{-st}}{s} G(s)$	$\left(\dfrac{z - 1}{z} \right) \mathfrak{z}_\mu \left[\dfrac{G(s)}{s} \right]$

From the definition in Eq. 7.79 it is also obvious that

$$G(z) = \lim_{\mu \to 0} G(z, \mu) \tag{7.80}$$

Let us now try to recapitulate what we have accomplished. The modified z-transform allows us to obtain the values of response $y(t)$ between sampling intervals. The sequence formed by the values of $y(t)$ at $t = \mu T$, $(1 + \mu)T$, $(2 + \mu)T, \ldots, (0 < \mu < 1)$ are related to the input sequence $f(k)$ by a transfer function $G(z, \mu)$ which can be found from Table 7.5 for a given $G(s)$. The output $Y(z, \mu)$ is given by (see Fig. 7.21a)

$$Y(z, \mu) = G(z, \mu)F(z)$$

---

†Some authors define a modified z-transform $G(z, m)$ as

$$G(z, m) = \mathfrak{z}[g(k + m - 1)T] = z^{-1} \sum_{k=0}^{\infty} g[(k + m)T]z^{-k}$$

Comparison of this definition with our definition (Eq. 7.79) shows that $G(z, \mu) = zG(z, m)$. The definition used in this book proves more convenient than $G(z, m)$.

and

$$y[(k + \mu)T] = \mathfrak{z}^{-1}[G(z, \mu)F(z)]$$

Hence the inverse z-transform of $G(z, \mu)F(z)$ yields the sequence $y(\mu T)$, $y[(1 + \mu)T], y[(2 + \mu)T], \ldots$ By assigning values to μ in the range 0 to 1, we can obtain all the information about $y(t)$. We shall now demonstrate the application of modified z-transform.

Example 7.6.   We shall find the response $y(t)$ of the system shown in Fig. 7.23a when

$$G_c(z) = \frac{z}{z - 1}$$

$$G(s) = \frac{1}{s + 4}$$

when the input $r(t)$ is a unit step and $T = 0.5$.

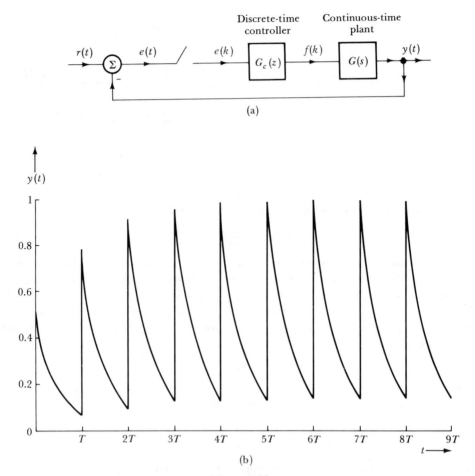

Figure 7.23

This system has a discrete-time controller and a continuous-time plant. The block diagram does not show the appropriate converters required at the interface of discrete-time and continuous-time system; these are implied. Thus the output of the sampler will be coded to convert it in a form suitable for a digital controller. Similarly, the output $f(k)$ of the controller will converted into sequence of impulses $f(k)$ before it is applied to the continuous-time plant. The input $r(t) = u(t)$, and $T = 0.5$. From Table 7.4, $G(z)$ for $G(s) = 1/(s + 4)$ is given by

$$G(z) = \frac{z}{z - e^{-4T}} = \frac{z}{z - e^{-2}}$$

Also,

$$\frac{Y(z)}{R(z)} = \frac{G_c(z)G(z)}{1 + G_c(z)G(z)}$$

$$= \frac{\left(\dfrac{z}{z - 1}\right)\left(\dfrac{z}{z - e^{-2}}\right)}{1 + \dfrac{z^2}{(z - 1)(z - e^{-2})}}$$

$$= \frac{z^2}{(z - 1)(z - e^{-2}) + z^2}$$

$$= \frac{z^2}{2(z - 0.394)(z - 0.174)}$$

Hence

$$Y(z) = \frac{z^2}{2(z - 0.394)(z - 0.174)} R(z)$$

From Table 7.4, for $r(t) = u(t)$, we have

$$R(z) = \frac{z}{z - 1}$$

and

$$Y(z) = \frac{z^2}{2(z - 1)(z - 0.394)(z - 0.174)}$$

and

$$\frac{Y(z)}{z} = \frac{z^3}{2(z - 1)(z - 0.394)(z - 0.174)}$$

$$= \frac{1}{z - 1} - \frac{0.583}{z - 0.394} + \frac{0.083}{z - 0.174}$$

Hence

$$Y(z) = \frac{z}{z - 1} - 0.583 \frac{z}{z - e^{-0.93}} + 0.083 \frac{z}{z - e^{-1.75}}$$

and

$$y(k) = 1 - 0.583e^{-0.93k} + 0.083e^{-1.75k} \tag{7.81}$$

This gives us the output at instants $t = 0, T, 2T, \ldots$, etc. In order to determine the output at all instants, we need to use a modified z-transform. The modified z-transform $G(z, \mu)$ for $G(s)$ is found from Table 7.5 as

$$G(z, \mu) = \frac{ze^{-2\mu}}{z - e^{-2}}$$

and

$$G(z) = \lim_{\mu \to 0} G(z, \mu) = \frac{z}{z - e^{-2}}$$

also

$$R(z) = \mathfrak{z}[u(k)] = \frac{z}{z - 1}$$

We also have

$$F(z) = \frac{G_c(z)}{1 + G_c(z)G(z)} R(z)$$

and

$$Y(z, \mu) = G(z, \mu)F(z)$$

$$= \frac{G(z, \mu)G_c(z)}{1 + G_c(z)G(z)} R(z)$$

$$= \frac{\dfrac{ze^{-2\mu}}{z - e^{-2}} \left(\dfrac{z}{z - 1}\right)\left(\dfrac{z}{z - 1}\right)}{1 + \left(\dfrac{z}{z - 1}\right)\left(\dfrac{z}{z - e^{-2}}\right)}$$

$$= \frac{z^3 e^{-2\mu}}{(z - 1)^2(z - e^{-2}) + z^2(z - 1)}$$

$$= \frac{z^3 e^{-2\mu}}{2z^3 - (3 + e^{-2})z + (1 + 2e^{-2})z - e^{-2}}$$

$$= \frac{z^3 e^{-2\mu}}{2(z - 1)(z - 0.3938)(z - 0.1738)}$$

Hence

$$\frac{Y(z, \mu)}{z} = \frac{z^2 e^{-2u}}{2(z - 1)(z - 0.3938)(z - 0.1738)}$$

$$= \frac{e^{-2\mu}}{z - 1} - \frac{0.583e^{-2\mu}}{z - 0.3938} + \frac{0.083e^{-2\mu}}{z - 0.1738}$$

and

$$Y(z, \mu) = e^{-2\mu}\left[\frac{z}{z - 1} - \frac{0.583z}{z - 0.3938} + \frac{0.083z}{z - 0.1738}\right]$$

$$= e^{-2\mu}\left[\frac{z}{z - 1} - 0.583\frac{z}{z - e^{-0.93}} + 0.083\frac{z}{z - e^{-1.75}}\right]$$

The inverse z-transform of $Y(z, \mu)$ is obtained from Table 7.3 to be

$$y[(k + \mu)T] = e^{-2\mu}[1 - 0.583e^{-0.93k} + 0.083e^{-1.75k}] \qquad 0 < \mu < 1 \qquad (7.82a)$$

Note that from Eq. 7.82a,

$$y(kT) = 1 - 0.583e^{-0.93k} + 0.083e^{-1.75k} \qquad (7.82b)$$

This is the same result as obtained earlier in Eq. 7.81. It is also obvious from Eqs. 7.82a and 7.82b that

$$y[(k + \mu)T] = y(kT)e^{-2\mu} \qquad 0 < \mu < 1 \qquad (7.82c)$$

Thus $y[(k + \mu)T]$ is an exponential† beginning at kT with value $y(kT)$ as shown in Fig. 7.23b. The output $y(t)$ therefore has a value $y(kT)$ at $t = kT$ and decays exponentially until $t = (k + 1)T$. The procedure is repeated again for $(k + 1)T < t < (k + 2)T$.

Example 7.7.   Find the response $y(t)$ of the system shown in Fig. 7.24a when the input $r(t) = u(t)$, $T = 1$, and under the conditions (a) no sampler and no hold circuit, (b) with sampler but no hold circuit, (c) with sampler and with hold circuit.

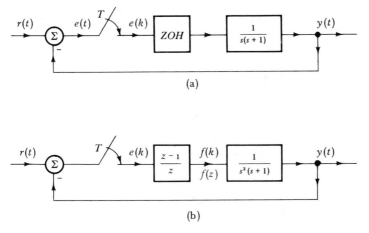

Figure 7.24

(a) No Sampler, No Hold Circuit. This is a continuous-time feedback system with transfer function

$$T(s) = \frac{1/s(s + 1)}{1 + 1/s(s + 1)} = \frac{1}{s^2 + s + 1}$$

---

†The time constant of this exponential is 0.25. This is because μ varies from 0 to 1 during the range of $t = T = 0.5$ sec.

and

$$y(t) = \mathcal{L}^{-1} \frac{1}{s(s^2 + s + 1)}$$

$$= \mathcal{L}^{-1} \left[\frac{1}{s} - \frac{0.58e^{j150°}}{s - 0.5 - j0.866} - \frac{0.58e^{-j150°}}{s - 0.5 + j0.866} \right]$$

$$= 1 - 1.16e^{-t/2} \cos(0.866t + 150°) \tag{7.83}$$

This function is plotted in Fig. 7.25.

(b) With samper, but no hold.

$$G(s) = \frac{1}{s(s + 1)}$$

$$= \frac{1}{s} - \frac{1}{s + 1}$$

From Table 7.5 we obtain ($T = 1$)

$$G(z, \mu) = \frac{z}{z - 1} - \frac{ze^{-\mu}}{z - 0.368}$$

$$= \frac{z[z(1 - e^{-\mu}) + (e^{-\mu} - 0.368)]}{(z - 1)(z - 0.368)} \tag{7.84a}$$

and

$$G(z) = \lim_{\mu \to 0} G(z, \mu) = \frac{0.632z}{(z - 1)(z - 0.368)} \tag{7.84b}$$

Also

$$E(z) = \frac{1}{1 + G(z)} R(z)$$

$$= \left(\frac{(z - 1)(z - 0.368)}{z^2 - 0.7363z + 0.368} \right) \left(\frac{z}{z - 1} \right)$$

$$= \frac{z(z - 0.368)}{z^2 - 0.7363z + 0.368} \tag{7.85}$$

and

$$Y(z, \mu) = G(z, \mu) E(z)$$

From Eqs. 7.84a and 7.85 we have

$$Y(z, \mu) = \frac{z^2[z(1 - e^{-\mu}) + (e^{-\mu} - 0.368)]}{(z - 1)(z^2 - 0.736z + 0.368)}$$

$$= \frac{z}{z - 1} - \frac{z(e^{-\mu}z - 0.368)}{z^2 - 0.7363z + 0.368}$$

$$= \frac{z}{z - 1} - \frac{e^{-\mu}(z - 0.368)z}{z^2 - 0.736z + 0.368}$$

$$+ 0.765(1 - e^{-\mu}) \frac{0.482z}{z^2 - 0.736z + 0.368}$$

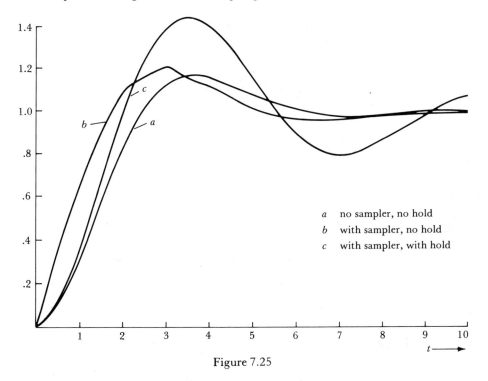

Figure 7.25

From Table 7.3 (pairs 2 and 10) we find

$$y[(k + \mu)T] = 1 + e^{-k/2}[0.765(1 - e^{-\mu}) \sin 0.93k - e^{-\mu} \cos 0.93k] \qquad (7.86)$$

By letting $k = 0, 1, 2, \ldots$, etc. and varying μ from 0 to 1, we get the plot of $y(t)$ as shown in Fig. 7.25.

(c) With Sampler and with ZOH. The ZOH circuit is replaced by $(z - 1)/z$ and $1/s$ (see Fig. 7.18). The transfer function $1/s$ is merged with $1/s(s + 1)$ as shown in Fig. 7.24b.

$$G_1(s) = \frac{1}{s^2(s + 1)} = \frac{1}{s^2} - \frac{1}{s} + \frac{1}{s + 1}$$

and

$$\begin{aligned}
G_1(z, \mu) &= \frac{z[\mu(z - 1) + 1]}{(z - 1)^2} - \frac{z}{z - 1} + \frac{ze^{-\mu}}{z - 0.368} \\
&= \frac{z[(\mu + e^{-\mu} - 1)z^2 + 2.368 - 1.368\mu - 2e^{-\mu})z + (0.368\mu + e^{-\mu} - 0.736)]}{(z - 1)^2(z - 0.368)}
\end{aligned}$$

and

$$G_1(z) = \lim_{\mu \to 0} G_1(z, \mu) = \frac{z(0.368z + 0.264)}{(z - 1)^2(z - 0.368)} \qquad (7.87a)$$

From Fig. 7.24b we observe that

$$F(z) = \frac{(z-1)/z}{1 + \frac{z-1}{z} G_1(z)} R(z)$$

$$= \frac{z-1}{z + \frac{(z-1)z(0.368z + 0.264)}{(z-1)^2(z-0.368)}} \frac{z}{z-1}$$

$$= \frac{(z-1)^2(z-0.368)}{(z-1)(z^2 - z + 0.632)}$$

and

$$Y(z, \mu) = G_1(z, \mu) F(z)$$

Therefore

$$\frac{Y(z, \mu)}{z} = \frac{(\mu - e^{-\mu} - 1)z^2 + (2.368 - 1.368\mu - 2e^{-\mu})z + (0.368\mu + e^{-\mu} - 0.736)}{(z-1)(z^2 - z + 0.632)}$$

$$= \frac{1}{z-1} - \frac{(2 - \mu - e^{-\mu})z - (1.368 - 0.368\mu - e^{-\mu})}{z^2 - z + 0.632}$$

and

$$Y(z, \mu) = \frac{z}{z-1} - (2 - \mu - e^{-\mu}) \frac{z(z - 0.5)}{z^2 - z - 0.632}$$

$$+ (0.595 + 0.214\mu - 0.81e^{-\mu}) \frac{0.617z}{z^2 - z - 0.632}$$

and $y[(k + \mu) T] = 1 - e^{-0.23k}$

$$\cdot [(2 - \mu - e^{-\mu}) \cos 0.89k - (0.595 + 0.214\mu - 0.81e^{-\mu}) \sin 0.89k] \quad (7.87b)$$

This function is plotted in Fig. 7.25.

It is interesting to compare the performance of the three cases studied in this example. First of all we observe that continuous-time system (without sampler and without hold) has the lowest overshoot and the lowest settling time.

Among the sampled-data systems [cases (b) and (c)], the transient behavior (overshoot, peak time, or settling time) of the system without hold is superior to that of the system with hold. The reason is that addition of hold adds one extra pole at the origin $(1/s)$. This is equivalent to adding a lag compensation. It was seen in Chapter 5 that addition of a lag generally causes deterioration in the transient behavior.

Unlike the system response in Example 7.6 (Fig. 7.23b) the response for the sampled-data system in this example is smooth and could have been obtained simply by joining the points at sampling instants. This is the result of the fact that the transfer function $G(s) = 1/s(s + 1)$, smoothens impulse or step inputs and the output has no jump discontinuities when either an impulse or a step input is applied to $G(s) = 1/s(s + 1)$. The system $1/(s + 4)$ in Example 7.6 does not have this property. It can smooth the step inputs but not the impulse input. If we had used a hold circuit in the system in Example 7.6, the input to $G(s) =$

$1/(s + 4)$, would be rectangular pulse instead of impulses and the responses $y(t)$ would be a smooth function without jump discontinuities. Without the hold circuit, however, the impulse input to $1/(s + 4)$ causes jump discontinuities. In general, if a hold circuit is used and if the denominator of $G(s)$ is of an order higher than that of the numerator, then the output $y(t)$ will be smooth without discontinuities. If a hold circuit is not used, then the order of the denominator of $G(s)$ must be at least 2 degrees higher than the order of the numerator in order to have smooth $y(t)$. This result follows from the initial-value theorem (Eq. 4.40). This is seen from the fact that the value of $sY(s)$, as $s \rightarrow \infty$, is 0 for the impulse input to $G(s)$ only if the order of the denominator of $G(s)$ is at least 2 degrees higher that the order of its numerator. Under this condition, the initial value of $y(t)$ has no jump discontinuity.

7.14   DESIGN OF SAMPLED-DATA SYSTEMS

Sampled-data systems in general are designed to meet certain transient (PO, t_r, t_s, etc.) and steady-state specification. The design procedure follows along lines similar to those used for continuous-time systems. In sampled-data systems we deal with a discrete-time model, such as shown in Fig. 7.20b. We begin with a second-order system. The relationship between the pole locations of open-loop transfer function and the corresponding PO, t_r, t_s, . . ., are determined Hence for a given transient specifications, an acceptable region in the z-plane is determined where the dominant poles of the closed-loop transfer function $T(z)$ should lie. Next we sketch the root locus for the system. The characteristic equation of the system in Fig. 7.20b is

$$1 + G(z) = 0$$

Hence we can use the root-locus techniques developed for continuous-time systems (Sec. 5.6). If the root locus passes through the acceptable region, the transient specifications can be met by simple adjustment of the gain parameter K. If not, we must use a compensator which will steer the root locus in the desired region.

From Fig. 7.7 we know that to reduce PO and t_s the dominant poles of $T(z)$ must move toward the origin. To decrease the rise time, the dominant poles must move anticlockwise (toward the left). The exact relationships are somewhat more complex that those for continuous-time systems. The reader is refered to any standard book of discrete-time systems.†

7.15   STATE-SPACE ANALYSIS OF DISCRETE-TIME SYSTEMS

We have shown that the nth-order differential equation can be expressed in terms of n first-order differential equations. Following analogous procedure, we can show that nth-order difference equation can be expressed in terms of

---

†B. C. Kuo, *Analysis and Synthesis of Sampled-Data Systems* (Englewood Cliffs, N.J.: Prentice-Hall, 1963).

n first-order difference equations. Consider a z-transfer function

$$H(z) = \frac{b_m z^m + b_{m-1} z^{m-1} + \cdots + b_1 z + b_0}{z^n + a_{n-1} z^{n-1} + \cdots + a_1 z + a_0} \tag{7.88}$$

If $f(k)$ and $y(k)$ are the input and the output respectively for this system, then $y(k)$ and $f(k)$ are related by a difference equation:

$$(E^n + a_{n-1} E^{n-1} + \cdots + a_1 E + a_0) y(k)$$
$$= (b_m E^m + b_{m-1} E^{m-1} + \cdots + b_1 E + b_0) f(k) \tag{7.89}$$

Using techniques developed in Sec. 4.14, we simulate this system in Kalman's first form (phase variables) as shown in Fig. 7.26. Let the signals appearing at the outputs of *n* delay elements be denoted by $x_1(k)$, $x_2(k), \ldots, x_n(k)$ as shown in Fig. 7.26. It is obvious that the input of the first delay is $x_n(k + 1)$. We can now write *n* equations at the input of each delay as

$$
\begin{aligned}
x_1(k + 1) &= x_2(k) \\
x_2(k + 1) &= x_3(k) \\
&\cdots\cdots\cdots\cdots \\
x_{n-1}(k + 1) &= x_n(k) \\
x_n(k + 1) &= -a_0 x_1(k) - a_2 x_2(k) - \cdots - a_{n-1} x_n(k) + f(k)
\end{aligned}
\tag{7.90}
$$

and

$$y(k) = b_0 x_1(k) + b_1 x_2(k) + \cdots + b_m x_{m+1}(k) \tag{7.91}$$

Equations 7.90 are *n* first-order difference equations in *n* variables $x_1(k)$, $x_2(k)$, and $x_n(k)$. These variables will be immediately recognized as state variables, since specification of initial values of these variables in Fig. 7.26 will uniquely determine the response $y(k)$ for a given $f(k)$. Thus Eqs. 7.90 represent state equations and Eq. 7.91 is the output equation. In matrix form we can write these equations as

$$
\underbrace{\begin{bmatrix} x_1(k+1) \\ x_2(k+1) \\ \cdots \\ x_{n-1}(k+1) \\ x_n(k+1) \end{bmatrix}}_{\mathbf{x}(k+1)}
=
\underbrace{\begin{bmatrix} 0 & 1 & 0 & \cdots & 0 & 0 \\ 0 & 0 & 1 & \cdots & 0 & 0 \\ \cdot & \cdot & \cdot & \cdots & \cdot & \cdot \\ 0 & 0 & 0 & \cdots & 0 & 1 \\ -a_0 & -a_1 & -a_2 & \cdots & -a_{n-2} & -a_{n-1} \end{bmatrix}}_{\mathbf{A}}
\underbrace{\begin{bmatrix} x_1(k) \\ x_2(k) \\ \cdots \\ x_{n-1}(k) \\ x_n(k) \end{bmatrix}}_{\mathbf{x}(k)}
+
\underbrace{\begin{bmatrix} 0 \\ 0 \\ \cdots \\ 0 \\ 1 \end{bmatrix}}_{\mathbf{B}} f(k)
$$

$$\tag{7.92a}$$

and

$$y(k) = \underbrace{[b_0 \quad b_1 \cdots b_m]}_{\mathbf{C}} \begin{bmatrix} x_1(k) \\ x_2(k) \\ \vdots \\ x_{m+1}(k) \end{bmatrix} \tag{7.92b}$$

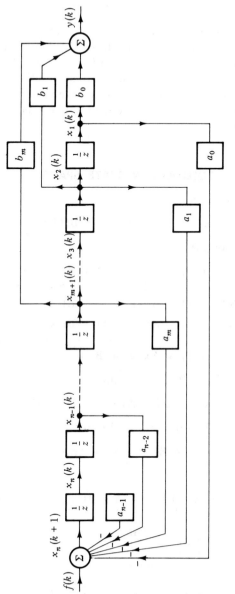

Figure 7.26

In general,

$$\mathbf{x}(k+1) = \mathbf{A}\mathbf{x}(k) + \mathbf{B}\mathbf{f}(k) \tag{7.93a}$$

$$\mathbf{y}(k) = \mathbf{C}\mathbf{x}(k) + \mathbf{D}\mathbf{f}(k) \tag{7.93b}$$

There may be more than one output, in general. Here we have represented a discrete-time system by state equations in phase variable form. There are several other possible representations, as discussed in Chapter 6. We may, for example, simulate the transfer function by a series, parallel, or other the Kalman's form. The output of each delay element qualifies as a state variable. We can now write the equation at the input of each delay element. The n equations thus obtained are the n state equations.

SOLUTION IN STATE SPACE

Consider the state equation

$$\mathbf{x}(k+1) = \mathbf{A}\mathbf{x}(k) + \mathbf{B}\mathbf{f}(k) \tag{7.94}$$

From this equation it follows that

$$\mathbf{x}(k) = \mathbf{A}\mathbf{x}(k-1) + \mathbf{B}\mathbf{f}(k-1) \tag{7.95a}$$

and

$$\mathbf{x}(k-1) = \mathbf{A}\mathbf{x}(k-2) + \mathbf{B}\mathbf{f}(k-2) \tag{7.95b}$$

$$\mathbf{x}(k-2) = \mathbf{A}\mathbf{x}(k-3) + \mathbf{B}\mathbf{f}(k-3) \tag{7.95c}$$

$$\cdots\cdots\cdots\cdots\cdots\cdots\cdots\cdots\cdots$$

$$\mathbf{x}(1) = \mathbf{A}\mathbf{x}(0) + \mathbf{B}\mathbf{f}(0)$$

Substituting Eq. 7.95b in Eq. 7.95a, we obtain

$$\mathbf{x}(k) = \mathbf{A}^2\mathbf{x}(k-2) + \mathbf{A}\mathbf{B}\mathbf{f}(k-2) + \mathbf{B}\mathbf{f}(k-2)$$

Substituting Eq. 7.95c in this equation, we obtain

$$\mathbf{x}(k) = \mathbf{A}^3\mathbf{x}(k-3) + \mathbf{A}^2\mathbf{B}\mathbf{f}(k-3) + \mathbf{A}\mathbf{B}\mathbf{f}(k-2) + \mathbf{B}\mathbf{f}(k-1)$$

Continuing this way, we obtain†

$$\mathbf{x}(k) = \mathbf{A}^k\mathbf{x}(0) + \mathbf{A}^{k-1}\mathbf{B}\mathbf{f}(0) + \mathbf{A}^{k-2}\mathbf{B}\mathbf{f}(1) + \cdots + \mathbf{B}\mathbf{f}(k-1)$$

$$= \underbrace{\mathbf{A}^k\mathbf{x}(0)}_{\substack{\text{zero-input} \\ \text{component}}} + \underbrace{\sum_{j=0}^{k-1} \mathbf{A}^{k-1-j}\mathbf{B}\mathbf{f}(j)}_{\text{zero-state component}} \tag{7.96}$$

The matrix $\mathbf{A}^k$ can be computed easily by the formula derived later in Eq. 7.98.

---

†The zero-state component in Eq. 7.96 is actually a convolution summation of matrices $\mathbf{A}^{k-1}u(k-1)$ and $\mathbf{B}\mathbf{f}(j)$.

z-Transform Solution

The z-transform of Eq. 7.94 is given by

$$z\mathbf{X}(z) - z\mathbf{x}(0) = \mathbf{A}\mathbf{X}(z) + \mathbf{B}\mathbf{F}(z)$$

Therefore

$$(z\mathbf{I} - \mathbf{A})\mathbf{X}(z) = z\mathbf{x}(0) + \mathbf{B}\mathbf{F}(z)$$

and

$$\begin{aligned}
\mathbf{X}(z) &= (z\mathbf{I} - \mathbf{A})^{-1}z\mathbf{x}(0) + (z\mathbf{I} - \mathbf{A})^{-1}\mathbf{B}\mathbf{F}(z) \\
&= (\mathbf{I} - z^{-1}\mathbf{A})^{-1}\mathbf{x}(0) + (z\mathbf{I} - \mathbf{A})^{-1}\mathbf{B}\mathbf{F}(z)
\end{aligned} \tag{7.97a}$$

Hence

$$\mathbf{x}(k) = \underbrace{\mathfrak{z}^{-1}[(\mathbf{I} - z^{-1}\mathbf{A})^{-1}]x(0)}_{\substack{\text{zero-input} \\ \text{component}}} + \underbrace{\mathfrak{z}^{-1}[(z\mathbf{I} - \mathbf{A})^{-1}\mathbf{B}\mathbf{F}(z)]}_{\substack{\text{zero-state} \\ \text{component}}} \tag{7.97b}$$

Comparision of Eq. 7.96 with Eq. 7.97 shows that

$$\mathbf{A}^k = \mathfrak{z}^{-1}[(\mathbf{I} - z^{-1}\mathbf{A})^{-1}] \tag{7.98}$$

The output equation is given by

$$\begin{aligned}
\mathbf{Y}(z) &= \mathbf{C}\mathbf{X}(z) + \mathbf{D}\mathbf{F}(z) \\
&= \mathbf{C}[(\mathbf{I} - z^{-1}\mathbf{A})^{-1}\mathbf{x}(0) + (z\mathbf{I} - \mathbf{A})^{-1}\mathbf{B}\mathbf{F}(z)] + \mathbf{D}\mathbf{F}(z) \\
&= \mathbf{C}(\mathbf{I} - z^{-1}\mathbf{A})^{-1}\mathbf{x}(0) + [\mathbf{C}(z\mathbf{I} - \mathbf{A})^{-1}\mathbf{B} + \mathbf{D}]\mathbf{F}(z) \\
&= \underbrace{\mathbf{C}(\mathbf{I} - z^{-1}\mathbf{A})^{-1}\mathbf{x}(0)}_{\text{zero-input response}} + \underbrace{\mathbf{H}(z)\mathbf{F}(z)}_{\substack{\text{zero-state} \\ \text{response}}}
\end{aligned} \tag{7.99a}$$

where

$$\mathbf{H}(z) = \mathbf{C}(z\mathbf{I} - \mathbf{A})^{-1}\mathbf{B} + \mathbf{D} \tag{7.99b}$$

Note that $\mathbf{H}(z)$ is the transfer function matrix of the system, and $H_{ij}(z)$, the ijth element of $\mathbf{H}(z)$, is the transfer function relating the output $y_i(k)$ to the input $f_j(k)$. If we define $\mathbf{h}(k)$ as

$$\mathbf{h}(k) = \mathfrak{z}^{-1}[\mathbf{H}(z)]$$

then $\mathbf{h}(k)$ represents a unit function response matrix of the system. Thus $h_{ij}(k)$, the ijth element of $\mathbf{h}(k)$, represents the zero-state response $y_i(k)$ when the input $f_j(k) = \partial(k)$, and all other inputs are zero.

7.16   STATE-SPACE ANALYSIS OF SAMPLED-DATA SYSTEMS

For sampled-data systems we may use a discrete-time model or a continuous-time model. The former is good only for the solution at sampling instants, whereas the latter can yield response for all values of t. We shall demonstrate the techniques for discrete-time as well as continuous-time model by an example.

Example 7.8.   Analyze the sampled-data system in Fig. 7.24 by using discrete-time and continuous-time models.

Discrete-Time Model.   In order to find the discrete model, let us refer to Fig. 7.24.   The transfer function $G_1(z)$ corresponding to $G_1(s)$ was found to be (Eq. 7.87a)

$$G_1(z) = \frac{z(0.368z + 0.264)}{(z - 1)^2(z - 0.368)}$$

Hence the open-loop transfer function (Fig. 7.24) is

$$\left(\frac{z - 1}{z}\right)\left(\frac{z(0.368z + 0.264)}{(z - 1)^2(z - 0.368)}\right) = \frac{0.368z + 0.264}{(z - 1)(z - 0.368)}$$

In order to write state equations for this system, we observe that the output of every first-order subsystem qualifies as a state variable.   Hence we break up this transfer function into two first-order subsystems as shown in Fig. 7.27.   This is the discrete-time model of the system in Fig. 7.24.   The variables $x_1(k)$ and $x_2(k)$ are the desired state variables.   From Fig. 7.27 we can immediately write state equations

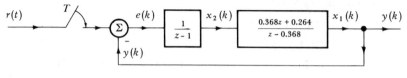

Figure 7.27

from the input-output relationship of these subsystems as

$$x_2(k + 1) = x_2(k) + e(k) = x_2(k) + r(k) - x_1(k) \qquad (7.100a)$$

$$x_1(k + 1) - 0.368x_1(k) = 0.368x_2(k + 1) + 0.264x_2(k)$$
$$= 0.368[x_2(k) + r(k) - x_1(k)] + 0.264x_2(k)$$

or

$$x_1(k + 1) = 0.632x_2(k) + 0.368r(k) \qquad (7.100b)$$

Equations 7.100a and 7.100b are the desired state equations.   We can express them in matrix form as

$$\begin{bmatrix} x_1(k + 1) \\ x_2(k + 1) \end{bmatrix} = \begin{bmatrix} 0 & 0.632 \\ -1 & 1 \end{bmatrix} \begin{bmatrix} x_1(k) \\ x_2(k) \end{bmatrix} + \begin{bmatrix} 0.368 \\ 1 \end{bmatrix} r(k) \qquad (7.101a)$$

and

$$y(k) = x_1(k) \qquad (7.101b)$$

The state equation can be now solved by methods discussed in Sec. 7.15.   Note that this model gives solution only at discrete instants of time.

Continuous-Time Model.   For continuous-time model analysis, we represent the system in Fig. 7.24 as shown in Fig. 7.28.   The state variables x_1 and x_2 are

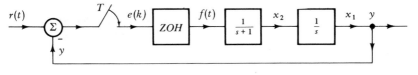

Figure 7.28

defined in the figure. We can write state equations by inspection of this figure as
follows:

$$\dot{x}_1 = x_2$$
$$\dot{x}_2 = -x_2 + f(t)$$

or

$$\begin{bmatrix} \dot{x}_1 \\ \dot{x}_2 \end{bmatrix} = \begin{bmatrix} 0 & 1 \\ 0 & -1 \end{bmatrix}\begin{bmatrix} x_1 \\ x_2 \end{bmatrix} + \begin{bmatrix} 0 \\ 1 \end{bmatrix}f(t)$$

$$\underbrace{\phantom{\dot{x}}}_{\dot{x}} \qquad \underbrace{}_{A} \quad \underbrace{}_{x} \quad \underbrace{}_{B} \quad f$$

$$\dot{x} = Ax + Bf$$

The input $f(t)$ is the output of the hold circuit. Hence

$$f(t) = e(k) \qquad kT \le t < (K + 1)T$$

The solution of the state equation is

$$x(t) = e^{A(t-t_0)}x(t_0) + \int_{t_0}^{t} e^{A(t-\tau)}Bf(\tau)\,d\tau \qquad (7.102)$$

letting $t_0 = kT$ and $t = (k + 1)T$, we obtain

$$x(k + 1) = e^{AT}x(k) + \int_{kT}^{(k+1)T} e^{A[(k+1)T-\tau]}Bf(\tau)\,d\tau \qquad (7.103)$$

During the interval $kT \le t < (k + 1)T, f(t)$ is given by

$$f(t) = [r(k) - x_1(k)]\delta(t - kT) \qquad \text{for no hold}$$
$$= r(k) - x_1(k) \qquad \text{with hold circuit}$$

Since we have a hold circuit in the system,

$$f(t) = r(k) - x_1(k)$$

and

$$x(k + 1) = e^{AT}x(k) + \int_{kT}^{(k+1)T} e^{A[(k+1)T-\tau]}B[r(k) - x_1(k)]\,d\tau$$

$$= e^{AT}x(k) + [r(k) - x_1(k)]\int_{kT}^{(k+1)T} e^{A[(k+1)T-\tau]}B\,d\tau$$

Letting $(k + 1)T - \tau = p$, we obtain

$$\mathbf{x}(k + 1) = e^{\mathbf{A}T}\mathbf{x}(k) + [r(k) - x_1(k)] \int_0^T e^{\mathbf{A}p}\mathbf{B}\, dp$$

Further, let

$$\int_0^T e^{\mathbf{A}p}\mathbf{B}\, dp = \boldsymbol{\beta}$$

This yields

$$\mathbf{x}(k + 1) = e^{\mathbf{A}T}\mathbf{x}(k) - \boldsymbol{\beta}x_1(k) + \boldsymbol{\beta}r(k)$$

If we define

$$\hat{\boldsymbol{\beta}} = [\boldsymbol{\beta} \quad 0]$$

then

$$\boldsymbol{\beta}x_1(k) = [\boldsymbol{\beta} \quad 0] \begin{bmatrix} x_1(k) \\ x_2(k) \end{bmatrix} = \hat{\boldsymbol{\beta}}\mathbf{x}(k)$$

and

$$\mathbf{x}(k + 1) = (e^{\mathbf{A}T} - \hat{\boldsymbol{\beta}})\mathbf{x}(k) + \boldsymbol{\beta}r(k) \qquad (7.104)$$

This is a state equation of the form of Eq. 7.93a, and can be solved by techniques discussed in Sec. 7.15. The solution of this equation yields the response only at the sampling instants. If a solution during a sampling interval $kT \le t < (k + 1)T$ is desired, we should use Eq. 7.102 with $t_0 = kT$. In this method we need the knowledge of $f(t) = r(k) - x_1(k)$ during this interval. Hence we must first solve Eq. 7.104 to obtain $x_1(k)$, and then substitute the value of $f(t)$ in Eq. 7.102.

PROBLEMS

7.1. A person borrows M dollars at an interest of β dollars/dollar/month from a bank. He pays regularly $x(k)$ dollars at the beginning of kth month. Write the equation for the outstanding mortgage $y(k)$ at the beginning of the kth month. Simulate this problem using delay elements.

7.2. In a simple model of a national economy, we shall assume

$$y(k) = c(k) + i(k)$$

where $y(k)$, $c(k)$, and $i(k)$ are the national income, the annual consumption, and the investment for the kth year respectively. The growth of the national income is assumed to be proportional to the investment. Thus

$$y(k + 1) - y(k) = \alpha i(k)$$

where α is a constant, known as the growth factor. In addition, it is assumed that the consumption is linearly related to the national income. Thus

$$c(k) = A + \beta y(k)$$

where A and β are constants. Find the equation for $y(k)$. Simulate this system using delay elements.

7.3. Solve the following difference equations.

(a) $y(k + 1) + 2y(k) = 0,$ $\quad y(0) = 1$
(b) $y(k + 2) + 3y(k + 1) + 2y(k) = 0,$ $\quad y(0) = 2,$ $y(1) = 1$
(c) $y(k + 2) + 2y(k + 1) + 2y(k) = 0,$ $\quad y(0) = 0,$ $y(1) = 1$
(d) $y(k + 2) + 2y(k + 1) + y(k) = 0,$ $\quad y(0) = y(1) = 1$

7.4. Find $y(k)$ in Prob. 7.1 if $M = 10,000,$ $\beta = .005,$ and $x(k) = 100$. Determine the value of k when the mortgage will be paid off.

7.5. In Prob. 7.2 find the solution for the national income $y(k)$.

7.6. Solve the following difference equations.

(a) $y(k + 1) + 2y(k) = f(k + 1);$ $\quad f(k) = e^{-k},$ $y(0) = 0$
(b) $y(k + 1) + 2y(k) = f(k);$ $\quad f(k) = e^{-k},$ $y(0) = 0$
(c) $y(k + 2) + 3y(k + 1) + 2y(k) = (3)^k;$ $\quad y(0) = y(1) = 0$
(d) $y(k + 2) + 2y(k + 1) + 2y(k) = f(k + 1) + 2f(k); f(k) = e^k, y(0) = y(1) = 0$

7.7. Solve Prob. 7.3, using the z-transform.

7.8. Solve Prob. 7.6, using the z-transform.

7.9. A digital filter in Fig. 7.11 has

$$G(z) = \frac{z^2 + 2z + 1}{z^2 - 1.142z + 0.4124}$$

The sampling rate is 10,000 samples/sec. Sketch the frequency response $|H(j\omega)|$ of this filter. Determine the filter bandwidth. What is the highest frequency that can be handled by this filter without running into aliasing effect? Simulate this filter using delay elements.

7.10. Simulate the following z-transfer functions using delay elements.

(a) $G(z) = \dfrac{z(z + 2)}{(z - 0.8)(z + 0.4)(z - 0.6)}$

(b) $G(z) = \dfrac{2z + 1}{z(z - 1)(z - 0.5)^2}$

(c) $G(z) = \dfrac{z^3}{(z - 0.4)(z^2 - 0.6z + 0.25)}$

7.11. Investigate the suitability of a cascade of two identical ZOH circuits for the hold operation (for reconstruction of a continuous signal from its sample train of impulses). Specifically, determine the unit impulse response of the circuit and sketch the output for a typical impulse train input. Determine and sketch the frequency response $|H(j\omega)|$ and comment on the filter characteristic of the circuit.

7.12. Figure P-7.12 shows a simplified model of a single-axis attitude control system of a lunar excursion module (LEM) in Prob. 5.17 (Fig. P-5.17). Draw the discrete

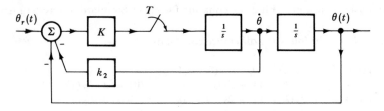

Figure P-7.12

time model of this system, and determine the z-transfer function relating the output samples $\theta(k)$ to the input samples $\theta_r(k)$.

7.13.   Repeat Prob. 7.12 for the system in Fig. P-7.13.

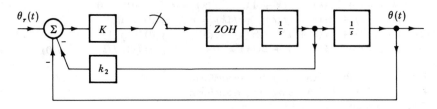

Figure P-7.13

7.14.   Determine the output $y(t)$ of systems in Fig. P-7.14, if the input $f(t) = e^{-2t}u(t)$.

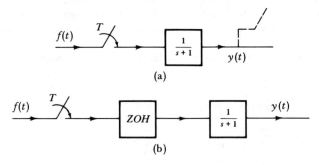

Figure P-7.14

7.15.   Write state equations for the system in Fig. P-7.12.
7.16.   Write state equations for the system in Fig. P-7.13.

A

Some Properties of Differential Operators

A.1 CANCELLATION OF COMMON FACTORS IN OPERATIONAL EQUATIONS

Consider the differential equation:

$$(p + a)N(p)x = (p + a)w \tag{A.1}$$

We shall now show that cancellation of the common factor $(p + a)$ in Eq. A.1 is not valid. We can rewrite Eq. A.1 as

$$(p + a)N(p)x - (p + a)w = 0$$

or

$$(p + a)[N(p)x - w] = 0 \tag{A.2}$$

If we let

$$N(p)x - w = u \tag{A.3}$$

then, from Eq. A.2, we have

$$(p + a)u = 0$$

This is a first-order differential equation of the form in Eq. 2.35 (chapter 2). Hence the solution for u is given by (see Eq. 2.36)

$$u = ce^{-at}$$

or

$$N(p)x - w = ce^{-at}$$

and

$$N(p)x = w + ce^{-at}$$
$$\neq w \tag{A.4}$$

If follows from Eq. A.4, that the cancellation of common factor in an operational equation is not valid.

A.2   CANCELLATION OF COMMON FACTORS IN THE NUMERATOR AND THE DENOMINATOR OF A RATIONAL FRACTION OPERATOR

We shall now show that if $D(p)$ is a polynomial in positive power of p, then

$$D(p)\frac{1}{D(p)} = 1 \tag{A.5}$$

But

$$\frac{1}{D(p)}D(p) \neq 1 \tag{A.6}$$

To prove these results, let us consider the differential equation

$$\frac{1}{D(p)}x = y \tag{A.7}$$

From the definition of inverse operator it follows that

$$x = D(p)y \tag{A.8}$$

Multiplying both sides of Eq. A.7 by $D(p)$, we obtain

$$D(p)\left[\frac{1}{D(p)}x\right] = D(p)y \tag{A.9}$$

Comparison of Eq. A.8 with Eq. A.9 yields

$$D(p)\left[\frac{1}{D(p)}x\right] = x \tag{A.10}$$

or

$$D(p)\frac{1}{D(p)} = 1$$

It is evident from Eq. A.10 that the common factor $D(p)$ in the numerator and the denominator can be canceled if it occurs in the form shown in Eq. A.10. On the other hand, if the common factor $D(p)$ occurs as shown in Eq. A.6, such a cancellation is not permitted. To prove this, let

$$\frac{1}{D(p)}D(p)x = u \tag{A.11}$$

Then

$$D(p)x = D(p)u$$

and

$$D(p)(x - u) = 0 \tag{A.12}$$

This is a homogeneous differential equation and $x - u \neq 0$. As an example if $D(p) = p + a$, then Eq. A.12 is given by

$$(p + a)(x - u) = 0$$

This is a first-order differential equation of the form in Eq. 2.35 (see Chap. 2). Hence the solution for $x - u$ is given by (see Eq. 2.36)

$$x - u = ce^{-at}$$

Hence

$$u = x - ce^{-at}$$

and

$$\frac{1}{D(p)}[D(p)x] = x - ce^{-at}$$

$$\neq x$$

It is evident that the common factor $D(p)$ cannot be canceled in this case. It follows from this discussion that

$$p\left(\frac{1}{p}\right) = 1 \tag{A.13}$$

but

$$\left(\frac{1}{p}\right)p \neq 1$$

A.3 COMMON FACTOR IN THE NUMERATOR AND THE DENOMINATOR OF $H(p)$

Occasionally we observe occurrence of a common factor in the numerator and the denominator of a transfer function $H(p)$. We saw in Appendix A.2 that a common factor may be canceled only under particular conditions.   For example, if

$$H(p) = G(p)\frac{N(p)}{G(p)D(p)} \tag{A.14}$$

Then we cancel $G(p)$ to yield

$$H(p) = \frac{N(p)}{D(p)}$$

However, if

$$H(p) = \frac{1}{G(p)}\frac{G(p)N(p)}{D(p)} \tag{A.15}$$

we cannot cancel the common factor $G(p)$.

It is interesting to understand the physical significance of these two possibilities. What do they mean in terms of the actual structure of the system? To answer this question we must first understand that canceling a common factor does not affect zero-state component (see Sec. 2.6). It will affect only zero-input component that arises due to common factor $G(p)$. It can be shown that when a transfer function is of the form in Eq. A.14, the system structure is as shown in Fig. A.1.

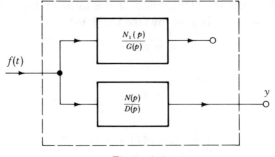

Figure A.1

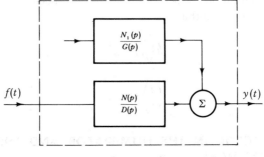

Figure A.2

The system can be represented as two decoupled subsystems as shown in Fig. A.1. The subsystem with transfer function $N_1(p)/G(p)$ does not contribute to the output. Consequently this system has no effect on the subsystem coupled to $f(t)$ and y. Consequently we can cancel $G(p)$.

Consider now the system in Fig. A.2. The upper subsystem $N_1(p)/G(p)$ has no input. Consequently its output is its zero-input response. This output x is being added to the output of the subsystem. Hence the output y actually has a contribution due to $G(p)$ in its zero-input component, and $G(p)$ cannot be ignored. Thus the situation in Eq. A.15 arises when the system structure is as shown in Fig. A.2. Note that in most problems the system structures are not so obvious as shown in Figs. A.1 and A.2. By special transformations, however, one can express a system with common factor in its numerator and denominator of $H(p)$ in one of the forms shown in Figs. A.1 and A.2. These ideas are closely connected with the concepts of controllability and observability of a system.

In analysis problems, if any doubt exists as to the correctness of cancellation one should include the term in the zero-input response arising due to the common factor. If the common factor is cancellable (structure as in Fig. A.1), the coefficient of the terms in the zero input response as found from Eq. 2.47c, will be zero for all physically possible initial conditions.

B

Partial-Fraction Expansion

In analysis of linear time-invariant systems we encounter functions that are the ratio of two polynomials (in p, $j\omega$, or s). Such functions are known as *rational algebraic fractions*. The function $F(x)$ below is such an example.

$$F(x) = \frac{b_m x^m + b_{m-1} x^{m-1} + \cdots + b_1 x + b_0}{x^n + a_{n-1} x^{n-1} + \cdots + a_1 x + a_0} \tag{B.1}$$

$$= \frac{N(x)}{D(x)} \tag{B.2}$$

There are two possibilities in regard to relative magnitudes of m and n: (1) $m \geq n$, and (2) $m < n$. The function is known as *improper fraction* if $m \geq n$, and is a *proper fraction* if $m < n$. If the fraction is improper, it can always be separated into a sum of a polynomial in x and a proper fraction. Consider for example, the function

$$F(x) = \frac{2x^4 + 3x^3 + x^2 + 2x}{x^2 + 4x + 3}$$

This is an improper fraction and can be expressed as

$$F(x) = \underbrace{2x^2 - 5x + 15}_{\substack{\text{polynomial in} \\ \text{positive powers of } x}} - \underbrace{\frac{43x + 45}{x^2 + 4x + 3}}_{\text{proper fraction}} \tag{B.3}$$

A proper fraction can be further expanded into partial fractions. The remaining discussion in this portion of the Appendix concerns with techniques of expanding a proper fraction into partial fractions.

B-1. ALL ROOTS SIMPLE (NO MULTIPLE ROOTS)

Consider the proper fraction

$$F(x) = \frac{b_m x^m + b_{m-1} x^{m-1} + \cdots + b_1 x + b_0}{x^m + a_{n-1} x^{n-1} + \cdots + a_1 x + a_0} = \frac{N(x)}{D(x)}$$

We shall factorize the denominator $D(x)$ into m first-order factors:

$$F(x) = \frac{N(x)}{(x - \lambda_1)(x - \lambda_2)\cdots(x - \lambda_n)} \qquad \text{(B.4a)}$$

It can be shown that $F(x)$ in Eq. B.4a can be expressed in partial fractions as

$$F(x) = \frac{k_1}{x - \lambda_1} + \frac{k_2}{x - \lambda_2} + \cdots + \frac{k_n}{x - \lambda_n} \qquad \text{(B.4b)}$$

In order to find coefficient k_1 we multiply both sides of Eq. B.4b by $x - \lambda_1$. This yields

$$(x - \lambda_1)F(x) = k_1 + k_2 \frac{x - \lambda_1}{x - \lambda_2} + \cdots + k_n \frac{x - \lambda_1}{x - \lambda_n}$$

Letting $x = \lambda_1$ or $x - \lambda_1 = 0$, we obtain

$$(x - \lambda_1)F(x)\,|_{x=\lambda_1} = k_1$$

or in general,

$$(x - \lambda_j)F(x)\,|_{x=\lambda_j} = k_j \qquad \text{(B.5)}$$

Note that $(x - \lambda_1)F(x)$ is the right-hand side of Eq. B.4a with the term $(x - \lambda_1)$ missing in the denominator. Thus to find coefficient k_1 corresponding to the term $(x - \lambda_1)$ we consider $F(x)$, ignore the term $(x - \lambda_1)$ in the denominator, and then let $x = \lambda_1$ in the rest of the expression. This yields k_1. This operation can be performed by concealing the term $(x - \lambda_1)$ by finger on the paper and letting $x = \lambda_1$ in the rest of the expression. Consider

$$F(x) = \frac{4x + 9}{x^2 + 5x + 6} \qquad \text{(B.6a)}$$

$$= \frac{4x + 9}{(x + 2)(x + 3)} \qquad \text{(B.6b)}$$

$$= \frac{k_1}{x + 2} + \frac{k_2}{x + 3}$$

To find the coefficient k_1 corresponding to the factor $(x + 2)$, conceal this factor $(x + 2)$ in Eq. B.6b by finger, and let $x = -2$ in the rest of the expression. This yields

$$k_1 = \frac{-8 + 9}{-2 + 3} = 1$$

Similarly, to find k_2, conceal $(x + 3)$ in Eq. 6b and let $x = -3$. This yields

$$k_2 = \frac{-8 + 9}{-3 + 2} = 3$$

and

$$F(x) = \frac{1}{x + 2} + \frac{3}{x + 3}$$

B.2   COMPLEX ROOTS

The procedure outlined above holds regardless of whether the roots of $D(x) = 0$ are real or complex. Consider for example

$$F(x) = \frac{2x + 3}{(x + 1)(x^2 + 4x + 5)}$$

$$= \frac{2x + 3}{(x + 1)(x + 2 - j1)(x + 2 + j1)} \tag{B.7}$$

In this case the roots of $D(x) = 0$ are -1, $-(2 + j1)$ and $-(2 - j1)$. The two complex roots are conjugates of each other. This is not a coincidence. If all the co-efficients of the polynomial $D(x)$ are real, then if there are any complex roots of $D(x) = 0$, they must occur in pairs of conjugates. Otherwise the product of all factors will not yield a polynomial with real coefficients. In all linear lumped systems, we come across functions (such as system functions) which are ratio of two polynomials with real coefficients.

We can now express $F(x)$ in Eq. B.7 by partial fractions as

$$F(x) = \frac{k_1}{x + 1} + \frac{k_2}{x + 2 - j1} + \frac{k_3}{x + 2 + j1}$$

As usual,

$$k_1 = (x + 1)F(x)\Big|_{x=-1} = \frac{2x + 3}{x^2 + 4x + 5}\Big|_{x=-1} = \frac{-2 + 3}{1 - 4 + 5} = \frac{1}{2}$$

$$k_2 = (x + 2 - j1)F(x)\Big|_{x=-2+j1} = \frac{2x + 3}{(x + 1)(x + 2 + j1)}\Big|_{x=-2+j1}$$

$$= \frac{-4 + j2 + 3}{(-1 + j1)(j2)}$$

$$= \frac{-1 - j3}{4} = \frac{\sqrt{10}}{4} e^{j(253°)} = \frac{\sqrt{10}}{4} \angle 253° \tag{B.8}$$

and

$$k_3 = (x + 2 + j1)F(x)\Big|_{x=-2-j1} = \frac{2x + 3}{(x + 1)(x + 2 - j1)}\Big|_{x=-2-j1}$$

$$= \frac{-4 - j2 + 3}{(-1 - j1)(-j2)}$$

$$= \frac{-1 + j3}{4} = \frac{\sqrt{10}}{4} e^{-j(253°)} = \frac{\sqrt{10}}{4} \angle -253° \tag{B.9}$$

and

$$F(x) = \frac{1/2}{x + 1} + \frac{-(1 + j3)/4}{2 + 2 - j1} + \frac{-(1 - j3)/4}{x + 2 + j1} \tag{B.10}$$

It is evident that k_2 and k_3, the coefficients corresponding to complex conjugate roots, are also conjugates of each other.

This result is general and can be easily proved.† Thus for complex conjugate roots it is necessary to compute the coefficient corresponding to one root only. The other coefficient is just a conjugate.

In our problems it proves most convenient to express the complex coefficients (corresponding to complex roots) in polar form. A complex number k can be expressed in polar form as

$$k = |k| e^{j\theta} = |k| \angle \theta$$

Thus Eq. B.10 can be expressed as

$$F(x) = \frac{1/2}{x + 1} + \frac{\frac{\sqrt{10}}{4} \angle 253°}{x + 2 - j1} + \frac{\frac{\sqrt{30}}{4} \angle -253°}{x + 2 + j1} \tag{B.11}$$

B.3   REPEATED (MULTIPLE) ROOTS

The case of repeated roots can be handled as follows. Let

$$F(x) = \frac{N(x)}{(x - \lambda_1)^r (x - \lambda_{r+1}) \cdots (x - \lambda_n)} \tag{B.12}$$

It can be shown that the partial fractions for this case are of the form

$$F(x) = \frac{a_0}{(x - \lambda_1)^r} + \frac{a_1}{(x - \lambda_1)^{r-1}} + \cdots + \frac{a_{r-1}}{x - \lambda_1}$$

$$+ \frac{k_{r+1}}{x - \lambda_{r+1}} + \cdots + \frac{k_n}{x - \lambda_n} \tag{B.13}$$

---

†This may be proved as follows. Let

$$F(x) = \frac{N_1(x)}{D_1(x)(x - \lambda)(x - \lambda^*)}$$

where $N_1(x)$ and $D_1(x)$ are polynomials in x with real coefficients. We can expand $F(x)$ into partial fractions as

$$F(x) = \frac{c_1}{x - \lambda} + \frac{c_2}{x - \lambda^*} + \frac{N_2(x)}{D_1(x)}$$

where

$$c_1 = \frac{N_1(x)}{D_1(x)(x - \lambda^*)} \bigg|_{x=\lambda} = \frac{N_1(\lambda)}{D_1(\lambda)(\lambda - \lambda^*)}$$

and

$$c_2 = \frac{N_1(x)}{D_1(x)(x - \lambda)} \bigg|_{x=\lambda^*} = \frac{N_1(\lambda^*)}{D_1(\lambda^*)(\lambda^* - \lambda)}$$

Since $N_1(x)$ and $D_1(x)$ are polynomial with real coefficients, it can be seen that c_1 and c_2 are conjugates of each other.

The coefficients $k_{r+1}, \ldots, k_n$ corresponding to simple roots $\lambda_{r+1}, \ldots, \lambda_n$ are found as usual (Eq. B.5). To find coefficients $a_0, a_1, \ldots, a_{r-1}$ corresponding to the root λ_1 of multiplicity r, we multiply both sides of Eq. B.13 by $(x - \lambda_1)^r$:

$$(x - \lambda_1)^r F(x) = a_0 + a_1 (x - \lambda_1) + \cdots + a_{r-1} (x - \lambda_1)^{r-1} + k_{r+1} \frac{(x - \lambda_1)^r}{x - \lambda_{r+1}}$$

$$+ \cdots + k_n \frac{(x - \lambda_1)^r}{x - \lambda_n} \quad \text{(B.14)}$$

Let $x = \lambda_1$ or $x - \lambda_1 = 0$ in Eq. B.14. This yields

$$(x - \lambda_1)^r F(x) \big|_{x = \lambda_1} = a_0$$

Next differentiate Eq. B.14 with respect to x and then let $x - \lambda_1$. This yields

$$\frac{d}{dx} [(x - \lambda_1)^r F(x)] \big|_{x = \lambda_1} = a_1$$

Proceeding in this manner, we obtain

$$a_j = \frac{1}{j!} \frac{d^j}{dx^j} [(x - \lambda_1)^r F(x)] \big|_{x = \lambda_1} \qquad j = 0, 1, 2, \ldots, (r - 1) \quad \text{(B.15)}$$

As an example, consider

$$F(x) = \frac{4x^3 + 16x^2 + 23x + 13}{(x + 1)^3 (x + 2)}$$

$$= \frac{a_0}{(x + 1)^3} + \frac{a_1}{(x + 1)^2} + \frac{a_2}{(x + 1)} + \frac{k}{x + 2}$$

The coefficient k, corresponding to simple root -2, can be found by Eq. B.5 as

$$k = \frac{4x^3 + 16x^2 + 23x + 13}{(x + 1)^3} \bigg|_{x = -2} = \frac{-32 + 64 - 46 + 13}{-1} = 1$$

To find coefficients a_0, a_1, and a_2 corresponding to the multiple root at -1, we use Eq. B.15:

$$a_0 = (x + 1)^3 F(x) \big|_{x = -1} = \frac{4x^3 + 16x^2 + 23x + 13}{x + 2} \bigg|_{x = -1}$$

$$= \frac{-4 + 16 - 23 + 13}{1}$$

$$= 2$$

$$a_1 = \frac{d}{dx} \left[\frac{4x^3 + 16x^2 + 23x + 13}{x + 2} \right] \bigg|_{x = -1}$$

$$= \left[\frac{12x^2 + 32x + 23}{x + 2} - \frac{4x^3 + 16x^2 + 23x + 13}{(x + 2)^2} \right] \bigg|_{x = -1} = 1$$

$$a_2 = \frac{1}{2} \frac{d^2}{dx^2} \left[\frac{4x^3 + 16x^2 + 23x + 13}{x + 2} \right] \bigg|_{x = -1} = 3$$

C

Bode Plots

C.1 FREQUENCY RESPONSE

The plots of $|H(j\omega)|$ and $\theta(\omega)$ as functions of ω give us the information about the steady-state response of the system to sinusoidal signals of various frequencies. For this reason these plots are called *frequency-response plots*. Sketching of these plots is considerably facilitated by using logarithmic units. Let us consider a system with the transfer function

$$H(s) = \frac{K(s + a_1)(s + a_2)}{s(s + b_1)(s^2 + b_2 s + b_3)} \tag{C.1a}$$

where the second-order factor $(s^2 + b_2 s + b_3)$ is assumed to have complex (conjugate) roots. We shall rearrange Eq. C.1a in the form

$$H(s) = \frac{Ka_1 a_2}{b_1 b_3} \frac{\left(\dfrac{s}{a_1} + 1\right)\left(\dfrac{s}{a_2} + 1\right)}{s\left(\dfrac{s}{b_1} + 1\right)\left(\dfrac{s^2}{b_3} + \dfrac{b_2}{b_3}s + 1\right)} \tag{C.1b}$$

and

$$H(j\omega) = \frac{Ka_1 a_2}{b_1 b_2} \frac{\left(1 + \dfrac{j\omega}{a_1}\right)\left(1 + \dfrac{j\omega}{a_2}\right)}{j\omega\left(1 + \dfrac{j\omega}{b_1}\right)\left[1 + j\dfrac{b_2\omega}{b_3} + \dfrac{(j\omega)^2}{b_3}\right]} \tag{C.1c}$$

It is obvious from this equation that $H(j\omega)$ is a complex function of ω. The magnitude function $|H(j\omega)|$ and the phase function $\theta(\omega)$ are given by

$$|H(j\omega)| = \frac{Ka_1 a_2}{b_1 b_3} \frac{\left|1 + \dfrac{j\omega}{a_1}\right|\left|1 + \dfrac{j\omega}{a_2}\right|}{|j\omega|\left|1 + \dfrac{j\omega}{b_1}\right|\left|1 + j\dfrac{b_2\omega}{b_3} + \dfrac{(j\omega)^2}{b_3}\right|} \tag{C.2a}$$

and

$$\theta(\omega) = \angle\left(1 + \frac{j\omega}{a_1}\right) + \angle\left(1 + \frac{j\omega}{a_2}\right) - \angle j\omega$$

$$- \angle\left(1 + \frac{j\omega}{b_1}\right) - \angle\left[1 + \frac{jb_2\omega}{b_3} + \frac{(j\omega)^2}{b_3}\right] \quad \text{(C.2b)}$$

It can be seen from Eq. C.2b that the phase function consists of addition of only three kinds of terms: (i) the phase of $j\omega$, which is $90°$ for all values of ω, (ii) the phase of the first-order term of the form $1 + \frac{j\omega}{a}$, and (iii) the phase of the second-order term

$$\left[1 + \frac{jb_2\omega}{b_3} + \frac{(j\omega)^2}{b_3}\right]$$

One can plot these three basic phase functions for ω varying from 0 to ∞ and then use these plots to construct the phase function of any transfer function. Note that if a particular term is in the numerator its phase is added, and if it is in the denominator its phase is subtracted. This makes it easy to plot the phase function $\theta(\omega)$ as a function of ω. We shall consider this topic a little later. First let us turn our attention to the problem of magnitude function $|H(j\omega)|$. Unlike the phase function the computation of $|H(j\omega)|$ involves multiplication and division of various terms. This appears quite a formidable task especially when we have to plot this function for the entire range of ω (0 to ∞). How much nicer it would be if we only had to add (or subtract) some basic forms of magnitude function! But why just wish when we can actually make it happen by a simple trick of logarithmic conversion? We know that a log operation converts multiplication, and division to addition and subtraction. So instead of plotting $|H(j\omega)|$, we shall plot log $|H(j\omega)|$ if it will simplify our task. We can also take advantage of the fact that logarithmic units are being used in many engineering areas. The logarithmic unit is the decibel and is twenty times the logarithm of the quantity (log to the base 10). Therefore $20\log_{10}|H(j\omega)|$ is simply the log magnitude in decibels (db). So instead of plotting $|H(j\omega)|$, we shall plot $20\log_{10}|H(j\omega)|$ as a function of ω. These plots (log magnitude and phase) are called *Bode plots*.

Let us again consider the transfer function in Eq. C.2a. For this function log magnitude is

$$20\log|H(j\omega)| = 20\log\frac{Ka_1a_2}{b_1b_3} + 20\log\left|1 + \frac{j\omega}{a_1}\right| + 20\log\left|1 + \frac{j\omega}{a_2}\right|$$

$$- 20\log|j\omega| - 20\log\left|1 + \frac{j\omega}{b_1}\right| - 20\log\left|1 + \frac{j\omega b_2}{b_3} + \frac{(j\omega)^2}{b_3}\right| \quad \text{(C.3)}$$

The term $20\log(Ka_1a_2/b_1b_3)$ is a constant. We then observe that the log magnitude is a sum of four basic terms corresponding to (i) a constant (ii) a pole or zero at the origin, $20\log|j\omega|$, (iii) a first-order pole or zero, $20\log|1 + j\omega/a|$, and (iv) complex conjugate poles or zeros, $20\log|1 + j\omega b_2/b_3 + (j\omega)^2/$

b_3 |. We can sketch these four basic terms as functions of ω and use them to construct the log-magnitude plot. Let us discuss each of the terms.

1. CONSTANT Ka_1a_2/b_1b_3

The log magnitude of this term is also a constant, 20 log (Ka_1a_2/b_1b_3). The phase contribution due to this term is zero.

2. POLE (OR ZERO) AT THE ORIGIN

Log Magnitude

Such a pole gives rise to the term -20 log $|j\omega|$. It is obvious that

$$-20 \log |j\omega| = -20 \log \omega$$

This function can be plotted as a function of ω. However, we can effect further simplification by using logarithmic scale for the variable ω itself. Let us define a new variable u as

$$u = \log \omega \qquad\qquad (C.4)$$

Hence

$$-20 \log \omega = -20u \qquad\qquad (C.5a)$$

The log-magnitude function $-20u$ is plotted as a function of u in Fig. C.1a. This is a straight line with a slope of -20. It crosses the u-axis at $u = 0$. The ω-scale ($u = \log \omega$) is also shown in Fig. C.1a. The semilog graphs can be conveniently used for this purpose. We can directly plot ω on semilog paper. Note that equal increments in u are equivalent to equal ratios on the ω-scale. Thus one unit along u is same as a decade along the ω-scale. Hence the log-magnitude plot has a slope of -20 *db* per decade (of the ω-scale), and it crosses the ω-axis at $\omega = 1$ ($u = 0$ is $\omega = 1$, since $u = \log_{10}\omega$). A ratio of 10 is a *decade* and a ratio of 2 is known as an *octave*. Ratio of 10 (a decade) along the ω-scale is 1 unit along the u-scale. It can be easily shown that a ratio of 2 (an octave) along the ω-scale is equal to 0.3010 ($\log_{10}2$) along the u-scale.† Hence a slope of 20 db/decade is same as a slope of 20 (0.3010) = 6.02 db/octave. This is commonly stated as 6 db/octave. Thus the slope -20 db/decade is the same as the slope -6/db octave.

---

†This can be shown as follows: Let ω_1 and ω_2 along the ω-scale correspond to u_1 and u_2 along the u-scale. Then log $\omega_1 = u_1$ and log $\omega_2 = u_2$, and

$$u_1 - u_2 = \log_{10} \omega_1 - \log_{10} \omega_2$$

$$= \log_{10} \frac{\omega_1}{\omega_2}$$

If

$$\frac{\omega_1}{\omega_2} = 2 \text{ (an octave)}$$

$$u_1 - u_2 = \log_{10} 2 = 0.3010$$

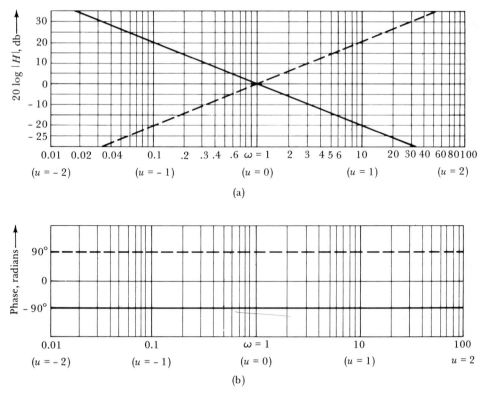

Figure C.1

For the case of a zero at the origin, the log-magnitude term is 20 log ω. This is a straight line passing through $\omega = 1$ and having a slope of 20 db/decade (or 6 db/octave). This plot is the mirror image about the ω-axis of the plot for the pole and is shown dotted in Fig. C.1a.

Phase

The phase function corresponding to the pole at the origin is $-\angle j\omega$ (see Eq. C.2b). Thus

$$\theta (\omega) = -\angle j\omega = -90° \tag{C.5b}$$

The phase is constant ($-90°$) for all values of ω. This is shown in Fig. 5.21b. For a zero at the origin the phase is $\angle j\omega = 90°$. This is the mirror image of the phase plot for the pole and is shown dotted in Fig. C.1b.

2. FIRST-ORDER POLE (OR ZERO)

The log magnitude due to first-order pole at $-a$ is $-20 \log \left| 1 + \dfrac{j\omega}{a} \right|$. Let us investigate the asymptotic behavior of this function for extreme values of $\omega (\omega \ll a$ and $\omega \gg a)$.

(a) For $\omega \ll a$,

$$-20 \log \left| 1 + \frac{j\omega}{a} \right| \cong -20 \log 1 = 0 \tag{C.6}$$

Hence the log-magnitude function $\to 0$ asymptotically for $\omega \ll a$ (Fig. C.2a). Let us consider the other extreme where $\omega \gg a$. In this case

$$-20 \log \left| 1 + \frac{j\omega}{a} \right| \cong -20 \log \left(\frac{\omega}{a} \right) \tag{C.7a}$$

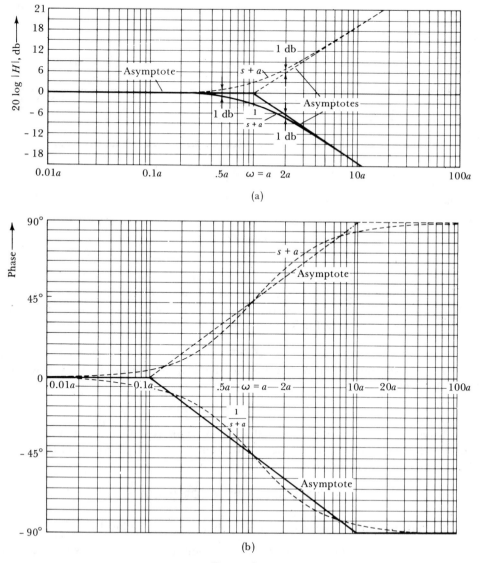

Figure C.2

$$= -20 \log \omega + 20 \log a \qquad \text{(C.7b)}$$
$$= -20u + 20 \log a \qquad \text{(C.7c)}$$

This represents a straight line (when plotted as a function of u, the log of ω) with a slope of -20 db/decade (or -6 db/octave).   When $\omega = a$, the log magnitude is zero (Eq. C.7b).   Hence this straight line asymptote crosses the ω-axis at $\omega = a$ as shown in Fig. C.2a.   The 2 asymptotes meet at $\omega = a$.   The exact log magnitude is

$$-20 \log \left| 1 + \frac{j\omega}{a} \right| = -20 \log \left(1 + \frac{\omega^2}{a^2} \right)^{\frac{1}{2}}$$

$$= -10 \log \left(1 + \frac{\omega^2}{a^2} \right) \qquad \text{(C.8)}$$

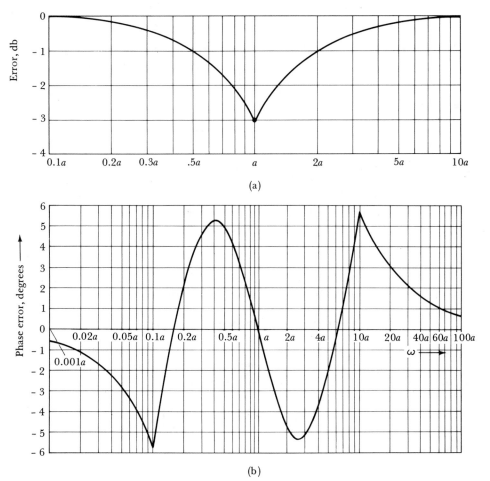

(a)

(b)

Figure C.3

This exact function is also shown in Fig. C.2a. Observe that the actual plot and the asymptotic plots are very close. A maximum error of 3 db occurs at $\omega = a$. This frequency is known as the *corner frequency* or *break frequency*. The error everywhere else is less than 3 db. The plot of error as a function of ω is shown in Fig. C.3a. It can be seen from this figure that the error at one octave above or below the corner frequency is 1 db and the error at two octaves above or below the corner frequency is 0.3 db. The actual plot is obtained by adding the error to the asymptotic plot.

For the first-order zero at $-a$, the log magnitude is 20 log $|1 + j\omega/a|$. This plot is the mirror image of the magnitude plot for the pole at $-a$ and is shown dotted in Fig. C.2a.

Phase

The phase for first-order pole at $-a$ is

$$\theta(\omega) = -\angle\left(1 + \frac{j\omega}{a}\right) = -\tan^{-1}\left(\frac{\omega}{a}\right) \tag{C.9}$$

Let us investigate the asymptotic behavior of this function. For $\omega \ll a$,

$$-\tan^{-1}\left(\frac{\omega}{a}\right) \cong 0$$

For $\omega \gg a$

$$-\tan^{-1}\left(\frac{\omega}{a}\right) \cong -90°$$

The actual plot along with the asymptotes is shown in Fig. C.2b. In this case we use three-line-segment asymptotic plot for greater accuracy. The asymptotes are (i) phase angle zero for $\omega < a/10$, (ii) a straight line with a slope $-45°$/decade and crossing the ω-axis at $\omega = a/10$, and (iii) phase angle $-90°$ for $\omega > 10\,a$. It can be seen from Fig. C.2b that the asymptotes are very close to the curve and the maximum error is 5.4°. The plot of the error as a function of ω is shown in Fig. C.3b. The actual plot is obtained by adding the error to the asymptotic plot.

The phase function for a zero at $-a$ is the mirror image about the ω-axis of the phase function for the pole at $-a$, and is shown dotted in Fig. C.2b.

3. SECOND-ORDER POLE (OR ZERO)

Let us consider the second order pole in Eq. C.1a. The denominator term is $s^2 + b_2 s + b_3$. We shall introduce an often-used standard form $s^2 + 2\zeta\omega_n s + \omega_n^2$ instead of $s^2 + b_2 s + b_3$. Hence the log magnitude function in Eq. C.3 becomes

$$-20 \log \left|1 + 2j\zeta\,\frac{\omega}{\omega_n} + \left(\frac{j\omega}{\omega_n}\right)^2\right|$$

and the phase function is

$$-\angle\left[1 + 2j\zeta\,\frac{\omega}{\omega_n} + \left(\frac{j\omega}{\omega_n}\right)^2\right]$$

The Log Magnitude

The log magnitude is given by

$$\text{log magnitude} = -20 \log \left| 1 + 2j\zeta\left(\frac{\omega}{\omega_n}\right) - \left(\frac{\omega}{\omega_n}\right)^2 \right| \tag{C.10}$$

For $\omega \ll \omega_n$ this becomes

$$\text{log magnitude} \cong -20 \log 1 = 0 \tag{C.11}$$

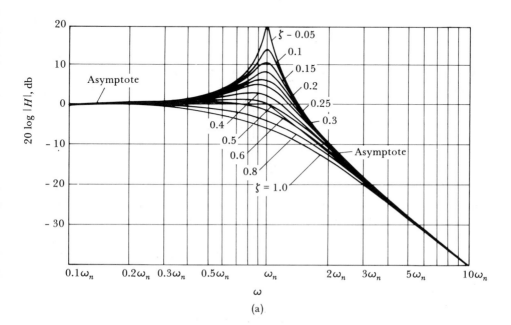

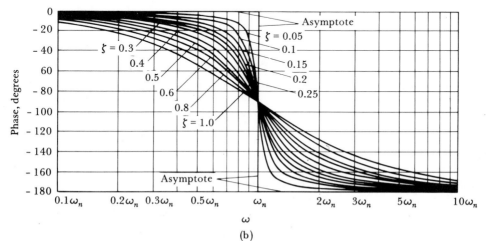

Figure C.4

For $\omega \gg \omega_n$ the log magnitude is

$$\text{log magnitude} \cong -20 \log \left| -\left(\frac{\omega}{\omega_n}\right)^2 \right|$$

$$= -40 \log \frac{\omega}{\omega_n} \tag{C.12a}$$

$$= -40 \log \omega - 40 \log \omega_n \tag{C.12b}$$

$$= -40u - 40 \log \omega_n \tag{C.12c}$$

The two asymptotes are (i) zero for $\omega < \omega_n$ and (ii) $-40u - 40 \log \omega_n$ for $\omega > \omega_n$. The second asymptote is a straight line with a slope of -40 db/decade (or -12 db/octave) when plotted against log ω scale. It begins at $\omega = \omega_n$ (see Eq. C.12b). The asymptotes are shown in Fig. C.4a. The exact log magnitude is given by (see Eq. C.10)

$$\text{log magnitude} = -20 \log \left\{ \left[1 - \left(\frac{\omega}{\omega_n}\right)^2\right]^2 + 4\zeta^2 \left(\frac{\omega}{\omega_n}\right)^2 \right\}^{\frac{1}{2}}$$

$$= -10 \log \left\{ \left[1 - \left(\frac{\omega}{\omega_n}\right)^2\right]^2 + 4\zeta^2 \left(\frac{\omega}{\omega_n}\right)^2 \right\} \tag{C.13}$$

It can be seen that the log magnitude in this case involves a parameter ζ. For each value of ζ, we have a different plot. For complex conjugate poles,† $\zeta < 1$. Hence we must sketch a family of curves for a number of values of ζ in the range 0 to 1. This is shown in Fig. C.4a. The error between the actual plot and the asymptotes is shown in Fig. C.5. The actual plot is obtained by adding the error to the asymptotic plot.

For the second-order zeros (complex conjugate zeros), the plots are mirror images about the ω-axis of the plots shown in Fig. C.4a. Note the resonance phenomenon of the complex conjugate poles (see Sec. 2.5). This phenomenon is barely noticeable for $\zeta > 0.707$ but becomes pronounced ‡ as $\zeta \to 0$.

Phase

The phase function for the second-order poles is

$$\theta(\omega) = -\angle \left[1 + 2j\zeta \left(\frac{\omega}{\omega_n}\right) - \left(\frac{\omega}{\omega_n}\right)^2\right]$$

---

†For $\zeta \geq 1$, the two poles in the second-order factor are no longer complex but real, and each of these two real poles can be dealt with as a separate first-order factor.

‡This can be seen from the fact that $s^2 + 2\zeta\omega_n s + \omega_n^2 = (s + \alpha - j\beta)(s + \alpha + j\beta)$, where $\alpha = -\zeta\omega_n$ and $\beta = \omega_n \sqrt{1 - \zeta^2}$. The natural mode corresponding to these poles is of the form $e^{-\alpha t} \cos(\beta t + \phi)$. As $\zeta \to 0$, $\alpha \to 0$, $\beta \to \omega_n$ and the natural mode approaches a sinusoidal signal $\cos(\omega_n t + \phi)$. Hence the response of the system to sinusoidal inputs of angular frequencies in the vicinity of ω_n is comparatively large.

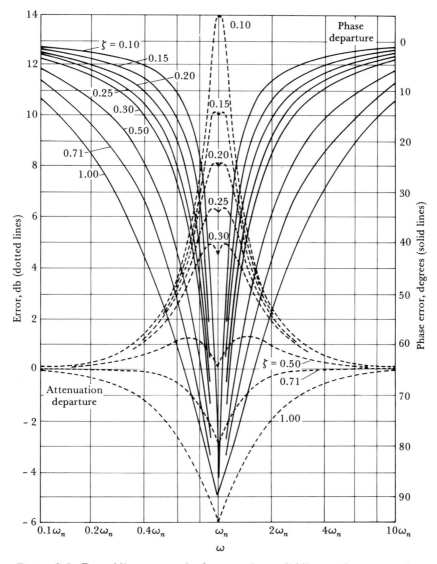

Figure C.5. Dotted lines—magnitude correction; solid lines—phase correction.

$$= -\tan^{-1}\left[\frac{2\zeta\left(\dfrac{\omega}{\omega_n}\right)}{1-\left(\dfrac{\omega}{\omega_n}\right)^2}\right] \tag{C.14}$$

For $\omega \ll \omega_n$,

$$\theta(\omega) \cong 0$$

For $\omega \gg \omega_n$,

$$\theta(\omega) \cong -180°$$

Hence the phase $\rightarrow -180°$ as $\omega \rightarrow \infty$. As for magnitude, we also have a family of phase plots for various values of ζ as shown in Fig. C.4b. A convenient asymptote for the phase of complex conjugate poles is a step function of magnitude $-180°$ and starting at $\omega = \omega_n$ as shown in Fig. C.4b. The error plot for such an asymptote is shown in Fig. C.5 for various values of ζ. The exact phase is the asymptotic value plus the error.

For complex conjugate zeros, the magnitude and phase plots are mirror images of those for the complex conjugate poles. We shall demonstrate the application of these techniques by two examples.

Example C.1.   Sketch Bode plots for the transfer function

$$H(s) = \frac{20s(s + 100)}{(s + 2)(s + 10)}$$

In the first step we write the transfer function in normalized form:

$$H(s) = \frac{20 \times 100}{2 \times 10} \frac{s\left(1 + \dfrac{s}{100}\right)}{\left(1 + \dfrac{s}{2}\right)\left(1 + \dfrac{s}{10}\right)}$$

$$= 100 \frac{s\left(1 + \dfrac{s}{100}\right)}{\left(1 + \dfrac{s}{2}\right)\left(1 + \dfrac{s}{10}\right)} \tag{C.15}$$

Here we have two first-order poles at -2 and -10, one zero at the origin and one zero at -100.

   Step 1.   For each of these terms, we draw an asymptotic plot as follows (see Fig. C.6a).
   (i) For the zero at the origin draw a straight line with slope 20 db/decade passing through $\omega = 1$.
   (ii) For the pole at -2, draw a straight line of slope -20 db/decade (for $\omega > 2$) beginning at the corner frequency $\omega = 2$.
   (iii) For the pole at -10, draw a straight line of slope -20 db/decade beginning at the corner frequency $\omega = 10$.
   (iv) For the zero at -100, draw a straight line of slope 20 db/decade beginning at the corner frequency $\omega = 100$.
   Step 2.   Add all the asymptotes as in Fig. C.6a.
   Step 3.   Apply corrections as follows. (see Fig. C.3a)
   (a) Correction at $\omega = 1$, due to corner frequency at $\omega = 2$ is -1 db. The corrections at $\omega = 1$ due to corner frequencies at $\omega = 10$ and $\omega = 100$ are quite small (see Fig. C.3a) and may be ignored. Hence the net correction at $\omega = 1$ is -1 db.

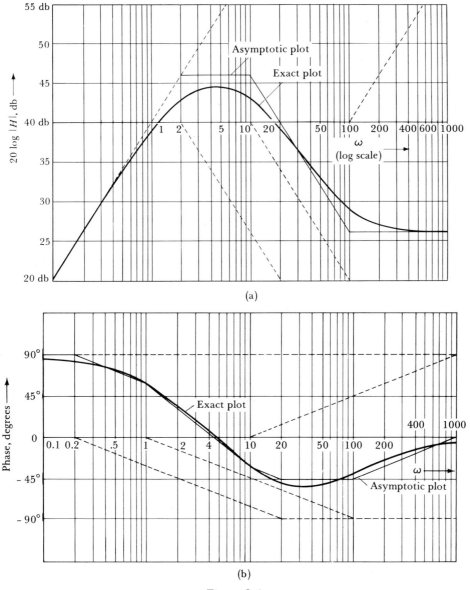

(a)

(b)

Figure C.6

(b) Correction at $\omega = 2$, due to corner frequency at $\omega = 2$ is -3 db and due to corner frequency at $\omega = 10$ is -0.4 db. Correction due to corner frequency $\omega = 100$ can be safely ignored. Hence the net correction at $\omega = 2$ is -3.4 db.

(c) Correction at $\omega = 10$, due to corner frequency $\omega = 10$ is -3 db and due to corner frequency $\omega = 2$ is -0.4 db. The correction due to $\omega = 100$ is 0.04 db and may be ignored. Hence the net correction at $\omega = 10$ is -3.4 db.

(d) Correction $\omega = 100$ due to corner frequency $\omega = 100$ is $+3$ db and corrections due to other corner frequencies may be ignored.

(e) Corrections at $\omega = 4$ and $\omega = 5$ may also be found to be about -1.75 db.

With these corrections, we can draw the magnitude plot as shown in Fig. C.6a. However, we have not yet taken into account the constant term 20 log 100 (see Eq. C.15). This is $20(2) = 40$ db. This term can be added to the plot by a simple trick of labeling the line from where the asymptotes began as the 40 db line as shown in Fig. C.6a. This implies shifting the horizontal axis upward by 40 db. This is precisely what is desired.

Phase Plots

We draw asymptotes corresponding to each of the four factors
  (i) Zero at the origin causes 90° phase shift.
 (ii) Pole at $s = -2$, gives rise to an asymptote of slope $-45°$/decade starting at $\omega = 0.2$, which goes up to $\omega = 20$. For $\omega < 0.2$, the asymptote has a zero value and for $\omega > 20$, the asymptotic value is $-90°$.
(iii) Similarly for the pole at $s = -10$, we have an asymptote with value zero for $-\infty < \omega < 1$ and a slope $-45°$/decade beginning at $\omega = 1$ and going up to $\omega = 100$. The asymptotic value for $\omega > 100$ is $-90°$.
(iv) For the zero at $s = -100$, the asymptote with slope $45°$/decade begins at $\omega = 10$ and goes up to $\omega = 1000$. For $\omega < 10$, asymptotic values is 0 and for $\omega > 1000$, the asymptotic value is 90°. All the asymptotes are added as shown in Fig. C.6b. The appropriate corrections are applied from Fig. C.3b. The exact phase plot is also shown in Fig. C.6b.

Example C.2. Sketch the magnitude and phase response (Bode plots) for the transfer function

$$H(s) = \frac{10(s + 100)}{s^2 + 2s + 100}$$

$$= 10 \; \frac{\left(1 + \dfrac{s}{100}\right)}{1 + \dfrac{s}{50} + \dfrac{s^2}{100}} \qquad\qquad (C.16)$$

Here we have a real zero at $s = -100$, and a pair of complex conjugate poles. When we express the second-order factor in a standard form

$$s^2 + 2s + 100 = s^2 + 2\zeta\omega_n s + \omega_n^2$$

we have

$$\omega_n = 10 \quad\text{and}\quad \zeta = 0.1 \qquad\qquad (C.17)$$

Step 1. Draw an asymptote of -40 db/decade (12 db/octave), starting at $\omega = 10$, for the complex conjugate poles, and another asymptote 20 db/decade, starting at $\omega = 100$, for the real zero.

Step 2. Add both the asymptotes.

Step 3. Apply corrections i) at $\omega = 100$, the correction due to corner frequency $\omega = 100$ is 3 db, and the correction due to corner frequency $\omega = 10$ may be ignored. At $\omega = 10$, the correction due to corner frequency 10 is 13.90 db (see

Fig. C.5a for $\zeta = 0.1$). We may find corrections at few more points, and construct the log magnitude plot as shown in Fig. C.7a. The constant factor 20 log 10 = 20 db can be accounted for by labeling the horizontal axis as the 20-db line (Fig. C.7a).

Phase Plot

The asymptotes for the complex conjugate poles is a step function of magnitude $-90°$ at $\omega = 10$, and the asymptote for the zero at $s = -100$ is a straight line

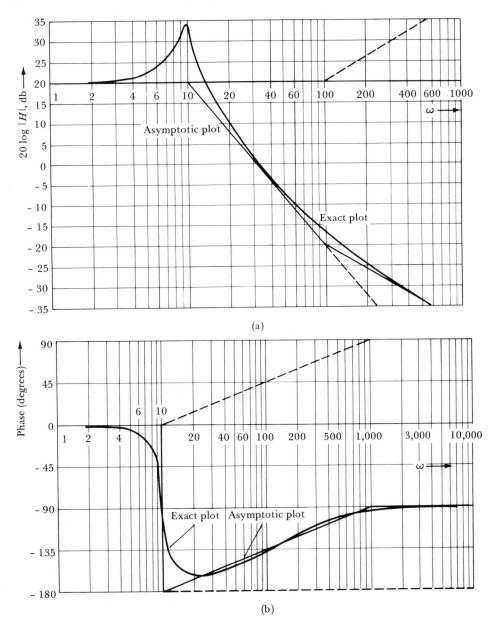

(a)

(b)

Figure C.7

with a slope of 45°/decade, and leveling off at $\omega = 10$, and $\omega = 100$ with values $0°$ and $90°$ respectively. The two asymptotes add to give a kind of sawtooth shown in Fig. C.7b. We now apply corrections from Fig. C.7b and Fig. C.3b to obtain the exact plot.

Poles and Zeros in RHP

In our discussion so far we have assumed poles and zeros of the transfer function to be in the LHP. What if some of the poles and/or zeros of $H(s)$ lie in the RHP. We can easily show that the magnitude plot depends only on the magnitude of the poles or zero values not their sign. Thus

$$\left| 1 + \frac{j\omega}{a} \right| = \left| 1 - \frac{j\omega}{a} \right| = \left(1 + \frac{\omega^2}{a^2} \right)^{\frac{1}{2}}$$

This is also true of the second-order function. Therefore log magnitude plots remain unchanged whether the pole(s) and/or zero(s) are in the LHP or RHP. The phase functions are slightly modified. We shall now show that the phase function of the RHP zero at $s = a$ is identical to the phase function of the LHP pole at $s = -a$, and the phase of the RHP pole at $s = a$ is identical to the phase of the LHP zero at $s = -a$.

Consider the phase of the RHP zero at $s = a$. This is

$$\angle \left(1 - \frac{j\omega}{a} \right) = -\tan^{-1} \left(\frac{\omega}{a} \right)$$

This is identical to the phase of the LHP pole at $s = a$. Similarly it can be shown that the phase of the RHP pole at $s = a$ is identical to that of the LHP zero at $s = -a$. The same conclusion applies to second order poles and zeros.

C.2 DETERMINATION OF TRANSFER FUNCTION FROM THE STEADY-STATE RESPONSE

In the previous section we were given the transfer function of a system. From the knowledge of the transfer function we developed techniques of determining steady-state response of a system to sinusoidal input signals. The other side of the coin is to determine the transfer function of the system if the steady-state response of the system to sinusoidal input signal is known. This problem has a significant practical utility. If we are given a system in a black box with only the input and output terminals available, the transfer function has to be determined by experimental measurements at the input and the output terminals. There are several possible ways of determining the transfer function by experimental measurements. The steady-state response to sinusoidal inputs is one of the possibilities that is very attractive because of the simple nature of measurements. One need to apply a sinusoidal signal at the input and observe the output. In steady state, the output signal is also a sinusoidal signal of amplitude $|H(j\omega)|$ times the input amplitude. The output and the input signals also have a phase difference $\angle H(j\omega)$ (see Fig. 2.22). We determine $|H(j\omega)|$ and $\angle H(j\omega)$ at several frequencies over the entire frequency range.   These yield the frequency response plots (or Bode plots) when plotted against log ω.   From these response curves we determine appropriate asymptotes by taking advantage of the fact that slopes of all asymptotes must be multiples of ± 20 db/decade.   From the asymptotes the corner frequencies are found.   This determines poles and zeros of the transfer function.   Caution, however, must be exercised in determining whether a given pole (or a zero) is in the LHP or RHP.   We have observed that a pole (or a zero) at $s = a$ or $s = -a$ both have identical magnitude plots but their phase plots are different.   From the phase plot, we can determine whether a given pole (or a zero) is in the LHP or RHP.

It can be further shown that when all of the poles and zeros lie in the LHP the magnitude $|H(j\omega)|$ and phase $\angle H(j\omega)$ are related.   The knowledge of one determines the other.†   Hence actually either the magnitude plot or the phase plot are sufficient to determine the transfer function if it is known that all of the poles and the zeros of the transfer function lie in the LHP.

---

†This is true also if all the poles and zeros lie in the RHP. See, for example, H. W. Bode, *Network Analysis and Feedback Amplifier Design* (New York: Van Nostrand, 1945), and E. A. Guillemin, *Theory of Linear Physical Systems* (New York: Wiley, 1963).

D

Vectors and Matrices

D.1 VECTORS

A system in general transforms a set of input signals and yields a certain set of output signals. Symbolically a system may be represented as shown in Fig. D.1. In general there may be m inputs (driving functions) and k outputs (response). In Fig. D.1 functions $f_1(t), f_2(t), \ldots, f_m(t)$ represent the m inputs and $y_1(t), y_2(t), \ldots, y_k(t)$ represent the k outputs. Thus an input at any instant is specified by m numbers. In dealing with entities characterized by more than one number, the use of vector representation proves very effective.

The reader is familiar with vectors in three-dimensional space. These vectors can be specified by three components or three numbers in a particular order. A vector in these cases is therefore a quantity specified by three numbers arranged in a particular order. The force vectors or electric or magnetic field vectors are specified by the three dimensions, and we need a maximum of three numbers to specify these vectors.

We can, however, extend the concept of a vector to quantities which are specified not necessarily by their components in physical space but by some n numbers and arranged in some order. Consider for example an electrical circuit with B branches with currents $i_1, i_2, \ldots, i_B$. For this network we may define a current vector $\mathbf{I}_B$ specified by B currents $i_1, i_2, \ldots, i_B$ in this particular order (or in any desired order). In general, therefore, a vector represents an entity specified by n numbers arranged in a definite order. The number n is the dimensionality of the vector. We shall represent a vector by a column. Thus a vector $\mathbf{x}$ specified by numbers $x_1, x_2, \ldots, x_n$ in that order, will be represented as

$$\mathbf{x} = \begin{bmatrix} x_1 \\ x_2 \\ \vdots \\ x_n \end{bmatrix} \tag{D.1}$$

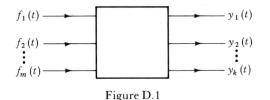

Figure D.1

and a vector $\mathbf{y}$ specified by numbers $y_1, y_2, \ldots, y_k$ will be represented by

$$\mathbf{y} = \begin{bmatrix} y_1 \\ y_2 \\ \vdots \\ y_k \end{bmatrix} \tag{D.2}$$

D.2   LINEAR VECTOR FUNCTIONS

The reader is familiar with the functions of scalars. When we say y is a function of x and represent the relation as

$$y = f(x) \tag{D.3}$$

we mean that for every value of x there exists a corresponding value of y which can be found from Eq. D.3. This concept of function can be easily extended to vectors. If corresponding to each vector $\mathbf{x}$ (in some set) there is a corresponding vector $\mathbf{y}$ found by some relationship $\mathbf{y} = f(\mathbf{x})$, we say that vector $\mathbf{y}$ is a function of vector $\mathbf{x}$. As an example, consider a linear two-part network shown in Fig. D.2. Let v_1 and i_1 represent the voltage and the current at the input terminals and v_2 and i_2 represent the voltage and the current at the output terminals. We may define vectors $\mathbf{v}$ and $\mathbf{i}$ as

$$\mathbf{v} = \begin{bmatrix} v_1 \\ v_2 \end{bmatrix} \quad \text{and} \quad \mathbf{i} = \begin{bmatrix} i_1 \\ i_1 \end{bmatrix}$$

Obviously the two vectors $\mathbf{v}$ and $\mathbf{i}$ are related in some manner depending upon the nature of the two-port. The matrix algebra deals with the problem of relating and manipulating the n-dimensional vectors.

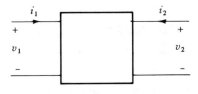

Figure D.2

Consider the two vectors $\mathbf{x}$ and $\mathbf{y}$ in Eqs. D.1 and D.2. Let these vectors be related by

$$\mathbf{y} = f(\mathbf{x}) \tag{D.4}$$

This relationship transforms (or maps) a given vector $\mathbf{x}$ into another vector $\mathbf{y}$. The most general form of relationship between $\mathbf{x}$ and $\mathbf{y}$ implies that each componant (element) of $\mathbf{y}$ is related to all the n components (elements) of $\mathbf{x}$ by some formula. This implies that

$$y_1 = f_1(x_1, x_2, \ldots, x_n)$$
$$y_2 = f_2(x_1, x_2, \ldots, x_n)$$
$$\cdots\cdots\cdots\cdots\cdots$$
$$y_k = f_m(x_1, x_2, \ldots, x_n)$$

For the sake of simplicity, let us restrict ourselves to the linear relationships. In such a case this equation takes the form

$$y_1 = a_{11}x_1 + a_{12}x_2 + \cdots + a_{1n}x_n$$
$$y_2 = a_{21}x_1 + a_{22}x_2 + \cdots + a_{2n}x_n$$
$$\vdots \tag{D.5}$$
$$y_k = a_{k1}x_1 + a_{k2}x_2 + \cdots + a_{kn}x_n$$

Relationships in Eq. D.5 are obviously linear.†

Relationship D.5 transforms each vector $\mathbf{x}$ into another vector $\mathbf{y}$. Such a transformation is called the *linear transformation of vectors*. If we are given the array of coefficients a_{ij} in Eq. D.5, then the relationship between vectors $\mathbf{x}$ and $\mathbf{y}$ is completely specified. The array of coefficients a_{ij} will be denoted by $\mathbf{A}$ for convenience.

$$\mathbf{A} = \begin{bmatrix} a_{11} & a_{12} & \cdots & a_{1n} \\ a_{21} & a_{22} & \cdots & a_{2n} \\ \cdot & \cdot & & \cdot \\ a_{k1} & a_{k2} & \cdots & a_{kn} \end{bmatrix} \tag{D.6}$$

The array of coefficients is called a *matrix*.

A matrix with m rows and n columns is called a matrix of the order (k, n), or a $(k \times n)$ matrix. For a special case $k = n$, the matrix is called a *square matrix* of order n.

_____

†This can be seen from the fact that if

$$\mathbf{y}_1 = f(\mathbf{x}_1) \quad \text{and} \quad \mathbf{y}_2 = f(\mathbf{x}_2)$$

then it follows from Eq. D.5 that

$$f(\alpha\mathbf{x}_1 + \beta\mathbf{x}_2) = \alpha\mathbf{y}_1 + \beta\mathbf{y}_2$$
$$= \alpha f(\mathbf{x}_1) + \beta f(\mathbf{x}_2)$$

See Chapter 1 for discussion of linearity.

It should be stressed at this point that matrix is not a number such as determinant, but is an array of numbers arranged in a particular order. It is convenient to abbreviate the representation of matrix $\mathbf{A}$ in Eq. D.6 by the form $(a_{ij})_{k \times n}$ implying a matrix of order $k \times n$ with elements a_{ij}'s. In practice, when the order $k \times n$ is either understood or need not be specified, the notation can be abbreviated to (a_{ij}). Note that the first index i of a_{ij} indicates the row and the second index j indicates the column of the element a_{ij} in the matrix $\mathbf{A}$.

The k simultaneous equations D.5 may now be expressed in a symbolic form as

$$\mathbf{y} = \mathbf{Ax} \tag{D.7}$$

or

$$
\begin{bmatrix} y_1 \\ y_2 \\ \cdot \\ y_k \end{bmatrix} =
\begin{bmatrix} a_{11} & a_{12} & \cdots & a_{1n} \\ a_{21} & a_{22} & \cdots & a_{2n} \\ \cdot & \cdot & & \cdot \\ a_{k1} & a_{k2} & \cdots & a_{kn} \end{bmatrix}
\begin{bmatrix} x_1 \\ x_2 \\ \cdot \\ x_n \end{bmatrix} \tag{D.8}
$$

It should be stressed here that Eq. D.8 is the symbolic representation of Eq. D.5. As yet we have not defined the operation of multiplication of a matrix and a vector. The quantity $\mathbf{Ax}$ is not meaningful until we define such an operation.

D.3   SOME DEFINITIONS AND PROPERTIES

DIAGONAL MATRIX

A square matrix whose elements are zero everywhere except on the main diagonal is a *diagonal matrix*. An example of a diagonal matrix is

$$
\begin{bmatrix} 2 & 0 & 0 \\ 0 & 1 & 0 \\ 0 & 0 & 5 \end{bmatrix}
$$

IDENTITY MATRIX OR UNIT MATRIX

A diagonal matrix with all its elements unity (along the diagonal) is called an *identity matrix* or a *unit matrix* and is denoted by $\mathbf{I}$. Note that the unit matrix is a square matrix:

$$
\mathbf{I} = \begin{bmatrix} 1 & 0 & 0 & \cdots & 0 \\ 0 & 1 & 0 & \cdots & 0 \\ 0 & 0 & 1 & \cdots & 0 \\ & & \cdots & & \\ 0 & 0 & 0 & \cdots & 1 \end{bmatrix} \tag{D.9}
$$

The order of the unit matrix is sometimes indicated by a subscript. Thus $\mathbf{I}_n$ represents the $n \times n$ unit matrix (or identity matrix). We shall however omit the subscript. The order of the unit matrix will be understood from the context.

ZERO MATRIX

A matrix having all its elements zero is a *zero matrix*.

SYMMETRIC MATRIX

A square matrix $\mathbf{A}$ is a *symmetric matrix* if $a_{ij} = a_{ji}$ (symmetry about the main diagonal).

EQUALITY OF TWO MATRICES

Two matrices of the same order are said to be equal if they are equal element by element. Thus if

$$\mathbf{A} = (a_{ij})_{m \times n} \quad \text{and } \mathbf{B} = (b_{ij})_{m \times n}$$

then $\mathbf{A} = \mathbf{B}$ only if $a_{ij} = b_{ij}$ for all i and j.

TRANSPOSE OF A MATRIX

If the rows and columns of an $m \times n$ matrix $\mathbf{A}$ are interchanged so that the elements in the ith row now become the elements in the ith column (for $i = 1, 2, \ldots, m$), the resulting matrix is called the transpose of $\mathbf{A}$ and is denoted by $\mathbf{A}'$. It is evident that $\mathbf{A}'$ is an $n \times m$ matrix. If

$$\mathbf{A} = \begin{bmatrix} 2 & 1 \\ 3 & 2 \\ 1 & 3 \end{bmatrix} \quad \text{then} \quad \mathbf{A}' = \begin{bmatrix} 2 & 3 & 1 \\ 1 & 2 & 3 \end{bmatrix}$$

Thus if

$$\begin{aligned} \mathbf{A} &= (a_{ij})_{m \times n} \\ \mathbf{A}' &= (a_{ji})_{n \times m} \end{aligned} \tag{D.10}$$

Note that

$$(\mathbf{A}')' = \mathbf{A} \tag{D.11}$$

D.4 MATRIX ALGEBRA

We shall now define matrix operations such as addition, subtraction, multiplication, and division of matrices. The definitions must be formulated so that they are useful in manipulation of matrices.

1. ADDITION OF MATRICES

For two matrices $\mathbf{A}$ and $\mathbf{B}$, both of the same order $(m \times n)$,

$$\mathbf{A} = \begin{bmatrix} a_{11} & a_{12} & \cdots & a_{1n} \\ a_{21} & a_{22} & \cdots & a_{2n} \\ \cdots & \cdots & \cdots & \cdots \\ a_{m1} & a_{m2} & \cdots & a_{mn} \end{bmatrix} \quad \text{and} \quad \mathbf{B} = \begin{bmatrix} b_{11} & b_{12} & \cdots & b_{1n} \\ b_{21} & b_{22} & \cdots & b_{2n} \\ \cdots & \cdots & \cdots & \cdots \\ b_{m1} & b_{m2} & \cdots & b_{mn} \end{bmatrix}$$

we define the sum $\mathbf{A} + \mathbf{B}$ as

$$\mathbf{A} + \mathbf{B} = \begin{bmatrix} (a_{11} + b_{11}) & (a_{12} + b_{12}) & \cdots & (a_{1n} + b_{1n}) \\ (a_{21} + b_{21}) & (a_{22} + b_{22}) & \cdots & (a_{2n} + b_{2n}) \\ \cdots\cdots\cdots\cdots\cdots\cdots\cdots\cdots\cdots\cdots\cdots\cdots \\ (a_{m1} + b_{m1}) & (a_{m1} + b_{m1}) & \cdots & (a_{mn} + b_{mn}) \end{bmatrix} \qquad (D.12)$$

or

$$\mathbf{A} + \mathbf{B} = (a_{ij} + b_{ij})_{m \times n}$$

Note that two matrices can be added only if they are of the same order.

2. MULTIPLICATION OF A MATRIX BY A SCALAR

We define multiplication of a matrix $\mathbf{A}$ by a scalar c as

$$c\mathbf{A} = c \begin{bmatrix} a_{11} & a_{12} & \cdots & a_{1n} \\ a_{21} & a_{22} & \cdots & a_{2n} \\ \cdots\cdots\cdots\cdots\cdots\cdots \\ a_{m1} & a_{m2} & \cdots & a_{mn} \end{bmatrix} \qquad (D.13)$$

$$= \begin{bmatrix} ca_{11} & ca_{12} & \cdots & ca_{1n} \\ ca_{21} & ca_{22} & \cdots & ca_{2n} \\ \cdots\cdots\cdots\cdots\cdots\cdots \\ ca_{m1} & ca_{m2} & \cdots & ca_{mn} \end{bmatrix} \qquad (D.14)$$

3. MATRIX MULTIPLICATION

For convenience, we shall first define multiplication of 2×2 matrices and then generalize the result. The product of matrices $\mathbf{A}$ and $\mathbf{B}$ is defined as

$$\underbrace{\begin{bmatrix} a_{11} & a_{12} \\ a_{21} & a_{22} \end{bmatrix}}_{\mathbf{A}} \underbrace{\begin{bmatrix} b_{11} & b_{12} \\ b_{21} & b_{22} \end{bmatrix}}_{\mathbf{B}} = \underbrace{\begin{bmatrix} (a_{11}b_{11} + a_{12}b_{21}) & (a_{11}b_{12} + a_{12}b_{22}) \\ (a_{21}b_{11} + a_{21}b_{21}) & (a_{21}b_{12} + a_{22}b_{22}) \end{bmatrix}}_{\mathbf{AB}} \qquad (D.15)$$

Observe that the 11 element in $\mathbf{AB}$ is formed by multiplying the elements of the first row in $\mathbf{A}$ with the corresponding elements in the first column of $\mathbf{B}$ and adding all the products. This is shown below.

$$\begin{bmatrix} a_{11} & a_{12} \\ a_{21} & a_{22} \end{bmatrix} \begin{bmatrix} b_{11} & b_{12} \\ b_{21} & b_{22} \end{bmatrix} \implies \begin{bmatrix} a_{11} & a_{12} \\ \downarrow & \downarrow \\ b_{11} & b_{21} \end{bmatrix} \implies a_{11}b_{11} + a_{12}b_{21}$$

Following along these lines, we can extend this multiplication rule to matrices of any order. It can be seen that this procedure will work only if the number of columns in $\mathbf{A}$ is equal to the number of rows in $\mathbf{B}$. Thus if $\mathbf{A}$ is an $m \times n$ matrix and $\mathbf{B} = n \times p$ matrix, we define the product

$$\mathbf{AB} = \mathbf{C}$$

where c_{ij} the element of $\mathbf{C}$ in the ith row and jth column is given by the sum of products of the elements of $\mathbf{A}$ in the ith row with the corresponding elements of $\mathbf{B}$ in the jth column. Thus

$$c_{ij} = a_{i1}b_{1j} + a_{i2}b_{2j} + \cdots + a_{in}b_{nj}$$

$$= \sum_{k=1}^{n} a_{ik}b_{kj} \tag{D.16}$$

This is shown below.

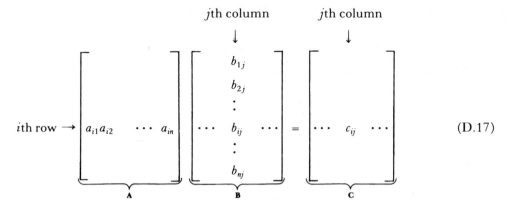

Note carefully again that the number of columns of $\mathbf{A}$ must be equal to the number of rows of $\mathbf{B}$ for this procedure to work. We stress here once again that $\mathbf{AB}$ the product of matrices $\mathbf{A}$ and $\mathbf{B}$ is defined only if the number of columns of $\mathbf{A}$ is equal to the number of rows of $\mathbf{B}$. If this condition is not satisfied the product $\mathbf{AB}$ is not defined and is meaningless. When the number of columns of $\mathbf{A}$ is equal to the number of rows of $\mathbf{B}$, matrix $\mathbf{A}$ is said to be **conformable** to matrix $\mathbf{B}$ for the product $\mathbf{AB}$. Observe that if $\mathbf{A}$ is an $m \times n$ matrix and $\mathbf{B}$ is an $n \times p$ matrix, $\mathbf{A}$ and $\mathbf{B}$ are conformable for the product and $\mathbf{C}$, the product $\mathbf{AB}$ is an $m \times p$ matrix.

We demonstrate the use of rule in Eq. D.16 by following examples.

Example D.1.

$$\begin{bmatrix} 2 & 3 \\ 1 & 1 \\ 3 & 1 \end{bmatrix} \begin{bmatrix} 1 & 3 & 1 & 2 \\ 2 & 1 & 1 & 1 \end{bmatrix} = \begin{bmatrix} 8 & 9 & 5 & 7 \\ 3 & 4 & 2 & 3 \\ 5 & 10 & 4 & 7 \end{bmatrix} \tag{D.18}$$

Example D.2.

$$\begin{bmatrix} 2 & 1 & 3 \end{bmatrix} \begin{bmatrix} 2 \\ 1 \\ 1 \end{bmatrix} = 8$$

In Eq. D.18 we see that the two matrices are conformable. If, however, we interchange the order of matrices as

$$\begin{bmatrix} 1 & 3 & 1 & 2 \\ 2 & 1 & 1 & 1 \end{bmatrix} \begin{bmatrix} 2 & 2 & 3 \\ 1 & 1 & 1 \\ 3 & 3 & 1 \end{bmatrix}$$

the matrices are no more conformable for the product. It is evident that, in general,

$$\mathbf{AB} \neq \mathbf{BA} \qquad (D.19)$$

Indeed $\mathbf{AB}$ may exist and $\mathbf{BA}$ may not exist as in Example D.2, or vice versa. We shall later see that for some special matrices

$$\mathbf{AB} = \mathbf{BA} \qquad (D.20)$$

When Eq. D.20 is true, the matrices $\mathbf{A}$ and $\mathbf{B}$ are said to be *commutable*. We shall stress here again that in general matrices are not commutable. Only for some special cases to be discussed later, the operation D.20 is valid.

In the matrix product $\mathbf{AB}$, the matrix $\mathbf{A}$ is said to be *postmultiplied* by $\mathbf{B}$ or the matrix $\mathbf{B}$ is said to be *premultiplied* by $\mathbf{A}$. The reader may also verify the following relationships.

$$(\mathbf{A} + \mathbf{B})\mathbf{C} = \mathbf{AC} + \mathbf{BC} \qquad (D.21)$$
$$\mathbf{C}(\mathbf{A} + \mathbf{B}) = \mathbf{CA} + \mathbf{CB} \qquad (D.22)$$

The reader can easily verify that any matrix $\mathbf{A}$ premultiplied or postmultiplied by identity matrix $\mathbf{I}$, remains unchanged

$$\mathbf{AI} = \mathbf{IA} = \mathbf{A} \qquad (D.23)$$

of course, we must make sure that the order of $\mathbf{I}$ is such that the matrices are conformable for the product. Thus if

$$\mathbf{A} = \begin{bmatrix} 2 & 3 \\ 1 & 1 \\ 3 & 1 \end{bmatrix}$$

Then

$$\mathbf{AI} = \begin{bmatrix} 2 & 3 \\ 1 & 1 \\ 3 & 1 \end{bmatrix} \begin{bmatrix} 1 & 0 \\ 0 & 1 \end{bmatrix} = \begin{bmatrix} 2 & 3 \\ 1 & 1 \\ 3 & 1 \end{bmatrix} = \mathbf{A} \qquad (D.24)$$

and

$$\mathbf{IA} = \begin{bmatrix} 1 & 0 & 0 \\ 0 & 1 & 0 \\ 0 & 0 & 1 \end{bmatrix} \begin{bmatrix} 2 & 3 \\ 1 & 1 \\ 3 & 1 \end{bmatrix} = \begin{bmatrix} 2 & 3 \\ 1 & 1 \\ 3 & 1 \end{bmatrix} = \mathbf{A} \qquad (D.25)$$

It is evident that $\mathbf{I}$ must be a 2×2 matrix in D.24 and a 3×3 matrix in D.25 to be conformable.

If $\mathbf{A}$ is an $n \times n$ square matrix, then for an $n \times n$ identity matrix

$$\mathbf{AI} = \mathbf{IA} = \mathbf{A}$$

Therefore matrices $\mathbf{A}$ and $\mathbf{I}$ are commutable.

4. MULTIPLICATION OF A MATRIX BY A VECTOR

Consider matrix Eq. D.8 which represent Eq. D.5. The right-hand side of Eq. D.8 is a product of a matrix $\mathbf{A}$ of the order $m \times n$ and a vector $\mathbf{x}$. If for the time being we treat vector $\mathbf{x}$ as if it were an $n \times 1$ matrix, then the product $\mathbf{Ax}$ according to matrix multiplication rule yields the right-hand side of Eq. D.5. It therefore follows that we may multiply a matrix by a vector as if the vector were an $n \times 1$ matrix. Note that the constrain of conformability still applies. Thus in this case $\mathbf{xA}$ is not defined and is meaningless.

5. MATRIX DIVISION

In order to define matrix division we must first define inverse of a matrix. To do this, consider the set of equations

$$
\begin{bmatrix} y_1 \\ y_2 \\ \cdots \\ y_n \end{bmatrix}
=
\begin{bmatrix}
a_{11} & a_{12} & \cdots & a_{1n} \\
a_{21} & a_{22} & \cdots & a_{2n} \\
\cdots & \cdots & \cdots & \cdots \\
a_{n1} & a_{n2} & \cdots & a_{nn}
\end{bmatrix}
\begin{bmatrix} x_1 \\ x_2 \\ \cdots \\ x_n \end{bmatrix}
\tag{D.26}
$$

We can solve this set of equations for $x_1, x_2, \ldots, x_n$ in terms of $y_1, y_2, \ldots, y_n$ by using Cramer's rule.† This yields

$$
x_1 = \frac{\begin{vmatrix}
y_1 & a_{12} & \cdots & a_{1n} \\
y_2 & a_{22} & \cdots & a_{2n} \\
\cdots & \cdots & \cdots & \cdots \\
y_n & a_{n2} & \cdots & a_{nn}
\end{vmatrix}}{\begin{vmatrix}
a_{11} & a_{12} & \cdots & a_{1n} \\
a_{21} & a_{22} & \cdots & a_{2n} \\
\cdots & \cdots & \cdots & \cdots \\
a_{n1} & a_{n2} & \cdots & a_{nn}
\end{vmatrix}}
$$

$$
= \frac{A_{11}}{|\mathbf{A}|} y_1 + \frac{A_{21}}{|\mathbf{A}|} y_2 + \cdots + \frac{A_{n1}}{|\mathbf{A}|} y_n
$$

where $|\mathbf{A}|$ is the determinant of the matrix $\mathbf{A}$ and A_{ij} is the cofactor of element a_{ij} in the matrix $\mathbf{A}$. The cofactor of element a_{ij} is given by $(-1)^{i+j}$ times the determinant of the $(n-1) \times (n-1)$ matrix that is obtained when the ith row and the jth column in matrix $\mathbf{A}$ are eliminated. We can obtain $x_2, x_3, \ldots, x_n$ in a

---

†Proof of Cramer's rule may be found in any textbook on determinants and matrices. See for example F. E. Hohn, *Elementary Matrix Algebra* (New York: MacMillan, 1958).

similar way by using the Cramer's rule. Thus

$$x_1 = \frac{A_{11}}{|\mathbf{A}|} y_1 + \frac{A_{21}}{|\mathbf{A}|} y_2 + \cdots + \frac{A_{n1}}{|\mathbf{A}|} y_n$$

$$x_2 = \frac{A_{12}}{|\mathbf{A}|} y_1 + \frac{A_{22}}{|\mathbf{A}|} y_2 + \cdots + \frac{A_{n2}}{|\mathbf{A}|} y_n \qquad (D.27)$$

$$\cdots\cdots\cdots\cdots\cdots\cdots\cdots\cdots\cdots\cdots$$

$$x_n = \frac{A_{1n}}{|\mathbf{A}|} y_1 + \frac{A_{2n}}{|\mathbf{A}|} y_2 + \cdots + \frac{A_{nn}}{|\mathbf{A}|} y_n$$

We can express Eq. D.27 in the matrix form as

$$
\begin{bmatrix} x_1 \\ x_2 \\ \vdots \\ x_n \end{bmatrix}
=
\begin{bmatrix}
\dfrac{A_{11}}{|\mathbf{A}|} & \dfrac{A_{21}}{|\mathbf{A}|} & \cdots & \dfrac{A_{n1}}{|\mathbf{A}|} \\[2mm]
\dfrac{A_{12}}{|\mathbf{A}|} & \dfrac{A_{22}}{|\mathbf{A}|} & \cdots & \dfrac{A_{n2}}{|\mathbf{A}|} \\[1mm]
\cdots\cdots\cdots\cdots\cdots\cdots\cdots \\[1mm]
\dfrac{A_{1n}}{|\mathbf{A}|} & \dfrac{A_{2n}}{|\mathbf{A}|} & \cdots & \dfrac{A_{nn}}{|\mathbf{A}|}
\end{bmatrix}
\begin{bmatrix} y_1 \\ y_2 \\ \vdots \\ y_n \end{bmatrix}
\qquad (D.28)
$$

We can express Eq. D.26 in matrix form as

$$\mathbf{y} = \mathbf{Ax} \qquad (D.29)$$

We shall now define $\mathbf{A}^{-1}$, the inverse of matrix $\mathbf{A}$ with the property

$$\mathbf{A}^{-1}\mathbf{A} = \mathbf{I} \qquad \text{unit matrix} \qquad (D.30)$$

Then premultiplying both side of Eq. D.29 by $\mathbf{A}^{-1}$, we obtain

$$\mathbf{A}^{-1}\mathbf{y} = \mathbf{A}^{-1}\mathbf{Ax} = \mathbf{Ix} = \mathbf{x}$$

or

$$\mathbf{x} = \mathbf{A}^{-1}\mathbf{y} \qquad (D.31)$$

Comparison of Eq. D.31 with Eq. D.28 shows that

$$
\mathbf{A}^{-1} =
\begin{bmatrix}
\dfrac{A_{11}}{|\mathbf{A}|} & \dfrac{A_{21}}{|\mathbf{A}|} & \cdots & \dfrac{A_{n1}}{|\mathbf{A}|} \\[2mm]
\dfrac{A_{12}}{|\mathbf{A}|} & \dfrac{A_{22}}{|\mathbf{A}|} & \cdots & \dfrac{A_{n2}}{|\mathbf{A}|} \\[1mm]
\cdots\cdots\cdots\cdots\cdots\cdots\cdots \\[1mm]
\dfrac{A_{1n}}{|\mathbf{A}|} & \dfrac{A_{2n}}{|\mathbf{A}|} & \cdots & \dfrac{A_{nn}}{|\mathbf{A}|}
\end{bmatrix}
\qquad (D.32a)
$$

$$
= \frac{1}{|\mathbf{A}|}
\begin{bmatrix}
A_{11} & A_{21} & \cdots & A_{n1} \\
A_{12} & A_{22} & \cdots & A_{n2} \\
\cdots\cdots\cdots\cdots\cdots \\
A_{1n} & A_{2n} & \cdots & A_{nn}
\end{bmatrix}
\qquad (D.32b)
$$

It should be noted that one of the conditions for the unique solution of Eq. D.26 is that the number of equations must be equal to the number of unknowns. This implies that the matrix $\mathbf{A}$ must be a square matrix. In addition, we observe from the solution as given in Eq. D.28 that for the solution to exist† $|\mathbf{A}| \neq 0$. Hence the inverse exists only for a square matrix and only under the condition that the determinant of the matrix be nonzero. A matrix whose determinant is nonzero is a *nonsingular* matrix. Thus an inverse exists only for a nonsingular (square) matrix. By definition, we have

$$\mathbf{A}^{-1}\mathbf{A} = \mathbf{I} \qquad\qquad\qquad (\text{D.33a})$$

One can easily show that

$$\mathbf{A}\mathbf{A}^{-1} = \mathbf{I} \qquad\qquad\qquad (\text{D.33b})$$

To do this we first postmultiply Eq. D.33b by $\mathbf{A}^{-1}$. This yields

$$\mathbf{A}^{-1}\mathbf{A}\mathbf{A}^{-1} = \mathbf{I}\mathbf{A}^{-1} = \mathbf{A}^{-1}$$

or

$$\mathbf{A}^{-1}(\mathbf{A}\mathbf{A}^{-1}) = \mathbf{A}^{-1}$$

evidently

$$(\mathbf{A}\mathbf{A}^{-1}) = \mathbf{I}$$

Thus $\mathbf{A}$ and $\mathbf{A}^{-1}$ are the inverse of each other—that is

$$(\mathbf{A}^{-1})^{-1} = \mathbf{A} \qquad\qquad\qquad (\text{D.34})$$

Note that matrices $\mathbf{A}$ and $\mathbf{A}^{-1}$ commute (see Eqs. D.33a and D.33b).

Example D.3.   Find $\mathbf{A}^{-1}$ if

$$\mathbf{A} = \begin{bmatrix} 2 & 1 & 1 \\ 1 & 2 & 3 \\ 3 & 2 & 1 \end{bmatrix}$$

Here

$$A_{11} = -4, \quad A_{12} = 8, \quad A_{13} = -4$$
$$A_{21} = 1, \quad A_{22} = -1, \quad A_{23} = -1$$
$$A_{31} = 1, \quad A_{32} = -5, \quad A_{33} = 3$$

and $|\mathbf{A}| = -4$. Therefore

$$\mathbf{A}^{-1} = \frac{1}{4} \begin{bmatrix} -4 & 1 & 1 \\ 8 & -1 & -5 \\ -4 & -1 & 3 \end{bmatrix}$$

---

†These two conditions imply that the number of equation are equal to the number of unknowns and all the equations are independent.

Observe that

$$\mathbf{AA}^{-1} = \frac{1}{4} \begin{bmatrix} 2 & 1 & 1 \\ 1 & 2 & 3 \\ 3 & 2 & 1 \end{bmatrix} \begin{bmatrix} -4 & 1 & 1 \\ 8 & -1 & -5 \\ -4 & -1 & 3 \end{bmatrix} = \begin{bmatrix} 1 & 0 & 0 \\ 0 & 1 & 0 \\ 0 & 0 & 1 \end{bmatrix}$$

and

$$\mathbf{A}^{-1}\mathbf{A} = -\tfrac{1}{4} \begin{bmatrix} -4 & 1 & 1 \\ 8 & -1 & -5 \\ -4 & -1 & 3 \end{bmatrix} \begin{bmatrix} 2 & 1 & 1 \\ 1 & 2 & 3 \\ 3 & 2 & 1 \end{bmatrix} = \begin{bmatrix} 1 & 0 & 0 \\ 0 & 1 & 0 \\ 0 & 0 & 1 \end{bmatrix}$$

D.5   DERIVATIVES AND INTEGRAL OF A MATRIX

When the elements of matrix $\mathbf{A}$ are functions of some variable t, we identify this fact by writing $\mathbf{A}$ as $\mathbf{A}(t)$. Thus a matrix $\mathbf{A}$ given by

$$\mathbf{A} = \begin{bmatrix} e^{-2t} & \sin t \\ e^t & e^{-t} + e^{-2t} \end{bmatrix} \tag{D.35}$$

has elements which are functions of t, and as result it is helpful to denote $\mathbf{A}$ by $\mathbf{A}(t)$.

The derivative of a matrix $\mathbf{A}(t)$ (with respect to t) is defined as a matrix whose ijth element is the derivative (with respect to t) of the ijth element of the matrix $\mathbf{A}$.

Thus if

$$\mathbf{A}(t) = [a_{ij}(t)]_{m \times n}$$

$$\frac{d}{dt}[\mathbf{A}(t)] = \frac{d}{dt}[a_{ij}(t)]_{m \times n} \tag{D.36a}$$

or

$$\dot{\mathbf{A}}(t) = [\dot{a}_{ij}(t)]_{m \times n} \tag{D.36b}$$

Thus the derivative of the matrix in Eq. D.35 is given by

$$\dot{\mathbf{A}}(t) = \begin{bmatrix} -2e^{-2t} & \cos t \\ e^t & -e^{-t} - 2e^{-2t} \end{bmatrix}$$

Similarly, we define integral of $\mathbf{A}(t)$ (with respect to t) as a matrix whose ijth element is the integral (with respect to t) of the ijth element of the matrix $\mathbf{A}$.

$$\int \mathbf{A}(t)\,dt = \left(\int a_{ij}(t)\,dt \right)_{m \times n} \tag{D.37}$$

For the matrix $\mathbf{A}$ in Eq. D.35 we have

$$\int \mathbf{A}(t)\,dt = \begin{bmatrix} \int e^{-2t}\,dt & \int \sin dt \\ \int e^t\,dt & \int (e^{-t} + e^{-2t})\,dt \end{bmatrix}$$

We can easily prove the following identities:

$$\frac{d}{dt}(\mathbf{A} + \mathbf{B}) = \frac{d\mathbf{A}}{dt} + \frac{d\mathbf{B}}{dt} \tag{D.38}$$

$$\frac{d}{dt}(c\mathbf{A}) = c\frac{d\mathbf{A}}{dt} \tag{D.39}$$

$$\frac{d}{dt}(\mathbf{AB}) = \frac{d\mathbf{A}}{dt}\mathbf{B} + \mathbf{A}\frac{d\mathbf{B}}{dt} = \dot{\mathbf{A}}\mathbf{B} + \mathbf{A}\dot{\mathbf{B}} \tag{D.40}$$

The proofs of identities D.38 and D.39 are trivial. We can prove Eq. D.40 as follows.

Let $\mathbf{A}$ be an $m \times n$ matrix, $\mathbf{B}$ an $m \times p$ matrix; then if

$$\mathbf{C} = \mathbf{AB}$$

from Eq. D.16, we have

$$c_{ik} = \sum_{j=1}^{n} a_{ij}b_{jk}$$

and

$$\dot{c}_{ik} = \underbrace{\sum_{j=1}^{n} \dot{a}_{ij}b_{jk}}_{d_{ik}} + \underbrace{\sum_{j=1}^{n} a_{ij}\dot{b}_{jk}}_{e_{ik}} \tag{D.41}$$

or

$$\dot{c}_{ik} = d_{ik} + e_{ik}$$

It is obvious from Eq. D.41 and multiplication rule, that d_{ik} is the ikth element of matrix $\dot{\mathbf{A}}\mathbf{B}$ and e_{ik} is the ikth element of matrix $\mathbf{A}\dot{\mathbf{B}}$. Hence Eq. D.40 follows.

If we let $\mathbf{B} = \mathbf{A}^{-1}$ in Eq. D.40, we obtain

$$\frac{d}{dt}(\mathbf{AA}^{-1}) = \frac{d\mathbf{A}}{dt}\mathbf{A}^{-1} + \mathbf{A}\frac{d}{dt}(\mathbf{A}^{-1})$$

But since

$$\frac{d}{dt}(\mathbf{AA}^{-1}) = \frac{d}{dt}\mathbf{I} = 0$$

We have

$$\frac{d}{dt}(\mathbf{A}^{-1}) = -\mathbf{A}^{-1}\frac{d\mathbf{A}}{dt}\mathbf{A}^{-1} \tag{D.42}$$

D.6   COMPUTATION OF $e^{\mathbf{A}t}$

Here we shall present without proof two additional methods of computing $e^{\mathbf{A}t}$

If the eigenvalues are distinct, then it can be shown that

$$e^{\mathbf{A}t} = c_0\mathbf{I} + c_1\mathbf{A} + c_2\mathbf{A}^2 + \cdots + c_{n-1}\mathbf{A}^{n-1} \tag{D.43}$$

where the constants $c_0, c_1, c_2, \ldots, c_{n-1}$ can be computed from

$$
\begin{bmatrix} c_0 \\ c_1 \\ \cdot \\ c_{n-1} \end{bmatrix} = \begin{bmatrix} 1 & \lambda_1 & \lambda_1^2 & \cdots & \lambda_1^{n-1} \\ 1 & \lambda_2 & \lambda_2^2 & \cdots & \lambda_2^{n-1} \\ \cdot & \cdot & \cdot & & \cdot \\ 1 & \lambda_n & \lambda_n^2 & \cdots & \lambda_n^{n-1} \end{bmatrix} \begin{bmatrix} e^{\lambda_1 t} \\ e^{\lambda_2 t} \\ \cdot \\ e^{\lambda_n t} \end{bmatrix}
\tag{D.44}
$$

Note that the matrix on the right-hand side is a form of Vandermonde matrix (transpose of the Vandermonde matrix encountered in Chapter 2.). For repeated eigenvalues, the procedure is slightly modified.†

Another formula is based on Sylvester's expansion theorem. If $\mathbf{A}$ has distinct eigenvalues, then it can be shown that

$$
e^{\mathbf{A}t} = \sum_{i=1}^{n} e^{\lambda_i t} \mathbf{F}_i
\tag{D.45a}
$$

where

$$
\mathbf{F}_i = \prod_{\substack{j=1 \\ j \neq i}}^{n} \left(\frac{\mathbf{A} - \lambda_j \mathbf{I}}{\lambda_i - \lambda_j} \right)
\tag{D.45b}
$$

---

†See for example K. Ogata, *State Space Analysis of Control Systems* (Englewood Cliffs, N.J.: Prentice-Hall, 1967).

E

Second-Order System with a Zero

From Eq. 5.16a

$$y(t) = \mathcal{L}^{-1} \frac{\omega_n^2}{a} \frac{s + a}{s(s + \zeta\omega_n - j\omega_n\sqrt{1 - \zeta^2})(s + \zeta\omega_n + j\omega_n\sqrt{1 - \zeta^2})}$$

$$= \mathcal{L}^{-1}\left[\frac{1}{s} + \frac{re^{j\psi}}{s + \zeta\omega_n - j\omega_n\sqrt{1 - \zeta^2}} + \frac{re^{-j\psi}}{s + \zeta\omega_n + j\omega_n\sqrt{1 - \zeta^2}}\right]$$

where

$$r = \frac{\sqrt{a^2 - 2a\zeta\omega_n + \omega_n^2}}{2a\sqrt{1 - \zeta^2}} \tag{E.1}$$

$$\psi = -\pi + \tan^{-1}\frac{\omega_n\sqrt{1 - \zeta^2}}{a - \zeta\omega_n} - \tan^{-1}\frac{\zeta}{\sqrt{1 - \zeta^2}} \tag{E.2}$$

Hence

$$y(t) = 1 + 2re^{-\zeta\omega_n t}\cos(\omega_n\sqrt{1 - \zeta^2}t + \psi) \tag{E.3}$$

To obtain t_p and PO, we find the solution of $dy/dt = 0$. This gives

$$\frac{dy}{dt} = 2re^{-\zeta\omega_n t}[-\omega_n\sqrt{1 - \zeta^2}\sin(\omega_n\sqrt{1 - \zeta^2}t + \psi) - \zeta\omega_n\cos(\omega_n\sqrt{1 - \zeta^2}t + \psi)]$$

$$= -2r\omega_n e^{-\zeta\omega_n t}\cos\left[\omega_n\sqrt{1 - \zeta^2}t + \psi - \tan^{-1}\frac{\omega_n\sqrt{1 - \zeta^2}}{\zeta\omega_n}\right]$$

$$= -2r\omega_n e^{-\zeta\omega_n t}[\cos(\omega_n\sqrt{1 - \zeta^2}t + \phi)]$$

where

$$\phi = \psi - \tan^{-1}\frac{\omega_n\sqrt{1 - \zeta^2}}{\zeta\omega_n} = \frac{\pi}{2} + \tan^{-1}\frac{\omega_n\sqrt{1 - \zeta^2}}{a - \zeta\omega_n} \tag{E.4}$$

At $t = t_p$, $dy/dt = 0$ and

$$\cos(\omega_n\sqrt{1 - \zeta^2}t_p + \phi) = 0$$

This yields

$$\omega_n \sqrt{1 - \zeta^2} \, t_p + \phi = \frac{\pi}{2} \tag{E.5}$$

Therefore

$$\omega_n t_p = \frac{\frac{\pi}{2} - \phi}{\sqrt{1 - \zeta^2}} = -\tan^{-1}\left(\frac{\omega_n\sqrt{1 - \zeta^2}}{a - \zeta\omega_n}\right) \Big/ \sqrt{1 - \zeta^2}$$

$$= -\tan^{-1}\left(\frac{\sqrt{1 - \zeta^2}}{\zeta\left(\frac{a}{\zeta\omega_n} - 1\right)}\right) \Big/ \sqrt{1 - \zeta^2} \tag{E.6}$$

To determine PO, we find y at $t = t_p$. This is given by

$$y(t_p) = 1 + 2re^{-\zeta\omega_n t_p}\cos(\omega_n\sqrt{1 - \zeta^2}\,t_p + \psi)$$

$$= 1 + \frac{\sqrt{a^2 - 2a\zeta\omega_n + \omega_n^2}}{a\sqrt{1 - \zeta^2}}e^{-\zeta\omega_n t_p}\cos(\omega_n t_p\sqrt{1 - \zeta^2} + \psi)$$

$$\text{PO} = [y(t_p) - 1] \times 100$$

$$= \frac{\sqrt{a^2 - 2a\zeta\omega_n + \omega_n^2}}{a\sqrt{1 - \zeta^2}}e^{-\zeta\omega_n t_p}\cos(\omega_n\sqrt{1 - \zeta^2}\,t_p + \psi) \tag{E.7}$$

We substitute the values of ψ and t_p (Eqs. E.2 and E.6) in Eq. E.7. Figure 5.16a shows PO as a function of $a/\zeta\omega_n$ for various values of ζ. Values of t_r are computed from Eq. E.3 using a digital computer. Figure 5.16b shows t_r as a function of $a/\zeta\omega_n$ for several values of ζ.

F

Nyquist Criterion
for Stability

In order to understand this criterion, it is helpful to introduce the concept of complex mapping.

F.1 COMPLEX MAPPING

The reader is familiar with the mapping of the functions of real variables. Mapping is a visual or graphical representation of the relationship between two variables which are related by a certain function. Thus when we say y is a function of x and represent the relation as

$$y = f(x) \qquad (F.1)$$

we mean that for every value of x there exists a corresponding value of y, and these values are related by Eq. F.1. This relationship is represented graphically in Fig. F.1. The graph is just a visual means of representing Eq. F.1 or of showing the values of y for the corresponding values of x.

The graphical or visual representation is quite simple for two real variables x and y. If, however, the two variables under consideration are complex, say s and W, then such a representation is not so easy. Each complex variable contains two variables within itself—the real variable and the imaginary variable. Thus

$$s = \sigma + j\omega \qquad (F.2)$$
$$W = u + jv$$

There are four variables in all, and a simple graphical representation in two- or even three-dimensional space is not possible. We overcome this difficulty by defining two planes: the s-plane and the W-plane. Each plane has two dimensions, and hence we obtain the required four dimensions to represent four variables. Let us write the functional representation

$$W = f(s) \qquad (F.3a)$$

or

$$u + jv = f(\sigma + j\omega) \qquad (F.3b)$$

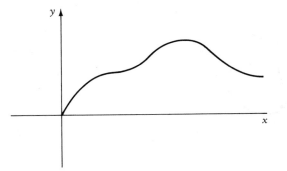

Figure F.1

Equation F.3a expresses the fact that for every value of s there is a corresponding value of W related through Eq. F.3a. This information can be represented graphically on two planes: the s- and W-planes. We represent a certain point s_1 in the s-plane and a corresponding point W_1 in the W-plane. Thus the point W_1 in the W-plane is a sort of image of point s_1 in the s-plane, and W_1 and s_1 are related by Eq. F.3:

$$W_1 = f(s_1) \tag{F.4}$$

Let us consider the simple function

$$W = s^2 \tag{F.5}$$

Then

$$u + jv = (\sigma + j\omega)^2 \tag{F.6a}$$
$$= \sigma^2 - \omega^2 + j2\sigma\omega \tag{F.6b}$$

It can be seen from Eq. F.6 that for every value of s there is a corresponding value of W. If, for example

$$s = 2 + j3$$

then

$$W = (2 + j3)^2$$
$$= -5 + j12$$

Similarly, for different values of s we get different values of W. If we consider a closed curve C_s in the s-plane, each point on this curve has an image in the W-plane, and thus the whole curve C_s in the s-plane maps into another closed curve C_w in the W-plane, provided that W is a continuous function of s along C_s. Thus C_w is an image of C_s, as shown in Fig. F.2. The points A, B, C, D, E, F, and G on C_s map into points A', B', C', D', E', F', and G' on C_w in the W-plane.

Now consider the transformation

$$W = (s - s_1) \tag{F.7}$$

We shall map the image in the W-plane of a closed curve C_s in the s-plane enclosing

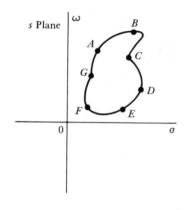 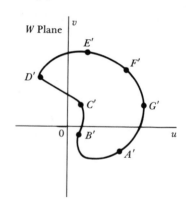

Figure F.2

the point s_1 (Fig. F.3a). This is done by computing the values of W for various values of s for points lying along C_s. The quantity $s - s_1$ is complex, in general, and can be conveniently visualized graphically in Fig. F.3a. For a point A on C_s, for example $s - s_1$, is represented by a phasor joining the point s_1 to point A. If we designate

$$s - s_1 = Me^{j\theta}$$

then, for each point along C_s, there is a certain value for M (magnitude of $s - s_1$) and θ (phase of $s - s_1$). We now map a point of magnitude M and argument θ in the W-plane. This point A' in the W-plane is the image of A in the s-plane. As we move along the entire contour C_s in the s-plane, θ goes through a net variation of 2π radians. Obviously the image in the W-plane of a closed curve C_s in the s-plane is also a closed curve C_w which encloses the origin of the W-plane once. Note that if we move along C_s in a clockwise direction, angle θ decreases, and hence the image C_w is also traversed in a clockwise direction.

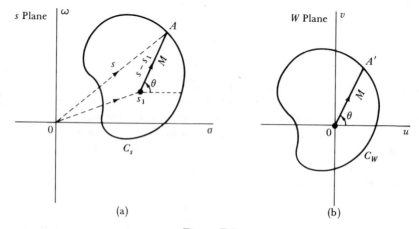

Figure F.3

Consider now another transformation:

$$W = \frac{1}{s - s_1} \tag{F.8}$$

$$= \frac{1}{M e^{j\theta}} = \frac{1}{M} e^{-j\theta}$$

Note that the magnitude of W is now $1/M$ and the argument is $-\theta$. Hence by the previous argument it follows that as we traverse the close curve C_s once in counterclockwise direction the phase of W goes through a net variation of -2π radians. In other words, C_w encloses the origin in the W-plane, but the corresponding direction of traverse is clockwise.

Next, consider the transformation

$$W = (s - z_1)(s - z_2) \tag{F.9}$$

We shall assume that C_s encloses z_1 and z_2 (Fig. F.4). We can interpret $s - z_1$ and

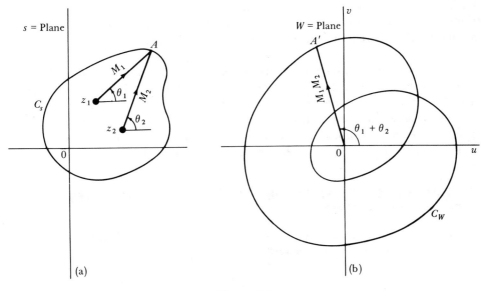

Figure F.4

$s - z_2$ as before. If

$$s - z_1 = M_1 e^{j\theta_1} \quad \text{and} \quad s - z_2 = M_2 e^{j\theta_2}$$

then

$$W = M_1 M_2 e^{j(\theta_1 + \theta_2)}$$

Note that when we traverse along C_s in a counterclockwise direction once the angles θ_1 and θ_2 each go through a variation of 2π radians. The argument of W goes through a change of 4π radians. Hence C_w encloses the origin in the W-plane twice in a counterclockwise direction.

Now consider the general transformation

$$W = \frac{(s - z_1)(s - z_2) \cdots (s - z_m)}{(s - p_1)(s - p_2) \cdots (s - p_n)}$$ (F.10)

If

$$s - z_k = M_k e^{j\theta_k}$$

and

$$s - p_i = N_i e^{j\phi_i}$$

then

$$W = \frac{M_1 M_2 \cdots M_m}{N_1 N_2 \cdots N_n} e^{j(\theta_1 + \theta_2 + \cdots + \theta_m - \phi_1 - \phi_2 - \cdots - \phi_n)}$$

As before, we shall assume that C_s encloses all of the m zeros and n poles. When we traverse once along C_s in a counterclockwise direction, each of the angles θ_k and θ_i goes through a variation of 2π radians. Hence the net variation of an argument of W is $2\pi(m - n)$, where m and n are the numbers of zeros and poles respectively of W enclosed by C_s. Thus C_w encloses the origin in the W-plane $(m - n)$ times, in a counterclockwise direction. If $m - n$ is negative, the enclosure is in a clockwise direction.

This discussion may be summarized by the following conclusion. A closed curve C_s in the s plane enclosing m zeros and n poles of $W(s)$ maps into a closed curve C_w in the W-plane encircling the origin of the W-plane $m - n$ times, in the same direction as that of C_s. If $m - n$ is negative, then the encirclement is in the opposite direction. Suppose $m = 2$ and $n = 5$, then C_w will encircle the origin in the W-plane three times in the direction opposite to that of C_s.

F.2 GRAPHICAL CRITERION FOR STABILITY IN FEEDBACK SYSTEMS (THE NYQUIST CRITERION)

We shall again consider a simple feedback system shown in Fig. 5.17. The transfer function $T(s)$ is given by

$$T(s) = \frac{G(s)}{1 + G(s)H(s)}$$

To investigate the stability of this system, it is sufficient to know whether $T(s)$ has any poles in the RHP. Since the poles of $T(s)$ are zeros of $1 + G(s)H(s)$, we need to know whether $(1 + GH)$ has any zeros in the RHP, if the function $(1 + GH)$ has m zeros and n poles in the RHP, then the image of a closed curve C_s in the s-plane, which encloses all RHP-poles and zeros of $(1 + GH)$, will encircle the origin of the $(1 + GH)$-plane $(m - n)$ times, in the same direction as that of C_s. It is evident if we draw the image of C_s in the GH-plane instead of in the $(1 + GH)$-plane, then the corresponding image of C_s will enclose $(m - n)$ times the point -1 in the same direction as that of C_s. [This is because a map of any curve in the GH-plane will be the same as that in the $(1 + GH)$ plane except that it will be shifted by -1.]

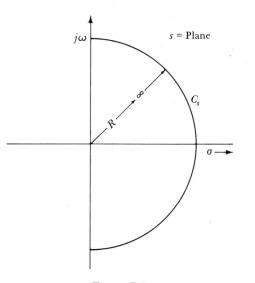

Figure F.5

Note that the curve C_s must enclose all poles and zeros of $G(s)H(s)$ in the RHP. This means that C_s must enclose all of the RHP. A suitable curve C_s, which will enclose all of the RHP, is shown in Fig. F.5. The curve encloses the region inside a semicircle of radius R as shown in Fig. F.5. The radius R is made arbitrarily large to enclose all of the RHP. In the limit, $R = \infty$. The curve C_s thus consists of the entire $j\omega$-axis, and a semicircle of radius $R = \infty$. To map this curve on the $G(s)H(s)$-plane we need to know the values of $G(s)H(s)$ along the $j\omega$-axis and along the semicircle. Along the $j\omega$-axis, $G(s)H(s)$ is given by $G(j\omega)H(j\omega)$. We may compute $G(j\omega)H(j\omega)$ (or obtain them from Bode plot) for $\omega = 0$ to ∞. In addition, we have (see Eq. 3.38)

$$G(-j\omega)H(-j\omega) = G^*(j\omega)H^*(j\omega)$$

Hence values of $G(j\omega)H(j\omega)$ for negative values of ω are simply conjugate of those for positive values of ω. Thus in mapping C_s in s plane from $s = -j\infty$ to $j\infty$, we need to map it only for $\omega = 0$ to ∞. The map for $\omega = 0$ to $-\infty$ will be the mirror image, about the horizontal axis, of the map for $\omega = 0$ to ∞.

We shall assume the denominator of $G(s)H(s)$ to be at least of the same degree as that of the numerator. This means $G(s)H(s)$ is either zero or a constant for $s = \infty$. Hence the entire semicircle of radius $R(R = \infty)$ in C_s maps into a point in the GH-plane.

Note that if the open-loop transfer function GH represents a stable system, then $(1 + GH)$ cannot have any poles in the RHP. This can be easily shown. Let

$$GH = \frac{N(s)}{D(s)}$$

Then

$$1 + GH = \frac{N(s) + D(s)}{D(s)} \tag{F.11}$$

Since by assumption the open-loop transfer represents a stable system, $D(s)$ has no zeros in the RHP. But zeros of $D(s)$ are the poles of $(1 + GH)$(Eq. F.11). Hence it follows that $(1 + GH)$ does not have poles in the RHP. This means that $n = 0$ and the number of encirclements of the critical point $(-1 + j0)$ in the GH-plane by the image of C_s is equal to m—that is, the number of zeros of $(1 + GH)$ enclosed in C_s.

We have shown above that if the open-loop transfer function represents a stable system, then the quantity $(1 + GH)$ has no poles in the RHP, and the total number of encirclement of the critical point $(-1 + j0)$ in the GH-plane, by the image of contour C_s (Fig. F.5), is equal to the number of zeros of $(1 + GH)$ in the RHP. If a closed-loop system is stable, then $(1 + GH)$ should have no zeros in the RHP. We are now ready to state Nyquist criterion for stability.

We shall first state the general Nyquist criterion. If $1 + G(s)H(s)$ has m zeros and n poles in the RHP, then the polar plot of the $GH(j\omega)$, the steady-state frequency response of the open-loop transfer function, as ω varies from $-\infty$ to $+\infty$ encircles the critical point $(-1 + j0)$, $(m - n)$ times in the clockwise direction.

In practice, however, most of the open-loop systems (with very few exceptions) are stable, and hence $n = 0$. In this case the Nyquist stability criterion is considerably simplified and the number of encirclements of point $(-1 + j0)$ by the plot of $GH(j\omega)$ is equal to the number of zeros of $(1 + GH)$. Hence for a stable open-loop transfer function the closed-loop system will be stable if the point $(-1 + j0)$ is not encircled by the polar plot of $GH(j\omega)$ as ω varies from $-\infty$ to ∞. This is Nyquist's abbreviated criterion for stability of a closed-loop system and applies to systems with stable open-loop transfer functions.

At times the plot may be rather complicated and the number of encirclements of the point $(-1 + j0)$ may not be obvious. In such cases, it may become necessary to trace out the net change in the angle subtended by $GH(j\omega)$ at point $(-1 + j0)$ as ω varies from $-\infty$ to $+\infty$. This may be easily done by actually drawing a vector from point $(-1 + j0)$ to $GH(j\omega)$ and moving the end of this vector along $GH(j\omega)$ from $\omega = -\infty$ to ∞ and noting the net angle generated by this vector at the point $(-1 + j0)$. Two such cases are shown in Fig. F.6.

The case of the open-loop transfer function containing poles on the $j\omega$-axis needs special handling. We must choose C_s so that no poles of $W(s)$ lies along its path, since $W = \infty$ at the pole. We must therefore make detours around the poles, if any, existing along the $j\omega$-axis.

Consider the function

$$GH = \frac{K}{s(s + 2)(s + 4)}$$

We shall consider the specific case of $K = 24$ and determine, using the Nyquist criterion, whether this open-loop transfer function represents a stable closed-loop system.

$$G(j\omega)H(j\omega) = \frac{24}{j\omega(j\omega + 2)(j\omega + 4)}$$

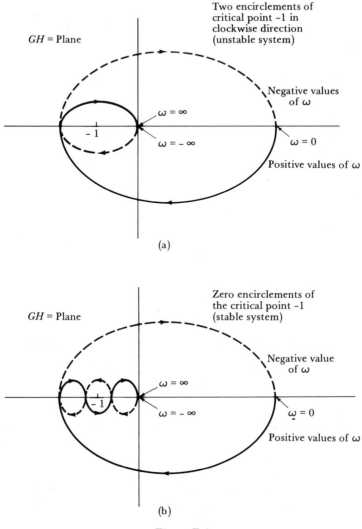

$GH = \text{Plane}$

Two encirclements of
critical point –1 in
clockwise direction
(unstable system)

Negative values
of ω

$\omega = \infty$

-1

$\omega = -\infty$

$\omega = 0$

Positive values of ω

(a)

$GH = \text{Plane}$

Zero encirclements of
the critical point –1
(stable system)

Negative value
of ω

$\omega = \infty$

-1

$\omega = -\infty$

$\omega = 0$

Positive values of ω

(b)

Figure F.6

Note that there is a pole on the $j\omega$-axis at $\omega = 0$. Hence we shall make a small
semicircular detour along C_s at $\omega = 0$ (Fig. F.7). Along the semicircular detour,

$$s = re^{j\theta} \tag{F.12}$$

where θ varies from $-\pi/2$ to $\pi/2$ radians. The function $G(s)H(s)$ along the detour
is given by

$$GH = \frac{24}{re^{j\theta}(re^{j\theta} + 2)(re^{j\theta} + 4)} \tag{F.13}$$

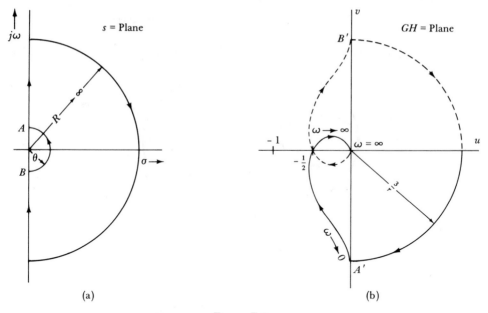

(a)   (b)

Figure F.7

The radius of the detour is arbitrarily small. If we let $r \rightarrow 0$ in Eq. F.13, we get

$$GH = \frac{24}{re^{j\theta}(2)(4)} = \frac{3}{r} e^{-j\theta}$$

As θ varies from $-\pi/2$ to $\pi/2$ radians along the detour, the corresponding map of GH traces out a semicircle of radius $3/r$ in the GH-plane with angle varying from $\pi/2$ to $-\pi/2$. We can use Bode plots to construct the Nyquist plot for $\omega = 0$ to ∞ (Fig. 5.44). The plot for $\omega = 0$ to $-\infty$ is mirror image (shown dotted) of the plot for $\omega = 0$ to ∞. The entire Nyquist plot is shown in Fig. F.7b. It can be seen that the Nyquist plot crossès the negative real axis at -0.5, and does not encircle the critical point -1. Hence the closed-loop system is stable for $K = 24$. If, however, we make $K > 48$, the plot will encircle the point -1. Therefore the closed-loop system is unstable for $K > 48$.

Bibliography

Balabanian, N., Bickart, T. A., and Seshu, S. *Electrical Network Theory* (New York: Wiley, 1969).

Bracewell, R. *The Fourier Transform and Its Applications* (New York: McGraw-Hill, 1965).

Brenner, E., and Javid, M. *Analysis of Electric Circuits* (New York: McGraw-Hill, 1959).

Cannon, R. H., Jr. *Dynamics of Physical Systems* (New York: McGraw-Hill, 1967).

Carlson, A. B. *Communication Systems* (New York: McGraw-Hill, 1968).

Cooper, G. R., and McGillem, C. D. *Methods of Signal and System Analysis* (New York: Holt, Rinehart, and Winston, 1967).

Cruz, J. B., and Van Valkenburg, M. E. *Introductory Signals and Circuits* (Waltham, Mass.: Ginn/Blaisdell, 1967).

D'Azzo, J. J., and Houpis, C. H. *Feedback Control System Analysis and Synthesis,* 2nd ed. (New York: McGraw-Hill, 1966).

De Russo, P. M., Roy, R. J., and Close, C. M. *State Variables for Engineers* (New York: Wiley, 1965).

Desoer, C. A., and Kuh, E. S. *Basic Circuit Theory* (New York: McGraw-Hill, 1969).

Dorf, R. C. *Modern Control Systems* (Reading, Mass.: Addison Wesley, 1967).

Frederick, D. K., and Carlson, A. B. *Linear Systems in Communication and Control* (New York: Wiley, 1971).

Freeman, H. *Discrete-Time Systems* (New York: Wiley, 1965).

Gold, B., and Rader, C. M. *Digital Processing of Signals* (New York: McGraw-Hill, 1969).

Guillemin, E. A. *Theory of Linear Physical Systems* (New York: Wiley, 1963).

Gupta, S. C., and Hasdorff, L. *Fundamentals of Automatic Control* (New York: Wiley, 1970).

513

Hohn, F. E. *Elementary Matrix Algebra* (New York: Macmillan, 1964).

Horowitz, I. *Synthesis of Feedback Systems* (New York: Academic Press, 1963).

Huang, T. S., and Parker, R. R. *Network Theory: An Introductory Course* (Reading, Mass.: Addison-Wesley, 1971).

Huelsman, L. P. *Circuits, Matrices, and Linear Vector Spaces* (New York: McGraw-Hill, 1963).

Huggins, W. H., and Entwistle, D. R. *Introductory Systems and Design* (Waltham, Mass.: Ginn/Blaisdell, 1968).

Jury, E. I. *Theory and Applications of the z-Transform Method* (New York: Wiley, 1964).

Kuo, B. C. *Analysis and Synthesis of Sampled-Data Control Systems* (Englewood Cliffs, N.J.: Prentice-Hall, 1963).

Kuo, F. F. *Network Analysis and Synthesis,* 2nd ed. (New York: Wiley, 1966).

Lathi, B. P. *Signals, Systems, and Communication* (New York: Wiley, 1965). *Communication Systems* (New York: Wiley, 1968). *An Introduction to Random Signals and Communication Theory* (New York and London: Intext Educational Publishers, 1968).

Leon, B. J. *Lumped Systems* (New York: Holt, Rinehart, and Winston, 1968).

LePage, W. R. *Complex Variables and the Laplace Transform for Engineers* (New York: McGraw-Hill, 1961).

Lighthill, M. J. *An Introduction to Fourier Analysis and Generalized Functions* (New York: Cambridge, 1958).

Lindorff, D. P. *Theory of Sampled-Data Control Systems* (New York: Wiley, 1965).

Mason, S. J., and Zimmerman, H. J. *Electronic Circuits, Signals, and Systems* (New York: Wiley, 1960).

Melsa, J. L., and Schultz, D. G. *Linear Control Systems* (New York: McGraw-Hill, 1969).

Mitra, S. K. *Analysis and Synthesis of Linear Active Networks* (New York: Wiley, 1969).

Newcomb, R. W. *Concepts of Linear Systems and Controls* (Belmont, Calif.: Brooks/Cole, 1968).

Ogata, K. *State Space Analysis of Control Systems* (Englewood Cliffs, N. J.: Prentice-Hall, 1967).

Papoulis, A. *The Fourier Integral and Its Applications* (New York: McGraw-Hill, 1962).

Perkins, W. R., and Cruz, J. B., Jr. *Engineering of Dynamic Systems* (New York: Wiley, 1969).

Ragazzini, G. R., and Franklin, G. F. *Sampled-Data Control Systems* (New York: McGraw-Hill, 1958).

Schwartz, M. *Information Transmission, Modulation and Noise,* 2nd ed. (New York: McGraw-Hill, 1970).

Schwartz, R. J., and Friedland, B. *Linear Systems* (New York: McGraw-Hill, 1965).

Šiljak, D. *Nonlinear Systems* (New York: Wiley, 1969).

Timothy, L. K., and Bona, B. E. *State Space Analysis* (New York: McGraw-Hill, 1968).

Van der Pol, B., and Bremmer, H. *Operational Calculus* (New York: Cambridge, 1955).

Index

Printer and Binder: Halliday Lithograph Corporation

80 81 82 9 8 7 6 5 4